D0143720

Mechanical Design in Organisms

Mechanical Design in Organisms

S. A. Wainwright

W. D. Biggs

J. D. Currey

J. M. Gosline

Princeton University Press
Princeton, New Jersey

Published by Princeton University Press, 41 William Street,
Princeton, New Jersey 08540
In the United Kingdom: Princeton University Press,
Oxford

First published 1976
by Edward Arnold (Publishers) Limited, London and by
Halsted Press, a Division of John Wiley & Sons, Inc., New York

LCC 82-47636
ISBN 0-691-08306-1
ISBN 0-691-08308-8 pbk.

Princeton University Press books are printed on acid-free paper,
and meet the guidelines for permanence and durability
of the Committee on Production Guidelines for Book Longevity
of the Council on Library Resources

Printed in the United States of America

7 6 5 4 3 2

Preface

We believe that the study of mechanical design in organisms using the approach of the mechanical engineer and the materials scientist can promote an understanding of organisms at all levels of organization from molecules to ecosystems.

Certain basic engineering concepts, such as strength, rigidity and viscoelasticity, can be understood in terms of interactions between atoms and molecules, and this understanding allows us to interpret the mechanical behaviour of skeletal *materials* in terms of their molecular organization. Building from this understanding of skeletal materials, we can apply engineering theory to consider the shape and distribution of *structural elements* within organisms and the overall design strategy of complete *skeletal systems*. Finally, knowing the mechanical design of an organism as a whole we can begin to consider its interaction with other organisms and with the physical environment. In travelling this road we have come across a number of design principles that appear to govern the structure-function relationships in organisms and interestingly in man-made structures as well.

This book is frankly evangelical: we wish to modify biologists' view of the world and to impress upon them the importance of mechanical design in all aspects of biology. It is no longer sufficient for biologists to study organisms at one level alone. Just as an engineer designing an engine cylinder head must be aware of the mechanical properties of alloys and the environmental consequences of exhaust emission, so too a biologist interested in the genetics of bristle patterns on a fly's wing ought to know why the bristles are stiff, how bristles modify the airflow over the wing, and how such aerodynamic modifications affect the behavioural patterns of the insect.

This book is also meant to be instructive, especially to the non-engineering biologist. The reader will, we hope, be given a fairly extensive understanding of an area of knowledge that is intellectually very satisfying. He should find here all he needs to gain a working foothold on the relevant aspects of mechanics.

We regard the book as a joint effort, but some apportionment of blame may be of interest to our readers. Roughly, Wainwright had the original idea, wielded the whip and wrote Chapters 1, 7, and 8. Gosline wrote Chapters 3 and 4, Currey wrote Chapter 5. Biggs wrote Chapter 2 and much of the engineering introduction in other chapters. Chapter 6 is a joint effort by Currey and Biggs. All authors have enclaves of writing in other chapters.

1975

S.A.W.
W.D.B.
J.D.C.
J.M.G.

Acknowledgement

The authors are grateful to the many authors and publishers who have allowed their illustrations to be reproduced in this book. All of the illustrations have been redrawn and in some cases modified, and there may be cases in which the original authors do not approve of the liberties we have taken with their material. In particular, Figure 7.23 has been redrawn and altered from Manton (1965); Figure 7.26 has been redrawn and substantially altered from Manton (1961a); Figures 7.17c and j have been redrawn and altered from Manton (1966) and Figure 7.17a has been created from inferences derived from the text of Manton (1950). In every case we are responsible for any inaccuracies we have wrought.

Contents

List of symbols

Pages numbered in italics include a figure; pages numbered in bold face contain definitions or equations; pages whose numbers are marked with an asterisk contain values of the designated property.

A = area, 9
a = length of a rigid segment in polymer, **53**
a_0 = critical crack length, 20, *21*
α = fibre angle, **294**, *294*
B = bulk compliance, 27
β_c = column buckling parameter, 267-268*
D = compliance, **27**
D = drag force, 248
δ = phase angle, **32**
ΔH = activation energy, **24**
E^* = complex modulus, **33**, *33*
E'' = loss modulus, **33**, *33*
E' = storage modulus, **33**, *33*
E = Young's modulus, **9**
$\dot{\epsilon}$ = flow rate, **23**
ϵ = strain, **7**, **9**
ϵ_t = true strain, **9**
η = efficiency of reinforcement, **151**-152*
η = viscosity, **23**
F = force, 9
F_E = critical (Euler) buckling force, **249**
f_0 = molecular friction coefficient, **58**
G = Gibbs free energy, **45**
G = shear modulus, **11**
γ = shear strain, **10**, *10*
H = Helmholtz free energy, **45**, 46
$H(\tau)$ = relaxation spectrum, **38**
h = elastic loss factor, **113**
I = second moment of area, **247**
J = polar second moment of area, **254**
J = shear compliance, 27
K = bulk modulus, **11**
k = Boltzmann's constant, **24**
L = length, 9
$L(\tau')$ = retardation spectrum, 38
l_c = transfer length, **149**
λ = extension ratio, **56**
M_c = average molecular weight between crosslinks in a polymer, 56
M = bending moment, 247, **265**
M = modulus, 33, 34
M_R = modulus of rupture, **166**
ν = Poisson's ratio, 10-**11**
Q = heat, **46**
R = larger or outer radius, 17
R = resilience, **15**
r = least radius of gyration, **250**
r = smaller or inner radius, 17
ρ = density, 56
S = applied stress, 16, 17
S = entropy, **45**
S = surface energy, **19**-*21*
σ = stress, **7**, **9**
σ_t = true stress, **9**
T = temperature, 24
T = torque, 253
tan δ = loss tangent = damping, **15**
τ = relaxation time, **24**-*25*
τ' = retardation time, **25**
τ = shear stress, **10**
U = internal energy, **45**
U = strain energy, **12**, *13*
V = volume fraction, 23
W = weight, 265
W = work, **20**
Z = section modulus, **247**, **256**, 257*

Chapter I
Introduction

This book deals with an interface between mechanical engineering and biology. The major strategic function of structural materials, elements and systems in organisms is that of mechanical support. The particular tactics of how each component or system carries out support depend on its mechanical properties. The theory and practice of materials science, mechanical engineering and biology tell us that mechanical properties of components and systems can be explained in terms of their structure. This, then, establishes a direct link between structure and mechanical function. This book looks at mechanical function in organisms and their components and attempts to correlate function with measurable mechanical properties and observed structure. Functions, properties and structures are discussed in terms of models and principles of design.

The idea that biological materials and structures have functions implies that they are 'designed'; hence the book's title. We run into deep philosophical waters here, and we can do little but give a commonsense idea of what we mean. In our view structures can be said to be designed because they are adapted for particular functions. They are not merely appropriate for these functions, because that could happen by chance.

In any species, at any moment in time, the organisms with structures more appropriate for particular functions will, on average be selected. This, along with the genetically controlled variation occurring in the population in each generation, will mean that the structure will change through the generations. The result, at any time, is that the structure is not only more or less appropriate for some function, but also that it has been adapted for that function in the past, and so can be considered to be designed. The designing is performed, of course, by natural selection. Natural selection takes account not only of how the structure performs a particular function, but also how this interacts with all the other processes the organism must carry out. Readers who are unhappy about using the word 'design' in such a context as that of this book may like to read WILLIAMS (1966), SOMMERHOFF (1950) and RUDWICK (1964).

A goal in making this book has been to state 'Principles of Design' for materials, skeletal elements and entire systems. These principles have been found lurking in a hundred different corners of the engineering and architectural design literature, often hidden by jargon, and we have selected those for which we have a biological illustration and present them with great enthusiasm and some anxiety to biologists. They appear on pages 22, 43, 268 and 297.

Chapters 2A, 2B, 3, 4 and 5 are about materials. This appears to be the first attempt to survey principles of mechanical design of biological structural materials. There is no attempt to survey *all* such materials: examples have been selected that tell us most about the appropriateness of those theories and models, borrowed from sister sciences, that are embodied in the design principles. Appropriate models, in turn, tell us much about the biomaterials we study.

Although it is the authors' intent that the reader should appreciate the mathematical basis of biological mechanical design features, the mathematically timorous reader should find it possible to skip over Chapter 2B the first time through. Perusal of the principles of the design of viscoelastic materials (Section 2.20) will provide the temporary handles necessary to get a feel for the materials discussed in Chapters 3 and 4. In fact, we hope that a biologist could find his pet subject via the index and, even if it were the kelps in Chapter 8, the reader should be able to backtrack, via the four sets of principles, to get as deep into the basics as he wants or as we can take him—whichever is reached first.

Chapter 6 on skeletal elements recognizes that, beyond the intrinsic properties of the materials themselves, the shape of individual bones, shells, etc., controls their mechanical properties. Chapter 7 sees the mechanical support systems of all macroscopic organisms as belonging to two types: rigid space frames and flexible, membranous cylinders. Chapter 8 tackles the largest subject in the fewest pages: the mechanical interactions between organisms and gravitational and fluid flow forces in their environment.

Thus, the book clearly reflects the pyramid of knowledge and changes its mission as it moves along. The shapes, sizes and behaviours of atoms and molecules can now be described thoroughly, accurately and precisely, by physical scientists. The larger the molecule or the more complex its interactions with others, the less can be said about it, and the less is the precision with which we say it. Principles of the structure and mechanical function of materials abound. Much can be, and is, said at this level.

At the next level, that of skeletal elements, there is a tremendous shrinkage of the total amount of hard information and rather fewer principles. On the other hand, although the number of types of elements (cylinders, rods, bricks, cables, membranes, etc.) is small, the number of specific variations on these that occurs among plant and animal species is apparently unlimited. Biologists know about diversity and deal with it every day. In this book we state principles and generalities as a background on which each biologist may construct patterns of specific information.

The number of types of systems into which the parts are assembled to give a functional organism is seen here to be only two: the membrane-bound hydrostat and the rigid framework, plus their various combinations. There are structural descriptions of more than a million species in the taxonomic literature of extinct and extant species. We say here, ever less about ever greater phenomena.

By the time we expose a system to external forces in its environment in Chapter 8, we are fresh out of principles and can only wave our hands at Nature and say that some organisms are more flexible than others. That may seem a meagre reward for such a long and sometimes arduous haul through thermodynamics, beam theory, Maxwell's lemma and Linnaeus' catalogue of species. But one is then at the top of the pyramid and, if one has consumed the substance of our message, the view from the top reveals a certain degree of order and some useful pathways through the multi-dimensional network of biological diversity in all levels of integration from the macromolecular to the ecological.

We have been led to our thinking by the work of many others. Primary sources, that we heartily recommend to the reader, include THOMPSON'S classic *On Growth and Form* (1917), especially the chapters entitled 'On Magnitude' and 'On Form and Mechanical Efficiency'. ALEXANDER'S *Animal Mechanics* (1968) is the most comprehensive treatment of mechanical phenomena in animals as well as being a model of brevity and clarity in scientific writing; his *Functional Design in Fishes* (1967) is narrower in scope; and his *Size and Shape* (1971) is a successful updating and, more importantly, a more functional view of many subjects first treated by THOMPSON (1917). The many published papers of PROF. SIR JAMES GRAY and the work of others in his group, as expressed in his book *Animal Locomotion* (1968), and the discussion of the evolutionary implications of mechanics in animals by CLARK in *Dynamics in Metazoan Evolution* (1964) are great synthetic works that repay continued visitation.

Many of the design principles we state are derived from various expressions by R. Buckminster Fuller whose insistence on a universal language of design was a key stimulus in the conception of this book. That such an approach to organisms is possible and fruitful is manifest in PICKEN'S book *The Organization of Cells and Other Organisms* (1960), which represents possibly the most thorough argument yet given for the structural basis of function in natural systems.

Finally, we believe that no one can read GORDON'S *The New Science of Strong Materials* (1968) and not be persuaded–and charmed–to drop everything and enter the field of materials science. All of these authors possess contagious enthusiasm for doing exciting things and for doing them well and we gratefully acknowledge having suffered this condition at their hands.

In setting limits for the present account, we have arbitrarily decided not to restate the accounts of the design of the vertebrate body (ALEXANDER, 1967, 1968; GRAY, 1968; NACHTIGALL, 1971). We have not considered fluid mechanics and this has led us to omit more than passing reference to a large literature on haemodynamics. Perhaps most difficult to justify is the omission of an account of muscle as a material or, in any detail, as a component in a support system. We have not considered surface tension as an environmental force and its importance to small organisms, nor have we asked just how far down the size scale our 'general' principles apply. Another book might be written about any of these subjects.

Part I
Materials

Chapter 2A
Principles of the strength of materials: Phenomenological description

2.1 Introduction

The central theme of this book is to describe and to discuss the various ways in which biological systems respond and adapt to an environment which involves mechanical stress. Since any such response or adaptation involves and may be determined by the nature and the properties of the materials which comprise the system, it is fitting that, in this chapter and the next, we should describe the behaviour of materials under applied stress. For this we do not need to be too specific. If we consider the whole spectrum of materials, both man-made and naturally formed, which are used to resist or to adapt to mechanical stresses, the list would be formidable and, even summarized, would fill several volumes. However, in terms of bulk properties, the number of patterns of behaviour is remarkably limited so that we can describe general features without troubling too much about the individual mechanisms which determine them. This is the purpose of the present chapter in which we shall deal largely with the observable phenomena, consider the relations between them, and show how we may construct models which may be used to describe the observed behaviour mathematically.

However, mere description does not, necessarily, explain the behaviour. For this we must go to the fundamental building blocks–namely the atoms and molecules of which the material is built. This is reserved for Chapter 2B where it will be seen that, in many cases, we are unable to offer a complete explanation and, once again, we must resort to models–more sophisticated ones to be sure, but still models.

This division into a phenomenological and a molecular approach is arbitrary and largely a personal choice but it seems to accord with experience. We tend, on the whole, to become familiar with the bulk properties of materials (iron is 'hard', glass is 'brittle', rubber is 'rubbery', timber is 'woody' and so on) long before we start to consider the reasons for these generalized observations.

Of course, in making generalized divisions between materials, we construct for ourselves a comparative language which involves the use of such terms as 'hard', 'soft', 'stiff', 'weak', 'ductile', 'brittle' and so on. Later we shall discuss some of these terms in more detail but, meanwhile, it is important to remember that they describe a set of properties which are not, in any sense, absolute, but which are relative to a set of given conditions. What is strong to you and me may be weak to the circus strong man. A 'weak' straw can be driven through a baulk of 'strong' timber by a hurricane and so on. Bone would be described, by most people as 'strong and brittle' and, by comparison, collagen is 'weak and ductile'. But collagen is, in fact, 'stiff and strong'. If it were not so it would not be so widely used as the basic component of all the tension members of the animal system—tendon, cartilage, etc.—all of which must withstand forces which are often considerably greater than those to which bone is subjected.

All of this may sound as though it should be in a text on metaphysics or semantics rather than in a text on biology and, indeed, this may be right. But to define the terms 'hard', 'soft', 'stiff', 'weak', etc. objectively and scientifically would lead to much hair splitting and to a terminology of wearisome complexity and prolixity. Thus, in what follows, the terms used to describe materials are used in their loose, but intuitively satisfying and acceptable, way. A 'strong' material is one which resists a force without too much change in shape; a 'weak' one does not. A 'hard' material resists indentation by a 'soft' material, and so on. Provided we remember that ranking materials in this way is dependent entirely upon the baseline which is chosen, the division of later topics into tensile, pliant and rigid materials will follow upon the general understanding and common experience of these terms.

2.2 Stress and Strain

Any attempt to quantify these properties must involve some definitions. When a force is applied to a body, the body responds in some way, usually by deforming and, if a greater force is applied, the deformation will be greater. However, it is clear that the situation will be different if the same force is applied to two bodies of different size so that, to avoid problems due to changing dimensions, the force is normally described in terms of the cross-sectional area over which it acts, this is a *stress* σ, defined simply as force/unit area and having units such as MN m^{-2}, N mm^{-2}, etc. The corresponding deformation is expressed as a *strain*—the ratio of the change in size to some basic size thus:

$$\epsilon = \mathrm{d}L/L,\ \mathrm{d}A/A,\ \mathrm{d}V/V \text{ etc.}$$

where L = length, A = area, V = volume. Being a ratio, strain has no dimensions although it is not unusual to find strain expressed as a percentage thus: $\epsilon = 0.1 \equiv 10\%$. Some texts even give strain in terms of units such as cm/cm, mm/mm, etc. There is no harm in this, and it does help to remind the reader that a strain expressed as, say, 0.1 cm/cm means that each cm of length has extended by 0.1 cm, so that a bar initially 10 cm long will extend by a total amount of 1 cm, whereas a bar 100 cm long will extend by 10 cm.

The discussion above has been written in terms of an externally applied force (stress) producing a deformation (strain). However, it must not be construed to imply that only external forces are relevant. If a material is heated it expands. This is a change of dimensions and is, therefore, a strain as defined above. Furthermore if the body is constrained in some way so that the expansion cannot occur physically, then stresses, in this case internal stresses, will be set up within the material. Any change in dimensions, no matter how caused, is a strain, and any force acting upon a point in the body produces a stress.

2.3 Linear Elasticity

Figure 2.1 (after BURSTEIN *et al.*, 1972) shows the relationship between an applied force and the corresponding extension when a sample of bone is pulled in a tension test. The portion OA of the force-extension curve is linear and this linear behaviour is exhibited initially by a very large number of materials which are called *linearly elastic* or *Hookean* (after Robert Hooke (1635–1702) who first proposed this linear relationship). Do, please, be quite clear about two things. The first is that engineers have acquired the very bad habit of describing materials simply as 'elastic' when they really mean 'linearly elastic'. This assumption is a great source of comfort to them and makes their sums very much easier than they would otherwise be. The second is that the term 'elastic', as used here, has little

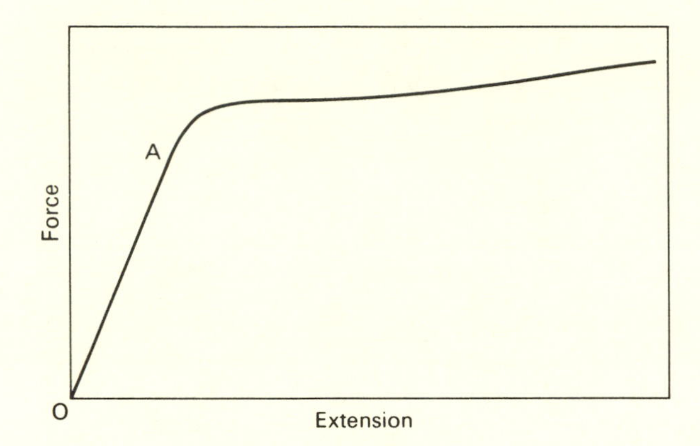

Fig. 2.1 Force-extension curve for bone (after BURSTEIN *et al.* 1972; courtesy of the *J. Biomech.*)

or nothing to do with the material 'elastic' used in your suspenders or your girdle. This is a substance which actually displays, over much of its working range, non-linear elasticity as we shall see in Chapter 4.

The assumption of linearly elastic behaviour has, in fact, led to the development of a whole branch of applied mathematics called the 'Theory of Elasticity' which sets out to examine all the possible consequences of the linear relationship at all points in a body under load. We need do no more here than establish a few of the more elementary definitions, but we may note one important feature which arises from Hooke's Law. It is that force and displacement are uniquely related by a single constant. Thus if a Hookean body is loaded with a force F_1 it will sustain a deflection, E_1 and if an additional force F_2 is now added, the deflection will become algebraically $(E_1 + E_2)$. Precisely the same deflection would have occurred if we had loaded it with a force $(F_1 + F_2)$ right at the start. This is the *principle of superposition* which forms the foundation of the majority of engineering design methods. The fact that it does not matter if we apply the second force a day, a month, or a year after the first implies the principle of *time independence* in the properties of Hookean materials.

2.4 The Elastic Moduli

The initial linear part, OA, of the force-deflection curve for bone in Fig. 2.1 will remain linear if the units are converted to stress and strain as defined above. The slope of the stress–strain curve is the ratio between the tensile stress and the strain and is called the *elastic modulus* or *Young's modulus*, after Thomas Young who first conceived of the idea in an almost totally incomprehensible paper in 1807. It is conventionally given the symbol E and has the units of stress (since it is a stress divided by a strain which is a dimensionless quantity)

$$E = \sigma/\epsilon \tag{2.1}$$

We should note here the convention which is normally followed in converting force and deformation into stress and strain. These quantities were defined in the days when most testing was applied to fairly hard, non-deformable metals which do not change their cross-sectional area greatly when stretched, so that the stress could, without much loss of accuracy, be referred to the original area of the specimen. Similarly the strain could be referred to the original length. Thus engineers have got into the way of defining these conventionally as:

$$\text{stress } \sigma = F/A_0 \quad \text{strain } \epsilon = \mathrm{d}L/L_0 \tag{2.2}$$

where F is force, A_0 is the original cross-sectional area and L_0 is the original length. For linearly elastic materials operating under small strains, this does not matter too much, but it is clearly a useless convention when one considers the case of a highly stretchable material like elastin. If the volume of the material is to remain constant, it follows that if we stretch it, the area decreases so that the force is acting over a smaller area than we started with and the stress is consequently higher. Thus we should define a *true stress* as

$$\sigma_t = F/A \tag{2.3}$$

where A is the actual, instantaneous area of the specimen. Similarly with the strain: as the material stretches, each incremental increase in length is being referred to a specimen which is getting progressively longer and we therefore define a *true strain* as:

$$\epsilon_t = \int_{L_0}^{L} \mathrm{d}L/L = \ln L/L_0 \tag{2.4}$$

where L is the actual length.

Because the integral leaves us with a logarithm, this is often called *logarithmic* or *natural strain*. It is of the greatest importance to distinguish carefully between conventional and true values when very extensible materials are being tested.

Even a linearly elastic material must show some change in area when it is extended. The relative decrease in area per unit increase in length is expressed by

Poisson's ratio which may be defined as:

$$\nu = -\epsilon_y/\epsilon_x = -\epsilon_z/\epsilon_x \qquad (2.5)$$

for a bar of material which is stressed in the x-direction and which contracts in either the y- or z-directions. Note that this definition is only true for isotropic materials, anisotropic materials (timber, bone, etc.) show different amounts of contraction in different directions and the equality of *Eq. 2.5* does not hold.

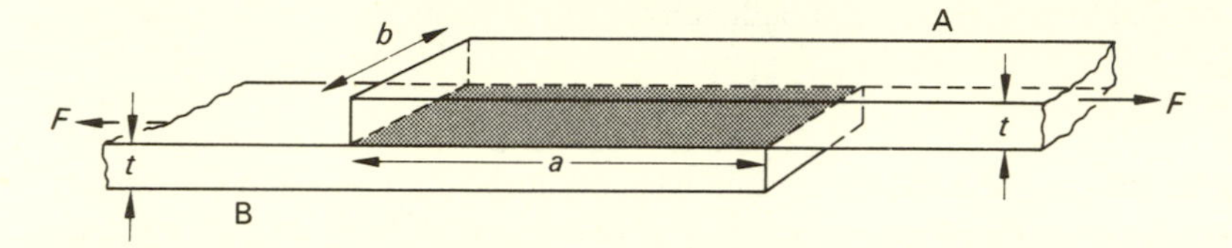

Fig. 2.2 Shear stress over area *ab*.

Shearing stresses (Fig. 2.2) are treated slightly differently. Here, again, the stress is defined as force per unit area, i.e., as

$$\tau = F/A \qquad (2.6)$$

but note that the area is now the *area over which the force is applied* and *not* the cross-sectional area of the member which is used to *apply* the shearing force. Thus in Fig. 2.2 the two members A and B are in *tension* but the *shear stress* is obtained by dividing the force F by the shaded area (ab).

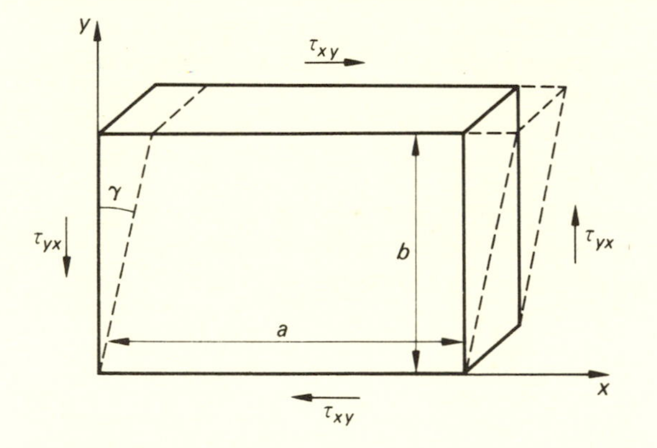

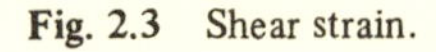
Fig. 2.3 Shear strain.

Shear stress produces *shear strain.* Figure 2.3 shows that an originally rectangular block is distorted into a parallelepiped by translation of the corners through an angle γ. If γ is measured in radians it is defined as the *shear strain* and is non-dimensional so that, for a linearly elastic material, we can write

$$\tau = G\gamma \qquad (2.7)$$

where G is the *shear modulus* of the material having the dimensions of stress (as with the Young's modulus E). Provided γ is small, shearing produces no change in volume.

Finally we can complete the definitions by considering the behaviour of a block of linearly elastic material under the action of a hydrostatic stress: usually pressure p. The strain is now most conveniently related to the volume change $\mathrm{d}V/V$ and stress is related to volume change via the *bulk modulus* K where

$$p = K\,\mathrm{d}V/V. \tag{2.8}$$

The various moduli may be shown to be related as follows:

$$G = \frac{E}{2(1+\nu)} \tag{2.9}$$

$$K = \frac{E}{3(1-2\nu)} \tag{2.10}$$

and consequently we can write

$$E = \frac{3G}{1 + G/3K}. \tag{2.11}$$

For most materials $3K \gg G$ so that *Eq. 2.11* approximates to

$$E \simeq 3G(1 - G/3K) \tag{2.12}$$

from which it is clear that Young's modulus is mostly controlled by the resistance to shearing (G) and not by change of volume (K). This is because in uniaxial stressing the surfaces are free to move inwards (or outwards) so that the resistance to volume change is not tested to any great extent in a tension or compression test. Thus uniaxial strength is quite a different property from hydrostatic strength.

2.5 Poisson's Ratio

Poisson's ratio is a much neglected property. One reason for this is that in non-isotropic materials it is formidably difficult to obtain reliable values for it and our knowledge, so far as biomaterials is concerned, is very limited. Another reason for neglect is that it can only show a very limited range of values. Table 2.1 shows how the ratios E/G and K/E (see *Eqs. 2.9* and *2.10* above) vary with Poisson's ratio. It can be seen, from *Eq. 2.10* that a material having $\nu = 0.5$ can undergo no volumetric strain whatever, irrespective of the stress, i.e., it is incompressible. This is characteristic of a liquid and Table 2.1 shows that at $\nu = 0.5, E = 3G$. At the other extreme for $\nu = 0, E = 2G = 3K$. Thus Poisson's ratio is a measure of the relative ability of a material to resist dilation (change of volume) and shearing (change of shape). A material with $\nu \to 0.5$ will resist change of volume but change shape easily and, as Table 2.2 shows, has a high value in those materials whose atoms can move freely past each other (e.g. rubber). In materials whose atoms are highly

Table 2.1

Poisson's ratio ν	Elastic E/G	Constants K/E
0.0	2.0	0.333
0.10	2.20	0.417
0.20	2.40	0.556
0.30	2.60	0.883
0.40	2.80	1.667
0.50	3.00	–

Table 2.2 Poisson's ratio for various materials

Material	ν	Reference
Sandstone	0.10	
Vitreous silica	0.14	
Bone	0.13–0.30	PIEKARSKI (1968); MCELHANEY (1966)
Apple flesh	0.21–0.34	CHAPPELL and HAMANN (1968)
Corn stalk (with nodes and pith)	0.23	PRINCE and BRADWAY (1969)
Glass	0.24	
Iron, Steel	0.30	
Timber	0.30–0.5	Various
Corn, horny endosperm	0.32	PRINCE and BRADWAY (1969)
Polyethylene	0.33	
Hard rubber	0.39	
Soft rubber	0.49	
Potato flesh	0.49	FINNEY and HALL (1967)
Gelatin gel	0.50	

bonded and therefore resist a change of shape $\nu \rightarrow 0$ (e.g., stone, ceramics, etc.).

Few reliable values for ν are available for biomaterials and those that are available should be treated with reserve. The anisotropy of the material makes accurate measurement difficult. Many of the published values are calculated from independent measurements. There is a distinct possibility that, in these materials, ν may vary with the stress level and with the rate of strain. If this were so, it would be of great importance for it would indicate that the material is changing its 'state', i.e., from a less 'fluid' to a more 'fluid' type of behaviour in its attempt to accommodate to the applied forces.

2.6 Elastic Resilience–Stored Energy

We have seen that a linearly elastic material deforms reversibly, i.e., it returns to its original length when the load is removed. The product of stress and strain gives energy, in this case *strain energy* and the amount is simply the area under the linear portion of the stress strain curve. i.e.,

$$U = \tfrac{1}{2}\epsilon\sigma = \tfrac{1}{2}\sigma^2/E = \tfrac{1}{2}\epsilon^2 E \tag{2.13}$$

per unit volume of material. Strain energy is well exemplified by a clock spring which is loaded by winding the clock and which then releases the stored energy to drive the mechanism. The capacity to absorb and release energy in this way is often called the *elastic resilience* of the material.

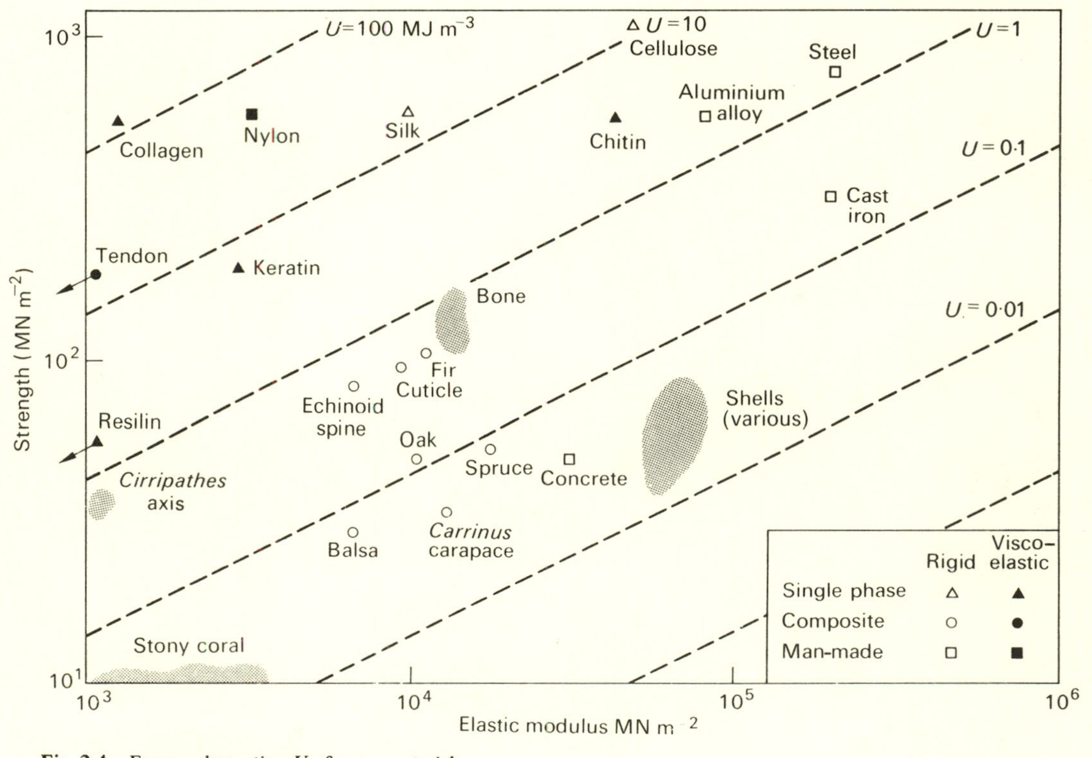

Fig. 2.4 Energy absorption U of some materials.

2.6 *Elastic Resilience–Stored Energy*

Many of the rigid biological materials are, substantially, elastic up to the point of failure and for these we can place a rather more specific interpretation on elastic resilience. Clearly the limiting amount of energy which can be absorbed per unit volume will be given by *Eq. 2.13* when the stress σ equals the fracture stress of the material. Note, however, that this would not necessarily be the maximum amount which could be absorbed because, if the stress built up slowly, it may be possible for the material to deform in other ways besides truly elastically so that additional energy would be absorbed. If we plot the strength and modulus of various materials such as shell, bone, etc., as shown in Fig. 2.4, the dashed lines give values of the strain energy per unit volume of the material from *Eq. 2.13.* Comparison with commonly used engineering materials shows that the energy absorption of shells is not good in absolute terms but that of bone and cuticle is considerably better and is, in fact, comparable with many of the commonly used engineering materials such as timber, cast iron and the harder aluminium alloys.

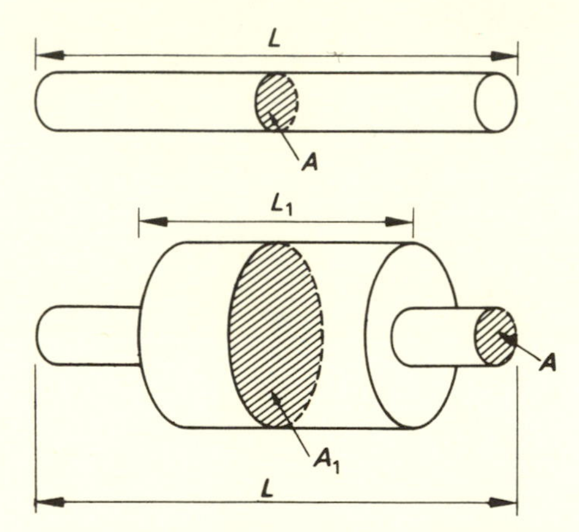

Fig. 2.5 The energy absorption of a body of length L and cross-sectional area A is compared with that for the same body with a swollen part of length L_1 and area A_1. This diagram is explained in the text.

The previous discussion has been concerned solely with the materials aspect of resilience but the actual physical shape of the specimen exerts a profound influence upon the capacity to store energy elastically. We shall consider this in more detail later, but meanwhile one observation is of interest here. It is that, under tensile loading, the shape of the cross-section, be it tube, rod or whatever, should be the same all along the length if the elastic resilience is to be a maximum. COTTRELL (1964) derives the following argument. Consider a uniform rod of length L and cross-sectional area A and let the material have a failing stress of σ_f. This can absorb energy amounting to $\sigma_f^2 AL/2E$ without failure. Now let us increase the area to A_1 over a limited length L_1 (Fig. 2.5). The energy absorbed by the unchanged part is

$$\sigma_f^2 A(L - L_1)/2E$$

and in the thickened part it is

$\cdot (\sigma_f A/A_1)^2 A_1 L_1/2E$

so that the total energy absorption is

$$\frac{\sigma_f^2 AL}{2E}\left[1-\frac{L_1}{L}\left(1-\frac{A}{A_1}\right)\right] \qquad (2.14)$$

and since $L_1 < L$ and $A_1 > A$ the resilience is *decreased* by increasing the cross-sectional area.

ALEXANDER (1968) points out that the term 'resilience' can have other meanings and uses it to mean the work recovered from an elastic recoil expressed as a percentage of the work put in. For a truly Hookean material this is unity, but for a material showing non-linear elasticity such as rubbers, the value is less than unity, so that in Fig. 2.6 it is the shaded area under the unloading curve (b)

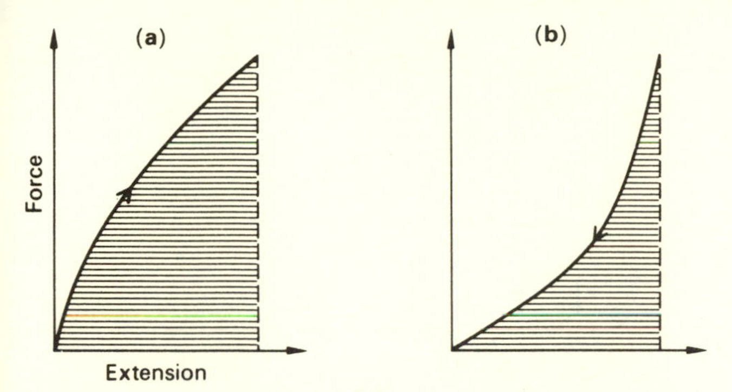

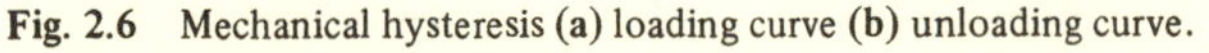
Fig. 2.6 Mechanical hysteresis **(a)** loading curve **(b)** unloading curve.

expressed as a percentage of area under the loading curve (a). An engineer would probably call this definition the 'rebound resilience', since it is conveniently measured by considering the magnitude of successive oscillations when a stretchable material like rubber, elastin, abductin, etc., is allowed to vibrate. We may define the term resilience R used in this way as

$$\ln\left(\frac{100}{R}\right) = \pi \tan \delta \qquad (2.15)$$

where $\tan \delta$ is the loss tangent defined in Section 2.15.

2.7 Elastic Stress Concentrations

It is common experience that if we continue to increase the load on a material it will eventually break. The applied stress is resisted by the material and when the limit is reached, we define this as the *strength* of the material, i.e., the maximum resistance to an applied force.

The concept of 'strength' as defined above assumes (though it does not actually say) that the force is applied uniformly over the whole cross-section of the specimen. We can represent this situation by considering the stress on each element of the body to be represented by a series of lines. These are often called 'force trajectories' and their density, over a given area, may be taken as representing stress. In a uniformly stressed body having a smooth, regular cross-section, the force trajectories are uniform (Fig. 2.7a), but the presence of a notch or hole, which supports no stress, causes the trajectories to bunch together so that the distribution of stress at the tip of the notch or hole is much more concentrated than it is in the bulk material. For this reason these discontinuities in the way of the force trajectories are called 'stress concentrations' or 'stress raisers'.

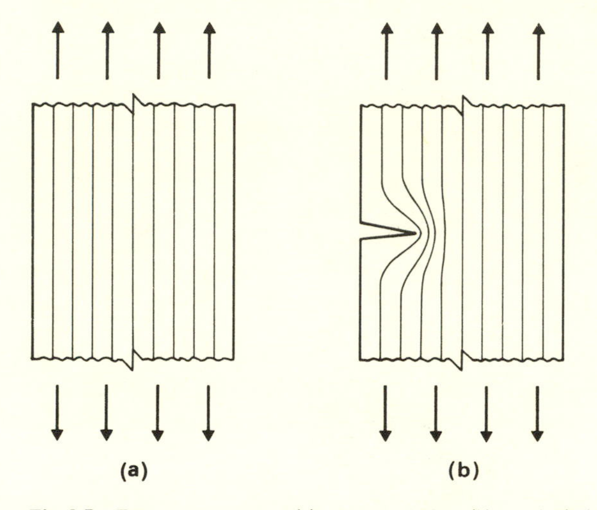

Fig. 2.7 Force trajectories in (**a**) unnotched plate (**b**) notched plate in simple tension.

The mathematical analysis and experimental verification of the magnitude of the local stresses produced occupies a fair proportion of engineering literature and a great deal of information is available (e.g., FAUPEL 1964; PETERSON, 1953; TIMOSHENKO, 1956; etc.). The analysis is too complex to be studied in detail here so that we will simply quote, without proof, one fairly general case—that of an elliptical hole in an elastic plate under an applied stress S (Fig. 2.8a). The full solution has been given by INGLIS (1913). For our present purpose we shall consider only the case where the stress is applied along either the major or the minor axis of the ellipse.

When loaded along the x-axis ($\beta = 0$) the stress at the ends of the minor axis is

$$\sigma_x = S_x(1 + 2b/a) \qquad (2.16)$$

and at the ends of the major axis it is $\sigma_y = -S_x$. Conversely when loaded along the y-axis ($\beta = \pi/2$) the stress at the ends of the minor axis is

$$\sigma_x = -S_y$$

and at the ends of the major axis it is

$\sigma_y = S_y(1 + 2a/b).$

Clearly a circular hole in an elastic plate is merely a special case where $a = b$ and

$$\sigma_x = 3S_x \quad \text{and} \quad \sigma_y = 3S_y \tag{2.17}$$

showing that the stress at the edge of the hole rises to a value which is three times that of the applied stress. Notice that if the sign of the applied stress is reversed, i.e., compression instead of tension, the maximum value of stress is reached along the compression axis as shown in Fig. 2.8b.

COTTRELL (1964) has pointed out that the magnitude of the disturbance in the stress pattern around a hole varies as d^{-2} where d is the distance—so that the effect is small at points remote from the hole. This result can be easily ex-

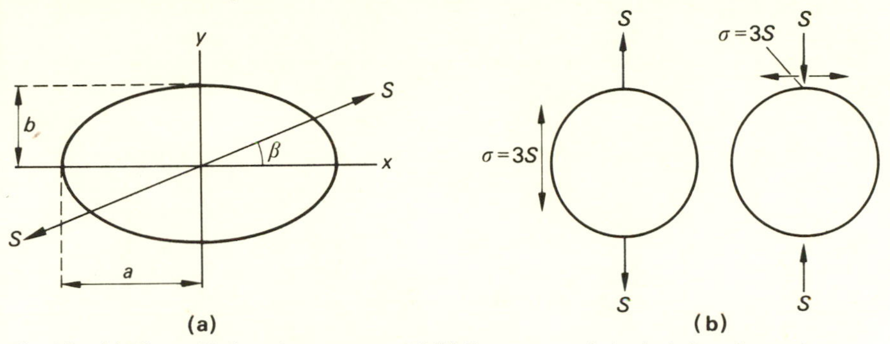

Fig. 2.8 (a) Elliptical hole under tension at *S-S* (b) Stresses around circular hole under tension and compression.

tended to a plate containing several holes or to holes near free surfaces, provided that the distances between them are not too small. If a small hole exists close to a large hole at $\beta = \pi/2$, the maximum stress at the edge of the small hole is then, approximately, 3^2S. If we extend this argument to a group of n successively small holes, each lying in the maximum stress field of the preceding one, the stress at the edge of the smallest is then of order 3^nS. Let the radii decrease by a factor of 10 from one to the next and let the radius of the largest be R and that of the smallest be r. Then $R = 10^n r$ and the concentrated stress σ_c at the smallest hole is approximately given by

$$\tfrac{1}{2}n = \tfrac{1}{2}\log_{10}(R/r)$$

i.e., by

$$\sigma_c \simeq S\sqrt{R/r}. \tag{2.18}$$

This is an extremely crude approximation but it serves to indicate the very large values of stress which may occur locally at a sharp corner or at the tip of a crack or notch.

2.8 Fracture of Linearly Elastic Solids

The basic theory of fracture in an isotropic elastic solid is due to GRIFFITH (1921) but before describing it we shall define what we mean by the term 'fracture'.

It is often used (generally by people who should know better) as being synonymous with the term 'failure' and it is true that, in most applications, fracture of the material does also mean that the artifact has failed in service. In exception to this, many pressure vessels are built with a weak plate which is deliberately designed to fracture if the pressure becomes too great. Fracture of the plate is a safety device. However, there are many cases where the structure (be it biological or man-made) may be deemed to have failed without fracture occurring. Thus materials may deform excessively (the car doors will not close or the animal feels pain or cannot co-ordinate), the cross-sectional area is reduced (by excessive wear or corrosion) or the material may deform permanently into a shape which is no longer capable of transmitting the load for which it was designed. All of these may be counted as failure in the sense that the structure is no longer able to fulfil its original function.

Fracture, quite specifically, requires that new surfaces shall be created. Since the creation of a new surface requires the supply of energy, we shall find that fracture is a process involving the balance of energies within the material. In the following section we shall only concern ourselves with one form of energy—namely strain energy from a source external to the system and, for simplicity, we shall only consider the case of uniaxial stressing—though the theory can quite easily be extended to cover the case of multiaxial stressing.

We can estimate the theoretical strength of a solid in a number of ways. The simplest method is to assume a totally homogeneous body which is, at all points, uniformly stressed to its breaking point. All the interatomic bonds must break simultaneously and the solid changes instantaneously to vapour. The energy of

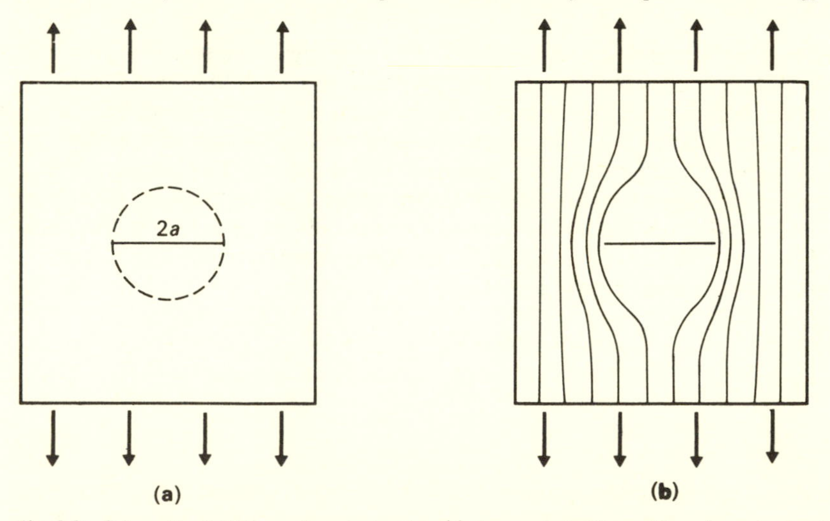

Fig. 2.9 Schematic Griffith crack under tension (**a**) Assumed situation with no force in circular area around crack. (**b**) More probable distribution of force trajectories.

vaporization, then, sets an upper limit to the fracture strength and this gives a theoretical value of about $E/5$, where E is the Young's modulus of the material. A more sophisticated treatment involving a detailed consideration of the interatomic forces leads to an estimate of about $E/20$. As a mean value then, we suggest that we could take a value of about $E/10$ as a working hypothesis.

Real solids fail at much lower values than this when they are tested in bulk form. 'Brittle' solids like glasses and ceramics break, with little or no measurable permanent deformation, at stresses of about $E/1000$ while 'ductile' materials like metals and linear polymers start to deform at about the same value and finally break at stresses of about $E/100$. Very strong metals may reach $E/50$ and a few, very carefully prepared materials are within sight ($E/20$) of the theoretical estimate.

The problem of this discrepancy was tackled in 1921 by A. A. GRIFFITH using glass as the archetypal 'brittle' material. It was suggested that the discrepancy could be explained if it were assumed that glass contained a number of tiny cracks which acted as stress concentrations.

Consider the crack of Fig. 2.9 and for simplicity assume the stress trajectories enclose a circular area as shown in a plate of unit thickness. The 'missing' strain energy of this region is

$$U = \left(\frac{\sigma^2}{2E}\right) \pi a^2$$

and if the crack now extends at each end by da the change in strain energy, i.e., the energy released is

$$\mathrm{d}U = \left(\frac{\sigma^2}{E}\right) \pi a \, \mathrm{d}a$$

if we neglect higher terms in da. Now the material opposes this. As the crack of Fig. 2.10 extends from (1) to (2), all the other bonds next in line must each stretch into the position occupied by its immediate predecessor and the total work done is exactly equal to the work required to break the bond, i.e., to the

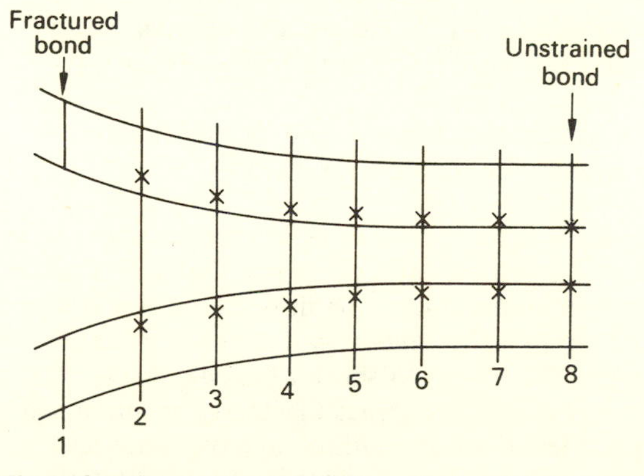

Fig. 2.10 Diagram of a crack tip in an elastic solid. As the crack extends from (1) to (2) each pair of atoms must move to the position shown by (x).

surface energy S per unit area. Thus for an extension of da at each end of the crack, the total work is

$$dW = 4S \, da \tag{2.19}$$

since four new surfaces are produced. The critical condition for crack propagation is therefore

$$4S \, da = \frac{\pi\sigma^2 a \, da}{E}$$

or

$$\sigma = \left(\frac{4ES}{\pi a}\right)^{1/2} \tag{2.20}$$

Actually, by using a more precise integration for the true shape of the stress trajectories (Fig. 2.9) Griffith obtained

$$\sigma = \left(\frac{2ES}{\pi a}\right)^{1/2} \tag{2.21}$$

but, for all practical purposes, and remembering that we are making quite a few assumptions already, we can take $\sigma \simeq (ES/a)^{1/2}$ as near enough.

Since the Griffith theory forms the cornerstone of fracture theory it is, perhaps, as well to digress a little here and to consider whether it provides both a necessary and a sufficient condition.

In order that any physical change may take place, it is necessary that two conditions be satisfied:

1 A *thermodynamic* condition which requires that the change must take place in order to minimize the total energy of the system. An automobile on a hill rolls down not up, and its potential energy is converted to kinetic energy. When the automobile comes to rest its kinetic energy is zero and its potential energy has been reduced. It can go on doing this until its potential energy is zero, at sea level for example.
2 A *mechanical* condition which requires that there should be a mechanism which permits (1) to occur. The automobile should have wheels.

The first condition is fairly easily seen. If we consider the two terms strain energy and surface energy as a function of the crack length, we obtain Fig. 2.11. Combining these, the dashed curve shows that an instability condition is reached at some critical length a_0 when the rate of release of strain energy equals the rate at which work is done in creating new surface. For all cracks shorter than a_0 energy must be supplied but beyond a_0 the energy available exceeds the energy required and a crack which has reached its critical length must continue to grow, since an increase in length can only *decrease* the stress which is needed to propagate the crack. In fact the crack will not only grow but its rate of growth will be accelerated.

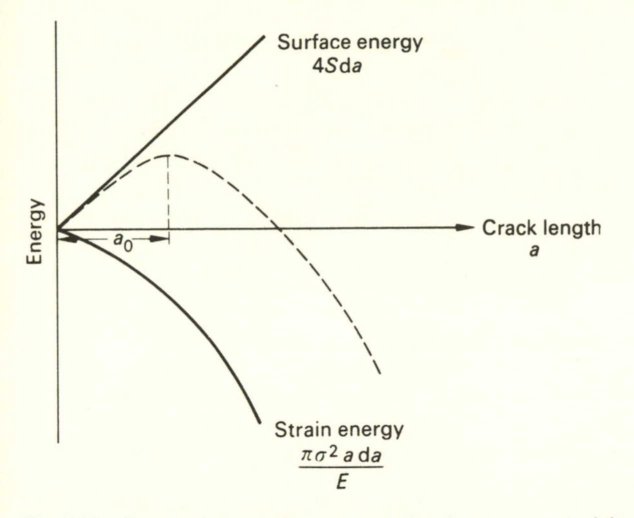

Fig. 2.11 Curves relating surface energy and strain energy to crack length (schematic). The resultant curve (dashed) has a maximum value at a_0 when the gradients of both curves are the same. The symbols are explained in the text.

In this way a fast, unstable crack is produced under a falling stress and, in extreme cases, the strain energy is so much in excess of that which is required that the crack can produce more and more surfaces by forking and splitting to produce a 'shatter' type of crack.

The 'mechanical' requirement is not quite so easy but can be seen from Fig. 2.10. In effect it requires that a relatively low level of strain energy in the bulk of the material shall be converted into a more concentrated form at the tip of the crack, i.e., that the surface energy term can be expressed in terms of purely local strains. The argument leading to *Eq. 2.19* and Fig. 2.10 shows that this can, indeed be so—though a detailed analysis is more difficult and takes us beyond the scope of the present book.

However, it is apparent that both conditions are satisfied so that, for truly elastic (i.e., Hookean) materials, the Griffith condition is both necessary and sufficient. We shall see later how this must be modified to take account of the possibility that some non-elastic deformation may (and, in real materials, usually does) occur but, meanwhile, we may note the following points:

1 That there exists, for a given material having fixed values of E and S and loaded to a given stress, a critical crack length a_0.
2 That if the size of the material be reduced below this length the fracture stress must increase, since it follows that a continuous fibre of diameter, for example 5 nm, cannot contain a crack of length 10 nm. Other things being equal, then, we expect fine fibres to be stronger than bulk solids.
3 That the Griffith theory is, in effect, a 'weakest link' theory and the fracture strength of a material will be determined by the size and orientation of the largest crack. The distribution of cracks will, in general, be random so that we expect a statistical spread in fracture strength as size or length increase.

4 That in a bundle of fibres, there is a higher probability of finding one which contains a crack which is close to the critical size. Thus one fibre may fail at a low stress and the remaining fibres must now each carry a slightly higher stress—load being fixed. This means that the fibre containing the next largest crack may fail, and so on. Thus in an extreme case, failure of the whole bundle may be determined by failure of the weakest fibre in it.
5 That the requirement of a balance between strain energy and surface energy represents only a *minimum* requirement. Strain energy may be used up in deformation, heat, or sound. We shall consider the first of these later.

2.9 Summary of Properties of Linearly Elastic Solids

1 Stress is linearly proportional to strain ($\sigma = E\epsilon$)—the proportionality constant being the elastic or Young's modulus (E), the shear modulus (G) or the bulk modulus (K) depending upon whether the stress applied is tension, shear or pressure.
2 The magnitude of the elastic modulus constant is a measure of the stiffness (or rigidity) of the material. The three moduli are simply related to each other via a constant, Poisson's ratio ν, whose magnitude lies between zero (for an ideally rigid material) and 0.5 for a fluid. The magnitude of ν is a measure of the relative capacity of the body to resist shear (change of shape) and dilatation (change of volume).
3 The relationships between the moduli are

$$G = \frac{E}{2(1+\nu)}, \qquad K = \frac{E}{3(1-2\nu)}$$

whence

$$E = \frac{3G}{1 + G/3K}.$$

4 Linear elasticity implies the ability to store energy reversibly. The capacity to store energy is a simple function of the stress and the modulus. It will be high in materials having a high fracture stress and a low modulus and low in materials with a low fracture strength and high modulus.
5 The fracture strength is not a material property but is a measure of the perfection of the structure, being determined by the size of the largest crack or defect in the material. For every elastic material, at any given stress level, there is a critical crack length beyond which fracture occurs in a rapid, thermodynamically unstable manner.
6 The defects which control the fracture strength of an elastic solid are statistically distributed. The chances of finding a defect of critical size therefore increase with the physical size of the specimen. The highest strengths will, therefore, be expected in fibres rather than in bulk solids.

2.10 Viscosity and Relaxation

We do not need to discuss the atomic structure of a linearly elastic solid in order to be able to visualize it as one which resists the application of an applied force by virtue of the strength of the bonds between the atoms of which it is composed. The fact that the material deforms under load and that the original dimensions are restored when the force is removed signifies, also, that no permanent or large scale movements, as of atoms or molecules translating past each other, occurs.

We now turn, briefly, to the opposite extreme of behaviour, that of viscous flow in which the applied energy is dissipated, within the material, by some process involving relative motion of the particles of which it is composed.

An ideal liquid is one which obeys Newton's law that the *flow rate* $\dot{\epsilon} = d\epsilon/dt$ is proportional to the applied stress, i.e., that

$$\sigma = \eta\dot{\epsilon} \qquad (2.22)$$

where the constant η is the viscosity. We have used the convention for *tensile* stress σ, above, despite the fact that the more conventional way of writing *Eq. 2.22* is in terms of shear stress. The concept of a liquid carrying a tensile stress may seem odd but the use of the conventional symbol for shear stress τ will land us in a lot of trouble with symbols later on. So assume that σ in *Eq. 2.22* refers to *any* stress which can produce an appropriate flow rate.

Pure Newtonian flow, as defined in *Eq. 2.22* rarely exists in fact, although the flow of gases and of low molecular weight liquids approximates very closely to it. The equation must be very significantly modified if there are suspended particles present as these disturb the flow pattern and energy is dissipated by friction between the fluid and the particles. The most general case requires the use of a polynomial

$$\eta = \eta_0(1 + \alpha V + \beta V^2 + \cdots)$$

where η_0 is the viscosity of the pure fluid, V is the volume fraction of particles and α, β, etc., are constants. Providing that the fraction of suspended particles is so small that they do not interact with each other this equation reduces to Einstein's equation

$$\eta = \eta_0(1 + \alpha V).$$

The viscosity of a fluid is a property resulting from attractive forces between particles of the fluid. These attractive forces are always present, but they are particularly important in liquids where they provide the binding force which keeps the particles in the liquid rather than the gaseous state. In a fluid at rest, the sum of all attractive forces acting on each particle must be zero, as any net force would cause motion. If, however, we consider a fluid moving through a tube, we observe that a velocity gradient is established with low velocity next to the tube wall and maximum velocity at the centre of the tube. In such a gradient the particles in one layer will be moving relative to those in another. We can find a model for this situation by considering a single particle moving under an external force past a set of stationary particles. The attractive interactions between neighbouring particles can be thought of as weak bonds, and when a particle moves relative to its neighbours, these bonds will be deformed. In the process of bond formation, energy will be released as heat. Thus, in a viscous process mechanical energy must be supplied to overcome attractive forces between particles, but this energy is dissipated as heat when new attractive interactions are formed. As long as there is a driving force there will be relative movement, and energy will be dissipated. The rate of energy dissipation is dependent on the relative velocity between the particles, and this, in turn, is dependent on the magnitude of the force.

This deformation requires energy which is supplied by the external driving force. The attractive forces in a fluid are effective only over a range of a few nm, so the bonds are readily broken by the driving force, but the moving particle soon approaches another stationary particle and new bonds will be established.

Thus we may visualize viscous flow as a process in which an atom or molecule moves into a 'hole' leaving behind it another 'hole' to be filled. In order that a 'hole' shall be filled, the atom must break clear of its neighbours which are holding it in a structural group. This happens when the amplitude of its vibration increases to the point where it can pull away. The vibrating atom must, therefore, overcome some sort of an energy barrier—applied stress will help lower the barrier and so will assist the effects of temperature fluctuations which increase the amplitude of vibration. The frequency of a given fluctuation is proportional to $\exp(\Delta H/kT)$, where ΔH is an activation energy, k is Boltzmann's constant and T is temperature, so that the rate of flow (and hence the viscosity) may be expressed as

$$\eta(T) = A \exp(\Delta H/kT) \qquad (2.23)$$

where A is a constant.

The distinction between solid and liquid behaviour is often drawn, rather arbitrarily, at a viscosity of 10^{15} poise (10^{14} kg m^{-1} s^{-1}), about the value for pitch, but a much more satisfactory way of expressing the difference is in terms of the ability of the material to relax under an applied load. An ideally elastic solid will sustain a given load for any length of time and when the load is removed the original dimensions are restored in very short time—of the order of the period of vibration of the atoms. A viscous liquid will continue to deform under the load and when the load is removed, it remains in its deformed shape. The elastic solid will have a *relaxation time* approaching infinity and the viscous liquid has a relaxation time close to zero. Many, perhaps most, real materials behave in a way that is somewhere between the two extremes and the relaxation time is a convenient way of describing their behaviour.

We derive it quite simply, as follows. For a Hookean body loaded to constant stress σ, the strain is ϵ. If we relax the stress over time t we obtain

$$\frac{d\sigma}{dt} = E\frac{d\epsilon}{dt}$$

where E is the modulus.

If some viscous flow can occur while the body is loaded the stress tends to decrease at a rate which depends on the initial value

$$\frac{d\sigma}{dt} = E\frac{d\epsilon}{dt} - \frac{\sigma}{\tau} \tag{2.24}$$

where τ is a constant having the dimensions of time and is the *relaxation time* of the material. If we extended the body by a fixed strain and then held it constant, we can obtain the stress after time t by integrating *Eq. 2.24* for $d\epsilon/dt = 0$ to obtain

$$\sigma(t) = \sigma_0 \exp(-t/\tau) \tag{2.25}$$

where σ_0 is the initial stress. Thus the relaxation time is the time required for the stress to fall to $1/e$ of its initial value. Clearly the relaxation time involves both the elasticity and the viscosity and we simply define it here, without any loss of generality as

$$\tau = \eta/E. \tag{2.26}$$

leaving the demonstration of the validity of this to Section 2.16.

Thus the relaxation time as defined in *Eq. 2.26* is a measure of the ratio of viscous to elastic behaviour. The really important feature, however, is the relationship of the relaxation time to the loading time t'. Silicone putty is a good example—under slow loading the time of load application $t' \gg \tau$ and the material flows like a liquid. Under impact loading $t' \ll \tau$, no relaxation occurs and the material bounces elastically. Pitch and asphalt are other classic examples. The importance of this will be shown in following chapters where it will be seen that virtually all biomaterials exhibit relaxation phenomena and it is essential that tests on these materials be carried out under conditions where the rate of loading is known, if the results are not to be totally misleading.

In the discussion so far, we have assumed that the extension is fixed and we allow the stress in the material to decay. The relaxation time τ is a measure of the rate of decay, as less stress is needed to sustain a given extension. However, we can reverse the situation by keeping the stress fixed and allowing the strain to vary. The argument is exactly the same and the rate of change of strain at constant stress is called the *retardation time* and has the symbol τ'.

2.11 Linear Viscoelasticity

A very wide range of materials show some combination of linearly elastic and viscous behaviour—many linear polymers and rubbers are familiar examples. However, this behaviour is especially noticeable in the pliant biomaterials and, to some extent also, in the tensile biomaterials. A discussion of the molecular theories is deferred to Chapter 2B, meanwhile we describe the experimentally verifiable behaviour of such materials.

First we must define the term 'linearly viscoelastic'. We saw, in Section 2.3 that a linearly elastic or Hookean material possesses a unique relationship between stress and strain so that if loaded to a stress σ_1, at time t_1, it develops a strain ϵ_1, given simply by $\epsilon_1 = \sigma_1/E$ and, moreover, if the stress is sustained for a period of time $t_2 - t_1$, the strain remains the same returning to zero only when the stress is removed. This sequence is shown, diagrammatically, in Fig. 2.12a. The response of a linear polymer such as rubber, polyethylene, or collagen to the same loading sequence is shown in Fig. 2.12b. Here the strain increases (usually asymptotically) as a function of time and, on unloading, decreases with

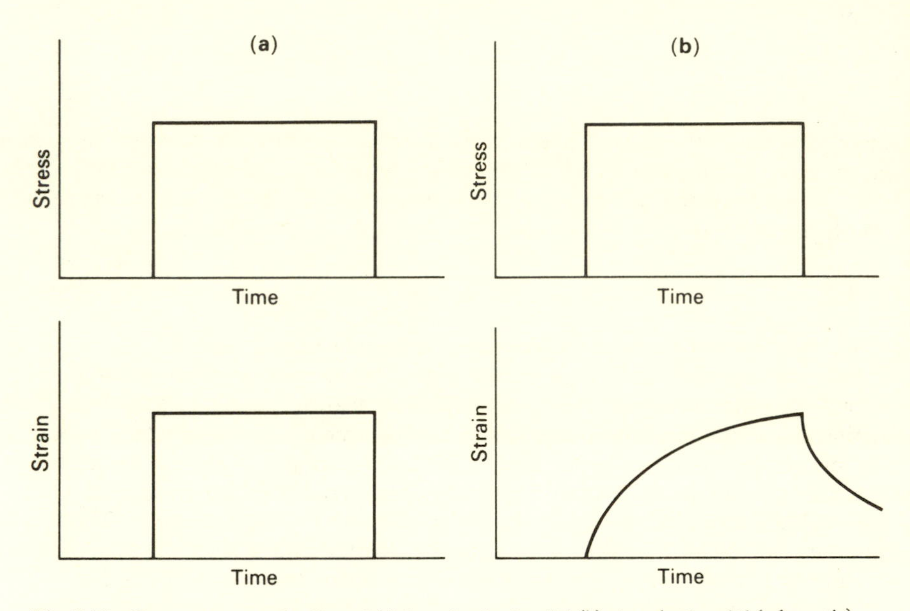

Fig. 2.12 Response to step loading of (a) linearly elastic solid (b) viscoelastic solid (schematic).

time. The test of *linear* viscoelasticity is that time is additive. Thus if a stress σ_1 produces a strain ϵ_1 in time t_1 and σ_2 produces ϵ_2 in t_1 then $(\sigma_1 + \sigma_2)$ will produce $(\epsilon_1 + \epsilon_2)$ in time t_1. More generally if a sequence of stresses $\sigma_1 \sigma_2 \ldots \sigma_n$ be applied at times $t_1 t_2 \ldots t_n$ then the strain at some time t is given by

$$\epsilon(t) = \sum_{i=1}^{n} \sigma_i D(t - t_i) \qquad (2.27)$$

where D is a material constant to be discussed later. This is a statement of *Boltzmann's superposition principle* and proper manipulation of this principle permits the analysis of time-dependent behaviour in much the same way as the analysis of the time independent behaviour of a Hookean solid. Materials which obey this principle are said to be linearly viscoelastic—though note that the stress–strain curve itself need not be linear. Indeed, it very rarely is.

2.12 Creep and Stress Relaxation

The experiment described above is known as a tensile creep experiment—the name is self explanatory. But it is clear that problems arise as soon as we start trying to define a modulus of elasticity. The stress is fixed but the strain at t_1 is less than that at t_2 so that the modulus, expressed as $E = \sigma/\epsilon$ would have different values if measured at t_1 and t_2 and would need to be defined in some other way such as, say

$$E(t) = \sigma/\epsilon(t)$$

where σ is the constant stress and $\epsilon(t)$ is the strain at some time t.

To avoid confusion with the phenomena of stress relaxation (to be described next) we usually express creep data in terms of a *compliance*, D, defined as the ratio strain/stress and having units of reciprocal stress, e.g., $m^2\ MN^{-1}$, etc. In these terms the behaviour is now defined as

$$\epsilon(t) = D(t)\sigma \qquad (2.28)$$

where $D(t)$ is the *tensile creep compliance*, and σ is the constant stress.

Similar behaviour occurs if the strain is fixed (by extending a specimen to a predetermined length). In this case a given initial stress is required and, over a period of time, the stress required to maintain that extension diminishes. This is known as *stress relaxation.* By analogy with *Eq. 2.28* we define a *relaxation modulus* $E(t)$ as

$$\sigma(t) = E(t)\epsilon \qquad (2.29)$$

where $\sigma(t)$ is now the time varying stress and ϵ the fixed strain.

The same phenomena occur if the test is made by shear or by volume deformation. Expressions analogous to *Eqs. 2.28* and *2.29* lead to the definition of shear relaxation modulus $G(t)$, the shear compliance $J(t)$, the bulk relaxation modulus $K(t)$ and bulk compliance $B(t)$ (often known as the compressibility). Note that, although the following discussion is in terms of $E(t)$ and $D(t)$ the appropriate symbols above may be substituted directly.

If either the tensile relaxation modulus or the tensile compliance are determined over a long enough period of time (usually several decades of time are needed) the results appear as shown in Fig. 2.13. Both curves become asymptotic at very short and very long times. The ratio stress/strain at $t \to 0$ is known as the *unrelaxed modulus* E_U—this corresponds to the *unrelaxed compliance* D_U. Similarly at $t \to \infty$ the asymptotes give the values of the *relaxed modulus* E_R and the relaxed compliance D_R respectively. Now it is very important to note that *only these* compliances and moduli are algebraically related, i.e.,

$$D_U = 1/E_U \quad \text{and} \quad D_R = 1/E_R.$$

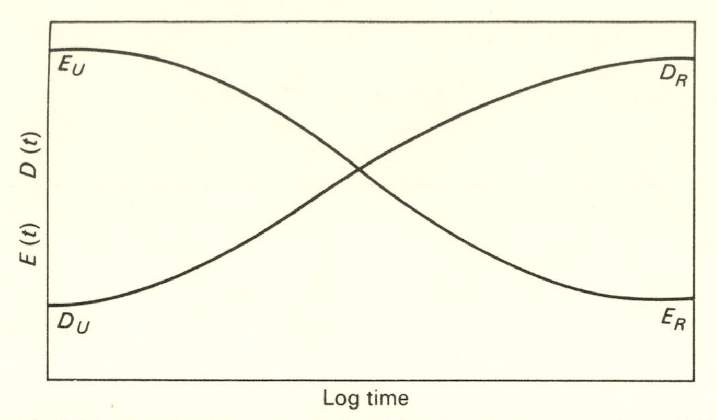

Fig. 2.13 Tensile relaxation modulus $E(t)$ and tensile creep compliance $D(t)$ from tests at constant temperature (schematic).

All other moduli and compliances are not simple reciprocals, i.e.,

$$D(t) \neq 1/E(t)$$

and in fact the exact relations between them may be highly complex. They are given in detail by FERRY (1970).

Summarizing the effects of time as discussed above we may note that:

1 Three possible types of behaviour may be exhibited by a material whose physical characteristics are described by a given relaxation time τ:
- (a) Hookean:—when the loading time $t' \ll \tau$. The modulus is now the unrelaxed modulus E_U; deformations are small and recoverable.
- (b) Viscous:—when $t' \gg \tau$. The modulus now tends to the relaxed modulus E_R and deformations are large and non-recoverable.
- (c) Intermediate:—when $t' \simeq \tau$. Deformations may be quite large but will not be immediately recovered on unloading. They will be recovered on resting in the unloaded state—this is often known as *delayed elasticity.*

2 The 'master curve' for total strain may be written in the form

$$\epsilon(t) = \epsilon_a + \epsilon_b + \epsilon_c$$

where the suffixes a, b, and c refer to the processes described in (1) above. The first, is purely Hookean, i.e., $\epsilon_a = D_U \sigma$. The second is purely Newtonian $\epsilon_b = D_R \sigma$ or $\epsilon_b = \sigma(t)/\eta$ while the third represents the effects of a delay time and is given by $\epsilon_c = D(t)\sigma$. Thus the total strain over time t is given by

$$\epsilon(t) = \sigma(D_U + D_R + D(t)).$$

Thus whether we describe a polymer as 'hard and elastic' or as 'soft and rubbery' depends on our time scale. As AKLONIS *et al.* (1972) put it ". . . in a test lasting two minutes the material would 'look like' a glass . . . in a test lasting

30 minutes it would look like a rubber. But in a test lasting 200 years it would look like a liquid." You pay your money, and you take your choice; but your terms of reference are highly significant when you come to describe a material as 'soft' or 'hard', 'stiff' or 'floppy', etc.

2.13 Effect of Temperature

Although most viscoelastic biomaterials work over a limited range of temperature, a discussion of the temperature dependence of the modulus and compliance is of great importance in understanding the molecular behaviour. Furthermore, an appreciation of these effects can provide a useful experimental tool in studying the time-dependent behaviour.

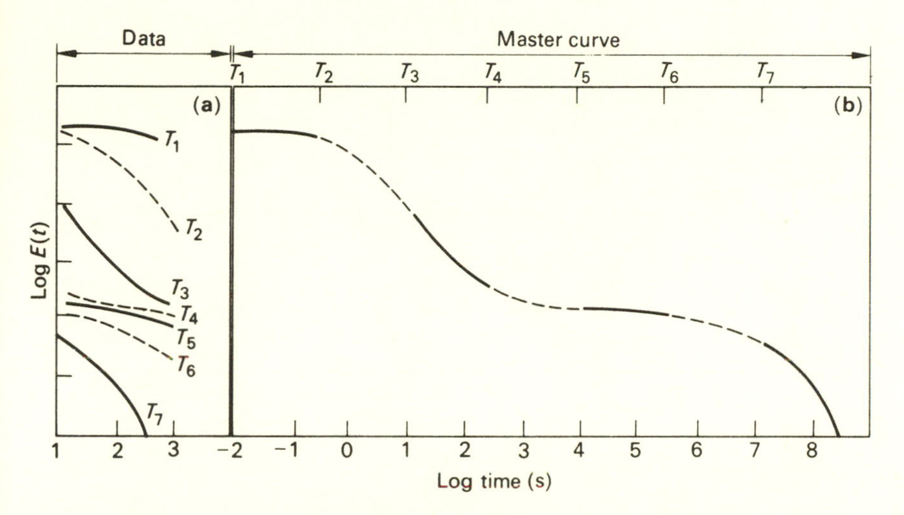

Fig. 2.14 Preparation of a stress relaxation master curve. (a) shows the experimental data, (b) shows the master curve obtained by 'shifting' the data from (a) relative to a reference temperature T_3. (After AKLONIS *et al.*, 1972. *Introduction to Polymer Viscoelasticity*, John Wiley & Sons, Inc.)

Tensile relaxation tests on a lightly crosslinked polymer of low molecular weight (i.e., similar in properties to a viscous liquid) might produce for the relaxation modulus $E(t)$ a family of curves such as those shown in Fig. 2.14 at elevated temperatures $T_1 < T_2 < T_3$ . . . etc. Since the moduli are obviously a function of temperature as well as of time, it appears possible that there may be some correspondence between time and temperature, i.e., that to achieve a value of the compliance which may require say 100 days at room temperature might be achieved in a more experimentally convenient time (say 1000 sec) at a higher temperature.

Experimentally this is indeed found to be so. Figure 2.14 (after AKLONIS *et al.*, 1972) demonstrates that a 'master curve' may be constructed showing the complete modulus–time behaviour relative to some arbitrarily chosen reference temperature, in this case T_3.

We may express these ideas as

$$E(t, T) = E(T_2, t/\alpha_T) \tag{2.30}$$

where the effect of changing the temperature is the same as applying a multiplying factor α_T to the time scale (i.e., an additive factor to the log time scale). Quantitative application of this 'shift' principle involves certain other corrections–in particular for the effect of temperature upon the inherent value of the modulus and also for its effect upon the molecular chain length. These are discussed more fully in the standard texts–meanwhile we accept the principle of data reduction in this way and proceed to the problem of selecting a reference temperature.

2.14 The Glass Transition

If short time tensile tests (each lasting, say, 10 seconds) be carried out on an amorphous linear polymer over a wide enough temperature range the results will appear as shown in Fig. 2.15.

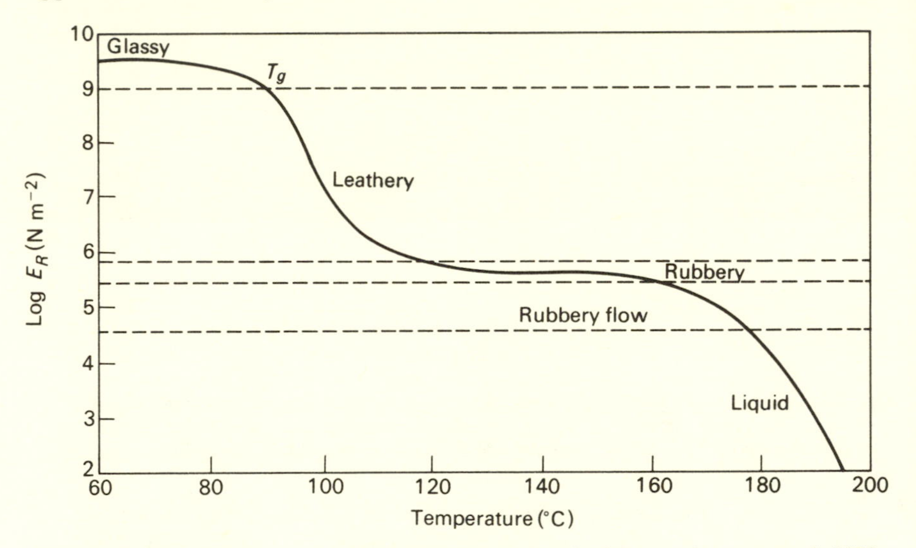

Fig. 2.15 Effect of temperature on the relaxed modulus E_R of polystyrene (AKLONIS *et al.*, 1972. *Introduction to Polymer Viscoelasticity*, John Wiley & Sons Inc.).

Below the inflection point at T_g the material is hard and stiff with moduli typically greater than 10^8 N m^{-2} but over a range of temperature (typically 5°–20°C) the modulus decreases by several orders of magnitude and the material becomes flexible and rubbery. At still higher temperatures the material starts to break down to become a viscous liquid.

The lower transition is called the glass transition T_g, it represents a region of temperatures below which the molecular chains are, essentially, 'frozen' in fixed positions to form a disordered 'lattice'. As the temperature increases, molecular vibration becomes more and more pronounced and the material starts to display

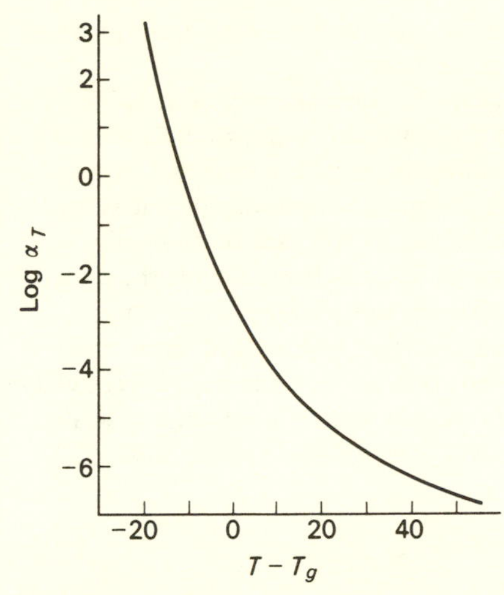

Fig. 2.16 The WLF equation for polystyrene, (after AKLONIS *et al.*, 1972. *Introduction to Polymer Viscoelasticity,* John Wiley & Sons Inc.).

some of the flow properties which characterize a viscoelastic solid. We shall discuss the reasons for this in Chapter 2B but, meanwhile, we may note that the glass transition provides a convenient reference temperature for carrying out the sort of reduction procedures described above.

It is now customary to reduce relaxation and creep data of polymers to T_g— this temperature being determined by some suitably slow testing technique. The function log α_T for this choice of reference temperature is as shown in Fig. 2.16 (FERRY, 1970) in which the points are chosen empirically to fit the equation

$$\log \alpha_T = \frac{-C_1(T - T_g)}{C_2 + (T - T_g)} \qquad (2.31)$$

known as the WLF (Williams Landel and Ferry) equation. All amorphous polymers exhibit similar behaviour when plotted in this way and the constants C_1 and C_2 vary only slightly from polymer to polymer. FERRY (1970) gives a list of variables which may be used for reduction of any of the various viscoelastic functions.

Thus the complete viscoelastic response of any polymer under any set of experimental conditions may be obtained if any two of the following functions are known:

1 the master curve at any temperature;
2 the modulus-temperature curve at any time;
3 the shift factors relative to some reference temperature.

2.15 Dynamic Behaviour

So far we have considered only those 'elastic' constants which may be determined under conditions of static stress or strain. Often more informative data are obtained when transient stresses or strains are applied and we now consider these situations as they apply to linearly viscoelastic solids.

Consider once again the stress–strain response of a Hookean material. Since the relationship between stress and strain is single-valued, a sinusoidally varying stress produces a strain pattern which is, at all times, in phase with the stress. Such a system is said to be conservative–i.e., energy is stored during a loading cycle and is released during an unloading cycle and, in the case of an ideally elastic solid, no energy is dissipated by plastic or viscous flow. This is shown in Fig. 2.17a. In a viscoelastic solid, however, there is some energy dissipation, since some flow must occur for all times greater than $t = 0$, so that stress and strain are no longer in phase. We may represent this as shown in Fig. 2.17b, where it is clear that the maximum stress does not coincide with the maximum strain and there is therefore a lag in the response. This is expressed as a lag angle δ.

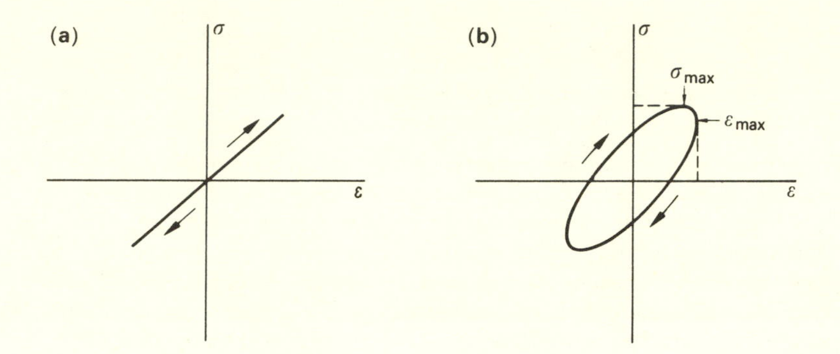

Fig. 2.17 Schematic diagram of reversed loading cycle on (a) linearly elastic (b) viscoelastic solid.

Since it is, experimentally, the easier way, let us imagine that we apply a sinusoidally varying strain

$$\epsilon = \epsilon_0 \sin \omega t$$

where ϵ_0 is the maximum strain in each cycle and ω is the circular frequency. The stress will vary as

$$\sigma = \sigma_0 \sin(\omega t + \delta) \tag{2.32}$$

where σ_0 is the maximum stress and δ is the phase angle. Now the modulus is $E = \sigma_0/\epsilon_0$ so that *Eq. 2.32* can be written

$$\sigma = \epsilon_0 E \sin(\omega t + \delta) \tag{2.33}$$

and, expanding

$$\begin{aligned}\sigma &= \epsilon_0(E \cos \delta) \sin \omega t + \epsilon_0(E \sin \delta) \cos \omega t \\ &= \epsilon_0 E' \sin \omega t + \epsilon_0 E'' \cos \omega t \qquad (2.34)\end{aligned}$$

The first term in *Eq. 2.34* represents that component of stress which is in phase with the strain and is a measure of the energy which is elastically stored in each cycle. Hence $E'(= E \cos \delta)$ is known as the *storage modulus.* The second term represents the out-of-phase component and is a measure of the energy which is dissipated (mostly as heat) so that $E''(= E \sin \delta)$ is called the *loss modulus.* The *phase angle* (loss angle) is usually expressed as its tangent

$$\tan \delta = E''/E'. \qquad (2.35)$$

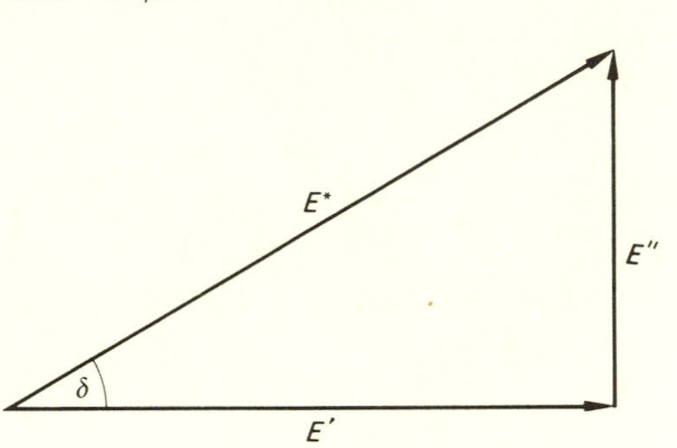

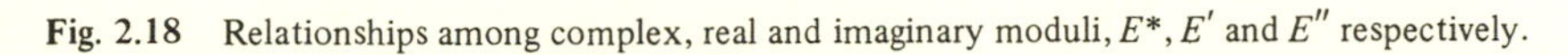

Fig. 2.18 Relationships among complex, real and imaginary moduli, E^*, E' and E'' respectively.

These relationships can also be deduced by assuming that the modulus is a complex quantity (Fig. 2.18), i.e., that

$$E^* = E' + iE'' \qquad (2.36)$$

and E^* is known as the *complex modulus* with E' and E'' described respectively as the *real modulus* and the *imaginary modulus* of the material.

By analogy with, say, the relaxation modulus (which is normally written $E(t)$ to show that it is a time-dependent property), it is good practice to write $E^*(\omega)$, $E'(\omega)$ and $E''(\omega)$ to show that these are functions also of time—but expressed in this case as frequency.

2.16 Viscoelastic Models

Visualization and analysis of viscoelastic effects are often simplified by considering mechanical or electrical analogues which demonstrate the type of behaviour which we have been discussing. We shall now consider some of these models but we must be careful always to remember that they are no more and no less than models, they demonstrate the behaviour but do not necessarily explain it at the molecular level, and we must be wary lest we assume too readily that the model is the real thing. Nonetheless, models do fulfil a useful role in helping to evaluate as well as to describe mechanical behaviour. They have been quite widely used in the study of such materials as bone (SEDLIN, 1965) and cytoplasm (HIRAMOTO, 1962).

We can define the two idealized types of behaviour—the Hookean 'spring' where $\sigma = M\epsilon$ and the Newtonian 'dashpot' where $\sigma = \eta\dot{\epsilon}$. Note that, for these models we are using M for the modulus, not E. This is because there is a real danger that one may too readily associate the spring constant with the actual modulus of a material. This may, indeed be admissible, but in the models M stands for the total elastic contribution from all sources. In the real material this may involve contributions from both linear and nonlinear elastic elements.

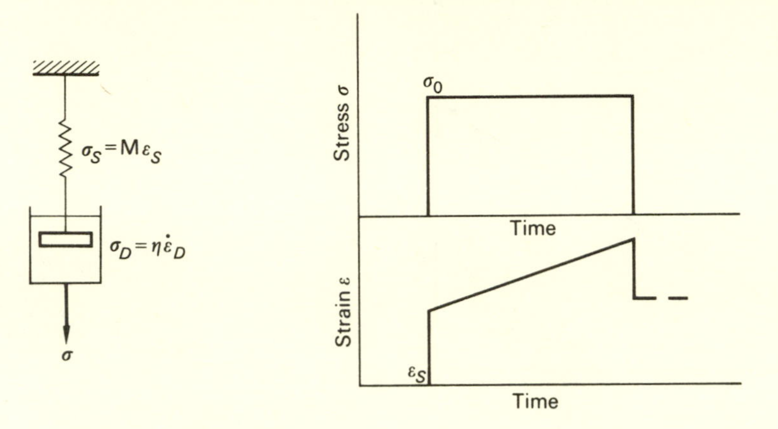

Fig. 2.19 Maxwell model for viscoelastic behaviour showing response to step loading.

The mechanical behaviour of many real materials can be represented by joining elastic and viscous elements in series. This is shown by the *Maxwell model* (Fig. 2.19) where both elements carry equal stress, so that the total stress is

$$\sigma = \sigma_S = \sigma_D$$

where the total strain is obtained from

$$\dot{\epsilon} = \dot{\epsilon}_S + \dot{\epsilon}_D$$

whence

$$\epsilon = \frac{\sigma}{M} + \int_0^t \frac{\sigma}{\eta}\,dt. \qquad (2.37)$$

For a step function stress $\sigma = \sigma_0$ = constant and

$$\epsilon = \sigma_0 \left(\frac{1}{M} + \frac{t}{\eta}\right)$$

or putting $\eta/M = \tau$

$$\epsilon = \frac{\sigma_0}{M}(1 + t/\tau). \qquad (2.38)$$

Similarly for constant strain $\epsilon = \epsilon_0$ we obtain, by differentiating

$$\frac{d\sigma}{dt} = -M\sigma/\eta$$

and the solution is

$$\sigma = \sigma_0 \exp(-Mt/\eta) = \sigma_0 \exp(-t/\tau) \tag{2.39}$$

so that this model exhibits a constant rate of flow under constant stress, but an exponential stress response to constant strain. This latter phenomenon describes stress relaxation and *Eq. 2.39* is the same as *Eq. 2.25.*

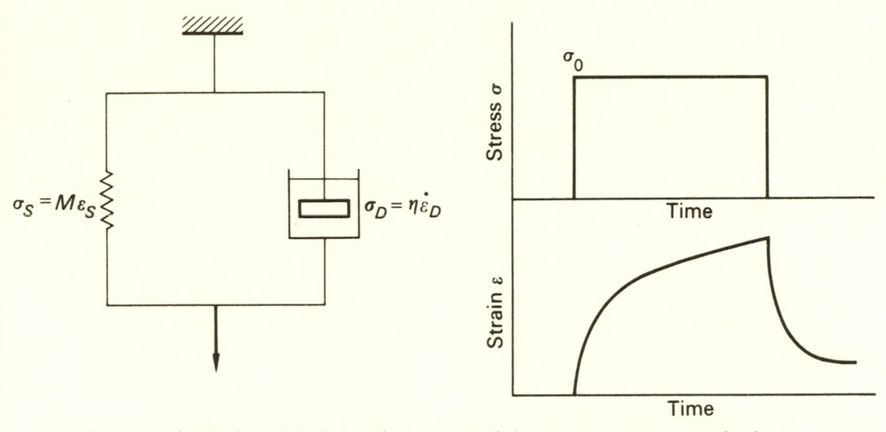

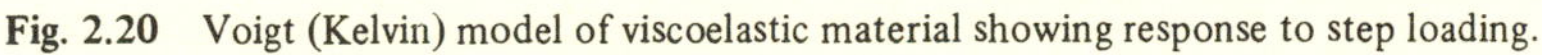

Fig. 2.20 Voigt (Kelvin) model of viscoelastic material showing response to step loading.

In the *Voigt* or *Kelvin model*, the two basic elements are joined in parallel (Fig. 2.20). This means that any changes in the strain of the elastic component are viscously damped. The total strain $\epsilon = \epsilon_S = \epsilon_D$, whereas the total stress $\sigma = \sigma_S + \sigma_D$ so that $\sigma = M\epsilon + \eta\dot{\epsilon}$ and if the stress varies as a function of time $\sigma(t)$ the solution, by standard methods, is

$$\epsilon = \exp(-t/\tau')\left[\epsilon_0 + \frac{1}{\eta}\int \sigma(t) \exp(t/\tau')\, dt\right] \tag{2.40}$$

where $\tau' = \eta/M$ and ϵ_0 is the strain at $t = 0$.

Under a constant stress $\sigma(t) = \sigma_0$ = constant and we obtain

$$\epsilon = \sigma_0/M + (\epsilon_0 - \epsilon_0/M) \exp(-t/\tau');$$

and if $\epsilon = 0$ at $t = 0$ then

$$\epsilon = \frac{\sigma_0}{M}(1 - \exp(-t/\tau')). \tag{2.41}$$

Figure 2.21 compares the behaviour of these two models when they are both subjected to the same step-loading function.

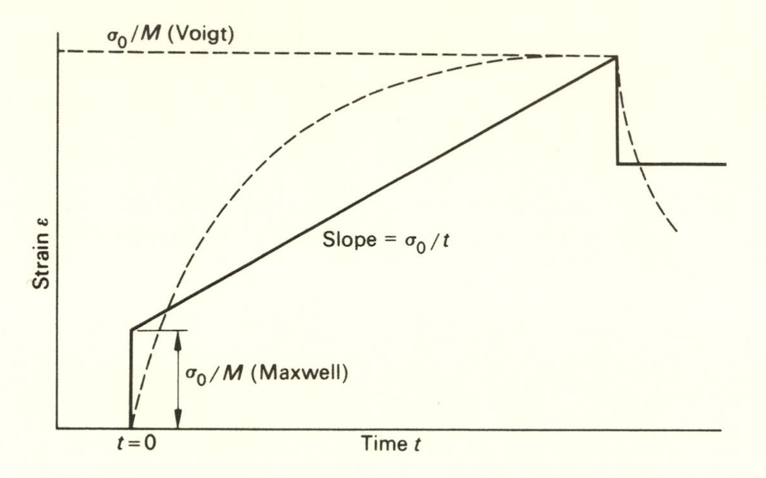

Fig. 2.21 Comparison of response to same constant stress of Maxwell model (solid curve) and Voigt model (dashed curve).

Note that the Maxwell solid shows a linear flow on loading but does not recover fully on unloading. By contrast the Voigt solid shows exponential creep (which becomes asymptotic to σ_0/M) and total recovery as a function of time after unloading.

2.17 Retardation and Relaxation Spectra

In this section we shall attempt to show how, by considering an array of viscoelastic models we arrive at the idea of a relaxation spectrum. This arises because not all of the molecular segments which go to make up a real polymer have the same relaxation behaviour so that when stressed different elements of the structure relax at different rates. The most noticeable effect of this, experimentally, is that the relaxation curve (defined by *Eq. 2.25*) is not linear when plotted on logarithmic coordinates. Therefore in order to obtain full relaxation data we need to be able to determine relaxation times at various temperatures and then to use the shift factor (*Eq. 2.30* or the WLF equation (*Eq. 2.31*)) to obtain the complete curve. The method is well illustrated by HAUGHTON *et al.* (1968) in relaxation tests on the cell wall of *Nitella opaca.*

We shall derive expressions for the relaxation spectrum in *Eqs. 2.42–2.44* but, meanwhile, as an illustration of the use of combined viscoelastic models we will show, without proof, how the relaxed and unrelaxed moduli (E_R and E_U) of real materials arise. Consider the model of Fig. 2.22 consisting of a Maxwell model with a parallel spring. This is one version of the so called 'standard linear solid' which is, in fact, quite adequate for describing the behaviour of many real materials. The reader may care to verify (either by inspection or analysis) that such a model can show linear elasticity (via $M_1 M_2$), exponential flow (via η and M_2),

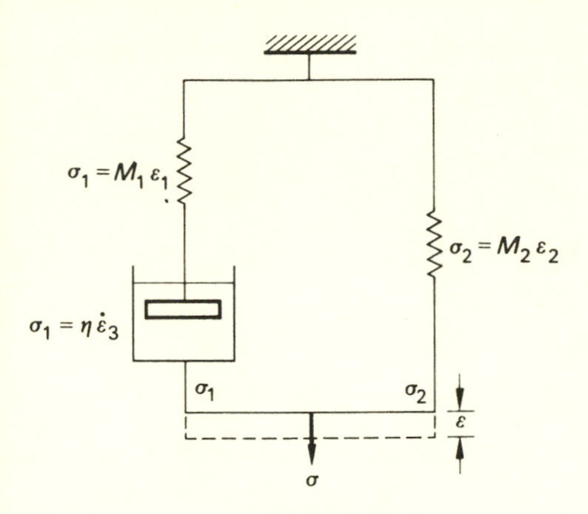

Fig. 2.22 'Standard Linear Solid'—this version consists of a spring in parallel with a Maxwell model. Other versions are possible.

the effect is small at points remote from the hole. This result can be easily ex-elastic and delayed elastic recovery. But it is the equation of motion which is of most interest here. By considering each element in turn we arrive at

$$\sigma + \tau\dot{\sigma} = M_2\epsilon + \eta\left(\frac{M_1 + M_2}{M_1}\right)\dot{\epsilon}.$$

Now when the model is in equilibrium $\dot{\epsilon} \to 0$, the second term on the left hand side disappears and the modulus is given only by M_2. At high rates of strain $\dot{\epsilon} \to \infty$ and the second term now dominates—the modulus is now $(M_1 + M_2)/M_1$. The first (M_2) corresponds to a relaxed modulus M_R, the second to an unrelaxed modulus $M_U = (M_1 + M_2)$ if we assume (as is generally the case) that $M_R \ll M_U$.

We now proceed to the concept of a relaxation or retardation spectrum and we consider first two Voigt models in series—this will require us to use τ' for the retardation time. The models are assumed to have different values of M and η. Under constant applied stress σ_0 the total strain is

$$\epsilon(t) = \frac{\sigma_0}{M_1}[1 - \exp(-t/\tau_1')] + \frac{\sigma_0}{M_2}[1 - \exp(-t/\tau_2')]$$

where $\tau_1' = \eta_1/M_1$ and $\tau_2' = \eta_2/M_2$. Putting this now in terms of moduli and compliances for a real material we can write, for n elements in series

$$\frac{\epsilon(t)}{\sigma_0} = D(t) = \sum\left(\frac{1}{E_n}\right)[1 - \exp(-t/\tau_n')]. \qquad (2.42)$$

Note that the contribution of each element to the equilibrium compliance is simply $1/E_n$. In other words each retardation time τ' is associated with a compliance $1/E$ giving a spectrum of retardation time.

2.17 *Retardation and Relaxation Spectra*

In real materials the retardation times are often so numerous and so closely spaced that the sum in *Eq. 2.42* may be replaced by an integral i.e.,

$$D(t) = \int \frac{1}{E} [1 - \exp(-t/\tau)] \, d\tau.$$

It is generally found convenient to plot the data as a function of $\ln \tau'$, and we now express the integral above in terms of $\ln \tau'$.

Since

$$d(\ln \tau') = 1/\tau' \, d\tau'$$

then

$$D(t) = \int \tau'/E [1 - \exp(-t/\tau')] \, d(\ln \tau')$$

where all values of τ' are covered by the integral. The quantity τ'/E is the contribution to the creep compliance of retardation times in the range $d(\ln \tau')$ and is called the *retardation spectrum* $L(\tau')$,

$$D(t) = \int L(\tau')[1 - \exp(-t/\tau')] \, d(\ln \tau') \qquad (2.43)$$

which defines the spectrum in terms of the tensile compliance.

Similarly for a series of n Maxwell models we obtain a *relaxation spectrum* $H(\tau)$ defined, in terms of the modulus as

$$E(t) = \int H(\tau)[\exp(-t/\tau)] \, d(\ln \tau) \qquad (2.44)$$

where

$$H(\tau) = \tau E(\tau).$$

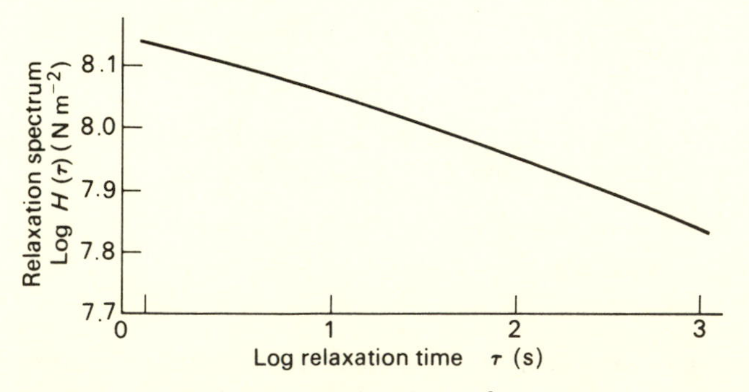

Fig. 2.23 Distribution of relaxation times at 0°C for cell wall of *Nitella opaca* (after HAUGHTON *et al.*, 1968; courtesy of the *J. exp. Bot.*).

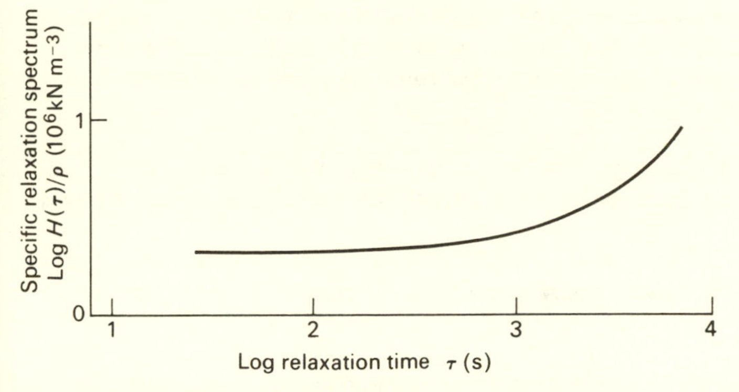

Fig. 2.24 Specific relaxation spectrum for wet wood (MORIIZUMI *et al.*, 1973*a*, *b*; courtesy of the *J. Japan Wood Res. Soc.*).

Typical distributions of relaxation spectra for *Nitella opaca* and wood are shown in Figs. 2.23 and 2.24. The implications of the shapes of these curves will be discussed in Chapter 5.

Thus for any given material either function $L(\tau')$ or $H(\tau)$ completely defines the behaviour and either may be obtained, in principle, from the inversion of the integrals defining $\epsilon(t)$ and $\sigma(t)$.

There are obvious difficulties in the way of checking the actual form of the distribution of relaxation times but, providing we accept the assumption that a continuous distribution exists, then any standard function may be used. FELTHAM (1955) shows that, for a set of parallel connected Maxwell elements the distribution of τ is log-normal (i.e., Gaussian against a logarithmic time base), and GROGG and HELMO (1958) have confirmed this for wheat dough. The same assumptions were made by MORROW (1960) to fit relaxation data from tests on bovine muscle.

Retardation and relaxation spectra are important for two reasons. The first is that they assist in the understanding of the molecular processes which underly the mechanical behaviour. Each molecular process has its own characteristic retardation time and the retardation spectrum represents the sum total of all the individual processes. It must be admitted that it is often difficult (and sometimes impossible) to interpret the spectra directly, but it is often possible to relate peaks in the spectrum to particular types of molecular movement, such as the rotation of side groups, etc.

The second use is that of allowing the results of different types of experiment to be directly compared. The results of any experiment over a range of temperature and frequency can be transformed into a spectrum to any degree of approximation which is justified by the experimental data. This spectrum can then, of course, be transformed into the results of any other experiment.

2.18 Fracture of Viscoelastic Materials

The atomic structure of the viscoelastic biomaterials will be described in detail in Chapter 2B and we shall not discuss it here beyond noting that, unlike crystalline solids, they consist essentially of loosely coiled and more or less entangled molecules whose actual physical length can vary quite widely within a given sample. Thus the application of an external force produces two classes of change in the atomic configuration:

1 A stretching of the loosely coiled longer molecules that accounts, very largely, for the linear or near linear elasticity of viscoelastic materials in the early stages of deformation.
2 Flow (i.e., relative movement) of the shorter coils past the longer ones and past each other. This flow is largely responsible for creep and relaxation phenomena.

Comparatively little work has been done on the fracture of polymeric biomaterials and we are, therefore, compelled to look for analogous behaviour in man-made polymers. Even here the situation is less well understood than it should be. The most comprehensive treatment of fracture in polymeric materials is due to ANDREWS (1968) and the present section largely follows and abridges his work.

The change from substantially elastic to substantially viscous behaviour is generally known as the '*yield point*' of the material. This is usually not well marked and is often defined as the stress which is required to achieve some level of permanent strain. In much mechanical testing a permanent deformation of 0.05% of the original length is often taken as an acceptable, if arbitrary, value.

The tensile yield stress in polymers is strongly dependent on both the temperature and the rate of strain and the subsequent deformation may exhibit either a rising (strain hardening) stress–strain curve or a falling (strain softening) curve (Fig. 2.25). In the former case the fracture stress is higher than the yield stress

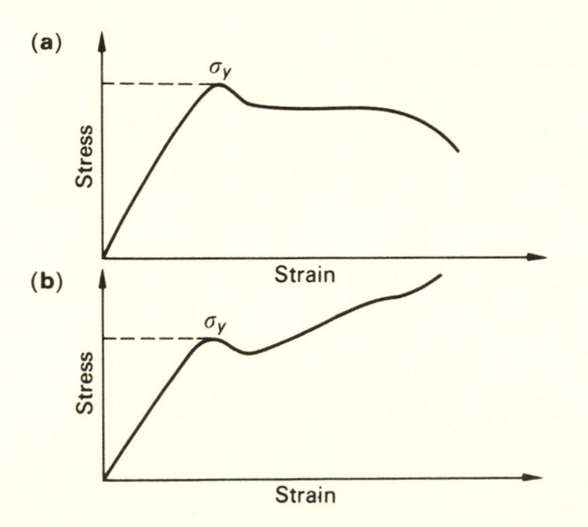

Fig. 2.25 Schematic stress-strain curves for (a) 'strain softening' and (b) 'strain hardening' polymers. In each case σ_y is the yield stress.

and, if the reduction in cross-section is taken into account it may well be considerably higher. This is of great importance in the absorption of strain energy. During post-yield extension the long molecules tend to become oriented along the stress axis and, as a result, a structure may be obtained which approaches that of a crystalline material. This is, in fact called 'strain-induced crystallization' and leads to a notable increase in the value of the instantaneous elastic modulus.

2.19 Generalization of the Griffith Theory of Fracture

It is fairly obvious that we cannot apply the Griffith theory directly to the fracture of viscoelastic materials. The theory is based upon the assumption of ideal linear elasticity in which the only effect of a flaw is to raise the stress locally without producing flow. Indeed, if flow did occur in a viscoelastic material, the radius of the crack tip would be changed by 'rounding off' and the basic equation for the stress distribution at the crack tip would need considerable modification.

OROWAN, (1950) found that metals, like plastics, needed unreasonably high values of the surface energy term S (*Eq. 2.21*) if the Griffith equation was to be used and proposed that the energy required to propagate the crack consisted of two parts. The first was the surface energy as already described–the second, and very much bigger term, is associated with the energy needed to produce plastic flow in the material. Thus we must replace S in the Griffith equation by some other term, say $\mathcal{G}$, which we shall not try to define too closely but which may be thought of as a 'characteristic energy' for crack propagation. It includes the surface energy, energy for plastic flow, any viscoelastic energy loss and even, if we wish, any heat and/or sound. All of these energies must be referred to unit area of crack since the dimensions of energy/unit area must be preserved.

With this assumption of a 'global' energy requirement we can now generalize the Griffith theory to those materials which are not linearly elastic. The argument here follows that of RIVLIN and THOMAS (1953). Following the 'thermodynamic' argument of Section 2.8 we can now generalize the Griffith theory to

$$-\left.\frac{\partial U}{\partial A}\right|_1 \geqslant \mathcal{G} \qquad (2.45)$$

where U is now the total stored energy in the body, A is the area of the crack and $\mathcal{G}$ is the 'characteristic energy'. The subscript 1 implies that no external work is done on the system during the interchange of energy between the body as a whole and the crack. This is to say that we are regarding the question as occurring with a simple system.

The above argument is perfectly general, but if we now move to particular cases the left hand side of the equation is expressible in terms of easily measurable quantities such as force, strain, etc. The right hand side is rather more difficult because it involves a number of variables–the most important is the nature of the energy density in the small volume of material at the tip of the crack. In the ideal linear solid this can be readily calculated by the various formulae dealing with elastic stress concentrations but, in the case of a non-elastic material it is impossible (or at least mathematically difficult) to be sure about the absolute value of the stress which is modified by plastic flow, visco-elastic recovery, etc. To get around this we introduce a constant $\mathcal{K}$. This is best thought of as a *stress intensity factor* which serves to multiply the applied stress by some constant which will define the efficiency by which the overall stored energy in the body is converted into available work at the crack tip.

From this point on the argument becomes increasingly complex and the original references (e.g., IRWIN, 1958; WELLS, 1953) should be consulted. However, for the present purpose we may state here the basic equation, rewritten in its most general form as

$$E\mathcal{G} = \pi \mathcal{K}^2 \tag{2.46}$$

where E is Young's modulus, $\mathcal{G}$ is essentially the same as the energy release rate $\mathrm{d}U/\mathrm{d}A$, and $\mathcal{K}$ is the factor which determines whether or not the stress at the crack tip will equal or become greater than $\mathcal{G}$ for, if it does, then it follows that propagation of the crack is mechanically and thermodynamically inevitable. This expression forms the basis of a quasi-science which is now known as 'fracture mechanics' which makes one important, and technically valid assumption. It is that unstable fracture occurs when the *total* situation at the tip of the crack reaches a critical value—either the energy release rate $\mathcal{G}$ becomes critical (i.e., reaches some value $\mathcal{G}_c$) or the geometry changes until some critical value of the stress intensity factor $\mathcal{K}_c$ is reached. From *Eq. 2.46* it is clear that either criterion will serve to define a property known as the 'fracture toughness' of the material.

The fracture toughness is, essentially, the resistance which a given material offers to the propagation of a crack. If the crack can propagate in a fast, unstable mode, the material is conventionally described as 'brittle' and has a low fracture toughness. If it can withstand the presence of a crack and if the further propagation of a crack requires that more energy be applied (in practical terms that the load must be increased), then the material has a high 'fracture toughness'.

Over the past three decades much study has been devoted to developing methods for measuring either $\mathcal{G}_c$ or $\mathcal{K}_c$. The original references (e.g., IRWIN, 1958; WELLS, 1953) should be consulted. For the present purpose we need only note that, in a material which is capable of yielding, we can write (ANDREWS, 1968)

$$\mathcal{K} = \frac{\sigma_y}{\sqrt{2r_p}} \tag{2.47}$$

where σ_y is the yield stress of the material and r_p is the radius of a zone, at the tip of the crack, where yielding can occur.

Now if r_p is small, the situation at the crack tip approximates to the Griffith situation for a linearly elastic solid, but if it is large, it is possible for gross deformation to occur, and the material behaves in a fully plastic (ductile) manner. Thus the fracture toughness depends, to some extent, upon the size and the shape of the plastic zone. In addition it depends upon the yield strength and the yield behaviour.

In polymeric materials the yield strength is a sensitive function of both time and temperature. For biomaterials *in vivo* we may ignore temperature effects–though we should note that tests *in vitro* should be carried out as close as possible to *in vivo* temperatures, since the yield stress is an exponential function of temperature; even the difference between 37°C and 20°C may be significant. The time effect is much more important. At very short times of loading, a viscoelastic material may behave like a linearly elastic solid and fracture, when it occurs, will be dictated by the Griffith mechanism. At slow rates of loading it behaves more like a viscous liquid which, in the limit, starts to flow at an infinitesimally small stress. Thus the yield stress is time dependent–the longer the load is applied the smaller the yield stress with consequent change in fracture toughness. The problem is mathematically complex, however, and will not be treated here.

The reader who has so doggedly pursued the argument to this point will doubtless be wondering what all this has to do with biomaterials. The most honest answer that we can give is that we do not know–simply because the question has not so far been studied. We shall return later to a discussion of the meaning and significance of the fracture stress and the toughness as 'design' parameters. What we are trying to show here is that careful, and quite sophisticated, ideas lie behind the apparently simple idea of fracture which must therefore guide not only selection of the test but also the interpretation of the results. The whole concept of fracture is a difficult one–even more difficult in biomaterials which are composed, in many cases, of combinations of linearly elastic and viscoelastic materials. Bone, shell, cuticle and timber all fall into this category and the amount of information which is currently available is so small as to be derisory. A careful study of fracture behaviour in rigid biomaterials, interpreted in the light of the findings of fracture mechanics, is vitally necessary if we are to understand the conditions under which an organism (especially one with a rigid skeleton) works.

2.20 Summary of Properties of Viscoelastic Materials

1 Viscoelastic materials do not possess a unique modulus–their atomic structure and mode of deformation is such that all properties, including modulus, must be related to the time over which the load was applied.
2 Mathematical models may be set up to describe the behaviour under load but there is no reason to suppose that these models can, necessarily, be interpreted directly in terms of atomic structure.
3 Although a viscoelastic solid may absorb more energy per unit volume than a linearly elastic solid, some of this energy will be dissipated by such processes as creep, stress relaxation, etc., and the energy returned to the system will be lowered thereby.
4 The most important single property is the relaxation time (or, for a real material the relaxation spectrum) since the value of this, relative to the time of load application, largely determines whether the behaviour is substantially elastic or substantially viscous.
5 Because of the absence of strong primary bonds resisting the external load, the modulus will be generally at least an order of magnitude less than that of a linearly elastic solid.

6 Fracture of a viscoelastic solid is attended by some dissipation of energy by yielding or relaxation. The Griffith criterion cannot therefore be applied, and the fracture characteristics must be expressed in terms of the size of the deforming zone as well as in terms of the material properties.
7 Because of the capacity for yield and the consequent 'rounding off' of stress raisers, viscoelastic materials are less susceptible to the presence of random defects than are linearly elastic materials. This may require some modification in the case of those viscoelastic materials in which the structure approaches that of a crystalline solid, e.g., by strain-induced crystallization.

Chapter 2B
Principles of the strength of materials: Molecular interpretation

2.21 Introduction

In the present chapter we attempt to interpret the phenomena previously described in terms of molecular structure. This presents little difficulty so far as linearly elastic materials are concerned–though we shall not attempt a molecular description of the process of fracture. A microstructural interpretation of viscoelastic properties leads, however, to a situation of formidable complexity. This arises partly because the study of these materials is still in its comparative infancy but, mostly, because of the inherent complexity of the molecules and of the various ways in which they can interact.

Thus, once again, we shall need to resort to simplified models and, in most cases, to a lack of rigour in the proofs. This is not to say that rigorous analysis is not possible (though we must admit that, in many cases it does not exist) but rather that, in a book of this type, such proofs are of a length and complexity as to preclude inclusion–if only for the peace of mind of the reader. There is, fortunately, a fairly comprehensive literature and the reader who feels that he has been short changed should consult it for a more complete understanding.

2.22 Thermodynamics of Mechanical Deformation

The previous chapter was concerned, mainly, with two quantities–force and displacement. The product of these is energy and, since thermodynamics concerns itself primarily with energy, it is convenient and conventional to start from a thermodynamic viewpoint.

We shall not attempt to introduce the topic here–this is better done in, for example, standard texts on physical chemistry such as GLASSTONE and LEWIS (1962). But it may not be out of place to mention a few of the more important thermodynamic statements which are of relevance to the present book.

1 Any spontaneous change involves an increase in entropy and, at equilibrium, entropy is a *maximum* and remains constant.

2 If we let U = internal energy, S = entropy, p = pressure, V = volume, and T = temperature, we can define the Helmholtz free energy

$H = U - TS$ (for constant V and p)

and the Gibbs free energy

$G = U + pV - TS$ (for constant T and p)

and *both* must be a *minimum* at equilibrium.

3 Since equilibrium involves the attainment of a completely random state, we can consider entropy as a function which measures the degree of randomness. This is given by Boltzmann's definition

$$S = k \ln N_a \tag{2.48}$$

where N_a is the number of possible arrangements of the particles which comprise the system and k is Boltzmann's Constant (1.38×10^{-23} J K^{-1}).
Consider now the first law:

$$dU = dQ + dW \tag{2.49}$$

where Q = heat and W = work

and the second law, written as

$$dQ = T\,dS.$$

Combining these we obtain

$$dU = T\,dS + dW.$$

If a bar, of initial length l is loaded by a force f, it extends by an amount dl and the work done is $dW = f\,dl$ so that

$$dU = T\,dS + f\,dl.$$

The Helmholtz free energy is $H = U - TS$ so that

$$dH = dU - T\,dS - S\,dT = f\,dl - S\,dT \tag{2.50}$$

and, if the conditions are such that temperature remains constant, the change in free energy is

$$dH = dU - T\,dS = f\,dl$$

so that

$$f = \left(\frac{\partial H}{\partial l}\right)_T = \left(\frac{\partial U}{\partial l}\right)_T - T\left(\frac{\partial S}{\partial l}\right)_T \tag{2.51}$$

showing that the force is distributed between a change in internal energy ($\partial U/\partial l$) and a change in entropy ($\partial S/\partial l$). We now proceed to consider the first of these terms, the change in internal energy.

2.23 Linear Elasticity

The first term in *Eq. 2.51* implies that part, at least, of the response of a material to an applied force is associated with a change in the internal energy of the body. In crystalline materials, where the atoms are arranged in orderly ranks in space, the capacity of the atoms for independent movement is limited and the bulk of the internal energy is found in the interaction energies between adjacent atoms. We suspect that these interaction energies must be both attractive and repulsive and we normally assume that the energy with which one atom attracts or repels another atom is a function of the distance between them—the energy of repulsion being much shorter ranged than that of attraction. We represent the total energy U by

$$U = Br^{-m} - Ar^{-n} \qquad (2.52)$$

where the first term represents the repulsive forces, the second term the attractive forces, r is the distance between the atoms and m and n are positive constants.

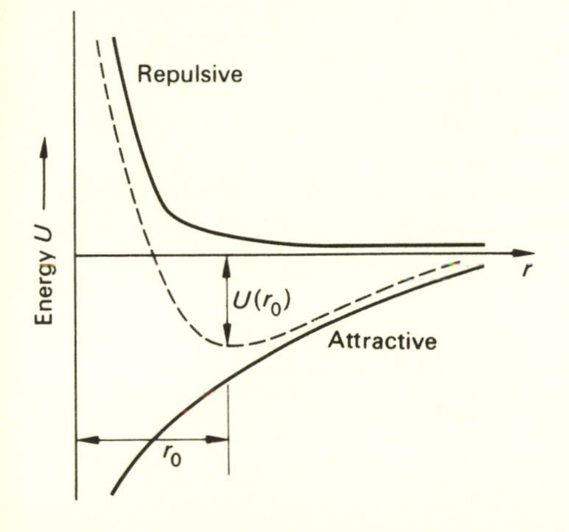

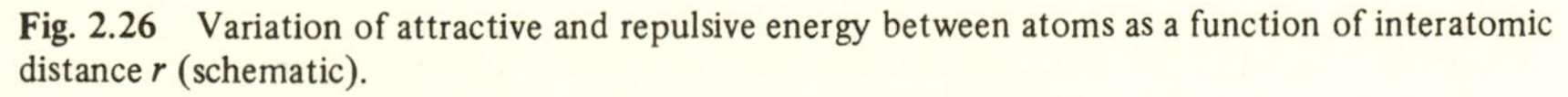
Fig. 2.26 Variation of attractive and repulsive energy between atoms as a function of interatomic distance r (schematic).

These curves are sketched in Fig. 2.26 and the net curve (shown dashed) has a minimum at $U(r_0)$. This is the point at which the attractive and repulsive forces balance and thus r_0 is the equilibrium spacing of the atoms.

The energy $U(r_0)$ at r_0 is the energy needed to dissociate the atom pair. Table 2.3 gives some typical data for this energy from which it is clear that the strength of the primary bonding mechanisms (covalent, electrovalent, etc.) is at least an order of magnitude greater than that of secondary bonding, and we generally expect to find that materials exhibit two types of behaviour. Materials which can be built up entirely from nondirectional primary bonds we expect to be hard and rigid—metals, oxides and ceramics generally fall into this group as do fully crosslinked polymers such as bakelite, ebonite, etc. The other group is

Table 2.3 Energy content of chemical bonds

Type	Bond	Interatomic distance nm	Energy kJ mole^{-1}
Electrovalent	Na Cl		494
Covalent	C=O	0.121	711
	C=C	0.133	
	H–O		463
	H–H	0.075	436
	C–O	0.142	351
	C–C (aliph)	0.454	348
Secondary	$=CH_2$		7.5
	$-CH_2-$	0.3–0.4	4.14
	–OH	between	30.3
	–COOH	groups	37.5
Hydrogen	e.g. >NH···O=C<		16.8

best represented by linear polymers, though we may include also such inorganic materials as asbestos, mica, etc. In these, the primary bonds are more or less directional and are usually saturated. Thus chains or plates are built up whose strength along the chain or in the plane of the plate is equal to that of the primary bond. However, bonding *between* completed chains or plates involves secondary forces only, so that the *intermolecular* strength may be low even though the *intramolecular* strength is high.

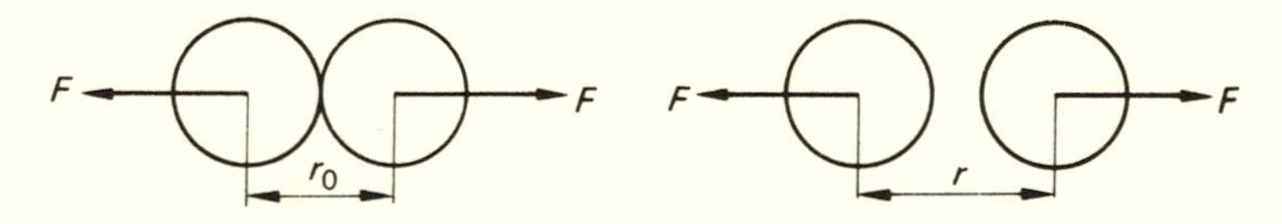

Fig. 2.27 Displacement of two atoms under external force.

We now consider the mechanical stability of a pair of atoms when an external force is applied. Consider two atoms in their equilibrium position and apply a force F (Fig. 2.27). Note that the force must be *balanced*, i.e., it must exert no resultant and no couple which might otherwise set the atoms in motion. When the force is applied the atoms find a new position r in which the interatomic and applied forces are balanced. If the displacement is ϵ

$$\epsilon = r - r_0$$

we find that the equilibrium condition is

$$F = \mathrm{d}U(r)/\mathrm{d}r$$

where $U(r)$ is the bond energy at the displacement r. If the force is tensile then $r > r_0$, if it is compressive then $r < r_0$.

The force in the bond is, therefore a function of the displacement and for each value of displacement there corresponds a particular force. The deformation is reversible—if the displacement returns to its original value then so does the force and vice versa. These are the features of *perfect elastic deformation.*

When the applied force is small the elastic displacement is directly proportional to the force. This follows from three things:

1 The bond energy $U(r)$ is a continuous function of r. This allows us to express the energy as a series

$$U(r) = U(r_0) + (\partial U/\partial r)r_0 + \tfrac{1}{2}(\partial^2 U/\partial r^2)r_0^2 + \cdots \qquad (2.53)$$

where $U(r_0)$ is the energy at r_0 and where the differential is taken at $r = r_0$.

2 The minimum in the curve at r_0 allows us to eliminate the second term in *Eq. 2.53* since $(\partial U/\partial r) = 0$ at a minimum.

3 The displacement is small so that we can ignore terms higher than r_0^2. Thus we find that

$$U(r) = U(r_0) + \tfrac{1}{2}(\partial^2 U/\partial r^2)r_0^2$$

whence

$$F = \mathrm{d}U/\mathrm{d}r = (\mathrm{d}^2 U/\mathrm{d}r^2)_{r_0}\, r \qquad (2.54)$$

i.e., the force is proportional to the displacement r via a constant $(\mathrm{d}^2 U/\mathrm{d}r^2)$. This of course is *Hooke's Law.*

The above argument enables us to draw certain conclusions about the magnitude of the constant which the reader will at once identify with the *elastic modulus*:

1 It is independent of the sign of r so that it should be the same in both tension and compression. This is, in fact, generally true of simple crystalline solids and small deformations.

2 The constant is a measure of the curvature of the U–r curve at the minimum point and it will be high when r is small. Since small atoms can pack more tightly than large ones, we expect a high modulus from materials such as carbon, boron, beryllium, etc., and a low modulus from lead and uranium. This is broadly true, though direct correlation between interatomic distance and modulus of the elements above would not be possible because of differences in crystalline structure. For elements (and compounds) having the same structure, however, close correlations may be obtained with the modulus varying as $1/r^4$.

3 We expect the moduli of amorphous polymers to be lower than those of crystalline solids because of the unoriented structure. Only when polymer molecules are aligned and extended will the modulus approach that of a crystalline material.

4 The minimum in the U–r curve is a consequence of the fact that $m > n$, i.e., that the repulsive forces dominate the behaviour. These are very short range forces, so we expect high moduli in those materials where we have strong, short range forces

operating between the atoms. In effect, this means solids which are bonded by covalent and electrovalent bonds. The elastic constant of diamond (covalent) or an oxide (electrovalent) is very much greater than that of a material containing, say, –CH groups.

If we derive the curve relating force and displacement from Fig. 2.26 we shall obtain something like Fig. 2.28. This, and our previous argument shows that linear behaviour can only exist over displacements where $r \ll r_0$, i.e., at infinitesimally small deformations.

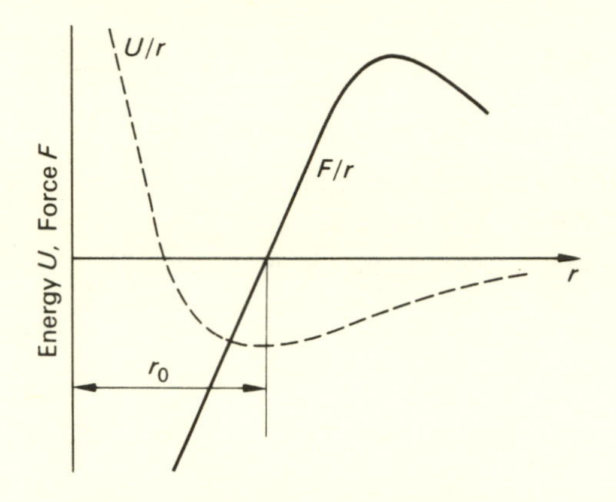

Fig. 2.28 Interatomic force as a function of atomic separation (solid curve). The dashed curve is the energy-separation curve of Fig. 2.26.

Why then do so many bulk solids display linear elasticity up to strains which are, typically, of the order of 0.001? The answer to this is complex and beyond our present scope, but the following points will serve to indicate the reasons:

1 Very pure specimens of stiff materials (such as SiO_2) are elastic in the sense that deformation is both recoverable and time-independent and are linearly elastic.

2 Single crystals of softer materials, such as lead, in a very high state of purity exhibit non-recoverable (i.e., plastic) deformation at vanishingly low stresses. The same material in polycrystalline form displays apparently linear behaviour over a measurable range of stress. This arises from the fact that, in a polycrystalline aggregate of randomly oriented crystals, there are restrictions to deformation of any one crystal due to the interaction with its neighbours. Elastic behaviour here is the result of an 'averaging' of many small inelastic deformations.

3 The assignment of linearity or non-linearity to a curve depends upon the sensitivity of the measurement—as the sensitivity increases, departures can be detected at lower values of stress.

2.24 The Structure of Polymers

The detailed structure of polymeric chains has been fully described in the standard texts (e.g., FLORY, 1953; 1969; CIFERRI, 1961; TOBOLSKY, 1960) and those of specific interest will be described in Chapters 3, 4 and 5. For our present purpose we may summarize the salient features as follows:

1 Polymeric materials are essentially composed of large molecules. These are frequently significantly larger in one dimension than they are in the other two. The length of the molecule is variable and is expressed in terms of the 'degree of polymerization' n, so that the average molecular weight of a polymer is nM_0 where M_0 is the molecular weight of the structural unit. Molecular weights are typically in the range 10^3–10^5.

2 In an *amorphous* polymer (either glassy or rubbery), the molecules are assumed to be entangled in a completely random manner. The evidence for total randomness is not completely unambiguous, but the explanation of the major change in properties (the glass transition—see Section 2.14) is unaffected by any uncertainty in the hypothesis.

3 In a *crystalline* polymer, regions of the tangled structure contain aligned segments of molecular chains—the 'crystallinity' being usually expressed in terms of the relative specific volumes of crystalline and amorphous regions.

In the following discussion we are concerned only with effects of stress, time and temperature on a randomly oriented structure.

2.25 Statistics of a Polymer Chain

Providing that the molecules are merely entangled and are not heavily cross-linked to other molecules, it is intuitively obvious that the application of an external force will cause changes in the internal configuration. A given atom in a molecular chain will not change its neighbours within the chain, but it will change its neighbours in adjacent chains several times during a loading/unloading sequence. Expressed in thermodynamic terms, we may state that the initial, unloaded configuration of any molecule is, or is close to being, the most probable one, so that this is an equilibrium configuration of maximum entropy. Under load the configuration changes to the one that is the most probable for the loaded condition and the entropy changes. Now the number of possible configurations available to a chain is a function of its end-to-end length r (i.e., whether convoluted or not—the straight chain is merely one of a number of possible configurations) and the number of conformations N_c decreases as r increases. Thus the *configurational entropy* (*Eq. 2.48*) $S_c = k \ln N_c$ must decrease as r increases.

Returning now to *Eq. 2.51*, the meaning of the second term, the change in entropy, is clear and we can extend the argument as follows. The right hand side of *Eq. 2.51* is a complete differential in l and T so that the order of differentiation is immaterial. Thus

$$\left(\frac{\partial f}{\partial T}\right)_l = -\left(\frac{\partial S}{\partial l}\right)_T$$

and, inserting this into *Eq. 2.51* we obtain

$$f = \left(\frac{\partial U}{\partial l}\right)_T + T\left(\frac{\partial f}{\partial T}\right)_l \tag{2.55}$$

which is the so called 'thermodynamic equation of state' for rubbers.

An important modification to *Eq. 2.55* is necessary for rubbers which do not deform at constant volume. The thermodynamic work W in *Eq. 2.49* must be amended to include pressure-volume work pV as, indeed, is customary in most classical thermodynamics. In this case *Eq. 2.55* becomes

$$f = \left(\frac{\partial U}{\partial l}\right)_{VT} + T\left(\frac{\partial f}{\partial T}\right)_{Vl}. \tag{2.56}$$

We shall consider this further in Chapter 4, but meanwhile we may note that it gives us an experimental method of some simplicity. The second term is the

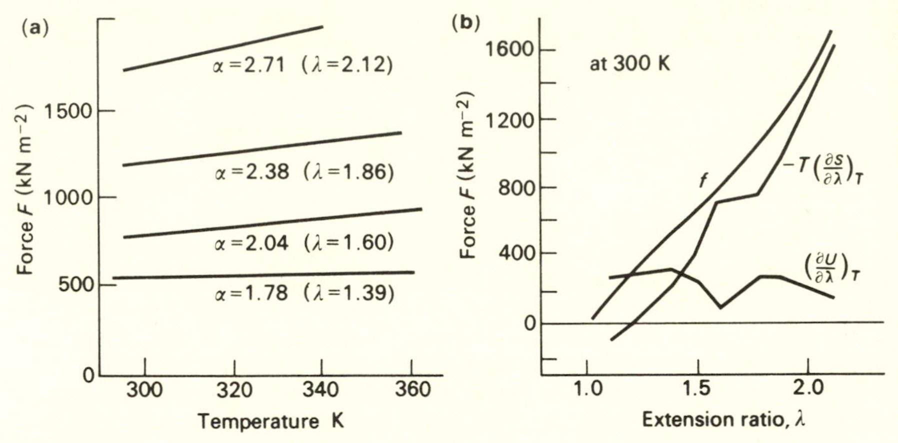

Fig. 2.29 (a) Force required to maintain constant extension for resilin tendon. Extension ratio λ refers to swollen unstrained length. (b) The contributions of entropy changes ($-T(\partial s/\partial\lambda)_T$) and internal energy changes $(\partial U/\partial\lambda)_T$ to the elastic energy of resilin. (After WEIS-FOGH, 1961 *a*; courtesy of the *J. molec. Biol.*).

variation of the isometric force (force needed to produce constant strain) with temperature, if this be determined then $(\partial U/\partial l)$ follows uniquely for a given temperature. There are corrections to be made, of course: varying temperature varies other factors as well but these are well discussed, e.g., by WEIS-FOGH, (1961*a*), TRELOAR, (1958) and ALEXANDER, (1966). Figure 2.29 shows the results for resilin and it will be seen that the force-extension curve is dominated by the entropy term.

It now remains to examine the nature of the relationship between force and entropy. For this we use the argument of TRELOAR (1958) which requires that we should consider the statistics of a single, flexible molecule and then use Boltzmann's equation (*Eq. 2.48*) to relate the number of possible configurations of such a chain to the entropy. The formal proof is quite complex and we do not include it here. Even so the reader who feels faint may skip to *Eq. 2.61*, although he should take particular note of *Eq. 2.58* which relates the end-to-end length of a flexible molecule to the number and size of the subunits.

Consider a long, flexible chain made up of a large number of stiff segments which can rotate freely at the joints and imagine it to be randomly oriented in space with one end fixed at the origin (Fig. 2.30). This is often called the 'random coil' model. KUHN (1934; 1939) and GUTH and MARK (1934) give the probability that the free end will be found at a point x, y, z in a volume (dx, dy, dz):

$$p(x, y, z)\ \mathrm{d}x, \mathrm{d}y, \mathrm{d}z = (b^3/\pi^{3/2}) \exp\left[-b^2(x^2 + y^2 + z^2)\right]\ \mathrm{d}x, \mathrm{d}y, \mathrm{d}z \qquad (2.57)$$

where $b^2 = 3/2na^2$. Here a is the length of the rigid segments and n is the number of such segments in the chain. Note that, for a molecular chain, the stiff segment

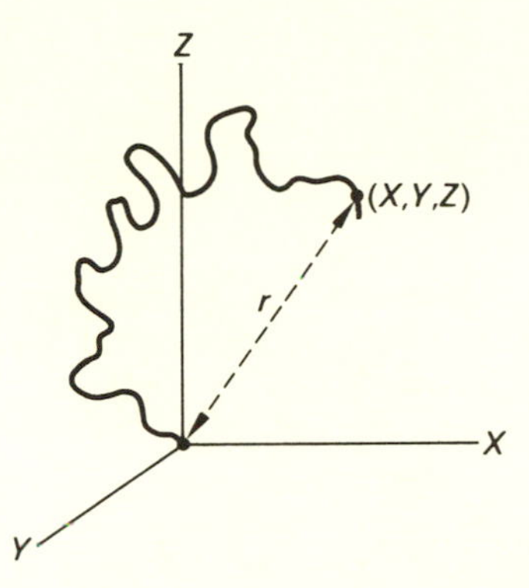

Fig. 2.30 Schematic diagram of randomly oriented flexible molecular chain. r = end-to-end distance.

does not necessarily correspond to a single monomer. We shall discuss this later in Section 2.27. TRELOAR (1958) has shown that, if the chain obeys Gaussian statistics the mean square end-to-end distance is given by

$$\bar{r}^2 = \frac{b^3}{\pi^{3/2}} \int_0^\infty r^2 \exp(-b^2 L^2) \cdot 4\pi r^2\ \mathrm{d}r$$

or

$$\bar{r}^2 = \frac{3}{2b^2} = na^2. \qquad (2.58)$$

The probability that the end-to-end length will be r is

$$p(r)\ \mathrm{d}r = \left(\frac{4b^3}{\pi^{1/2}}\right) r^2 \exp(-b^2 r^2) \cdot \mathrm{d}r \qquad (2.59)$$

Now since the configurational entropy S_c is related to the number of possible configurations by Boltzmann's equation (*Eq. 2.48*) and we use *Eq. 2.59* to obtain

$$S_c = \text{const.} - kb^2r^2 \qquad (2.60)$$

i.e., the entropy decreases as r increases as noted above. *Eq. 2.60* gives the entropy of a chain whose ends are held apart by a distance r. If they are now moved by dr, the entropy change will be

$$\frac{dS_c}{dr} = -2kb^2r$$

and, if internal energy and temperature are constant, the force required is

$$f = -T\left(\frac{dS}{dr}\right)_T = 2kTb^2r \qquad (2.61)$$

which is the equation of state for a single polymer chain freely oriented in space.

The argument above is not rigorous and is given more fully in the standard texts (e.g., BUECHE, 1962; TRELOAR, 1958, etc.). However, it serves to indicate the following principles:

1 If the ends of the molecule are moved from their most probable position, the change in entropy causes a force to act *along the line joining the ends of the molecule.* In fact the molecule acts as if it were a spring and, although there is no contribution from the internal energy, *Eq. 2.61* shows that it behaves like a Hookean spring having a spring constant $2kTb^2$.

2 The entropy force increases as temperature increases and is proportional to the absolute temperature. The force is inversely proportional to the mean square end-to-end length of the molecule since $b \propto 1/r$ from *Eq. 2.58.* In this it differs significantly from the short range forces which determine Hookean elasticity.

2.26 Rubber Elasticity

We shall now derive an expression for the equation of state for rubber elasticity and to show how the applied stress is related to the measured strain in a bulk solid. This seems pretty obvious but, in this case we shall see that the elastic constant is related to the number of chains in the network. This approach was, in fact, used by WEIS-FOGH (1961*a*, *b*) to evaluate the structure of resilin (Chapter 4). As before, the complete arguments go far beyond our present scope, the original texts should be consulted or an extremely clear presentation by AKLONIS *et al.* (1972). Meanwhile the mathematically faint hearted may skip to *Eqs. 2.69* and *2.70* noting, en route, the definition of extension ratio λ below *Eq. 2.65.*

The derivation of the equation of state for bulk solids requires a number of simplifying assumptions:

1 The internal energy of the system is independent of the conformation of the individual chains.

2 An individual chain obeys Gaussian statistics.

3 The total number of conformations possible in an isotropic network is the product of the number of conformations available to the individual chains.

Thus, N_a, the total number of conformations available to the chain is given by rewriting *Eq. 2.57* as

$$N_a(x_i y_i z_i) = (b^3/\pi^{3/2}) \exp[-b^2(x_i^2 + y_i^2 + z_i^2)]$$

where x_i, y_i, z_i are the end coordinates of a chain of end-to-end length r_i in the unstrained condition. By analogy with *Eq. 2.60* the entropy for N such unstrained chains is

$$S_c = 3k \ln(b/\pi^{3/2}) - k \sum_{i=1}^{N} b^2(x_i^2 + y_i^2 + z_i^2)$$

giving a Helmholtz free energy for the unstrained state

$$H_u = H_0 + kT \sum_{i=1}^{N} b^2(x_i^2 + y_i^2 + z_i^2), \qquad (2.62)$$

where H_0 is that part of the free energy which is not related to conformational changes.

In the strained state the chain deforms to a length r_i' having coordinates x_i', y_i', z_i'. We can relate this to the macroscopic strains by assuming that a cube of unit volume deforms, under a homogeneous stress, into a rectangular parallelepiped having dimensions $\lambda_1, \lambda_2, \lambda_3$. These are called the *principal extension ratios* ($\lambda = 1 + \epsilon$, where ϵ is the strain as conventionally measured dl/l_0). If we choose the coordinate axes to coincide with the principal strain axes

$$x_i' = \lambda_1 x_i, \quad y_i' = \lambda_2 y_i, \quad z_i' = \lambda_3 z_i$$

and the free energy for the deformed state is now

$$H_d = H_0 + kT \sum (\lambda_1^2 x_i^2 + \lambda_2^2 y_i^2 + \lambda_3 z_i^2). \qquad (2.63)$$

In order to allow for the fact that, in a real network, there are limitations (e.g., crosslinks, partial extensions) on the number of conformations it is customary to write

$$\overline{b^2 r^2} = \frac{3}{2} \frac{(\bar{r}_0^2/\bar{r}_f^2)}{N}$$

where $\bar{b}^2$ is averaged over all the free chains, $\bar{r}_0^2$ is the mean square end-to-end distance of the chain *in the network* and $\bar{r}_f^2$ is the mean square length of the *free chain.* The parameter $\bar{r}_0^2/\bar{r}_f^2$ is often called the *front factor* and represents the average deviation of the chains from the dimensions they would assume if they were free. For an ideal rubber the front factor is unity.

The most important single correction now to be made is that which takes into account the fact that the deformation is not volume-constant but involves some dilation. This requires a correction to the definition of the extension ratios after which the change in free energy obtained by subtracting *Eq. 2.62* from *Eq. 2.63* may be differentiated with respect to L_0 the sample length, to give the equation of state of the form

$$f = \frac{NkT}{L_0}\left(\frac{\bar{r}_0^2}{\bar{r}_f^2}\right)\left[\alpha - \frac{1}{\alpha^2}\right]\left(\frac{V}{V_0}\right)^{2/3} \tag{2.64}$$

where α is the 'corrected' extension ratio.

To express this as stress, we define the area A_0 of the undeformed sample ($V_0 = A_0L_0$) and let the number of network chains per unit volume be $N_0 = N/V_0$ whence

$$\sigma = \frac{f}{A_0} = N_0kT\left(\frac{\bar{r}_0^2}{\bar{r}_f^2}\right)\left(\lambda - \frac{V}{V_0\lambda^2}\right)$$

or, expressed in terms of moles and noting that V/V_0 is close to unity

$$\sigma = N_0RT\left(\frac{\bar{r}_0^2}{\bar{r}_f^2}\right)\left(\lambda - \frac{1}{\lambda^2}\right). \tag{2.65}$$

Since the extension ratio $\lambda = 1 + \epsilon$ we can expand

$$\lambda^{-2} = (1 + \epsilon)^{-2} = 1 - 2\epsilon + \cdots$$

neglecting higher terms. The time independent tensile modulus is now

$$E_0 = \frac{\sigma}{\epsilon} = 3N_0RT\left(\frac{\bar{r}_0^2}{\bar{r}_f^2}\right) \tag{2.66}$$

or, expressed in terms of the average molecular weight between crosslinks M_c as

$$E_0 = \frac{3\rho RT}{M_c}\left(\frac{\bar{r}_0^2}{\bar{r}_f^2}\right) \tag{2.67}$$

where ρ is the density ($= N_0M_c$).

But rubbers have Poisson's ratio $\nu \rightarrow 0.5$ and we noted in Chapter 2A that $E = 3G$ for this case so that we can rewrite *Eq. 2.66* without much loss of accuracy as

$$G_0 = N_0RT\left(\frac{\bar{r}_0^2}{\bar{r}_f^2}\right) \tag{2.68}$$

i.e.,

$$f = G_0A_0\left(\lambda - \frac{1}{\lambda^2}\right) \tag{2.69}$$

or

$$\sigma = G_0\left(\lambda - \frac{1}{\lambda^2}\right) \tag{2.70}$$

in which forms it is commonly used. As before, the treatment is not rigorous but once again it explains the observations that the elastic restoring force is proportional to the temperature and to the number of chains in the network. But it is clear that, in this case, the strain dependence of the force is not Hookean. Up to strains of about 0.5 (λ = 1.5), there is good agreement between *Eq. 2.70* and experiment—thereafter marked divergencies occur. These are due to such factors as strain-induced crystallization, limiting extension of chains, swelling in solvents, etc., and they have been extensively discussed, e.g., by TRELOAR (1958), CIFERRI (1961), FLORY *et al.* (1959).

2.27 Molecular Interpretations of Rubbery Polymers

An ideal flexible long-chain polymer made up of a large number n of segments, each of length a, will take on a random-coil conformation provided the following three conditions are met. (1) The junctions between segments must allow completely free rotation; (2) there must be sufficient thermal energy available to keep all segments of the molecule in motion; (3) the molecule must be physically unrestrained. Under these conditions the mean square end-to-end distance (*Eq. 2.58*) is

$$\bar{r}^2 = na^2$$

In the derivation of the equation of state for rubber-like elasticity, it was assumed that conditions (1) and (2) held, and the effect of physical restraint in the form of an applied force was analysed. We will now consider conditions (1) and (2), to get an idea of how real polymers behave.

When applying the argument in *Eqs. 2.57–2.61*, it is tempting to relate the stiff segment of the ideal random-coil molecule to the spacing between atoms in a polymer chain (i.e., the bond length). By definition the segment is a unit about which there is completely free rotation, and chemical bonds impose considerable limitation on rotational freedom between atoms. In the case of a simple hydrocarbon chain, the fixed bond angle between carbon atoms restricts rotation to a single, conical shell, and if we assume free rotation within this conical shell, it is possible to calculate a corrected value for $\bar{r}^2$ (see TANFORD, 1961). With the tetrahedral bond angle of a carbon chain, the result is $\bar{r}^2 = 2nB$, where B is the bond length. The effective segment length is thus $2^{1/2}B$. Even in very simple hydrocarbons, there is not free rotation around a bond angle; side chain interactions create a number of most probable rotation shapes around any bond, and the bulky side chains of proteins and polysaccharides may severely reduce rotational freedom. Other interactions which reduce chain flexibility include the peptide bond between amino acids in a protein, which has partial double bond character that virtually eliminates rotation, and the sugar rings of polysaccharides which also act as rigid units, allowing rotation only around the oxygen in the glycoside bond. The net result of these restrictions on rotational freedom is that the effective segment length must include several bond lengths. It is possible to describe real polymer molecules in terms of the random-coil model, but only if we use an empirical term for the segment length which reflects the flexibility of the polymer chain.

Condition (2) specifies that the polymer segments be in constant motion due to thermal energy in the system. This means that the thermal energy available to each segment must be large relative to the frictional forces which tend to retard segmental motion. By considering the relative magnitude of the frictional forces and of thermal energy, we may reach an understanding of the molecular interactions which create the time-dependent, viscous properties of polymeric materials.

In the previous chapter we considered liquid flow as a 'jump' process in which bonds are continually broken and reformed, as long as there is some driving force. The relationship between the applied force and the velocity of flow may be expressed in terms of a molecular friction coefficient (f_0) which is characteristic of the particular fluid system and is a measure of the force (F) required to move a single particle at unit velocity through its environment thus

$$F = vf_0 \tag{2.71}$$

when f_0 can also be defined in terms of a segmental friction coefficient

$$f_0 = nf_s$$

where n is the number of segments.

Polymer molecules in dilute solution are assumed to act independently, but as the concentration is increased, we must take into account interactions between the polymer molecules. These interactions can take on a number of forms, but it will be convenient for us to consider only one of the more important interactions, namely polymer chain entanglements. Figure 2.31 shows two closely placed random-coil molecules which have looped together to form an entanglement. A force F, acting on one of the molecules will move that molecule at a lower velocity than would be predicted by *Eq. 2.71* because any motion of this molecule will tend to drag the other along with it. In effect, the molecular weight is increased; thus, the segment number n must be replaced by some larger number n^*, resulting in either a lower velocity for a given force or requiring a larger force for any given velocity. A quantitative treatment of chain entangle-

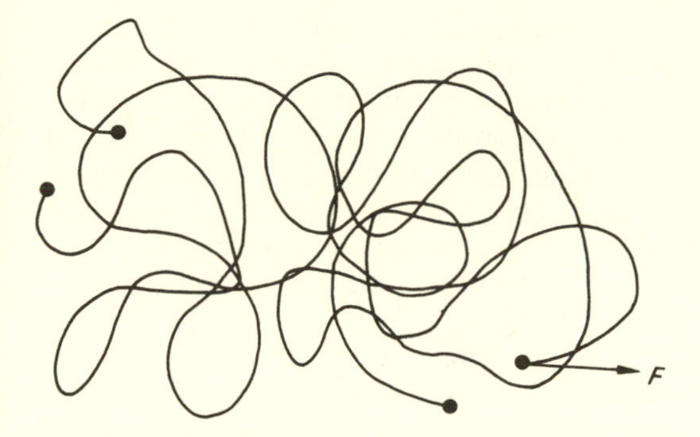

Fig. 2.31 Two entangled randomly-coiled molecules. F is a deforming force.

ment and polymer viscosity (BUECHE, 1962) predicts that at fixed polymer concentration (in weight per unit volume), viscosity will vary with molecular weight to the first power up to some critical molecular weight (M_c). Above this critical value, viscosity increases as a function of (molecular weight)$^{3.5}$, as shown in Fig. 2.32. According to Bueche, the critical value is equivalent to twice the molecular weight between entanglements and is the molecular weight at which entanglements begin to form. The rapid increase in viscosity above M_c reflects the importance of entanglements to the viscous interactions of polymeric materials. The actual value of M_c will depend on the weight concentration of the polymer solution tested and on the nature of the polymer itself. Viscosity measurements on a number of different polymeric materials both in solution and in bulk have shown that real polymeric materials do behave in the manner predicted above (see FERRY, 1970).

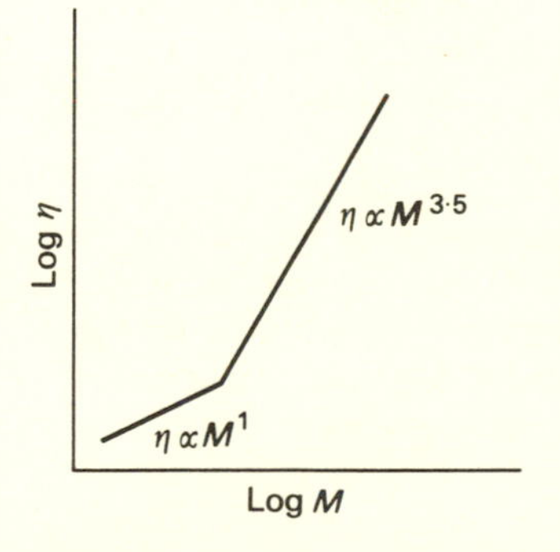

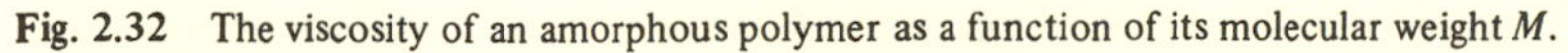

Fig. 2.32 The viscosity of an amorphous polymer as a function of its molecular weight M.

Finally we consider the random thermal motion of individual molecules as opposed to the motion of whole segments. This produces relative movements between neighbouring particles that will be retarded by frictional forces. In fact, Einstein's description of diffusion, the process of molecular motion due to thermal kinetic energy, states that,

$$D = kT/f_0 \tag{2.72}$$

where D is the diffusion constant, k the Boltzmann constant, T the absolute temperature, and f_0 the molecular friction coefficient as defined above. Thus, the random thermal movements of a polymer segment are limited by the same segmental friction coefficient (f_s) used to describe viscous flow of polymer solutions. The theory of rubber elasticity predicts that a force which tends to move the chain ends of a random-coil molecule apart will cause the molecule to take on a new, less random conformation. The rate at which this conformation change takes place will be proportional to the ratio of the thermal energy available (kT), and the segmental friction coefficient of the particular polymer system in question. That is, it will be proportional to some segmental diffusion constant. In an ideal rubber it is assumed that this diffusion constant is large, allowing the conformational changes to take place at a rate equal to or greater than the rate at which the applied force moves the chain ends apart. In the ideal rubber system, all the work put into deforming the molecule is stored as a change in the conformational entropy and no energy is dissipated as heat. If, however, f_s is large or kT small, the rate of conformational change due to thermal kinetic energy may not be able to keep pace with the rate at which the applied force moves the chain ends apart. Under these circumstances the applied force will in effect have to drag some of the polymer segments through their environment, and some of the mechanical energy put into the system will be dissipated as heat in this viscous process. In this case the additional force is required to move the polymer segments at a rate that is Δv faster than the random thermal movements. On the basis of the relationship expressed in *Eq. 2.71* the force and hence the energy dissipated will increase as Δv increases.

2.28 Molecular Structure and the Master Curve

We have described two apparently distinct viscous modes associated with the deformation of polymeric materials, one involving the flow of polymer molecules relative to one another, and the other involving conformational changes of these folded molecules. Now consider how these viscous interactions are coupled with the rubber-like elastic process described previously to give viscoelastic behaviour. Let us think in terms of a stress-relaxation experiment in which a test sample is deformed at time zero and the stress required to maintain this strain followed with time. From this experiment we can derive the time-dependent modulus $E(t)$ which will be characteristic of the molecular structure of the test sample (Fig. 2.33). First consider a polymer network in which the individual molecules are bound together by chemical crosslinks. During the initial deformation, the configuration of the polymer molecules will be altered and mechanical energy will be stored as a decrease in conformational entropy. Because all the molecules are linked together into a single unit, there will be no relative movements of the polymer molecules, and any time-dependent properties will arise from the viscous modes associated with conformational changes. As previously mentioned, in ideal systems this viscous contribution is zero, and all of the energy put into the system is recoverable. The behaviour of such an ideal rubber is indicated in Fig. 2.33, curve A (including the dashed portion). $E(t)$ is constant over the entire time range, indicating that none of the energy put into

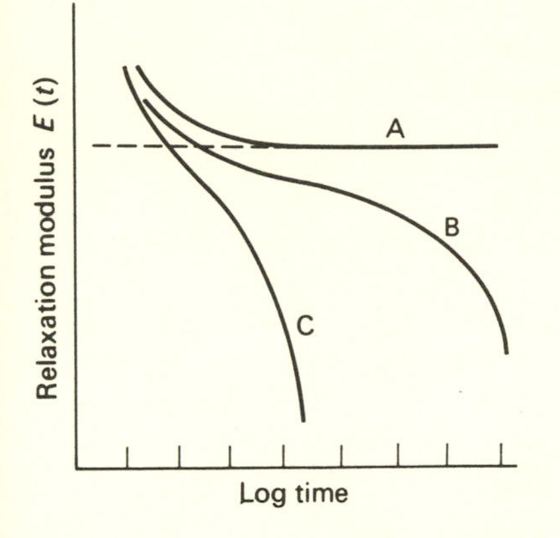

Fig. 2.33 The stress relaxation properties of (A) a crosslinked network polymer (B) a noncross-linked high molecular weight polymer and (C) a noncrosslinked low molecular weight polymer.

the initial deformation is lost. The solid line portion of curve A, represents the behaviour of a real crosslinked rubber. At long times, $E(t)$ is constant, showing that strain energy is indeed stored and the presence of this equilibrium modulus at long times is often used as an indication of a crosslinked polymer network. At short times, however, the sample appears to be more rigid (short times are equivalent to high rates of deformation). Thus, more energy is required for the initial, rapid deformation than is recoverable at longer times and, as we saw previously, this extra energy is required because random thermal segmental motion is unable to keep pace with the rate of deformation.

A low molecular weight, noncrosslinked polymer will be mechanically similar to a very viscous liquid (curve C). Such a material will deform under stress as long as the stress is applied, and $E(t)$ will decrease rapidly to zero. This rapid decrease in $E(t)$ reflects the ease with which the polymer molecules can flow past one another. However, a high molecular weight, noncrosslinked polymer will behave in a manner that suggests that there are temporary crosslinks retarding viscous flow (curve B). There is a pseudo-equilibrium modulus or plateau where $E(t)$ decreases gradually with time, because entanglements and other intermolecular interactions increase the effective viscosity and thus slow down the flow phenomena. The plateau indicates that at short times a high molecular weight polymer will show rubber-like properties, as an applied force will alter the molecular conformation and elastic energy will be stored. But given sufficient time, the temporary crosslinks will break down and allow the polymer chains to return to their unstrained conformation. By this process, the stored strain energy will be dissipated as heat. The viscous or time-dependent modes are associated with regions of the response curves in Fig. 2.33, where $E(t)$ changes rapidly with time. Regions of the curves where $E(t)$ is constant with time indicate time-independent, rubber-elastic behaviour.

Since the basic event in all the flow processes is the movement of a molecular segment we expect "... a priori, that the several time, temperature and frequency parameters should be interrelated" (McCRUM *et al.*, 1967). This is found to be so, but, as we have seen, the relationships are usually fairly complex. However, for the simplest case possible—that of a model having a single relaxation time, we expect that, as a consequence of thermally activated process

$$\tau = \tau_0 \exp(\Delta H/kT)$$

as is, indeed, the case.

To summarize, let us consider the total response or master curve shown in Fig. 2.34. Here storage modulus (E') and loss angle ($\tan\delta$) from a test at constant frequency are plotted against temperature. The shape of the curve indicates that the material tested is a high molecular weight, noncrosslinked polymer like the one tested in Fig. 2.33 (curve B). A similar response curve could be obtained if E' were measured at constant temperature over a frequency range. However, the frequency range needed would be of the order of 15 to 20 logarithmic decades

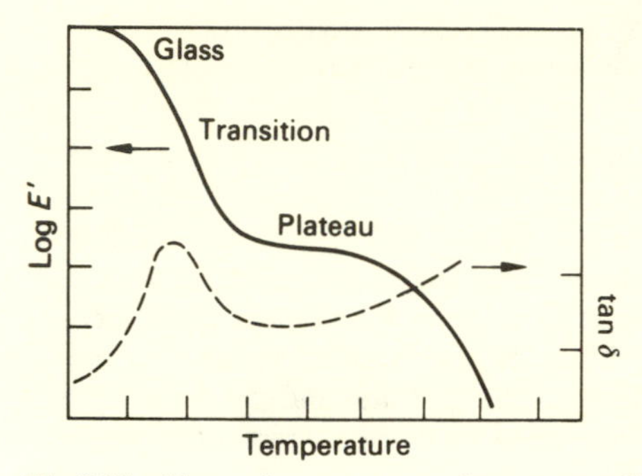

Fig. 2.34 The total response curve for a noncrosslinked high molecular weight polymer. Storage modulus (E') and damping ($\tan\delta$) are measured at constant frequency over a range of temperatures.

of frequency. Such a curve would have to be constructed from the results of a number of experiments carried out at different temperatures using the time-temperature superposition principle. One difference that would be noted between the two plots is that the transition region at constant temperature would be much broader than the one at constant frequency. This difference reflects the increase in the viscous activation energy in the transition region due to the reduction in free volume as the material contracts thermally.

Figure 2.34 shows the four characteristic regions of the viscoelastic response curve. At the low temperature end, the material is a rigid glass, like Perspex (Plexiglass) or polystyrene at room temperature, in which the folded polymer molecules are 'frozen' into a fixed conformation. The value of E' in the glassy region will be of the order of 10^9 N m^{-2}. The damping is low in the glass region because there is no segmental motion and, hence, no viscous dissipation associated with conformational changes.

The transition region corresponds to the onset of segmental motion. The damping is very high and goes through a maximum here because of the large viscous forces which retard the segmental movements required for conformational changes. Polymeric materials in the transition region of their response curves are often used as vibration dampers because of this large energy dissipation.

The plateau is the region of rubbery behaviour. Damping is low and passes through a minimum here because segmental motion due to thermal energy is keeping pace with the deformation rate, and because intermolecular crossbridges in the form of entanglements or chemical crosslinks prevent the flow of molecules relative to one another. The applied force alters the polymer conformation and strain energy is stored. If the material had been a crosslinked polymer network, the plateau would have extended to much higher temperatures and flow phenomena would not be observed. In fact, the modulus of a network polymer in the plateau region will increase slightly with increased temperature. In the flow region of an uncrosslinked polymer, the entanglements break down and the molecules are able to move relative to one another so that the modulus drops quickly and damping increases because of the viscous dissipation in the flow process.

Chapter 3
Tensile materials

3.1 Introduction to Crystalline Polymers

The basic design feature of tensile materials involves the parallel arrangement of high modulus fibres to form rope-like structures that are relatively inextensible when stressed by tensile forces in the fibre direction, but that are readily deformed by compression forces and forces normal to the fibre direction. This tensile rigidity is due to the high degree of preferred orientation of the fibres. The flexibility is due to the ability of these fibres to slide along one another and relieve shear stresses that arise in the bending of a solid structure. In virtually all cases, the tensile structures in organisms can be described as parallel arrays of crystalline polymeric fibres, where the polymer may be either protein or polysaccharide. The mechanical properties of biological tensile materials made from these fibres can be attributed almost entirely to the fibres alone. As we shall see later, the fibrous proteins and polysaccharides are also found in a wide variety of rigid and pliant composite materials whose properties are very different from those of the component fibre, and which are attributable to structural arrangement and interfibrillar bonding and the nature of the matrix. In this chapter we will be concerned with the structure and properties of crystalline polymeric fibres as pure substances, and the properties of the tensile materials in which they are found.

The key to the mechanical properties of these fibrous materials lies in their crystalline nature. The primary fibrous structures in biological materials, the silks and other extended protein chains, collagen, cellulose, and chitin, belong to a range of polymeric substances whose molecular chains aggregate into regions of crystalline order. Such synthetic materials as polyethylene, nylon, and polytetrafluoroethylene (Teflon) also belong to this range of crystalline polymers. Crystallinity in a polymeric material has very important mechanical consequences. In general, the Young's modulus of a bulk crystallized polymer is two to three orders of magnitude greater than that of an amorphous polymer above its glass transition temperature (e.g., rubber), and the modulus of oriented, crystalline fibres is another order of magnitude greater still. In recent years significant advances have been made in our understanding of the relationship between structure and mechanical properties of crystalline polymers. So before looking at specific, biological fibres, it will be useful to consider the principles governing the crystallization of all polymers.

3.1.1 Factors Affecting Crystallinity in Polymers

The term 'crystal' refers to material whose atoms or molecules are ficed into a regular, ordered, three-dimensional array. Thus, the structure of a crystal can be described in terms of a repetitious space lattice where the basic repeating unit of structure is the crystal unit cell. Coherent scattering of X-rays by the regularly placed atoms in the space lattice (i.e., X-ray diffraction) is commonly used as a means of detecting crystalline organization in materials, and X-ray diffraction studies have shown that regions of crystalline organization exist in some polymeric materials. The crystalline regions in linear polymers are usually interpreted to be areas where extended or perhaps helically coiled polymer chains are closely packed in parallel arrays (see Figs. 3.3 and 3.4). Within these crystalline regions the atoms that make up the polymer chains are fixed in a regular, three-dimensional array, and this array is stabilized by a large number of attractive forces between neighbouring atoms. Although the unit cell of a simple organic crystal such as that of glucose contains several complete molecules, the unit cell of most linear polymer crystals contains only a small portion of the entire polymer molecule. Thus, it is possible for a single polymer chain to be incorporated into several crystalline regions and into the intervening amorphous regions as well.

Two major factors determine whether or not a polymer will crystallize. First, polymer chains can only be incorporated into crystalline regions if they are linear. That is, branched polymers will not crystallize. Second, the crystallization of a

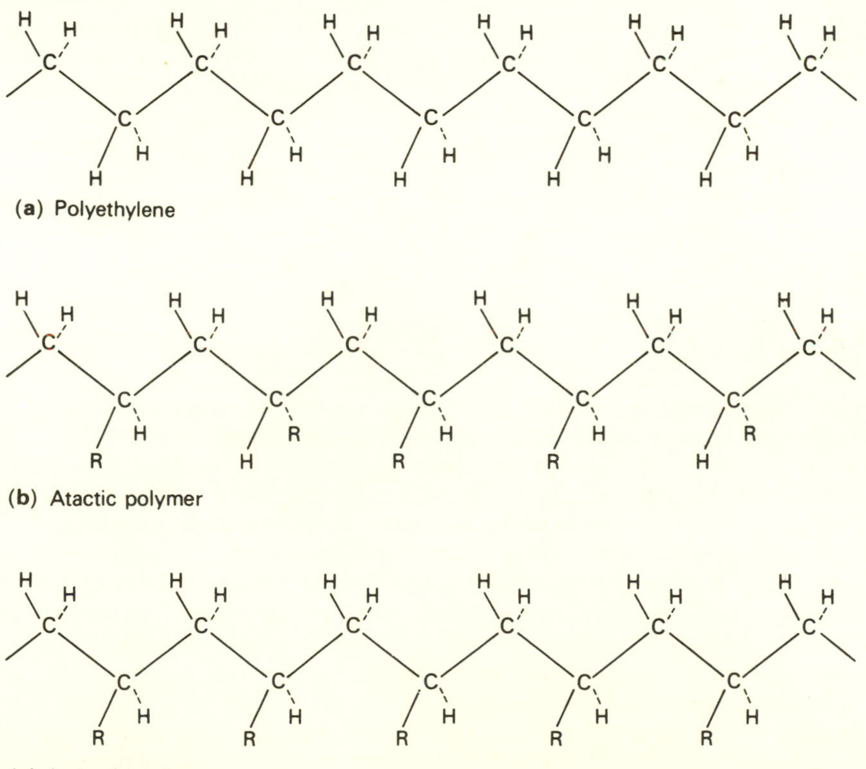

Fig. 3.1 Molecular structure of (**a**) linear polyethylene, (**b**) an atactic polymer and (**c**) an isotactic polymer.

linear polymer requires great regularity in the chain structure. Linear polyethylene provides an excellent example of the type of regularity necessary for crystallization. The chemical structure of polyethylene is very simple, $CH_3(CH_2)_nCH_3$, and as the side groups are all small and identical, there are no possible stereo-isomers to create irregularities in the structure (Fig. 3.1a). In polymers with several types of side groups, stereo-isomers will play an important role in crystallization. If the synthesis of the polymer does not involve a stereo-specific mechanism, the polymer produced will be a random sequence of the different stereo-isomers. This type of polymer, referred to as *atactic* or without order (Fig. 3.1b), will not crystallize. In some cases it is possible to synthesize *isotactic* polymers in which all of the monomers have the same stereo-configuration (Fig. 3.1c). Isotactic polymers are usually crystalline, even when the side groups are quite bulky.

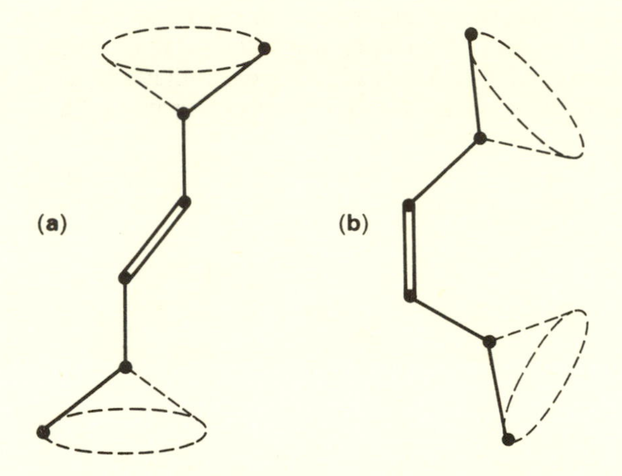

Fig. 3.2 The (a) *trans* and (b) *cis* configurations of a molecule.

An excellent example of the effect of chain conformation on crystallinity is provided by a comparison of natural rubber and gutta-percha (BUNN, 1955). Natural rubber contains only the *cis* isomer of polyisoprene, while gutta-percha contains only the *trans* form of the same chemical compound. As shown in Fig. 3.2, the only difference between the two materials is in the disposition of the C—C bonds on either side of the double bond. In the *cis* form there is an awkward 'kink', while the *trans* configuration is very similar to the extended carbon chain found in crystalline polyethylene. Natural rubber, the *cis* isomer, is an amorphous polymer and crystallizes to an appreciable extent only when subjected to large strains. Gutta-percha is a highly crystalline polymer.

Another form of irregularity in the chain structure is that caused by co-polymerization. In a copolymer two or more different types of monomer are combined to form a polymer. Most of the synthetic processes used to create copolymers produce random sequences of the different monomers and such random sequences lack the regularity necessary for crystallization. Indeed, co-polymerization is often used to inhibit crystallization and enhance rubbery properties in synthetic polymers. A copolymer will crystallize only if there is a specific sequence that is repeated for a substantial distance along the polymer chain.

It is important to note that all of the chemical processes involved in the synthesis of biological polymers contain stereo-specific mechanisms. Two major factors contribute to crystallinity in proteins. Only one of the stereo-isomers of the 21 amino acids, the *laevo* form, is present in proteins, and the synthetic mechanism of proteins can be programmed through the DNA base sequence to produce a unique repeating sequence of amino acids. Thus, it is possible to produce isotactic copolymers that will crystallize. Both of these factors are important in determining the structure of silks, keratin, and collagen. The structural polysaccharides, cellulose and chitin, are both made from a single stereo-isomer of the monomer unit, D-glucose and N-acetyl-D-glucosamine respectively, and all of these units are linked by the same β(1-4)glycoside bond. In this case, the β-linkage plays the same role as the *trans* configuration of gutta-percha. Polysaccharides linked by the α (1-4) glycoside bond (e.g., glycogen) do not form stable crystalline structures like cellulose and chitin.

3.1.2 The Structure of Polymer Crystals

Early X-ray diffraction studies of a wide variety of polymeric materials indicated the presence of crystalline ordering in a number of these materials, but until quite recently we have had no clear idea of the structure of these crystalline regions. Early attempts to define the structure of these regions resulted in the 'fringed micel' model as shown in Fig. 3.3. The model proposed that crystalline regions were areas where polymer chains were closely packed in parallel arrays, and the polymer chains were thought to run straight through these crystalline regions into adjoining amorphous regions and then possibly, into other crystalline regions.

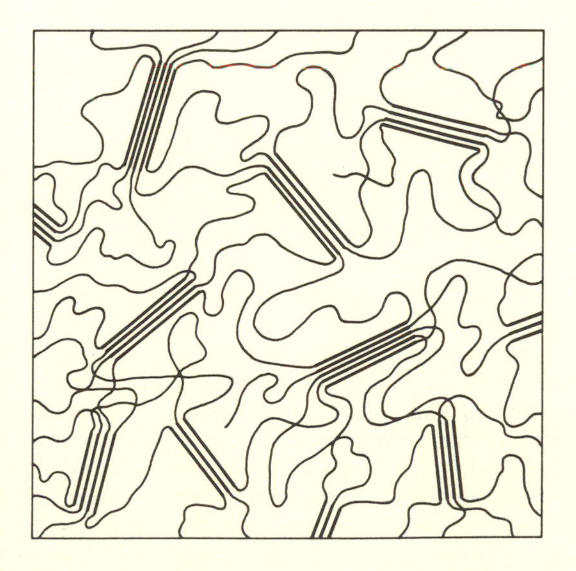

Fig. 3.3 A 'fringed micel' model of a crystalline polymer.

The discovery of polyethylene single crystals (FISCHER, 1957; KELLER, 1957) brought about a marked change in our thinking about crystalline polymers. These crystals, when observed under the electron microscope, appear to be flat, diamond-shaped structures several micrometers on a side and about 10 nm thick. Electron diffraction studies (KELLER, 1957) revealed the startling fact that the *b* axis of the crystal (the axis parallel to the polymer chains) was at right angles to the flat surface of the crystal; that is, the polymer chains run through the thickness of the crystal and not parallel to the surface. The molecular weight of the polyethylene used in these studies indicated that all of the polymer chains were many times longer than the thickness of the crystal. The conclusion drawn was that each polymer molecule was folded back on itself many times within the crystal,

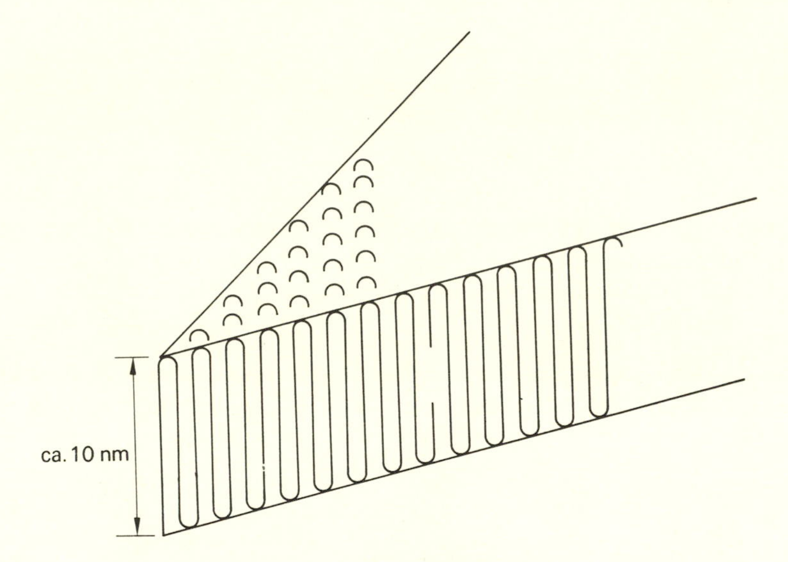

Fig. 3.4 A folded-chain model in a polyethylene crystal.

and the 'folded-chain' model (Fig. 3.4) was constructed to describe the structure of such polymer crystals. Since the first observations of polyethylene single crystals, many other linear polymers have been shown to form crystals in a similar way, and the folded-chain crystal model is now accepted as a very general crystalline arrangement for polymeric materials.

The studies of ANDERSON (1964) and LINDENMEYER (1965) indicate that the folded-chain structures characteristic of polyethylene single crystals are also present in bulk-crystallized polyethylene and other linear polymers. The process of bulk-crystallization involves the cooling of concentrated polymer solutions or pure polymer melts to temperatures below the crystal melting point, and the crystals that form during such a bulk process are in some ways different from the polymer single crystals which form very slowly from dilute solutions. Folded-chain crystal structures do form, but they are not found as discrete, perfect single crystals. Instead, crystalline spherulites are formed in which folded-chain lamellae are radially oriented around centres of crystallization. Figure 3.5 presents a schematic representation of several lamellae in a bulk-crystallized, linear polymer. The lamellae are seen to contain a number of imperfections and are separated from one another by amorphous regions.

Study of the kinetics of crystallization in polyethylene has shown that the thermodynamically most stable crystal form is an extended-chain lamella in which the parallel polymer molecules are fully extended without any chain folding (LINDENMEYER, 1966). Extended-chain structures are energetically favoured because they allow the maximum number of stabilizing interactions to form between adjacent molecules in the crystal lattice, but extended-chain lamellae are not normally found in polymers crystallized from the melt or from solution.

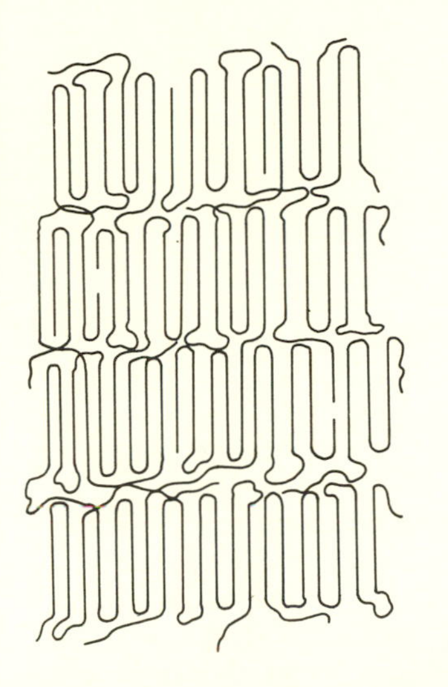

Fig. 3.5 Four lamellae in a bulk-crystallized linear polymer.

The folded-chain, lamellar structure represents a meta-stable state that is favoured under most conditions because the polymer chains lack the mobility to form extended-chain lamellae (LINDENMEYER, 1966). Extended-chain lamellae will form only if low molecular weight polymers are used (MANDELKERN, 1966) or if crystallization takes place under extreme conditions.

As the crystalline polymers occurring in biological materials are fibrous in nature, it may be of some value to consider the structure of fibres drawn from synthetic polymers. X-ray diffraction evidence from a wide range of polymer fibres indicates that the polymer chains become aligned with the fibre axis during extension. This evidence was initially interpreted to mean that the lamellar structure was drawn out into an extended-fibre form. DISMORE and STATTON (1966) proposed that the structure of cold drawn nylon 6-6 was an extended-chain form as shown in Fig. 3.6a. They observed on annealing (holding at a temperature slightly below the crystal melting temperature) that the fibres became more dense (i.e., more crystalline) and that the tensile strength of the fibres

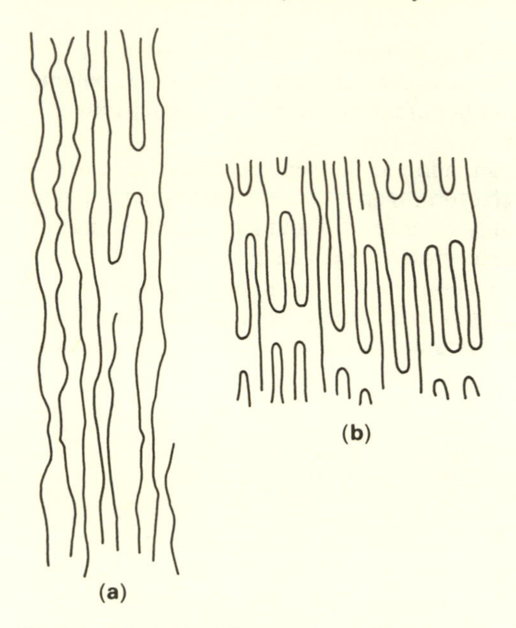

Fig. 3.6 (a) Extended-chain model and (b) lamellar structure model of polymer fibres (after DISMORE and STATTON, 1966; courtesy of the *J. Polymer Sci.*).

decreased. They proposed that a lamellar structure formed during annealing similar to that shown in Fig. 3.6b.

More recently it has been recognized that the lamellar structure is not completely drawn out into extended chains. PETERLIN (1969) concluded that the structure of cold-drawn polyethylene was very similar to that shown in Fig. 3.6b. The lamellae are partially drawn out, but they still remain and are oriented perpendicular to the fibre axis. The amorphous regions between the lamellae are extended to form crosslinks which hold the lamellae together. When these drawn polyethylene fibres break it is the crosslinks that break. According to this model, annealing would allow the lamellae to recrystallize, thus removing some crystal imperfections. Annealing would also relieve the stresses in the extended crosslinks that would then become amorphous regions between the lamellae.

Also of interest is the structure of polymers that have crystallized while under stress. The work of ANDREWS (1964) with natural rubber and KELLER and MACHIN (1968) with polyethylene, gutta-percha, nylon and polypropylene indicates that when polymers crystallize under stress, folded-chain lamellar structures are formed just as in bulk-crystallized polymers, but instead of forming spherulites the lamellae form in rows that are perpendicular to the direction of the stress. X-ray diffraction patterns of stress-crystallized polymers are virtually identical to those from drawn fibres, and it is reasonable to assume that stress-crystallized polymers can be described as parallel lamellae separated by amorphous regions. The process of crystallization under stress should be kept in mind as a model for the formation of such materials as the silks that are drawn from the spinnerettes as they are formed.

3.1.3 Mechanical Properties of Crystalline Polymers

Crystallinity in polymeric materials has a very important effect on the mechanical properties. When the degree of crystallinity is less than 20%, crystalline regions act as crosslinks in an amorphous polymer network and such slightly crystalline materials are mechanically similar to crosslinked rubbers. At degrees of crystallinity above 40%, crystalline regions begin to impinge on one another, and the material becomes quite rigid. The Young's modulus of a highly crystalline polymer can be as much as 10^3 times that of a comparable amorphous polymer. The biological, crystalline polymers that will be discussed in this chapter, namely the silks, collagen, cellulose and chitin, all fall into the category of highly crystalline polymers.

In the previous chapter, it was shown that an amorphous polymer, consisting of randomly oriented flexible molecules showed time-dependent elastic behaviour associated with entropy changes. The elastic moduli were generally low except

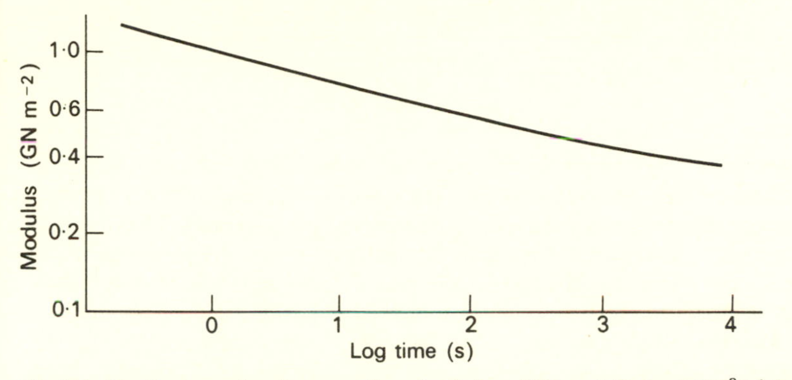

Fig. 3.7 Tensile stress-relaxation curve for polyethylene, 85% crystallinity at 40.7°C (after BECKER, 1961).

at temperatures below the glass transition T_g. Crystalline materials are characterized by time-independent elasticity associated with changes in the internal energy due to stretching of the chemical bonds. The crystalline polymers fall somewhere in between. In the case of highly crystalline polymers, it is reasonable to assume that the energy elastic mechanism predominates and, thus, that the mechanical properties of these crystalline polymers are quite time-independent.

The time-independence of mechanical properties in crystalline polymers is best demonstrated by the behaviour of these materials under long-term stresses. Figure 3.7 shows the results of a stress-relaxation experiment with high density (high degree of crystallinity), linear polyethylene. The stress-relaxation modulus is seen to drop off very gradually in a nearly linear manner for about 5 logarithmic decades of time. The rate at which the stress decays is very low, indicating that the assumption of time-independence is probably valid. Broad relaxation-time and retardation-time spectra that extend to very long times are characteristic of crystalline polymers.

There are two types of molecular movements that contribute to the broad, gradual relaxation seen in Fig. 3.7: those that take place in the crystal lattice and those involving the amorphous areas (FERRY, 1970). As the crystalline organization of a polymer crystal is stabilized by a large number of very weak bonds, it is possible for the polymer chains to be shifted very slightly and very gradually within the crystal lattice in response to applied stresses. The process is analogous to the crystal dislocations that occur during the plastic deformation of metals. A bond is broken between two groups, the crystal plane (or in the case of polymer crystals, the polymer chain) slides until groups are in close register again, and a new bond forms between adjacent chains. Deformation of the amorphous regions results in movement of the lamellae as units, either by slipping along one another or by rotation.

Table 3.1 The effect of orientation on the tensile modulus of several crystalline, polymeric materials. (From NIELSEN, 1962, *Mechanical Properties of Polymers*, Van Nostrand Reinhold, New York)

Material	$\frac{E \text{ of drawn fibre}}{E \text{ of unoriented polymer}}$
Polyethylene Terepthalate	17.8
Nylon 6-6	19.2
Viscose Rayon	9.3
Polyacrylonitrile	5.8

The process of drawing was previously seen to have a dramatic effect on the orientation of polymer chains and lamellae in crystalline polymers. The orientation of chains and lamellae has an equally dramatic effect on the mechanical properties. Both tensile strength and tensile modulus are increased in polymers with a high degree of preferred orientation, but most striking is the effect of orientation on the tensile modulus. Table 3.1, taken from NIELSEN (1962), demonstrates the effects of the degree of preferred orientation of polymer chains on the tensile modulus of crystalline polymers. In each case the effect of orientation is expressed in terms of the ratio of the tensile modulus of drawn fibres with a high degree of preferred orientation to that of the undrawn material. Recalling Peterlin's model for polymer fibres, we can see that the high modulus values in the drawn polymers are due to the extended nature of the crosslinks between crystalline lamellae. The load is applied directly to the covalently-bonded, polymer chains.

Table 3.2 provides an example of the effect of chain orientation on tensile strength. In this case, tensile strength of cellophane films has been measured in

Table 3.2 The effect of polymer chain orientation on the tensile strength of cellophane films (MARK, 1932)

	Tensile strength ($N\ m^{-2}$)	
Test angle	Before stretching	After stretching
0°	11.6×10^7	37.5×10^7
30°	11.2×10^7	22.3×10^7
60°	11.4×10^7	14.5×10^7
90°	11.5×10^7	6.4×10^7

different directions both before and after uniaxial stretching. The test angle given in this table is the angle between the direction of stretch and the direction in which the tensile strength test was carried out. The unstretched film has uniform tensile strength in all test directions, as would be expected if the crystalline regions are randomly oriented in the film. In the stretched film, however, there is a high degree of preferred orientation parallel to the direction of stretch, and the tensile strength is clearly dependent on the test angle. The tensile strength of the film parallel to the preferred axis is about 3.3 times that of the isotropic, unstretched film, but the tensile strength perpendicular to the preferred axis is much lower, lower even than the tensile strength of the isotropic film.

The long term mechanical properties of drawn fibres are very similar to those of undrawn, crystalline polymers with the exception that the relaxation-time spectra of drawn fibres are usually broader and are shifted to longer times (FERRY, 1970). That is, the mechanical properties of drawn fibres are even less time-dependent than those of bulk crystallized polymers. The molecular relaxation mechanism involving the amorphous regions is relatively unimportant in drawn fibres because the lamellae are already aligned perpendicular to the stress direction and the amorphous regions are already extended. The relaxation mechanism involving dislocations in the crystal lattice must account for most of the long term stress–relaxation. The shift of relaxation-time spectra to longer times probably reflects the absence of the amorphous mechanism.

3.2 Silk

The term 'silk' is most commonly associated with textile fibres derived from the cocoon of the silk moth, *Bombyx mori*, but there are many other arthropod species that produce extra-cellular, fibrous proteins that can also be classed as silks. The silks differ from other fibrous proteins in that they are formed by spinning from concentrated protein solutions and are used wholly outside the animal in the form of cocoons, webs, etc. The variety of these extra-cellular products is such that it is impossible to characterize all of the silks with any one protein conformation, but the parallel-β structure characteristic of *Bombyx mori* silk is the predominant form and the one most often associated with the silks. In this discussion we will be primarily concerned with the parallel-β silks, their composition, structure and mechanical properties, but some of the other forms of silk will be mentioned as well.

3.2.1 The Structure of Parallel-β Silks

The silk fibre as it is formed by the larva of the silk moth, *Bombyx mori*, is a complex structure made of two strands of fibroin surrounded by an amorphous sheath of sericin. The discussion that follows deals only with the crystalline proteins that make up the fibroin strands. The sericin, which plays a role in sticking fibres together is relatively unimportant in the tensile properties of silk.

Both X-ray diffraction studies and chemical analyses have been used to investigate the nature of silk. It has long been recognized that fibroin is a crystalline protein and early attempts to characterize the crystal structure interpreted it as extended linear polymer chains (MEYER, 1942). Chemical analyses indicated an abundance of the small amino acids, glycine, alanine and serine. In recent years more sophisticated analyses have provided a fairly complete picture of the structure of silk fibroin.

Table 3.3 Amino acid composition of several arthropod silks

	Bombyx mori[1]	*Antheraea*[2] *pernyi*	*Araneus*[3] *diadematus*, drag line	*Araneus*[4] *diadematus*, flagelliform gland
Glycine	445	265	320	442
Alanine	293	441	363	83
Valine	22	7	16	67
Leucine	5	} 8	16	14
Isoleucine	7		13	10
Serine	121	118	59	31
Threonine	9	1	17	25
Aspartic acid	3	47	8	27
Glutamic acid	10	8	121	29
Lysine	3	1	12	13
Arginine	5	26	18	11
Histidine	2	8	–	7
Tyrosine	52	49	10	26
Phenylalanine	6	6	–	11
Proline	3	3	27	205
Tryptophan	2	11	–	–
Methionine	1	–	–	–
Cystine (half)	2	–	–	–
X-ray group	1	3a	4	–

(Sources: 1, LUCAS and RUDALL, 1968; 2, SCHROEDER and KAY, 1955; 3, PEAKALL, 1964; 4, ANDERSEN, 1970.)

Table 3.3 shows the complete amino acid analysis of a number of different silks. Note that about 86% of the composition of the silk of *Bombyx mori* is accounted for by the amino acids glycine, alanine and serine, and that the number of glycine residues is roughly equal to the number of alanine and serine residues together. This composition suggests a possible repeating structure, poly-(gly-X), where X is either alanine or serine. As the side chains of both alanine and serine are small and of similar dimensions, a polymer of this nature would certainly be regular enough (see section 3.1.1) to aggregate into regions of crystalline order. MARSH *et al.* (1955*a*) carried out an extensive X-ray diffraction analysis of *Bombyx mori* silk and concluded that the structure could be modelled by the synthetic polypeptide poly-L-alanyl-glycine in the form of anti-parallel, pleated sheets of extended protein chains linked by inter-chain hydrogen bonds. The anti-parallel arrangement indicates that adjacent folded protein chains run in opposite directions; pleated sheets indicate that the chains form layers in the plane of the inter-chain hydrogen bonds and that these layers are stacked one upon the other. The amino acid side chains extend into the spaces between the pleated sheets in the manner shown in Fig. 3.8. Because of the precise alternating amino acid sequence, the pleated sheets could be arranged in such a manner that the

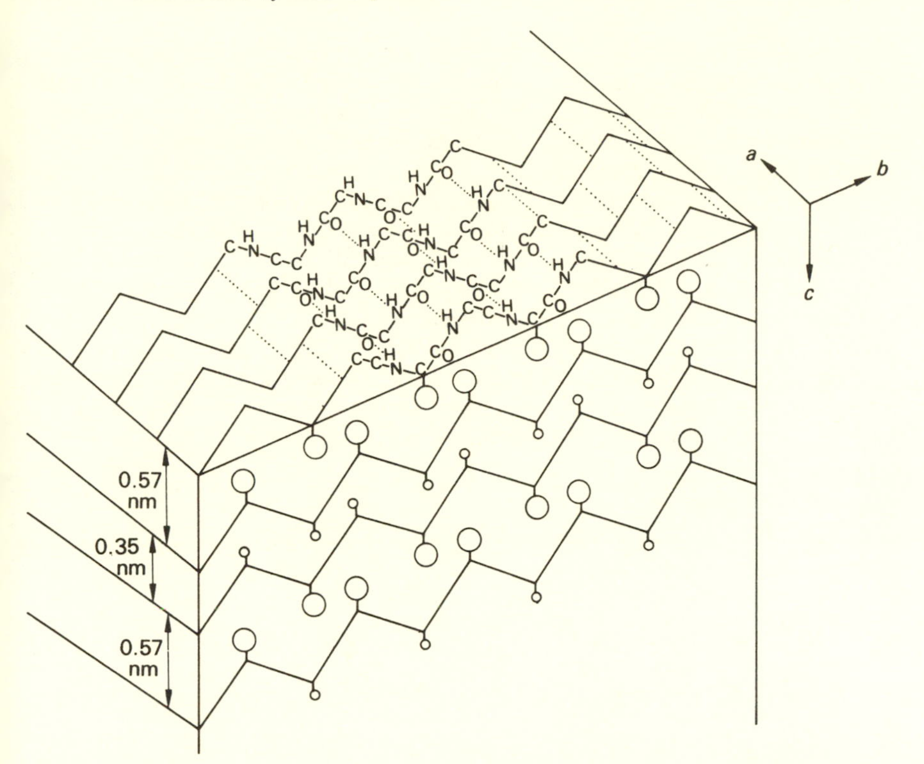

Fig. 3.8 Pleated sheet model of *Bombyx mori* silk (after MARSH *et al.*, 1955*a*; courtesy *Biochim. Biophys. Acta*).

smaller side chains of glycine in one sheet oppose only glycine side chains in the next sheet and the larger alanine side chains oppose only similar side chains. This type of structure gives a characteristic *c*-axis spacing (inter-sheet or side chain spacing) of 0.93 nm; 0.35 nm for the glycine side chains and 0.57 nm for the alanine side chains. FRASER *et al.* (1965) made X-ray diffraction studies of the synthetic polypeptide poly-L-alanylglycine and found that its structure was very similar to that predicted by MARSH *et al.* (1955*a*). In a later study (FRASER *et al.*, 1966) the synthetic polypeptide poly-L-alanylglycyl-L alanylglycyl-L-serylglycine was used as a model for the crystalline regions of *Bombyx mori* silk. X-ray diffraction studies indicated that this polypeptide more nearly matches the structure of native silk than does poly-L-alanylglycine.

DOBB *et al.* (1967) studied the fine structure of silk fibroin and observed microfibrils that were roughly rectangular in cross-section, 2.0 nm in the *c*-axis (inter-sheet) direction and 6.0 nm in the *a*-axis (inter-chain) direction. This suggests a basic unit of structure which is four pleated-sheets thick (LUCAS and RUDALL, 1968). KORATKY *et al.* (1964) observed that particles could be obtained in solution from *Bombyx mori* silk with dimensions of 21.2 x 6.6 x 1.9 nm suggesting that the extent of the crystalline regions along the *b*-axis (fibre-axis) is roughly 21.0 nm. Amino acid analyses (Table 3.3) suggest that not all of the material in fibroin is crystalline. There must be sizeable amorphous regions between the crystallites containing the larger amino acids. The general picture arising from available information is one of anti-parallel, extended-chain crystals (i.e., chain lamellae) oriented with the polymer chains parallel to the fibre axis and separated by amorphous regions. This is essentially the same structure as found in drawn polyethylene fibres.

MARSH *et al.* (1955*b*) studied the silk of the moth *Antheraea pernyi* and found an anti-parallel, pleated sheet structure that was slightly different from that of *Bombyx mori* silk. The difference was in the disposition of the side chains between the pleated sheets. *Bombyx mori* silk has alternating layers of glycine side chains and alanine side chains, while in *Antheraea pernyi* silk the structure is essentially that of poly-L-alanine. The layers are all the same, giving a *c*-axis spacing of about 1.06 nm or 0.53 nm between each pleated sheet. These differences are illustrated in Fig. 3.9. The larger inter-sheet spacing in *Antheraea pernyi* silk reflects the relatively greater proportion of larger amino acids present (see Table 3.3). The

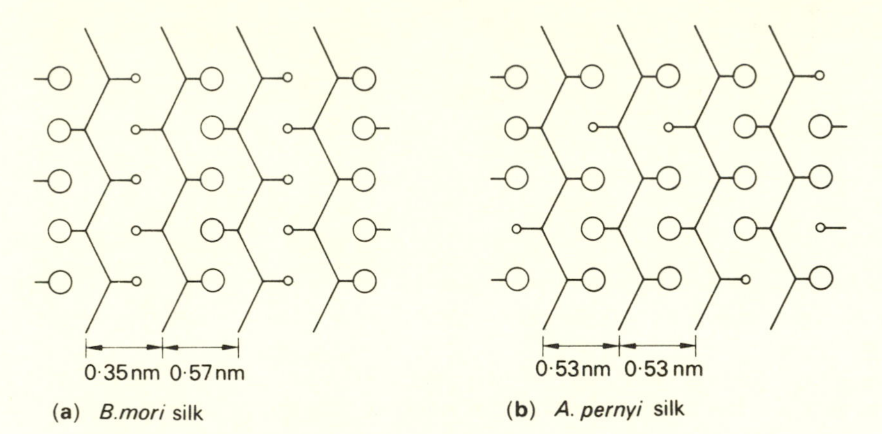

Fig. 3.9 Comparison of side chain disposition in *B. mori* and *A. pernyi* silks. Small circles represent hydrogen of glycine side chain; large circles represent CH_3 or OH side chains of alanine or serine (after MARSH *et al.*, 1955*a*; courtesy *Biochim. Biophys. Acta*).

amino acids glycine, alanine and serine account for 82% of the total, and the number of alanine and serine residues is more than twice the number of glycine residues.

X-ray diffraction and amino acid studies of a number of arthropod silks indicate that the basic structure of most parallel-β silks is similar to that of *Antheraea pernyi* silk. Differences that occur are in the side chain spacing of the various silks, and WARWICKER (1960) has proposed a classification for the parallel-β silks based on these (see Table 3.4). Amino acid analyses of the silks from these various groups indicate that those silks having a higher proportion of the more bulky amino acids generally have a larger side chain spacing (LUCAS *et al.*, 1960).

An interesting aspect of the formation of silk fibres is the need to keep the concentrated protein solutions in the silk gland in the liquid state. The protein concentration can range from 15% to 30% in these glands (IIZUKA, 1966), and the tendency of these proteins to form polymer crystals poses a serious problem

Table 3.4 Dimensions of unit cell of silk fibroins in nm (WARWICKER, 1960; courtesy of *J. Molec. Biol.*)

Group	Origin	*a*	*b* (fibre axis)	*c* (side chain)
1	*Bombyx mori*	0.944	0.695	0.93
2a	*Anaphe moloneyi*	0.944	0.695	1.00
2b	*Clania* sp.			
3a	*Antheraea mylitta*	0.944	0.695	1.06
3b	*Dictyoploca japonica*			
4	*Thaumetopoea pityocampa*	0.944	0.695	1.50
5	*Nephila senegalensis*	0.944	0.695	1.57

to the animal. LUCAS and RUDALL (1968) report that it is not uncommon for moribund individuals to be found with 'solid' glands, but in this case it is difficult to distinguish the cause from the effect. Does solidified silk kill, or do dying insects lose the ability to keep the silk liquid? The presence of amino acids other than glycine, alanine and serine probably plays an important role in preventing coagulation. The larger amino acids that do not fit the crystal lattice form amorphous regions that can isolate crystallites from each other and thus prevent them from precipitating (LUCAS and RUDALL, 1968).

RAMSDEN (1938) observed that shearing of the viscous contents of silk glands between glass slides would cause it to coagulate. IIZUKA (1966) studied the effect of shear rate on the coagulation of fibroin solutions and also analyzed the dimensions of the spinneret of the silk worm *Bombyx mori* with regard to the shearing forces arising in this tube. He concluded that the shearing forces in the spinneret align molecules thus enhancing crystallization and also align the randomly oriented crystallites thus promoting coagulation. The greater the shearing forces, the higher the degree of crystallinity of the product. Another important aspect of silk formation is drawing of the fibres as they are spun. As the fibre is extruded from the spinneret it is stretched or drawn, and this drawing will both increase the degree of preferred orientation of the crystalline regions and extend the chains in the amorphous regions between crystalline lamellae. This provides a second mechanism for increasing crystallinity and, we shall soon see, for enhancing the mechanical properties of the fibre.

3.2.2 The Mechanical Properties of Silk

Because the silk of *Bombyx mori* provides an important textile fibre there is a relative abundance of information on its mechanical properties, but with the exception of a few isolated studies on the mechanical properties of spiders' silk, there is virtually no information on the properties of the other silks. The data which are available are presented in Table 3.5. The starred values were given in the literature in grams per denier and have been converted to more useful units, $N\,m^{-2}$, by assuming a density of 1.35 for *Bombyx mori* silk and of 1.26 for spider silk. The values for Young's modulus should only be taken as approximations, as the stress-strain curves from which the values were taken are usually not linear, and we often do not know what portion of the curves these values represent. There are also variations in tensile strength and extensibility values which are probably due to differences in testing method (i.e., rate of extension,

Table 3.5 The mechanical properties of some arthropod silks. The starred values were given in the literature in the units of grams per denier. They were converted to N m^{-2} by assuming a density of 1.35 for *B. mori* silk and 1.26 for spider silks.

Silk	Young's modulus (N m^{-2})		Tensile strength (N m^{-2})	Extensibility (%)	Reference
1 *Bombyx mori*					
Cocoon thread	10^{10}	@ 47% RH			
	8.9×10^{9}	@ 60% RH			
	7.1×10^{9}	@ 70% RH	6×10^{8} @ 65% RH	18% @ 65% RH	1
	6.3×10^{9}	@ 80% RH			
	5.1×10^{9}	@ 90% RH			
			3.5 to 4.4×10^{8}		2
	10^{10}		4.8×10^{8}	20%	3
	1.1×10^{10}	@ 0.028 Hz, 60% RH*	4.5×10^{8}*	18%	4
	1.6×10^{10}	@ 170 HZ, 60% RH*			
			6.2×10^{8} @ 80% RH	21% @ 80% RH	5
			3.5×10^{8}*		6
			4.4×10^{8}	16%	7
2 *Aranea diadematus*					
Drag line	4.7×10^{9}		14.2×10^{8}	30%	6
	2.8×10^{9}	(1.2×10^{10})*	8.7×10^{8}*	31%	7
Cocoon thread	2.2×10^{9}		5.4×10^{8}	24%	6
	5.5×10^{8}	(6×10^{9})*	2.5×10^{8}*	46%	7
Sticky spiral thread				517%	6
3 *Meta reticulata*					
Sticky spiral thread			$2.3 \times 3.3 \times 10^{8}$	600 to 1600%	6

(References. 1. DENHAM and LONSDALE, 1933; 2. MARK, 1933; 3. SHIMIZU, FUKUDU and KIVIMURA, 1957; 4. IIZUKA, 1965; 5. MEYER, 1942; 6. DE WILDE, 1943; 7. LUCAS, 1964.)

temperature, relative humidity, etc.) or to inaccurate determinations of specimen thickness.

DENHAM and LONSDALE (1933) studied the effect of relative humidity on the modulus, tensile strength, and extensibility of *Bombyx mori* silk. In general, their results indicated that modulus and tensile strength decrease while extensibility increases with increased humidity. The greatest change takes place between about 70% relative humidity and saturated conditions. Water molecules probably penetrate into the amorphous regions between adjacent crystallites and loosen the structure by competing for hydrogen bond sites on the exposed protein chains. The study of IIZUKA (1965) provides perhaps the best measurements of the mechanical properties of *Bombyx mori* silk. He used a dynamic testing technique and gave values at two frequencies, 170 and 0.028 Hz. The decrease in modulus over this nearly four logarithmic decade time range is only 32%, similar to the very gradual decrease in modulus with time observed for a number of synthetic, crystalline polymers. Iizuka found that modulus, tensile strength and extensibility varied with the thickness of the test sample, all decreasing as the sample thickness was increased. The degree of crystallinity, as measured by an X-ray diffraction method, was also found to vary with thickness. Crystallinity ranged from 40% in the thickest fibres to 47% in the thinnest and was observed to increase with stretching, but never to more than 50%. These observations suggest that the mechanical properties of silk fibres are closely related to the degree of crystallinity.

DE WILDE (1943) and LUCAS (1964) report data for the mechanical properties of spider silk. It appears from Table 3.5 that the Young's modulus of *Bombyx mori* cocoon silk is considerably greater than that of spider silk, but the modulus figures for spider silk were calculated directly from the tensile strength and extensibility data. This calculation is based on the assumption that the stress–strain curve is linear, and as can be seen in Fig. 3.10 these curves are definitely not linear. The modulus figures given in parentheses in Table 3.5 have been calculated using the initial slopes (lines OA and OB) of the stress–strain curves in Fig. 3.10. The modulus of spider drag line is nearly the same as that of *Bombyx mori* silk. Spider cocoon silk appears to have a lower modulus than either of the other silks.

Comparison of tensile strength and extensibility data suggests that *Bombyx mori* and spider cocoon silk have about the same tensile strength with the spider silk being a bit more extensible. The tensile strength of spider drag line silk is considerably greater than either of the other silks. This is probably associated with the fact that drag line silk is more extensible than *Bombyx mori* silk. The drag line is the thread which is used by the spider to suspend or lower itself during its normal movements. It is produced by the same glands that produce the frame fibres and radii of the orb web (ANDERSEN, 1970). These fibres are extremely important tensile structures. LUCAS (1964) calculates that the drag line of an adult female spider weighing 0.65 g can support a load of 1.0 g. This allows only a small safety factor considering the drag line not only has to support the weight of the spider but also must overcome the inertia of the spider when it drops. Lucas guesses that the extensibility of the drag line allows for a gradual deceleration of the dropping spider, and thus, keeps the load within the critical

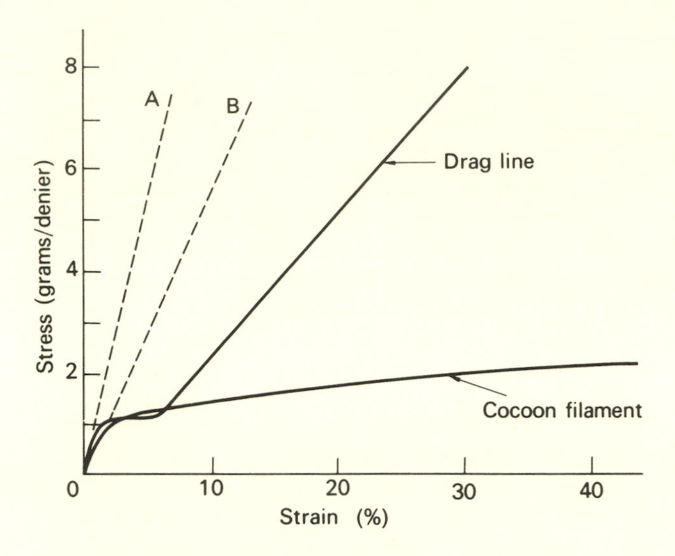

Fig. 3.10 Stress–strain curves for drag line silk and cocoon silk from the spider *Araneus diadematus.* To convert stress values to N m^{-2}, multiply by 8.8 x 10^7 ρ, where ρ is the density of silk (*ca.* 1.26). Lines OA and OB are the initial slopes of the two stress-strain curves (after LUCAS, 1964).

limits. The web requires these flexible, strong fibres in order to deal with the movement of the supports to which the web is attached, to restrain struggling prey, and to keep the web from tearing in the wind.

3.2.3 Other Types of Silk

A number of unusual protein structures have been observed in the wide variety of silks that have been examined. LUCAS and RUDALL (1968) review the literature on these structures so we will only mention some examples in passing. X-ray diffraction patterns of the silk from the lacewing fly *Crysopta* appears to be nearly identical to the pattern of parallel-β silks except that the X-ray pattern is rotated 90° with respect to the fibre axis (PARKER and RUDALL, 1957). GEDDES *et al.* (1968) interpreted the structure of this silk as shown in Fig. 3.11. The structure is an anti-parallel, pleated sheet structure of the *Antheraea pernyi* type, but the protein chains run at right angles to the fibre axis. They suggest that the protein chains fold back on themselves every 2.5 nm, or at every eighth amino acid residue. These silks, known as cross-β silks, can be extended up to six times initial length, and the extended silks give parallel-β X-ray patterns. The mechanical significance of the cross-β structure is completely unknown.

Other silks have been found which have an α-helical structure (LUCAS and RUDALL, 1968). The α-helix, described by PAULING and COREY (1953*a*), is a stable protein conformation characteristic of a wide variety of structural proteins, including keratin, myosin and fibrinogen. In the α-helix, each amide group in the protein chain is hydrogen bonded to the third carbonyl group from it. The helix has 3.7 residues per turn and a pitch of 0.52 nm to 0.56 nm. According to LUCAS and RUDALL (1968) α-helical silks are produced by all aculeate Hymenoptera

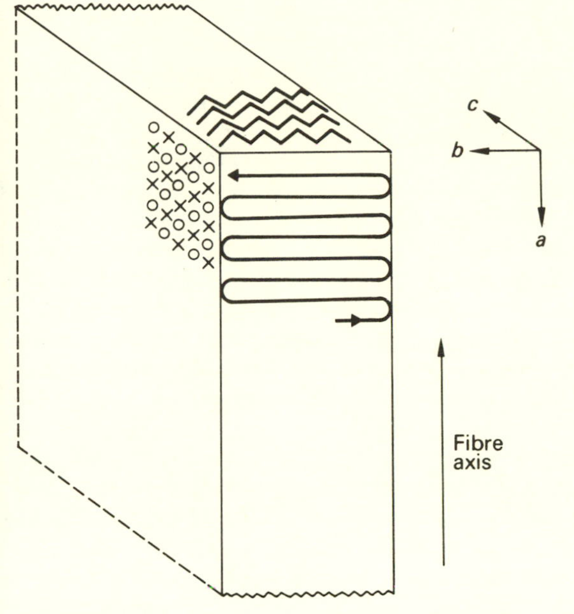

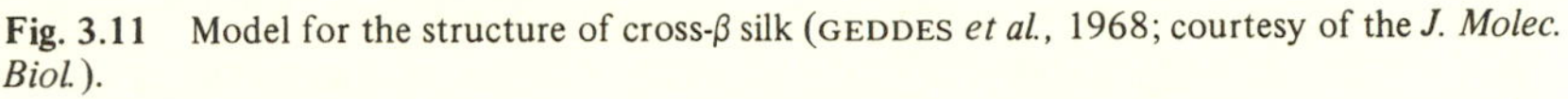

Fig. 3.11 Model for the structure of cross-β silk (GEDDES *et al.*, 1968; courtesy of the *J. Molec. Biol.*).

(bees, wasps, and ants). ATKINS (1967) proposes that these helices are wound together into four-chain ropes.

The silk of the sticky spiral of orb webs should really be discussed in the section on pliant materials, but because it is one of the silks used by spiders in their webs it will be included here. DE WILDE (1943) reports that the sticky spiral of *Meta reticulata* can be extended up to 16 times its initial length before it breaks, and that it can be extended reversibly up to 14 times its initial length. These are clearly not the properties of a crystalline polymer, but rather those of an amorphous one. ANDERSEN (1970) gives an amino acid analysis for the silk from the flagelliform gland which is reported to produce the sticky spiral thread of the spider *Araneus diadematus* (see Table 3.3). The most notable feature of this analysis is the high proline content. Proline is known to play an important role in protein conformations because it imposes a 'kink' in the protein chain (SZENT–GYÖRGYI and COHEN, 1957). Andersen suggests that the high levels of proline in this silk inhibit crystallization, thus keeping the material completely amorphous, and by virtue of its amorphous nature the silk takes on rubbery properties. The mechanical properties of amorphous polymers will be dealt with in greater detail in a later section.

3.3 Collagen

Collagen is the basic structural fibre of the animal kingdom. It is found in virtually all of the animal phyla, usually as a component of a complex, pliant connective tissue, but it is also found in a number of tensile structures, such as tendon, where it is present as parallel arrays of nearly pure collagen fibres. In most animals, excepting the arthropods, collagenous tissues provide the majority of the passive structural elements (e.g., skin, cartilage, tendon, and bone), and collagen is often the most abundant protein in the animal. Because of the wide distribution and mechanical importance of collagenous tissues, a great deal of research has been carried out on the biology and the chemistry of collagen, and this has resulted in an immense literature. Fortunately, a number of reviews covering many aspects of collagen have been published in the past few years, including GOULD (1968), a three volume *Treatise on Collagen*; BAILEY (1968*a*), a general treatment of the structure and chemistry of collagen; and ELDEN (1968), on the physical properties of collagen fibres. In this discussion we will only outline the concepts relating to the structure of the collagen molecule and the arrangement of these molecules into fibres. For more complete coverage of the structure, chemistry, metabolism and pathology of collagen the above reviews should be consulted.

3.3.1 The Structure of Collagen

The term collagen does not specify a single protein or a unique amino acid sequence. It is rather, the general description of a class of proteins which, through similarities in amino acid sequence, have similar conformation and physical properties. Several of the characteristics commonly used to identify collagens are listed below.

1 Amino acid analysis. Collagens contain the amino acid hydroxyproline which is not found in most other proteins. Proline and hydroxyproline together account for about 20% of the total amino acid residues. Glycine makes up one-third of the total amino acids.

2 Collagen shows a characteristic 64 nm banding pattern when observed with the electron microscope.

3 X-ray diffraction patterns of stationary fibres have a 0.29 nm meridional reflection corresponding to the distance between residues along the helices, a 1.0 nm layer line corresponding to the pitch of the minor helix, and a 1.1 to 1.3 nm equatorial reflection corresponding to the diameter of the coiled-coil.

4 Upon heating, collagen fibres shrink to about one-third of their original length and become rubbery.

The amino acids mentioned in item 1 of this list play a major role in determining the three-dimensional conformation of collagen, and items 3 and 4 are indications of this conformation. Item 2 reflects the manner in which collagen molecules aggregate to form fibres.

The structure of collagen is based on the helical arrangement of three, non-coaxial, helical polypeptides, stabilized by inter-chain hydrogen bonds. Three helical polypeptides are thrown into a second order helix with a long pitch around a central axis to form a coiled-coil structure (RAMACHANDRAN, 1963). Figures 3.12a and b show this organization and Fig. 3.12c shows an end-on view of the coiled-coil with the helices extending out of the plane of the page towards you. It is to the α-carbon that the side chains are attached. Every third amino acid in the proposed structure is glycine, and all the glycines face the centre of the triple-helix. As the glycine side chain is only a single hydrogen atom, the minor

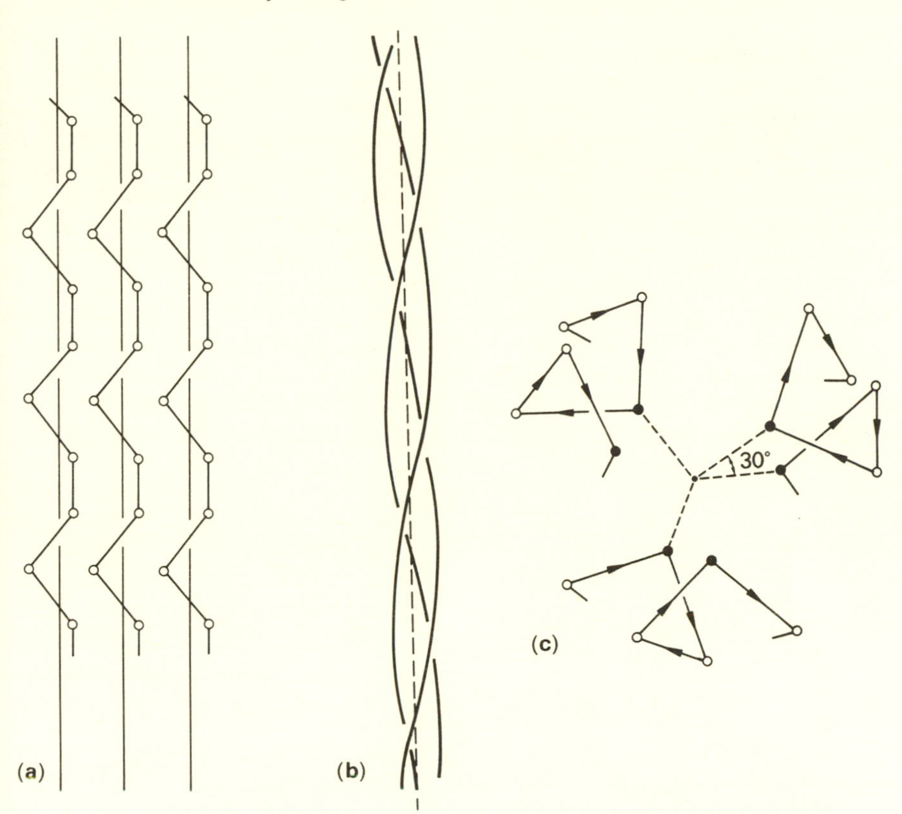

Fig. 3.12 The coiled-coil in collagen structure. (a) Three polypeptide chains coiling around three separate (minor) axes, (b) the three minor axes are shown coiled around a central axis (dashed line), (c) end-on view of the coiled-coil. Each circle represents the α-carbon of an amino acid. Solid circles are glycine. Solid line in (c) is the polypeptide chain (RAMACHANDRAN, 1963).

helices can be packed closely together to form the tight, triple-helix. For each turn of the minor helix, the major helix rotates 30° around the central axis. A number of models have been constructed which specify in more detail the structure of the collagen triple-helix, and the RICH and CRICK (1961) Collagen II model and the model of RAMACHANDRAN and SASISEKHARAN (1961) have emerged as the two most likely structures. The models are quite similar, embodying the basic features outlined above, but Collagen II allows only one inter-chain hydrogen bond per three amino acid residues while the Ramachandran model allows two stabilizing bonds. There is considerable evidence for each model (see BAILEY, 1968*a*) but the two-bonded structure is usually the preferred one.

That the collagen molecule is truly a crystalline structure is indicated by the thermal shrinkage noted in characteristic 4. This transition from an extended, triple-helical structure to a random coil is a thermodynamically first-order transition, characteristic of melting in crystalline materials (FLORY and GARRETT, 1958). The temperature at which this transition takes place (T_D) varies in collagens from different animals. The denaturation temperature has been correlated with the pyrrolidine content (i.e., the amount of proline and hydroxyproline) and found to be higher in collagens with a higher proportion of these amino acids (HARRINGTON, 1964; RAO and HARRINGTON, 1965).

As stated previously, the amino acids glycine, proline and hydroxyproline play a major role in determining the conformation of collagen. The polypeptides *poly* (gly-pro-pro) and *poly* (gly-pro-OHpro) have been synthesized and found to have coiled-coil structures virtually identical to that of native collagen (ENGEL *et al.*, 1966; TRAUB and YONATH, 1966). It is now believed that the 'collagen fold' or triple-helical structure will be formed if any polypetide contains enough tripeptides with the amino acid sequences (gly-pro-pro), (gly-pro-OHpro), or (gly-pro-X-).

If collagenous tissue is extracted in weak, neutral salt solutions or in weak acids (pH 3 to 4) a small amount of soluble collagen is obtained which retains the triple-helical structure. BOEDETKER and DOTY (1956) carried out a number of experiments with weak acid-extracted collagen and concluded that the molecules

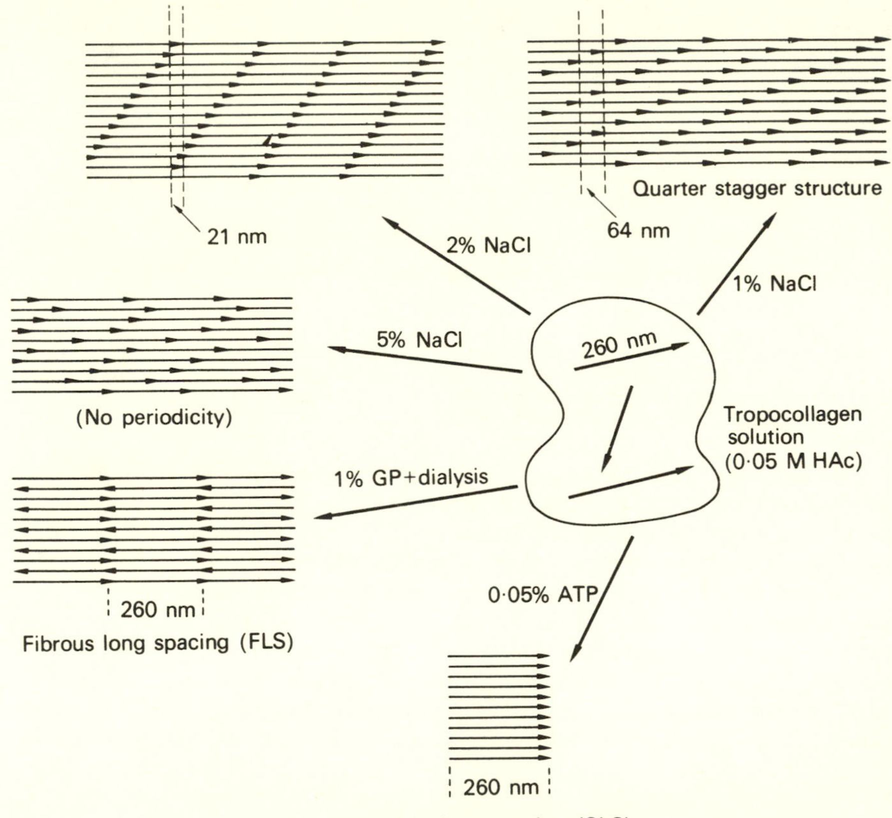

Fig. 3.13 Patterns of aggregation of tropocollagen molecules that are precipitated by various treatments of tropocollagen in weakly acidic solution. Each arrow represents the 300 nm long tropocollagen molecule. (After SCHMITT, 1956; courtesy of the American Philosophical Society).

had dimensions 1.36 by 300 nm and a molecular weight of 345 000. This molecule has roughly the same diameter as the triple-helix seen by X-ray diffraction methods. The shrinkage at elevated temperatures that is characteristic of insoluble collagens was also observed with the soluble form, as indicated by a major shift in the specific rotation at the denaturation temperature. More recent estimates of the dimensions of soluble collagen molecules (see BAILEY, 1968*a*) suggest that it is a bit shorter (280 nm) and the molecular weight a bit less (265 000 to 310 000) than suggested by BOEDETKER and DOTY (1956). GROSS *et al.* (1954) termed this soluble form of collagen 'tropocollagen' and proposed that it was the basic subunit for the assembly of collagen fibres.

Under certain conditions tropocollagen in solution will precipitate to form fibrous structures, and the arrangement of the tropocollagen molecules in these aggregates depends on the chemical composition of the precipitating medium. Figure 3.13 summarizes much of the work that has been carried out on the structure of precipitated collagens (SCHMITT, 1956). The form with 64 nm periodicity is typical of vertebrate collagens and numerous invertebrate collagens. It is referred to as the *quarter-stagger* structure because adjacent molecules are transposed one-quarter of their length in the axial direction. The 21 nm spacing, characteristic of some invertebrate collagens (PIEZ and GROSS, 1959), arises in a similar manner, but the shift between adjacent molecules is smaller. Fibres without any obvious periodicity, similar to fibres seen in the cuticles of *Ascaris* and *Lumbricus*, arise by a random transposition of adjacent molecules. The two long

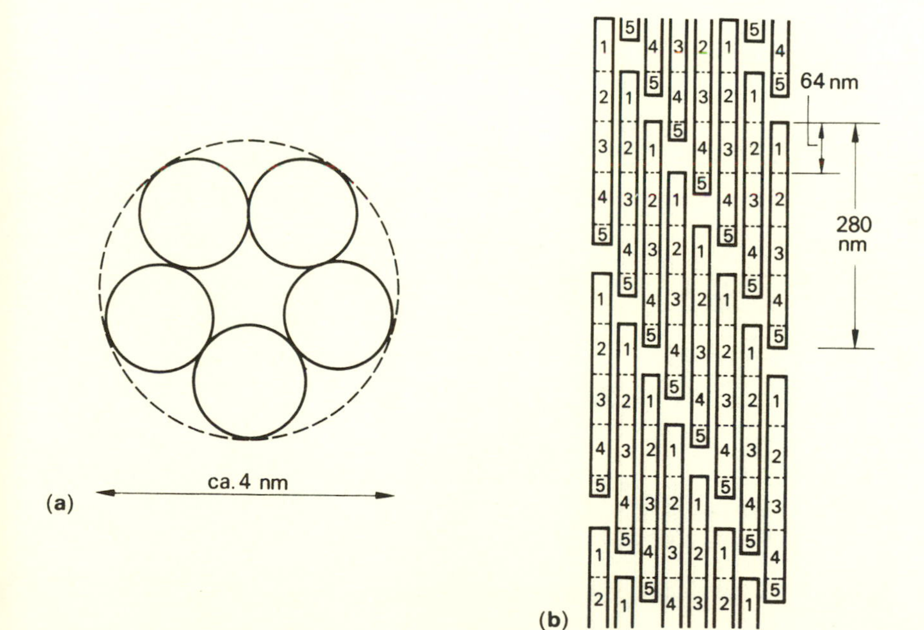

Fig. 3.14 After SMITH'S (1968) revised quarter-stagger model for collagen (**a**) in fibrillar form (cross-section) and (**b**) in planar form. Courtesy of *Nature*.

spacing forms, fibrous long spacing (FLS) and segment long spacing (SLS), are not represented in natural collagens, but they have played an important role in our understanding of collagen fibres. Close examination of fine banding patterns on these forms has shown that the collagen molecule is directional. That is, it has a head and a tail which are distinct from each other.

Detailed analysis of the quarter-stagger arrangement indicated that the tropocollagen molecule is actually 4.4 times the 64 nm period observed in electron micrographs. This observation led to a slight modification of the quarter-stagger model (SMITH, 1968). Figure 3.14 shows the revised edition. The molecule is broken up into five zones, instead of the four quarters of the earlier model. The first four zones are of equal length (64 nm), and the 5th zone is only about 25 nm. The new arrangement requires a 30 to 40 nm space between the ends of each molecule. These spaces have been observed and are thought to play an important role as mineral nucleating sites in the mineralization of collagenous tissues such as bone.

SMITH (1968) realized that this modified quarter-stagger model could be used to construct a two-dimensional sheet of collagen, but would not easily fit into a three-dimensional array. He postulated that fibrils were formed by folding a sheet five molecules wide into a hollow tube (Fig. 3.14a). Such a fibril would have a diameter of about 4 nm. MILLER and WRAY (1971) and MILLER and PARRY (1973) made a careful assessment of the small angle reflections in the collagen X-ray diffraction pattern, and found an equatorial reflection indicating a 3.8 nm

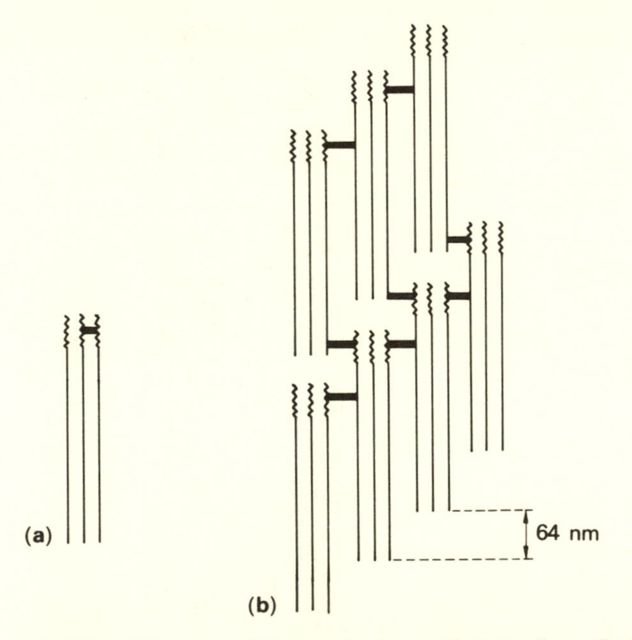

Fig. 3.15 (a) Intra- and (b) intermolecular crosslinks in collagen. Straight lines represent polypeptide chains in the helical region of tropocollagen molecules; wavy lines represent the *N*-terminal telopeptide (BAILEY *et al.*, 1970*a*).

spacing. This is very close indeed to the 4 nm diameter of the fibril proposed by Smith. The X-ray data also suggest that the molecules twist around the long axis of the fibril to form helices. BOUTEILLE and PEASE (1971) observed 3 to 3.5 nm filaments in thin sections of collagen fibres. These filaments, presumably the fibrils of SMITH (1968) appeared to spiral around the fibres to form helices with a pitch of about 1 μm. However, a detailed polarized light study by DIAMANT *et al.* (1972) suggests that the 4 nm fibrils are organized into a planar crimped arrangement rather than a helix.

The precipitated forms of collagen seen in Fig. 3.13 are held together by numerous, weak interactions between bonding sites located in specific regions along the tropocollagen molecule. The characteristic aggregations and banding patterns of collagen fibres are due to the specificity of these bonding sites, and the instability of these interactions is indicated by the ease with which the precipitated collagens can be brought back into solution. The insoluble collagens

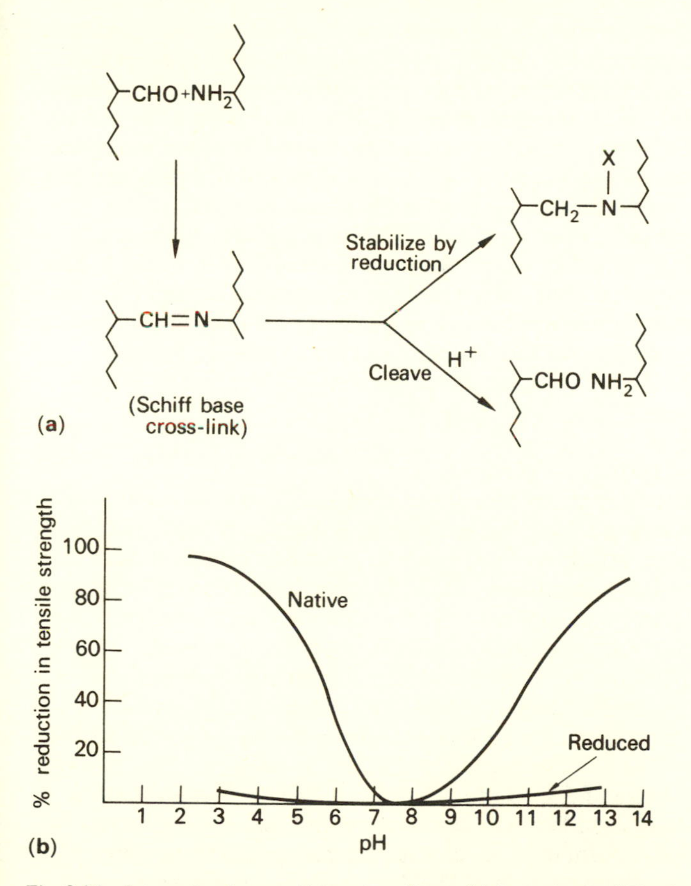

Fig. 3.16 Intermolecular crosslinkage in collagen. (a) formation, stabilization and cleavage and (b) relation between tensile strength and stability of the link at varying pH (BAILEY, 1968 *b*; courtesy of *Biochim. Biophys. Acta*).

found in animal tissues must contain other, more stable bonds, and indeed, both intra-molecular and inter-molecular, covalent bonds have been found (see Fig. 3.15). These bonds are located in the *N*-terminal telopeptide, a non-helical segment of the polypeptide chain (see BAILEY *et al.*, 1970*a*). Intra-molecular bonds link individual polypeptide chains within a single molecule, and inter-molecular bonds link adjacent molecules together into a single, insoluble unit.

BAILEY (1967; 1968*b*) described a labile, intermolecular crosslink in rat-tail tendon. This link involves the formation of a Schiff base between a lysine amino group in the telopeptide region of one molecule and an aldehyde on another (see Fig. 3.16a). This crosslink is easily broken in weak acid solutions, but it can be stabilized by reduction with lithium borohydride to a bond that is not cleaved in weak acid. This system provides an excellent example of the importance of crosslinking to the mechanical properties of collagen fibres. Figure 3.16b shows the per cent reduction in the tensile strength of rat-tail tendon at various pH values as compared to the tensile strength at pH 7. The strength drops off to near zero at both low and high pH in the untreated tendon, but if the tendon is treated with borohydride to stabilize the links, the tensile strength remains essentially unchanged over the entire pH range. BAILEY *et al.* (1970*a, b*) have also isolated a very stable, covalent, inter-molecular crosslink. The distribution of these two crosslinks varies from tissue to tissue in a single species and also between species. The labile, Schiff-base crosslink is quite common in tissues from young animals, but it is often replaced by more stable links as the animal matures.

The impression one gets is that mechanical failure in collagen fibres results not from the breaking of rod-like tropocollagen molecules, but from the pulling apart of adjacent molecules. Thus, intermolecular crosslinking is one of the most important factors affecting the mechanical properties of collagen fibres. Of course, leather chemists have recognized this for years and are constantly trying to develop new methods for crosslinking animal skins to produce strong, durable leather.

3.3.2 Mechanical Properties of Collagen Fibres

The only mechanical information available on collagen as a tensile material is that on the properties of vertebrate tendon and ligament. Tendon is the structure which provides the rigid attachment of muscle to bone, and as such it must transmit the muscle force with a minimum of loss. This is achieved through the parallel arrangement of collagen fibres to form rope-like structures with a high modulus of elasticity and high tensile strength. Collagen is the major component of tendon, making up 70 to 80% of the dry weight, but also present are fibroblasts and other cells, non-collagenous protein, polysaccharide, and inorganic salts (ELLIOTT, 1965).

ELLIOTT (1965) lists tensile strength data for a number of vertebrate tendons from several sources. The values range from 2×10^7 to 1.4×10^8 N m^{-2}, with most of them in the range of 5×10^7 to 10^8 N m^{-2}. Tensile strength testing involves testing to failure and failure is associated with flaws in the material. Because flaws are not uniform in all specimens, one always obtains a range of tensile strength values. Quite apart from the data scatter due to experimental technique, it is clear that the tensile strength also depends on the origin of the tendon (BENEDICT *et al.*, 1968) and on the age of the animal from which it is taken (BLANTON and BIGGS, 1970). These systematic differences probably reflect the differences in the nature of the crosslinks in tendon from different sources and the changes in crosslinking with age that were mentioned previously.

ELLIOTT (1965) correlated tendon size and the maximum isometric tetanic force for six muscles from the hind limb of the rabbit (Fig. 3.17). By using his values for the water content (53%) and for the collagen content (71% of the dry weight) of the rabbit tendon, and by assuming that the density of tendon is 1.25, one obtains values of 1.2×10^7 N m^{-2} and 2.2×10^7 N m^{-2} as the maximum stress applied by penniform and fusiform muscles respectively to their tendons. Thus the tensile strength is roughly 3 to 6 times the maximum isometric tetanic muscle stress. HARKNESS (1968) predicts that greater stresses can arise in tendons due to very rapid muscle contractions which may be twice the isometric muscle tension, but this still leaves a safety factor of around 2.

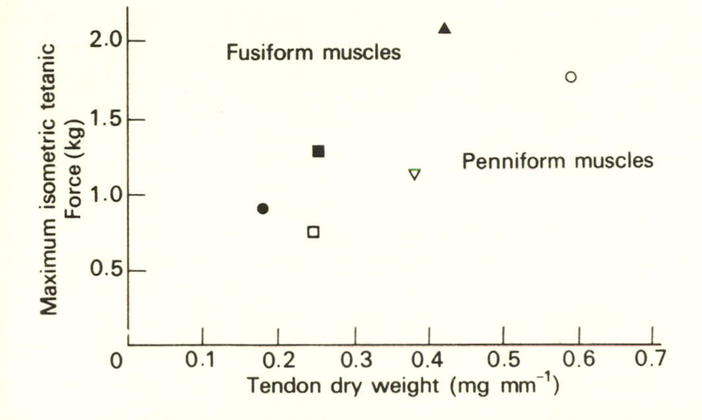

Fig. 3.17 The maximum isometric tetanic force developed by six muscles from the rabbit as a function of the size of the tendon which attaches the muscle to bone. Muscle tension creates a stress of 1.2×10^7 N m^{-2} in the tendons of penniform muscles and 2.2×10^7 N m^{-2} in tendons of fusiform muscles. (From ELLIOTT, 1965, Structure and function of mammalian tendon. *Biol. Rev.*, **40**, 392–421).

Figure 3.18 shows a typical stress–strain curve for tendon. The tendon can be extended reversibly to strains of about 4%, but strains above this level result in irreversible changes. The tendon breaks at strains of around 8 to 10% (RIGBY *et al.*, 1959). The modulus (slope of the stress–strain curve) increases gradually over the first few per cent extension, and then reaches a value of about 10^9 N m^{-2}. RIGBY *et al.* (1959) studied rat-tail tendon with polarized light and observed that unstretched tendon has a wavy appearance that disappears on stretching to strains of 3% or above. They concluded that the tendon was composed of slightly 'crimped' collagen fibres, and that on extension the fibres straighten out and become parallel. The initial rounded portion of the stress-strain curve reflects the straightening of these 'crimps', and the linear portion of the curve indicates that the force is being applied directly to extended collagen fibres.

As previously derived, the maximum isometric muscle tension creates a stress of roughly 20 MN m^{-2} in tendons attached to fusiform muscles. This means that

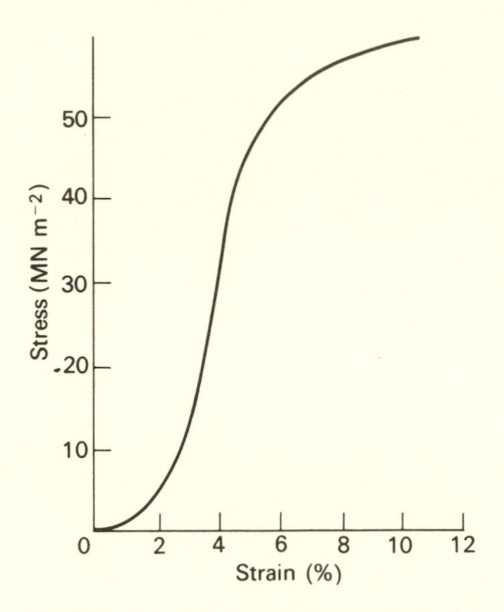

Fig. 3.18 A typical stress–strain curve for tendon.

under normal working conditions the tendon is strained within the limits of 0 to 3%. That is, the normal working range falls within the part of the stress–strain curve that is dominated by the 'crimped' fibres. This suggests that the 'crimped' fibres play a role in damping sudden loads. The tendon would appear to deform in a controlled manner to reduce the shock of rapidly applied load. In fact, the tendon functions in just the opposite manner. Figure 3.19 shows the results of two different experiments assessing the effect of strain rate on the stress–strain properties of collagen fibres. Figure 3.19a shows the initial portion of stress–strain curves from four experiments run at different strain rates (HAUT and LITTLE, 1969). It is clear that the rounded portion of the curve is reduced as the strain rate is increased, until in curve A it is nearly eliminated. At the relatively high strain rate of 19.7% per minute (curve A) it would take six seconds to reach 2% strain. This is slower than the strain rates that must occur in running and jumping animals. The apparent shock-absorbing effect is mostly an artifact of experiments carried out at unusually slow strain rates. At strain rates which approximate to those found in living creatures the stress–strain curves will be very nearly linear with only slight rounding at low strains.

MASON (1965) measured both the static modulus (derived from the slope of stress–strain plots carried out at low strain rates) and the sonic modulus (derived from the sonic velocity of a 20 kHz sound pulse passed through the tendon) for rat-tail tendon at various extensions (Fig. 3.19b). He observed some important differences between the high strain rate measurements (sonic modulus) and the static measurements. At zero strain the sonic modulus was high, but the static modulus approached zero. He concluded that some material between the 'crimped' fibres carried the stress at low extensions, and that the properties of this inter-fibrillar material were very strain rate-dependent. As tendon is extended, the load

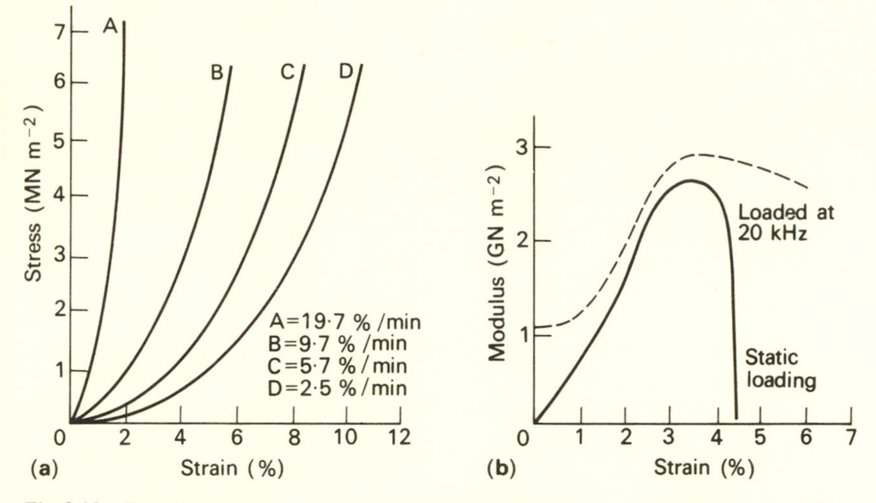

Fig. 3.19 The effect of strain rate on the mechanical properties of tendon and ligament. (a) the initial part of 4 stress–strain curves of canine anterior cruciate ligament run at different strain rates. (b) The Young's Modulus of rat-tail tendon as measured statically and at a very high strain rate (20 kHz sonic modulus). (a) HAUT and LITTLE (1969); reproduced with permission from the *J. Biomech.* (b) after MASON (1965).

is transferred from the inter-fibrillar material to the collagen fibres; thus, both the static and sonic moduli are high and increase as the 'crimps' are drawn out. The properties of the collagen appear not to be very strain rate-dependent. When the tendon yields at strains around 4 to 5%, the static modulus drops quickly indicating a breakdown of the structural continuity of the collagen fibres, but the sonic modulus remains high, indicating that substantial regions of the collagen fibres remain intact and well aligned. This is consistent with the idea that collagen fibres break not because the tropocollagen molecules are disrupted but because adjacent molecules pull apart.

It is difficult at this time to get an accurate picture of this proposed inter-fibrillar material, but the presence of glycosaminoglycan matrix materials in tendon may provide an important clue. Studies have been carried out to determine the role of non-collagen components in tendon (see PARTINGTON and WOOD, 1963), but they have dealt mainly with tensile strength measurements and have generally been inconclusive. It seems likely that the tensile strength of collagen fibres is determined by considerations of crosslinking alone and that the non-collagen components play a more subtle role in determining some of the time-dependent mechanical properties.

One further but often neglected aspect of the mechanical properties of tendon is the effect of long-term forces on collagen fibres. As previously mentioned, the long-term properties of crystalline polymers are characterized by broad distributions of retardation-time and relaxation-time spectra that extend to very long times. That is, crystalline polymers show gradual, long-term creep and stress-relaxation. All the available structural information points toward collagen being a crystalline material, and on this basis one expects that the long-term properties of collagen fibres would fit within the general pattern for crystalline polymers. This is indeed the case.

HALL (1951) studied the creep of collagen fibres over long periods of time. His testing procedure involved following the extension of collagen fibres under constant load for 200 minutes, removing the load for 21 hours to allow the fibres to relax, and then applying the load again. He carried out nine such cycles on a single sample and found that the length of the fibres at the end of each successive test was greater than at the end of the preceding test. That is, even when fibres are allowed to relax between successive loadings there is an irreversible creep that continues for long periods. Hall attributes this creep to 'slipping of secondary cross-bonds'.

The results of stress-relaxation tests carried out on rat-tail tendon (RIGBY *et al.*, 1959) are shown in Fig. 3.20. In this figure the stress-relaxation of tendon held at strains of 3.5% (well within the limits of reversible stress–strain behaviour)

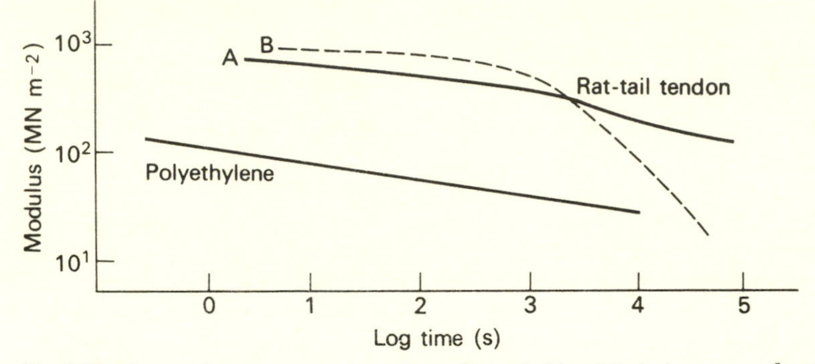

Fig. 3.20 Stress-relaxation curves for collagen fibres. Solid and dashed curves are for rat-tail tendon strained 3.5% and 7.5% respectively (RIGBY *et al.*, 1959); with permission of the *J. Gen. Physiol.* The stress-relaxation curve of bulk crystallized polyethylene is included for comparison (after BECKER, 1961).

and 7.5% (well above the safe limit for reversible extension) are compared with the stress–relaxation behaviour of bulk-crystallized polyethylene (BECKER, 1961). The modulus of tendon within the safe limits (curve A) is roughly an order of magnitude greater than that of the bulk-crystallized polyethylene. This is a reflection of the high degree of preferred orientation of crystalline regions in the collagen fibres. The shapes of the stress–relaxation curves are very similar, showing nearly linear, gradual decreases in modulus over a broad time range. The tendon strained to 7.5% (curve B) has clearly been damaged in some manner. Although the modulus decreases gradually for the first 10^3 seconds, it drops rapidly at longer times indicating a break in the structural continuity of the fibres.

The molecular relaxation mechanism that accounts for the long-term properties of collagen fibres must be similar to those described earlier for drawn, crystalline polymeric fibres. Gradual shifts of the numerous weak interactions, both those within the triple helix and those between molecules, must play an important role. But collagen fibres differ from other crystalline fibres in that they are secondarily crosslinked. Surely, these covalent crosslinks limit the extent of creep and stress-relaxation. According to our present understanding of the disposition of crosslinks in collagen fibres there are relatively few crosslinks per tropocollagen molecule. The arrangement shown in Fig. 3.15b indicates only three crosslinks per 300 000 molecular weight unit, which leaves considerable freedom for the minor adjustments in the crystal' structure that result in the observed long-term properties.

One may well ask, what relevance these long-term properties have to the functioning of tendons and ligaments in animal supportive systems? In the case of tendons that are stressed only occasionally for short periods of time, these properties are probably of little importance, but in the case of tendons under constant stress for extended periods of time the answer may be different. Take, for example, bovine Achilles tendon. The skeletal structure of ungulates is such that they literally stand on their toes. Thus, the Achilles tendon attaches the extensor muscle of the 'ankle' joint to the tarsal bone and is under constant stress as long as the animal remains standing. Cows spend a lot of time standing up—what happens to their Achilles tendon? The answer is, probably nothing. Although the experiment has not yet been carried out, we suspect that collagen fibres in bovine Achilles tendon are more highly crosslinked than tendons which are not subjected to such long-term stresses, and that this increased crosslinking reduces the rate of creep or stress-relaxation to insignificant levels.

Perhaps the best example of long-term properties in collagen fibres is provided by the cuticle collagen of the nematode *Ascaris*. This animal is supported by a hydrostatic skeleton made up of an internal fluid surrounded by an external container, the cuticle. The details of this structural system will be discussed more fully in Chapter 7, but essentially the external collagen fibres act as tensile structures which restrain the internal hydrostatic pressure. HARRIS and CROFTON (1957) measured the internal pressure of this animal and found it to vary between 253 and 2400 N m^{-2}. The pressure rarely fell below 334 N m^{-2} and was greater than 867 N m^{-2} for about 40% of the time. Here is a tensile collagen that is under *continual* tensile stress. Recent studies of the structure of the collagen that makes up this cuticle (McBRIDE and HARRINGTON, 1967*a, b*) have revealed some rather interesting anomalies. The molecular unit of *Ascaris* cuticle collagen has a diameter similar to that of other collagens, but it is considerably shorter, having a molecular weight of only 62 000. Unlike other collagens, this collagen is made from a single polypeptide chain that is folded back on itself to form the characteristic triple-helix. Most important, *Ascaris* cuticle collagen contains 27 residues per 1000 of 1/2-cystine. Cystine, which is known to form disulphide (—S—S—) crosslinks between protein chains, is not found in other collagens. If all cystine residues in cuticle collagen participate in inter-molecular crosslinks there will be a total of 36 such crosslinks per 300 000 molecular weight unit. (This figure was arrived at by using a value of 112 for the average molecular weight of amino acids in *Ascaris* cuticle collagen.) Thus, *Ascaris* cuticle collagen has the potential for being a very highly crosslinked system. That the cystine residues do in fact participate in intermolecular crosslinks is demonstrated by the fact that normally insoluble cuticle collagen is readily dissolved in reagents that break the disulphide bond. It seems reasonable that this high density of crosslinks provides the means to deal with the continual tensile stresses that arise in the normal functioning of the collagen in nematode cuticle.

3.4 Cellulose

Cellulose is by far the most abundant of the natural, fibrous substances, but its occurrence in tensile materials is very limited. Cellulose is most often found as the high modulus, fibrillar component of rigid composites, such as wood, where it is found in association with lignin, hemicellulose and a variety of other materials (see SIEGEL, 1968). Because plant structures are formed by the assembly of cellular units, tensile materials made from cellulose do not contain the long, continuous fibres characteristic of such tensile materials as tendon. Rather, tensile plant materials consist of linear aggregates of cellular units which form fibrous structures. The cells found in these fibrous structures are normally elongated with their long axes parallel to the direction of the tensile force, and the cellulose microfibrils within the walls of these fibre cells have a high degree of preferred orientation parallel or nearly parallel to the long axis of the cell. Thus, the mechanical properties of plant materials in general and tensile plant materials in particular, depend both on the properties of the cell wall (i.e., the properties of cellulose) and on the matrices that bind the cells together into larger units. In this section we will be concerned primarily with the structure and mechanical properties of individual cellulose fibrils, and with the properties of the relatively few tensile plant materials that seem to depend mainly on the parallel arrangement of

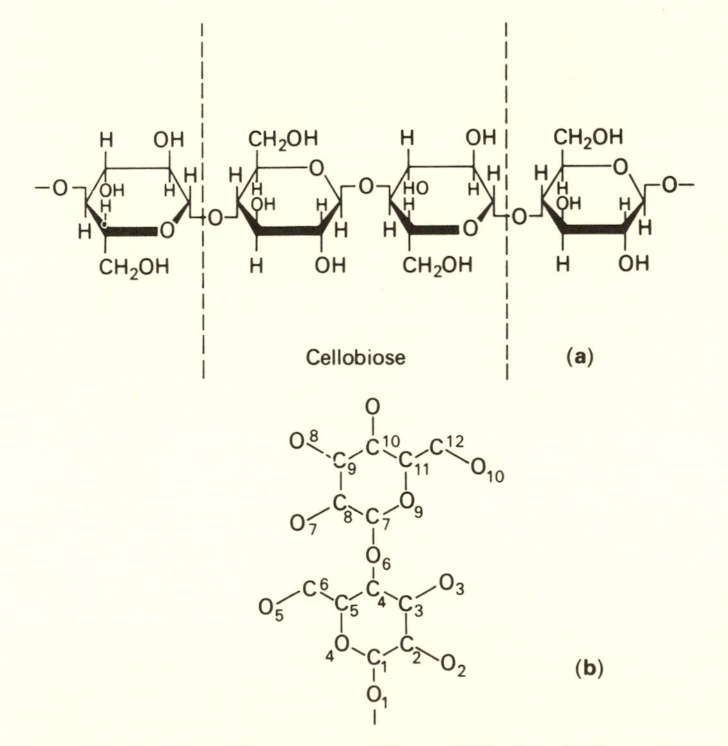

Fig. 3.21 The chemical structure of cellulose. (**a**) The disaccharide repeating unit cellobiose and (**b**) the numbering of carbon and oxygen atoms in cellobiose.

cellulose fibres. The general structure and properties of the plant cell wall will be discussed in the section on wood in Chapter 5 where it will be seen that lignin and hemicellulose play an extremely important role.

3.4.1 The Structure of Cellulose

Cellulose is an isotactic polymer of D-glucose linked by β (1–4) glycoside bonds. By virtue of the β-linkage, each successive glucose ring is rotated 180° around the chain axis, making the disaccharide cellobiose the true repeating unit of cellulose (see Fig. 3.21). The accepted crystal structure for native cellulose is that proposed by MEYER and MISCH (1937) and later modified by FREY-WYSSLING (1955). The unit cell has the following dimensions:

a = 0.835 nm
b = 1.030 nm
c = 0.79 nm
β = 84°

and consists of anti-parallel, extended cellulose chains, stabilized by inter-chain hydrogen bonds. Figure 3.22 is a diagram of the Meyer and Misch structure, in which the inter-chain hydrogen bonds are in the 002 plane. The modification of

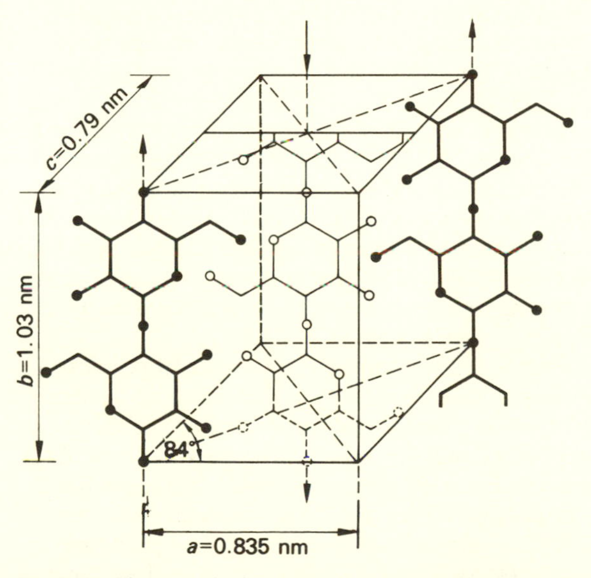

Fig. 3.22 The unit cell of crystalline cellulose proposed by MEYER and MISCH (1937). Courtesy of *Helv. Chem. Acta.*

FREY-WYSSLING (1955) involves a 0.25 shift of the anti-parallel chains along the b-axis. That is, the downward pointing chain in Fig. 3.22 (open circles) is shifted 0.26 nm to bring the C-6 hydroxyl of one chain into close proximity with the bridge oxygen of the adjacent chain. In terms of the numbering scheme shown in Fig. 3.21b, this allows hydrogen bonds $O_1 \rightarrow O_5$ and $O_6 \rightarrow O_{10}$ to form in the

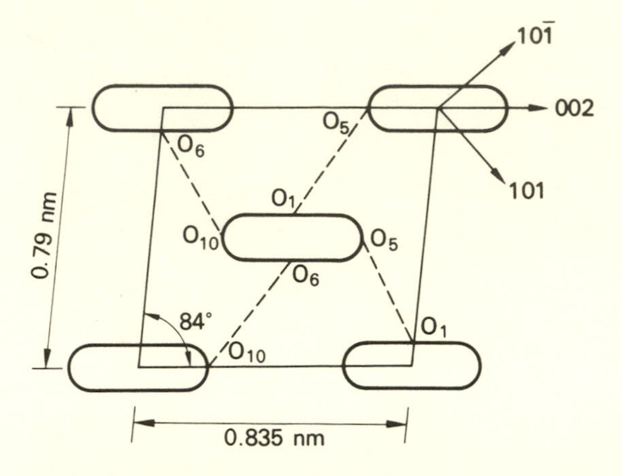

Fig. 3.23 A view down the *b*-axis (fibre axis) of the cellulose unit cell showing distribution of hydrogen bonds proposed by FREY-WYSSLING (1955); reproduced from *Biochim. Biophys. Acta.*

101 and $10\bar{1}$ planes between cellulose chains. Figure 3.23 shows a view looking down the cellulose chains (*b*-axis) which indicates the disposition of the inter-chain hydrogen bonds.

Recently VISWANATHAN and SHENOUDA (1971) have proposed a helical structure for cellulose. The proposal is based on the difference between the *b*-axis repeat (1.030 nm; the length of the cellobiose unit projected on the *b*-axis) and the length of cellobiose (1.039 nm; from X-ray diffraction studies of crystalline cellobiose). They assume that the disaccharide cellobiose unit is canted at an angle of 7° 36′ to the fibre axis ($\cos \alpha = 1.030 \div 1.039; \alpha = 7° 36'$), giving a helix with a radius of 0.158 nm and a pitch of 7.2 nm or seven cellobiose units per turn. The diffraction intensities calculated for a left-handed helix with a shift of 0.25 between the anti-parallel chains were found to provide the best agreement with measured intensities. At present there is no further evidence to support this hypothesis, but it will be interesting to see what effect such a helical structure has on the disposition of interchain hydrogen bonds.

In electron micrographs the plant cell wall appears to be made of cellulose fibrils about 20 to 25 nm in width, but when the cell wall is disrupted by sonication or other means, microfibrils 3.5 nm in diameter are observed (FREY-WYSSLING, 1959). The constancy of the 3.5 nm dimension and the discovery of these fibrils in a wide range of cellulosic materials has led to the suggestion that these fibrils are the basic crystalline unit of cellulose. Thus, the 3.5 nm structure has come to be known as the elementary fibril. Figure 3.24 shows the structure proposed by FREY-WYSSLING and MÜHLETHALER (1963) for the elementary fibril, in which the 101 plane is parallel to the surface of this crystalline unit. The larger 25 nm fibrils presumably contain 40 to 50 of these elementary crystalline units separated by non-crystalline regions.

The question now arises, can these elementary fibrils have the folded-chain crystal structure characteristic of many crystalline polymers? Or is native cellulose an extended-chain crystalline polymer as some workers propose? One must specify 'native cellulose' because there are a number of regenerated celluloses,

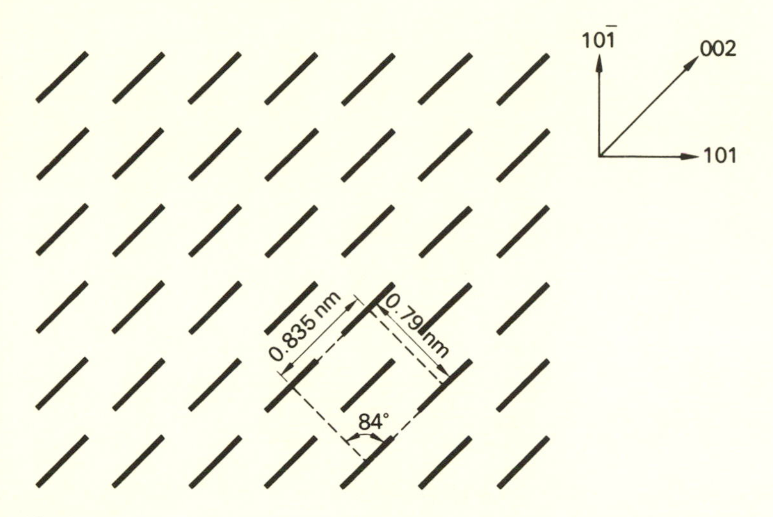

Fig. 3.24 The arrangement of cellulose chains in a transverse section of an elementary microfibril. Heavy lines represent glucose rings. Chains extend in and out of the page (FREY-WYSSLING and MÜHLETHALER, 1963). Reproduced from *Makromol. Chem.* **62**, 25, Fig. 3, Hüthig and Wepf Verlag, Basel.

such as rayon, which may be quite different from the native form. Folded-chain single crystals of various cellulose derivatives have been obtained from dilute solution and they are very similar in appearance to the polyethylene single crystals mentioned previously (MANLEY, 1963; BITTIGER *et al.*, 1969). Cellulose precipitated from cadmium ethylene-diamene solution forms 3.5 nm fibrils which appear in electron micrographs to be very similar to the elementary fibrils. The ratio of the extended chain length (from molecular weight measurements) to the fibril length (from the electron microscope) is roughly 10 to 1, indicating a folded-chain structure within these fibres (MANLEY, 1971). On the basis of this observation and on the appearance of 4.0 nm periodicity in cellulose fibrils, MANLEY (1971) proposed the model shown in Fig. 3.25a for the elementary fibril. It consists of a ribbon of folded-chains about 3.0 nm wide pleated into a flattened, helical fibre with a pitch of 4.0 nm. Other more conventional folded-chain structures (Fig. 3.25b) have been proposed by MARX-FIGINI and SCHULZ (1966) and others.

The folded-chain models have been attacked because they do not account for the high modulus and tensile strength of cellulose fibres (MARK *et al.*, 1969); while extended-chain structures can more easily account for the mechanical properties (MARK, 1967). It is now generally accepted that native cellulose is an extended-chain polymer crystal. Perhaps the best evidence for the extended-chain form is provided by the experiment reported by MUGGLI *et al.* (1969) and MÜHLETHALER (1969). The experiment is explained diagrammatically in Fig. 3.26. Sections of thickness equal to the extended length of the cellulose chains (2 μm) were cut from native cellulose fibres, and the molecular weight of the cellulose taken from these sections was compared to that of cellulose from the unsectioned fibres. The cellulose from one thousand 2 μm sections was pooled,

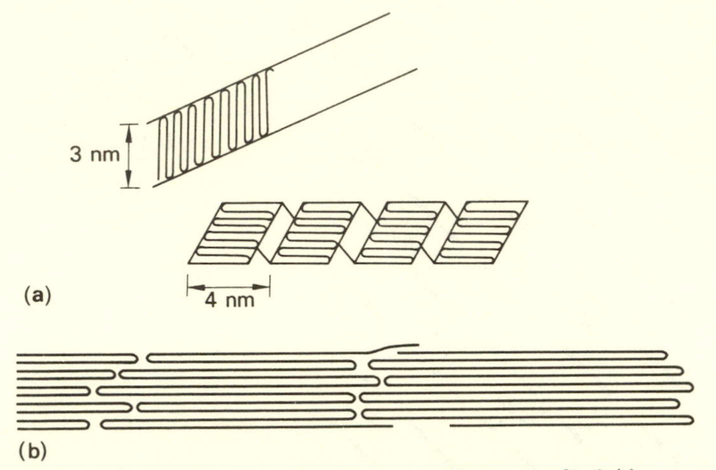

Fig. 3.25 Folded-chain models of the structure of cellulose microfibril. **(a)** MANLEY (1971); courtesy of the *J. Polymer Science*; **(b)** Modified from MARX-FIGINI and SCHULZ (1966); courtesy of *Biochim. Biophys. Acta.*

and the molecular weight determined by gel-filtration chromatography. With this technique high molecular weight fractions pass through the column more quickly and hence appear to the left in the plots shown in Fig. 3.26b and c. Case A shows the expected results for a folded-chain structure; case B for an extended-chain structure. In each graph the solid line represents cellulose from unsectioned fibres and the dashed line represents cellulose from sections. The molecular weight distribution of cellulose from sections of folded-chain fibres should be only slightly changed from that of the unsectioned material as only a few of the cellulose molecules will be cut (case A); whereas the average molecular weight of cellulose from sections of extended-chain fibres should be roughly one-half that of the unsectioned cellulose (case B). Figure 3.26c shows the results of the experiment. The number average molecular weight of cellulose before sectioning was 2.03×10^6, while that after sectioning was 0.83×10^6. These results are in complete agreement with the expected results for an extended-chain cellulose fibre.

How then can we explain the apparently folded-chain structures observed by MANLEY (1963, 1971) and others? We have stated that extended-chain polymer crystals provide the thermodynamically most stable form because each fold represents a segment of the polymer chain that will not fit into the crystal lattice and hence is energetically unfavourable. But when polymers crystallize from solution it is very unlikely that the random-coil molecules in solution will be able to extend fully to fit into an extended-chain crystal. Thus, the folded-chain crystal is an energetic compromise that arises in crystallization from solution. CHANZY *et al.* (1967) and CHANZY and MARCHESSAULT (1969) report that polyethylene synthesized in a *non-solvent* medium spontaneously forms extended-chain crystals. That is, when polymerization and crystallization take place simultaneously, the polymer chains do not have the opportunity to become folded because there is no solution phase. SARKO and MARCHESSAULT (1969)

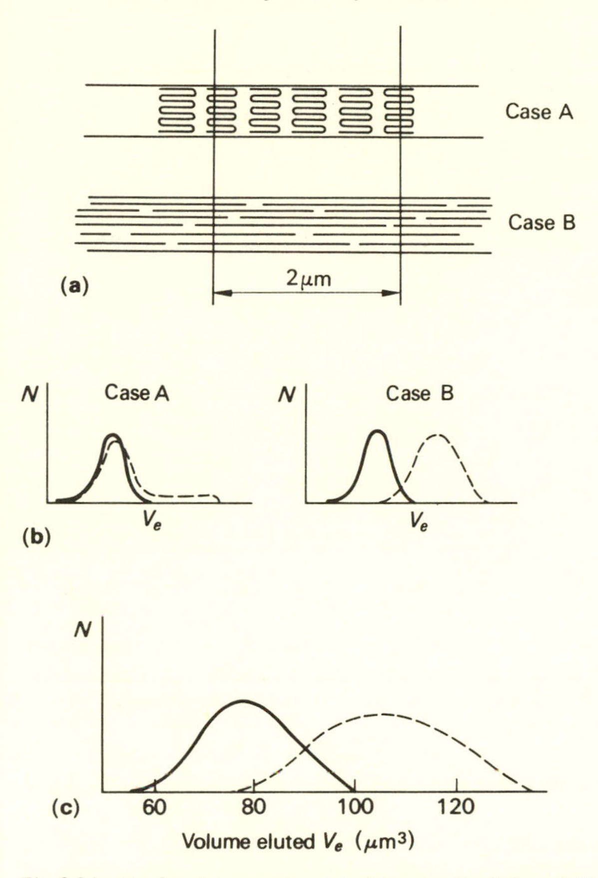

Fig. 3.26 Results of an experiment to determine if cellulose chains in native microfibrils are folded or extended (MUGGLI *et al.*, 1969; MÜHLETHALER, 1969; courtesy of the *J. Polymer Science*). *N*, concentration. See text for explanation.

suggest that simultaneous polymerization and crystallization provide an appropriate model for the synthesis of native cellulose fibres. Cellulose, which is very insoluble in aqueous media, is incorporated directly in the crystal lattice as fast as it is synthesized, leaving no opportunity for the formation of folded-chain crystals. All observations of folded-chain cellulose structures have been made on material precipitated or crystallized from solution. It is no wonder that there has been so much confusion.

3.4.2 Mechanical Properties of Cellulose Fibres

The tensile cellulose materials provide a number of economically important textile fibres, and as a result, a great deal of information is available on the mechanical properties of these fibrous materials. Many studies have been carried out on dried or chemically treated fibres and, although this information is of great value to the textile industry, it is less so for us. In this discussion we will be primarily concerned with the properties of wet, natural plant fibres as they are

Table 3.6 The dimensions of plant fibre cells

Fibre source	Cell width in μm (1)	Cell length in cm (1)	Ratio (1/w)	Fibre angle (avg. of S_1 and S_2)	Degree of polymerization (5)
Bast Fibres					
hemp (*Cannabis*)	16–50	1.2–2.5	500–700	2.3° (2)	9300
jute (*Corchorus*)	20–25	0.15–0.5	75–200	7.9° (2)	–
flax (*Linum*)	11–20	1–3	900–1500	6° (3)	7000–8000
ramie (*Bohemeria*)	20–75	12–15	2000–6000	6° (3)	10 800
bamboo fibres (*Bambusa*)	–	–	–	5° (2)	–
Seed Hairs					
cotton (*Gossypium*)	15–30	2.5–6	1700–2000	28°–44° (4)	15 300
Tracheids from woody plants	–	–	–	30°–60° (2)	7500–10 000

References. (1) BAILEY, TRIPP and MOORE, 1963; (2) PRESTON, 1952; (3) FREY-WYSSLING, 1952; (4) MARK, 1967; (5) GORING and TIMMELL, 1962.

found in the living plant, but we will also consider the properties of pure, crystalline cellulose.

Bast fibres, derived from the stems of dicotyledonous plants, provide perhaps the best example of tensile cellulose materials, and it is just this tensile nature that makes the bast fibres so useful as textile fibres. As can be seen in Table 3.6, these fibres are composed of very elongated cells, with the ratio of cell length to width reaching into the thousands for flax and ramie fibres. The fibre angle (the angle between the cellulose fibrils and the long axis of the cell) of the secondary cell wall is always very small, and the fibre angle of the S_2 layer, which makes up the majority of the cell wall (see Fig. 5.31), is in all cases less than the average value given in Table 3.6. Another important tensile plant material is the reaction wood produced by woody angiosperms. The fibre angle in the secondary cell wall of tracheids (the fibre cells of woody plants) usually falls within the range of 30 to 60°, but in 'tension wood' a thick layer is formed in the secondary cell wall which has a very low lignin content and a fibre angle of about 5° (CÔTÉ and DAY, 1965; WARDROP, 1965). The formation and function of tension wood will be discussed at greater length in Section 7.6.3.

Table 3.7 provides some data on the mechanical properties of tensile cellulose materials, and for comparison, selected data for collagen tendon and *Bombyx mori* silk. The cellulose fibres are seen in all cases to have a higher Young's modulus and tensile strength than either the collagen or silk. The tensile modulus of wet bast fibres is roughly an order of magnitude greater than that of collagen

3.4.2 *Mechanical Properties of Cellulose Fibres*

Table 3.7 The mechanical properties of tensile cellulose materials

	Tensile Modulus ($N\ m^{-2}$)	Ratio $\frac{E_{dry}}{E_{wet}}$	Tensile Strength	Ratio $\frac{\sigma_{dry}}{\sigma_{wet}}$	Extensibility
Bast Fibres					
hemp wet	3.5×10^{10} (2)	2	–	–	–
dry	7×10^{10} (1)		9.2×10^{8} (1)		1.7% (1)
jute wet	–	–	–	–	3.0% (4)
dry	6×10^{10} (2)		8.6×10^{8} (4)		–
flax wet	2.7×10^{10} (3)	3-3.7	8.8×10^{8} (6)	0.96	2.2% (6)
dry	$8\text{-}11 \times 10^{10}$ (2)		8.4×10^{8} (6)		1.8% (6)
ramie wet	1.9×10^{10} (2)	4.2	1.08×10^{9} (6)	0.89	2.4% (6)
dry	8×10^{10} (1)		$9.0\text{-}9.4 \times 10^{8}$ (6)		2.3% (6)
Seed Hairs					
cotton wet	7.5×10^{9} (3)	3.6	$2.4\text{-}8.3 \times 10^{8}$ (6)	ca. 0.9	–
dry	2.7×10^{10} (1)		$2\text{-}8 \times 10^{8}$ (6)		6-12% (6)
Tracheids wet	–	–	8.2×10^{7} (5)	1.5	–
dry	–		1.2×10^{8} (5)		–
Collagen					
Tendon wet	2×10^{9}	–	10^{8}	–	8-10%
Silk					
Bombyx mori	10^{10}	–	6×10^{8}	–	18-20%

References. (1) FREY-WYSSLING, 1952; (2) MEYER and LOTMAR, 1936; (3) MEREDITH, 1946; (4) MARK, 1967; (5) KLAUDITZ, 1957; (6) MEYER, 1942.

and several times that of silk. Differences in tensile strength are not so pronounced, but bast fibres are much less extensible than either of the protein fibres, suggesting that the bast fibres may not be as tough.

When considering the Young's modulus of the individual cellulose microfibrils in the cell wall of a living plant, one must recognize that the properties of the microfibril are not identical to that of pure, crystalline cellulose. Although cellulose is a highly crystalline polymer, a sizeable portion of the cellulose microfibril is non-crystalline, and the properties of the microfibril must depend on both the crystalline and amorphous regions. A value of 3 to $4 \times 10^{10}\ N\ m^{-2}$, which represents the upper range of modulus values for wet bast fibres, probably provides a reasonable value for the cellulose microfibrils in the cell wall of living plants. The value for wet fibres was chosen because the cell wall material in a living plant is always wet.

Note that the modulus of dry bast fibres is three to four times that of wet fibres. Water penetrates the amorphous regions and competes for potential hydrogen bonding sites and loosens the interactions between adjacent crystalline regions. Thus, the modulus values for dry bast fibres will approach that of crystalline cellulose, and the highest value in Table 3.7 (1.1×10^{11} N m^{-2} for flax fibres) must be quite close. But amorphous regions, even though bridged by bonding between adjacent crystalline regions, must still reduce the stiffness of the fibres. SAKURADA *et al.* (1962) followed changes in the X-ray diffraction pattern of ramie fibres while the fibres were being stretched. They used the change in the *b*-axis (fibre axis) spacing as a measure of strain and obtained a value of 1.37×10^{11} N m^{-2} for the Young's modulus of crystalline cellulose. Attempts have also been made to calculate the tensile modulus of crystalline cellulose from a knowledge of the crystal structure. TRELOAR (1960) based his calculations on the force required to straighten the primary valence bond angle of the bridge oxygen between glucose rings. He obtained a value of 5.6×10^{10} N m^{-2}, much lower than the experimental values. More recently, GILLIS (1969) obtained a value of 2.5 to 3.0×10^{11} N m^{-2} by considering the stretching of inter-chain hydrogen bonds as well as the straightening of the primary valence chain bonds. If we accept Gillis' value as an upper limit, the tensile modulus of crystalline cellulose lies between 1.4 and 3.0×10^{11} N m^{-2}.

The high tensile strength of native cellulose fibres is largely a result of the extended-chain structure of crystalline cellulose. The large overlap between adjacent, extended cellulose chains creates a total inter-chain interaction which is of the same order of magnitude as the valence-bond strength of the cellulose chain. In fact, MEYER (1942) and later MARK (1967) have proposed that the failure of cellulose fibres involves the scission of the polymer chains rather than the slipping of adjacent chains due to the breaking of inter-chain hydrogen bonds. According to the calculations of MARK (1967) a stress of 1.9×10^{10} N m^{-2} is required to break the cellulose chains and a stress of at least 3.7×10^{10} N m^{-2} is required to cause failure by chain slipping. For both calculations it was assumed that the stress is applied uniformly over the test sample. The figure for failure due to chain slipping was calculated for an elementary microfibril containing 36 cellulose chains with a single overlap area 750 cellobiose units long and 6 chains wide containing 4500 hydrogen bonds. This represents the weakest overlap arrangement in an elementary microfibril containing cellulose with a degree of polymerization (DP) of 3000 (i.e., 3000 glucose units per cellulose molecule). From Table 3.7 it is clear that the tensile strength of real cellulose fibres is roughly one-twentieth of these calculated values. This discrepancy is due to a number of factors which act to create a non-uniform stress distribution within the test sample. As these stress concentrating effects should be equally important for both modes of failure, it is reasonable to conclude that the weaker of the two modes, namely chain scission, should be the cause of failure.

A number of studies have shown that the tensile strength of cellulose fibres is increased with increased molecular weight up to a degree of polymerization of about 2500. Increased degree of polymerization above this value has no effect on the tensile strength (see MARK, 1967). This suggests that below DP = 2500 the overlap length is shorter than some critical length and the fibres fail by chain slipping. Above DP = 2500 the fibres must yield by chain scission. The degree of polymerization in natural cellulose materials (Table 3.6, column 6) is several times greater than this critical value, again indicating that natural cellulose fibres fail by the scission of cellulose chains.

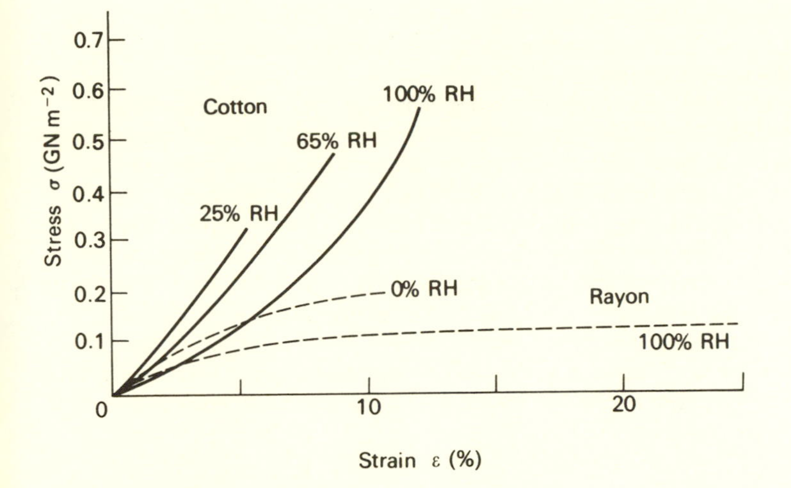

Fig. 3.27 Stress–strain curves for cotton and rayon fibres at different humidities. Each curve ends at breaking point (MEREDITH, 1946).

Increased water content does not appear to reduce the tensile strength of bast fibres (Table 3.7); in fact, wet fibres appear to be stronger than dry ones. MEYER (1942) used this fact as evidence that tensile failure in native cellulose fibres was due to chain scission rather than chain slipping. Water will loosen inter-chain interactions by competing for the inter-chain hydrogen bonds and should reduce the tensile strength if the fibres fail by chain slipping. MEREDITH (1946) studied the effect of water content on the properties of cotton fibres (Fig. 3.27), and found that both extensibility and strength were increased with increased water content. He concluded that plastic deformation of the amorphous regions in wet fibres relieved some of the stress concentrations and thus increased the tensile strength. Reconstituted cellulose fibres such as rayon are characterized by folded-chain structure and show a decrease in tensile strength with increased water content. In this case failure probably results from an unfolding of the folded-chains; a chain slipping process.

The high tensile modulus and tensile strength of bast fibres must be directly correlated with the low fibre angle in the secondary cell wall. In other plant cells where the fibre angle in the secondary cell wall is much greater, these tensile properties must be sacrificed for strength and rigidity in other directions. PRESTON (1963) measured the properties of a number of sisal fibres whose fibre angles in the secondary cell wall ranged from 10° to 50°. The results are shown below:

Fibre angle	50°	10°
Tensile strength ($N\ m^{-2}$)	8.3×10^7	5×10^8
Extension to break	14.5%	2%
Tensile modulus ($N\ m^{-2}$), initial	3×10^9	10^{11}
Tensile modulus ($N\ m^{-2}$), average	5×10^8	2.5×10^{10}

The properties of sisal fibres with a fibre angle of 10° closely resemble those of the bast fibres, but increased fibre angle greatly reduces tensile strength and stiffness while increasing extensibility.

Cellulose fibres under long-term stresses behave in the manner previously described for other crystalline, polymeric materials. PASSAGLIA and KOPPEHELE (1958) carried out stress-relaxation tests on oriented (drawn) and unoriented viscose rayon monofilaments. Modulus decreased gradually over nearly ten logarithmic decades of time, with the rate of decrease of the oriented fibres less than that of the unoriented fibres. As mentioned previously, the reconstituted celluloses (e.g., rayon) probably have a folded-chain crystal structure rather than the extended-chain structure of native cellulose. Fortunately, HAUGHTON *et al.*, (1968) have made stress-relaxation measurements on the cell wall of the alga *Nitella*, which contains cellulose as the major structural component. Their results are plotted in Fig. 3.28. The rate at which the stress-relaxation modulus drops is

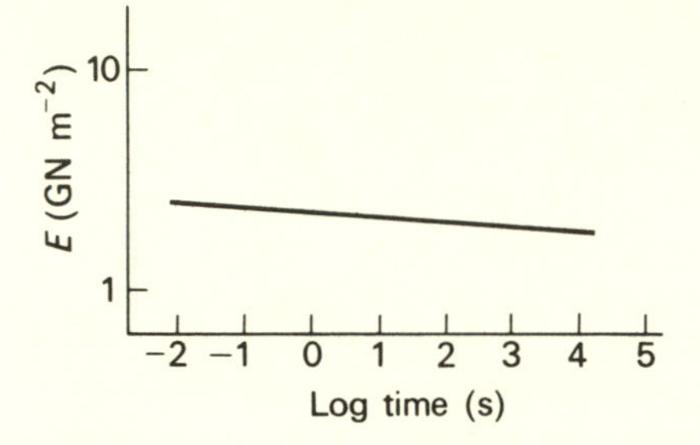

Fig. 3.28 Stress-relaxation of *Nitella* cell wall (HAUGHTON *et al.*, 1968; courtesy of the *J. exp. Bot.*).

extremely low, much lower than that of the oriented rayon sample and lower than either polyethylene or collagen (see Fig. 3.20). This unusually low rate of relaxation is a result of the extended-chain crystal form of native cellulose and the large number of hydrogen bonds that stabilize the extended chains. In folded-chain polymer crystals (i.e., polyethylene and rayon) and in collagen the overlap region is much smaller than in native cellulose. This reduces the number of inter-chain stabilizing bonds between adjacent molecules and makes stress-relaxation phenomena more likely.

3.5 Chitin

Chitin is the second most widely distributed of the basic fibrous materials in the animal kingdom. Although its phylogenetic distribution is not as broad as that of collagen, chitinous structures are found in numerous invertebrate phyla as well as in a number of plants (see RUDALL, 1955; JEUNIAUX, 1971). The greatest development of chitin-containing structures has taken place in the phylum Arthropoda, where chitin-protein complexes in the form of cuticle provide the structural basis for the characteristic exoskeleton of this animal group. Like cellulose, chitin is found primarily as the high modulus, fibrillar component in rigid composite materials (e.g., arthropod cuticle) and is not frequently found in tensile materials. Indeed, arthropod tendon or apodeme is the only recognizably tensile structure made from chitin.

3.5.1 *The Structure of Chitin*

Chitin is nominally isotactic poly-N-acetyl-D-glucosamine linked by β (1–4) glycoside bonds (see Fig. 3.29). Nominally because there is some evidence to

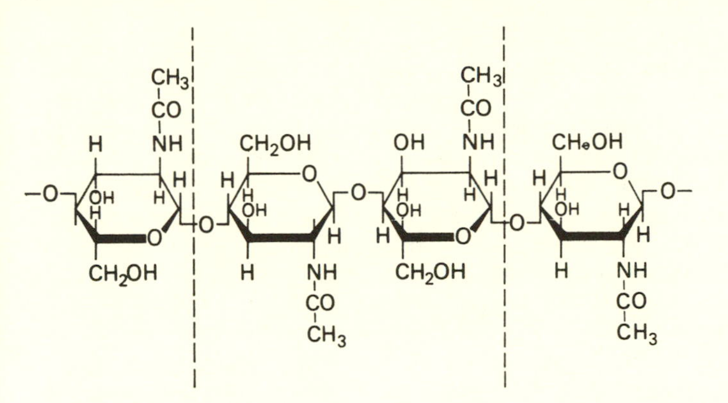

Fig. 3.29 The chemical structure of chitin, indicating the disaccharide repeating unit chitobiose.

suggest that every sixth or seventh residue is the non-acetylated form, D-glucosamine (RUDALL, 1963). Chitin is a highly crystalline polymer with a crystal structure quite similar to that of cellulose, but detailed X-ray diffraction studies of chitin from a variety of sources (RUDALL, 1963) have revealed that it exists in three distinct crystal forms. α-chitin is the only form present in the Arthropods

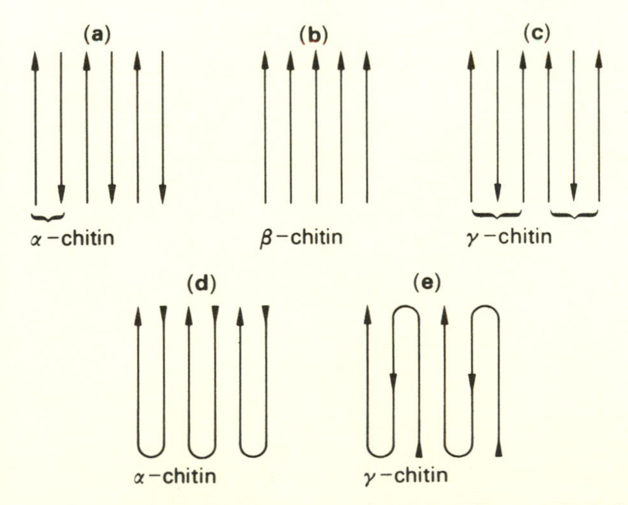

Fig. 3.30 Diagram showing the direction of the polymer chains in the three crystallographic forms of chitin. (**a**) α-chitin. (**b**) β-chitin. (**c**) γ-chitin. (**d** and **e**) folded-chain models for α- and γ-chitin respectively. (RUDALL, 1963, 1969).

and has also been identified in the molluscs and possibly in the Cnidaria. β-chitin has been identified in the Brachiopods, the Annelids, cephalopod Molluscs, and in pogonophoran tubes. The less common γ-form is known only in the Brachiopods and cephalopod Molluscs. JEUNIAUX (1971) reviews the literature on the distribution of the various chitins.

According to RUDALL (1963) the three forms of chitin differ only in the direction of adjacent polymer chains within the crystal unit cell (Fig. 3.30). The unit cell of α-chitin contains two, anti-parallel chitin chains. β-chitin is made up of parallel chains, hence the unit cell contains a single chain. γ-chitin has a unit cell containing three chains with the central chain running anti-parallel to the other two. Detailed crystal structures for α-chitin and β-chitin have been worked out by CARLSTRÖM (1957) and BLACKWELL (1969) respectively, as shown in Fig. 3.31. This figure shows a view looking down the chitin chain axes (*ac* plane) and a view from the side (*bc* plane). Both the α and the β crystal forms can be described as piles of chitin chains linked in the *a* direction by CO · · · NH, inter-chain hydrogen bonds. The view of the *ac* plane in Fig. 3.31 shows the disposition of these inter-chain bonds in the two crystal forms.

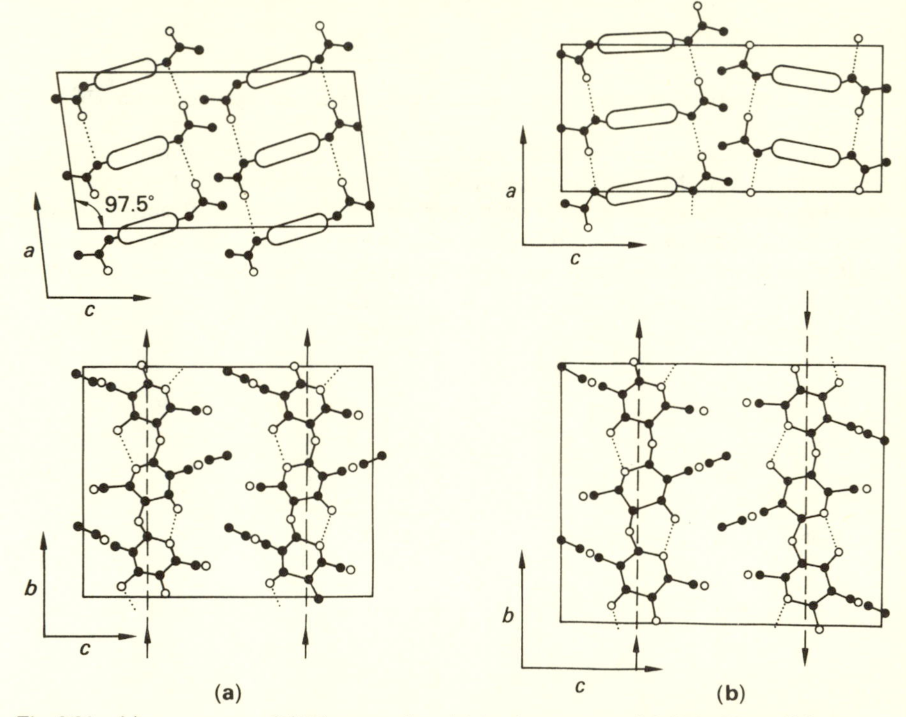

Fig. 3.31 (a) BLACKWELL'S (1969) proposed model for the structure of β-chitin. Unit cell dimensions are $a = 0.485$ nm, $b = 1.038$ nm, $c = 0.926$ nm, $\beta = 97.5°$. Dotted lines represent hydrogen bonds; open circles represent oxygen atoms; solid circles represent carbon and nitrogen atoms; large ovals in top diagram represent space occupied by glucose rings. (b) CARLSTRÖM'S (1957) model for the structure of α-chitin. Unit cell dimensions are $a = 0.476$ nm, $b = 1.028$ nm, $c = 1.895$ nm. Other conventions as in Fig. 3.31a. (a) Courtesy of *Biopolymers*; (b) reproduced with permission from *J. Biophys. Biochem. Cytol.*

RUDALL (1963, 1969) has proposed folded-chain structures for both α- and γ-chitin, as shown in Fig. 3.30d, e. α-chitin is seen to consist of extended chitin chains with a single fold, and the γ-form is a chain with two folds. β-chitin, in which the polymer chains are parallel, must be a fully extended-chain crystal, as any chain folding would result in an anti-parallel structure. The evidence cited to support chain folding in native chitin is the transition of β-chitin to α-chitin which takes place in certain reagents (LOTMAR and PICKEN, 1950; RUDALL, 1963). When β-chitin is held in 6N HCl it is converted to the α-form, and this conversion is accompanied by a contraction to about one-half of the original length. This process takes place not in solution but in the 'swollen, but solid state' (RUDALL, 1963). The shortening by one-half certainly suggests the folding of extended chains to form the α-chitin structure proposed by Rudall, but this alone is not sufficient evidence to conclude that native α-chitin is a folded-chain crystal. The $\beta \rightarrow \alpha$ transition indicates that the anti-parallel structure is the energetically more favourable form, and this is supported by the fact that α-chitin is much less soluble than β-chitin. But the α-chitin formed during the $\beta \rightarrow \alpha$ transition need not be the same as native α-chitin. Native α-chitin and perhaps native γ-chitin may well be extended-chain, anti-parallel, crystalline polymers like native cellulose.

Electron micrographs of cuticle-secreting cells in the locust show chitin fibrils virtually touching the cell membrane (WEIS-FOGH, 1970), and as no fibrillar material is observed within these cells, Weis-Fogh proposes that chitin synthesis takes place at the cell surface. Thus, the condition for extended-chain crystal formation, that of simultaneous polymerization and crystallization, is apparently met in the case of chitin as well as cellulose.

Chitin observed in electron micrographs is always seen as long, thin fibrils, but the dimensions of these fibrils vary according to the origin of the specimen (NEVILLE, 1967). There does not appear to be any single crystalline unit in the sense of the elementary fibril of cellulose, or at least such a unit has not been discovered. The following list from NEVILLE (1967) gives the observed widths of the chitin fibrils:

Crustaceans	10 nm
Insect endocuticle	2.5 nm
Insect trachea	7.5 to 30 nm
Diatoms	20 to 30 nm

3.5.2 Mechanical Properties of Chitin Fibres

The only known tensile structure made from chitin is arthropod tendon or apodeme. As shown in Fig. 3.32a and b, these muscle attachments are formed by invaginations of the cuticle making up the exoskeleton. Very little structural or mechanical information is available on apodeme, but electron microscopic and X-ray diffraction studies of the tendon from the hind leg of the locust (A. C. Neville, personal communication) indicate a very high degree of preferred orientation with the chitin fibrils parallel to the long axis of the tendon. Polarized light studies of lobster tendon (CLARK and SMITH, 1936) indicate that the chitin fibrils in this structure are parallel to the long axis of the tendon.

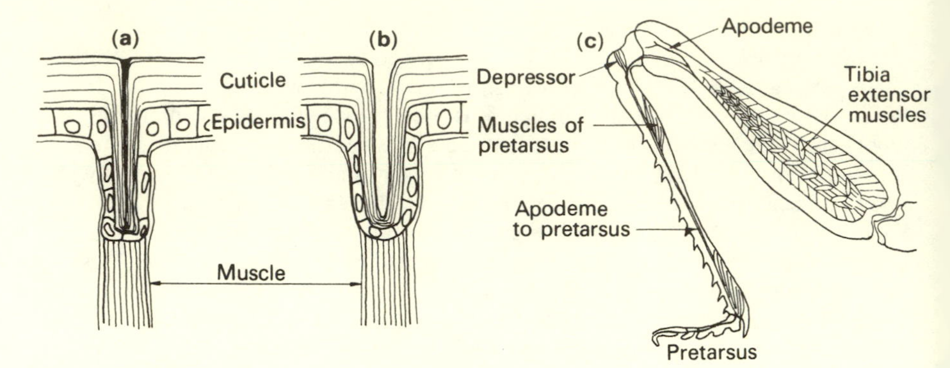

Fig. 3.32 (a) and (b) are longitudinal sections showing the structure of apodeme and apophysis as solid and hollow invaginations of cuticle. (c) is a see-through diagram of the back leg of a locust. Note the stout apodeme attaching tibial extensor muscles to the tibia and the long, flexible apodeme that depresses the pretarsus.

The length and thickness of chitin tendon depends on the particular functional role that each tendon plays in the skeleto-muscular system. Figure 3.32c shows two tendons in the jumping leg of the locust. The apodeme attaching the powerful tibial-extensor muscle is short and very stout; while the apodeme which articulates the pretarsus is long and thin, and as the diagram suggests quite flexible. In sclerotized cuticle the crosslinked proteins found in association with chitin fibrils create a very tough and rigid material. The flexibility of apodeme, inferred from Fig. 3.32c, suggests that the properties of this structure depend primarily on the parallel arrangement of chitin fibrils, and that the crosslinked proteins are relatively unimportant.

Information on the mechanical properties of chitin fibres is very sparse indeed, and all available data are given in Table 3.8. One would expect on the basis of similarities in the chemical and crystal structures of chitin and cellulose that these materials would have similar mechanical properties as well. We find that the

Table 3.8 The mechanical properties of chitin fibres

Chitin		Young's modulus N m^{-2}	Tensile strength N m^{-2}	Extensibility	Reference
spider apodeme *Tegenaria agrestis*		–	$2.7\text{–}6.8 \times 10^7$	–	HOMANN, 1949
Purified chitin	dry	–	9.5×10^7	–	THOR and HENDERSON, 1940
	wet	–	1.8×10^7	–	THOR and HENDERSON, 1940
Beetle, '*Balkenlage*'	dry	4.5×10^{10}	5.8×10^8	1.3%	HERZOG, 1926
Cellulose fibres (ramie)	dry	8.0×10^{10}	9.0×10^8	2.3%	MEYER, 1942

3.5.2 Mechanical Properties of Chitin Fibres

Young's modulus of chitin, as calculated by JENSEN and WEIS-FOGH (1962) from the stress–strain data of HERZOG (1926), is quite similar to that of cellulose. On the other hand, the tensile strength of chitin appears to be considerably lower than that of well oriented cellulose fibres (e.g., ramie). At present there is no information on the degree of polymerization of any of the samples tested, but the fact that wet chitin has a lower tensile strength than dry chitin (THOR and HENDERSON, 1940) suggests that the degree of polymerization is not as high in chitin as in most cellulose fibres. If the degree of polymerization is low, the number of overlapping residues between adjacent chains is quite low, and water molecules competing for the inter-chain hydrogen bonding sites will loosen the structure and allow the fibres to yield by chain slippage. The unusually high tensile strength observed by HERZOG (1926) indicates either a higher degree of preferred orientation, as suggested by JENSEN and WEIS-FOGH (1962), or a higher degree of polymerization, or both.

Chapter 4
Pliant materials

4.1 Introduction

This chapter deals with the class of materials which function in supportive systems by being deformed, and we choose to call them pliant materials. It is necessary to specify that the concept of a pliant material is based on functional considerations because all materials are deformed under stress, but pliant materials function by being deformed. Mechanically the pliant materials have a low modulus, can be deformed to large strains without breaking, and often show long-range, reversible elastic properties similar to those of rubber. These properties, which are very different from those of the high modulus solids found in tensile and rigid materials can be understood in terms of the properties of amorphous (i.e., non-crystalline) polymeric materials. Tensile and rigid materials are made from crystalline or highly crosslinked solids whose elastic properties arise from energy changes due to the deformation of chemical bonds. The amorphous polymers, however, can be described in terms of flexible, long-chain molecules that store elastic energy as entropy changes due to alterations of the polymer chain conformation.

There appear to be three different groups of pliant biomaterials. The first are proteins having mechanical properties very similar to those of rubber. These materials can usually be described as single phase amorphous polymers and are usually found in mechanical systems requiring long-range elastic energy storage for relatively high-speed cyclic processes. Included in this group are resilin, abductin, and elastin. The second group is an extremely diverse one that includes the materials commonly referred to as soft connective tissues. However, we shall call them pliant composites, as they contain some amorphous polymer component in association with a relatively inextensible, high modulus fibre. The amorphous component is usually a hydrated protein-polysaccharide complex but can also be one of the rubbery proteins mentioned above. Collagen provides the fibre component in virtually all the animal phyla. In general, the use of multiphase composites rather than single phase amorphous systems makes it possible for a small number of amorphous components to provide a wide range of pliant materials that are very finely tuned to their function in the organism. The third group is at the border line between the pliant and rigid materials. It contains materials that function by giving shape, but in a pliant sort of way. The group includes materials such as cartilage, which seem to depend on the water-binding properties of hydrated, amorphous polymers for their particular mechanical properties.

4.2 The Protein Rubbers

The protein rubbers provide a good starting point for this discussion of pliant biomaterials because of the relative simplicity of their organization and because we are able to give a good account of their mechanical properties in molecular terms. The materials to be discussed, resilin, abductin, and elastin, are the only known long-range elastic biomaterials that exist as one-phase amorphous polymer systems. As we shall see, other pliant biomaterials are known which have similar properties based on other amorphous components, but these materials are all pliant composites.

4.2.1 Resilin

Resilin is a protein rubber first described by WEIS-FOGH (1960) as a cuticular component of the insect flight system. The several bits of elastic cuticle described by Weis-Fogh function as energy storing elastic elements that counteract the inertial forces generated by the reciprocating wing movements during flight. According to WEIS-FOGH (1960) it is the use of this elastic element in the flight system of insects that gives this group such aerobatic success. The distribution of resilin in flight systems and other insect structures is discussed by ANDERSEN (1971) and WEIS-FOGH (1972).

Dry resilin is a hard, essentially glassy polymer, but when swollen with water, resilin shows all of the properties of an amorphous polymer network above its glass transition temperature. Resilin will swell in a number of solvents, including water, formamide, formic acid and other protein solvents, but is completely insoluble in all solvents that do not cleave peptide bonds. It can be extended reversibly to several times its resting length and will not creep under constant load for extended periods of time (WEIS-FOGH, 1960). In the unstrained, swollen state resilin appears structureless in light microscopic, electron microscopic, and X-ray diffraction studies (ELLIOTT, HUXLEY, and WEIS-FOGH, 1965). However, upon stretching, the isotropic material becomes birefringent (WEIS-FOGH, 1960, 1961*b*).

The study by WEIS-FOGH (1961*a*) on the thermodynamics of the elastic process of resilin clearly shows that this material is a three-dimensional network of crosslinked, random-coil molecules. As shown previously the thermodynamic description of elasticity predicts that the tensile force required to maintain a fixed extension will equal:

$$f = (\partial U/\partial l)_{VT} + T(\partial f/\partial T)_{Vl} \qquad (4.1)$$

The first term on the right corresponds to the force which results from changes in internal energy with extension, and the second to the force due to changes in conformational entropy. For an ideal rubber $(\partial U/\partial l)_{VT} = 0$, and all the elastic force is due to entropy changes. The form of this equation allows us to determine experimentally the relative contributions of internal energy and entropy changes in any elastic material. This is done by measuring the force required to maintain constant extension over a range of temperatures. The slope of such a plot times the absolute temperature gives the entropy contribution; the intercept of this curve at $T = 0^\circ K$ gives the internal energy contribution. It should be noted that this equation contains the constant volume coefficients $(\partial U/\partial l)_{VT}$ and $(\partial f/\partial T)_{Vl}$. It is usually convenient to carry out these experiments at constant pressure rather than at constant volume. Constant pressure measurements can be used if the volume of the material remains fairly constant over the temperature range used because when $(\partial V/\partial T)_{Pl} = 0$, $(\partial f/\partial T)_{Pl} = (\partial f/\partial T)_{Vl}$. WEIS-FOGH (1961*a*) made thermo-elastic measurements as described above on resilin tendon from the dragonfly *Aeshna*. The thermo-elastic data are plotted in Fig. 2.29a, and the calculated entropy and internal energy contributions are shown in Fig. 2.29b. The entropy component follows the force-extension values very closely over extensions of $\lambda = 1.1$ to 2.2, and the internal energy component remains low over this range. These results are very similar to those obtained for a number of synthetic rubbery materials, indicating that resilin is indeed a rubber elastomer.

WEIS-FOGH (1961*b*) was able to specify further the nature of this polymer network. We have seen (*Eq. 2.65*) that the force-extension relationship of a polymer network can be described as,

$$f = NkT\,(\lambda - \lambda^{-2}) \qquad (4.2)$$

where N is the number of polymer chains per unit volume of the material, k the Boltzmann constant, and T the absolute temperature. This equation is based on the assumption that the number of segments n between crosslinks in the network is large. If, however, n is relatively small *Eq. 4.2* will have to be modified because the force will increase more rapidly at high extensions than predicted. Weis-Fogh found that the force-extension curve of resilin tendon was similar to that of a rubber with roughly 15 freely orienting segments between crosslinks. The elastic modulus (G) of the resilin samples tested by Weis-Fogh was 6.4×10^5 N m^{-2} (For rubbers, when $\lambda = 1$, $G = E/3$. According to WEIS-FOGH (1960) $E = 1.9 \times 10^6$ N m^{-2}.) This value can be used in the equation derived from the network theory of rubber elasticity (TRELOAR, 1958 and *Eqs. 2.67, 2.68*, etc.) to give a measure of the molecular weight of polymer chains between crosslinks, M_c:

$$G = \rho RT/M_c \qquad (4.3)$$

where ρ is the density of the polymer in weight of dry polymer per unit volume, R the molar gas constant, and T the absolute temperature. The front factor of *Eqs. 2.67, 2.68* is here set as unity for an ideal rubber. The elastic modulus of resilin corresponds to a value of $M_c = 5100$, or roughly 60 amino acid residues. Thus, each of the 15 freely-orienting segments is made up of approximately 60 amino acids.

The mechanical information presented here provides direct evidence for the presence of a crosslinked network in resilin. Amino acid analyses (see Table 4.1) indicate that resilin is high in glycine and acidic amino acids but has no cystine (BAILEY and WEIS-FOGH, 1961). Clearly, disulphide bridges are not providing the crosslinks. Andersen's work on the crosslinks of resilin (reviewed in ANDERSEN, 1971) has shown that the protein chains are linked together through di- and tri-tyrosine residues. From chemical analyses it was calculated that there are enough di- and tri-tyrosine groups present to give a molecular weight between

Table 4.1 Amino acid composition of the rubbery proteins abductin, resilin and elastin. Values are numbers of residues per 1000 residues of the material (1, KELLY and RICE, 1967; 2, BAILEY and WEIS-FOGH, 1961; 3, ROSS and BORNSTEIN, 1969.)

	Abductin[1] (*Pecten irradians*)	Resilin[2] (Prealar arm)	Elastin[3] (Ligamentum nuchae)
Aspartic acid	18	108	6
Threonine	11	30	9
Serine	69	81	10
Glutamic acid	12	48	15
Proline	13	79	120
Glycine	630	367	324
Alanine	33	106	223
½Cystine	–	–	4
Valine	6	32	135
Methionine	90	–	–
Isoleucine	8	17	26
Leucine	2	24	61
Tyrosine	2	31	7
Phenylalanine	93	24	30
Lysine	8	–	7
Histidine	6	6	1
Arginine	3	–	5
Hydroxyproline	–	–	11
Desmosine/4	–	–	8

crosslinks of M_c = 3200. This is in good agreement with the value M_c = 5100 derived from mechanical measurements. It indicates that these compounds do, in fact, crosslink the resilin network, and that they are probably the only crosslinks present (ANDERSEN, 1971).

Resilin functions in the living insect by providing an elastic mechanism in the flight system that stores the kinetic energy of the moving wing, and in so doing, helps to decelerate the wing at the end of its stroke and then accelerate it in the opposite direction. In order for a rubber to function efficiently in this manner, it must have a low energy loss in the frequency range required for the flight system, i.e., tan δ must be kept very small. JENSEN and WEIS-FOGH (1962) measured the dynamic mechanical properties of the prealar arm (a structure made of resilin with some chitin lamellae) from the flight system of the locust *Schistocerca.* They carried out forced oscillation measurements over a frequency range of 10 to 200 Hz. The sample was deformed in a manner such that the chitin fibres were flexed and not directly pulled upon. The data are plotted in Fig. 4.1 as the elastic loss factor (h) against frequency. This loss factor is the ratio of the energy dissipated per half cycle to the maximum energy stored per half cycle. In very slightly damped materials such as resilin $h = \pi \tan \delta$. The loss factor is clearly frequency- and amplitude-dependent, and in the biological range of frequency the loss factor appears to be very low. Although Jensen and Weis-Fogh do not specify the biological range of strain amplitude, the value of 0.1 mm is probably

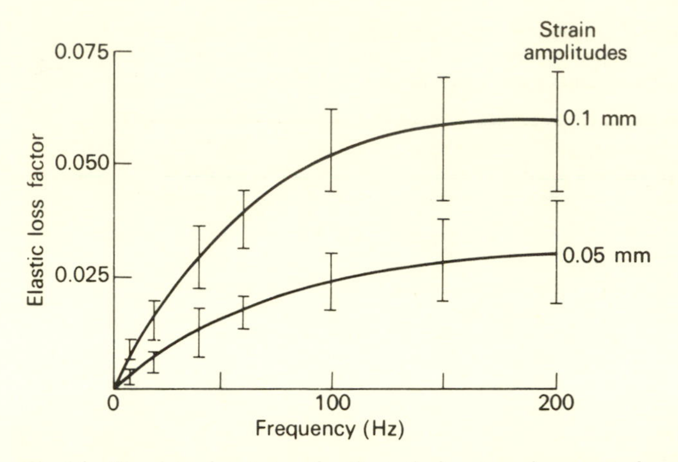

Fig. 4.1 The elastic loss factor of resilin in the locust prealar arm as a function of frequency. The elastic loss factor is explained in the text (after JENSEN and WEIS-FOGH, 1962; reproduced from *Phil. Trans. roy. Soc. Lond.*).

reasonably close. Jensen and Weis-Fogh interpret these results in terms of an elastic efficiency, $\eta_e = 1 - h$, which is the ratio of the energy stored per cycle to the total energy input. They conclude that the elastic efficiency in the biological range is 96 to 97%. The term resilience, $R = 1 - 2h$, can also be used as a measure of the efficiency of a rubber. Resilience is the ratio of the energy recoverable in one cycle to the total energy input. The difference between elastic efficiency and resilience arises because energy can only be stored during one half of the cycle, during extension; while energy is dissipated during the entire cycle. The resilience in the biological range is 92 to 94%. This is equal to or better than the resilience of the very best synthetic rubbers. Resilin is clearly well suited for its role in the insect flight system.

4.2.2 Abductin

Abductin is the protein rubber found in the inner hinge-ligament of the bivalve mollusc schell. The bivalve shell is composed of two rigid shell valves articulated along one edge by a flexible ligament (Fig. 4.2). The adductor muscle pulls the two valves together stretching the outer portion of this ligament and compressing the inner portion. In the shell of the scallop *Pecten* this outer part contains a fibrous protein which resists tensile strains. Thus, the abductin in the inner ligament of *Pecten* is compressed when the shell is closed, and it is this compressed bit of rubbery material that provides the force to reopen the shell when the adductor muscles relax. Most bivalve molluscs are sedentary beasts, and the rate at which the shell opens is of little importance. Of course, the animals must be able to close the shell rapidly in dangerous situations, but it will not matter if it takes two or even ten seconds for the shell to reopen. However, the scallops are very active bivalves (Pectinidae) that swim by rapid adductions of the shell. The animals are jet propelled by water expelled during shell closing and ALEXANDER (1966) reports that the shell opens and closes about three times a second in a

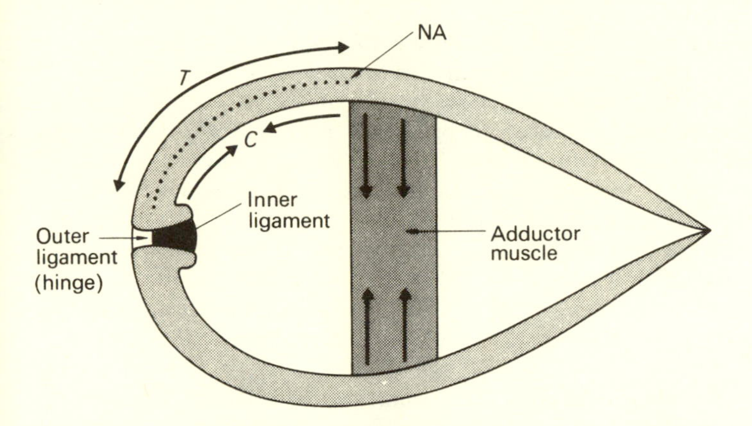

Fig. 4.2 Diagrammatic section through a scallop. *T*, tensile and *C*, compressive forces in shell are external and internal respectively to neutral axis, NA.

swimming *Pecten*. This is another system where the efficiency of the elastic element is very important.

TRUEMAN (1953) measured the hysteresis of ligaments in a number of bivalve shells by measuring the force required to hold the shell at a given angle of gape over an opening and closing cycle. The plots are in the form of loops, and the area within the loop indicates the amount of energy dissipated in the cycle. Figure 4.3 shows data for two of the swimming bivalves, *Pecten maximus* and *Chlamys opercularis*, and one of the more sedentary forms, *Lutraria lutraria*. These static measurements indicate that the ligament of the more active forms is more efficient. but measurements at the frequency and over the angles of gape actually attained in swimming are needed before we can be certain of this. The inner ligament of

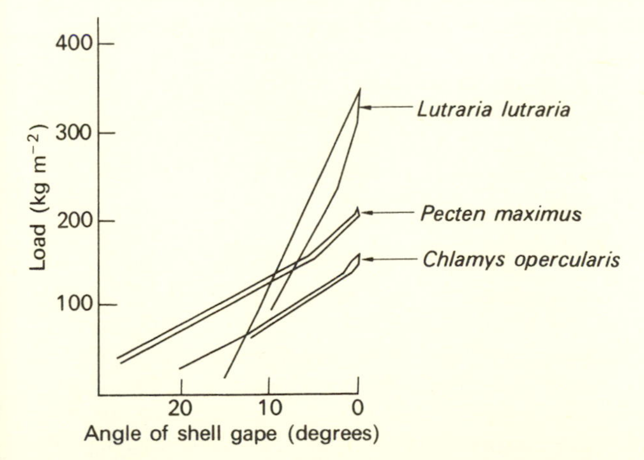

Fig. 4.3 Hysteresis (the area enclosed in each curve) of the ligaments of three species of bivalved molluscs. *Pecten* and *Chlamys* are swimmers in the family Pectinidae; *Lutraria* burrows in marine sediments (after TRUEMAN, 1953; courtesy of the *J. exp. Biol.*).

bivalves like *Lutraria* and *Mytilus*, which show a marked hysteresis during the opening and closing cycle, contains calcium carbonate crystals interspersed within the abductin. In the Pectinidae, however, the inner ligament is free of crystalline inclusions, and the mechanical properties observed for these ligaments can be attributed to the abductin alone.

ALEXANDER (1966) made thermo-elastic measurements on abductin from *Pecten* and found that it, like resilin, is a true rubber elastomer, with conformational entropy changes providing the major contribution to the elastic force. Abductin has a Young's modulus of 4×10^6 N m^{-2}. Alexander also measured the elastic efficiency of this protein rubber at 4 Hz by observing the decrement of free oscillations of abductin within the shell. The resilience was found to be 91%.

The mechanical evidence indicates that abductin is a crosslinked polymer network, but amino acid analyses of abductin from *Pecten irradians* (see Table 4.1) give no indication of the crosslinking agent (KELLY and RICE, 1967). However, ANDERSEN (1967) isolated a compound from abductin hydrolysates which was very similar to the di-tyrosine crosslink he had previously found as a crosslink in resilin. The compound was identified as 3,3′-methylene-bistyrosine, and differs from di-tyrosine in that the two phenol groups are linked together through a CH_2 group.

4.2.3 Elastin

Elastin is the protein rubber of the vertebrates. Although it is usually found in association with collagen and glycosaminoglycan in pliant composites, it does occur in nearly pure forms. The *ligamentum nuchae* of the ungulates contains 80% elastin and only a small amount of collagen and other materials. Elastin is never found in large blocks, but exists as very fine fibres. In the *tunica media* of the aorta, for example, elastin is found as sheets 2.5 μm thick, and in the *ligamentum nuchae* the elastic fibres are about 7 μm in diameter (FRANZBLAU, 1971). Mechanically elastin is very similar to the other protein rubbers. The elastic modulus (G) of around 6×10^5 N m^{-2} is very similar to that of resilin and abductin. Elastin shows long-range reversible extensibility and will not creep when loaded for extended periods of time. Purified samples of *ligamentum nuchae* break at roughly twice their resting length. Like resilin, elastin is a rigid, glassy solid when dry, but it becomes rubbery when swollen with water or other protein solvents. Elastin is completely insoluble in all solvents that do not cleave peptide bonds.

MEYER and FERRI (1936) made thermo-elastic measurements on water-swollen elastin and found large negative values for the internal energy contribution to the elastic force. This, they concluded, resulted from stress-induced crystallization. HOEVE and FLORY (1958), however, explained this anomalous internal energy change in terms of volume changes due to alterations in the degree of swelling with temperature. MEYER and FERRI (1936) measured the constant pressure coefficient $(\partial f/\partial T)_{Pl}$ and obtained misleading values because of the large negative coefficient of thermal expansion. HOEVE and FLORY (1958) found the volume of elastin to be independent of temperature in a 30% solution of ethylene glycol in water, and thermo-elastic measurements carried out in this solvent suggest that elastin is a true rubber elastomer. Recently, VOLPIN and CIFERRI (1970) have shown that a correction for the volume change can be applied to the thermo-elastic measurements of elastin in water. They found that elastin behaves like a rubber in that entropy changes account for 90% of the elastic force, but this rubbery behaviour is superimposed on the swelling changes with temperature.

The amino acid composition provides an important key to the molecular structure of elastin. Over 95% of the amino acids in elastin have non-polar side chains (see Table 4.1), and non-polar amino acids in a protein are known to aggregate in the centre of the molecule and exclude water by forming what is often called a hydrophobic bond but more properly called a hydrophobic interaction. The number of hydrophobic amino acids suggests that the hydrophobic interaction in elastin is the major factor determining the molecular structure of this material. The magnitude of this interaction has led workers such as PARTRIDGE (1967, 1970) to conclude that elastin can be described as a two-phase system of compact protein spheres with free solvent in the intervening spaces. This is quite a different structure from the single phase polymer network usually associated with rubbery materials. In support of this model, PARTRIDGE (1967) has shown that elastin can be used as a gel-filtration medium to separate solute molecules of different molecular size. The results of such gel-filtration studies suggest that elastin is a network of protein globules with pores of about 3.2 nm diameter running through it. Partridge suggests that the globules have a molecular weight of about 50 000 and are about 5 nm in diameter.

WEIS-FOGH and ANDERSEN (1970) suggested that the unfolding of the hydrophobic region of these protein spheres could contribute to the elastic force by creating a larger surface where the non-polar side chains could interact with water. According to the theory of NEMETHY and SCHERAGA (1962*a*, *b*, *c*), water forms a highly ordered, hydrogen-bonded cage structure around a non-polar molecule in solution. Thus, the process of moving a non-polar group out of the hydrophobic interior of an elastin molecule into water will involve a large positive free energy change because of this ordering effect of the solute on the water. Thermodynamic measurements made by stretching elastin samples in a microcalorimeter (WEIS-FOGH and ANDERSEN, 1970) have shown that there is an extremely large internal energy change associated with the elastic mechanism of elastin. (Remember that the internal energy change in other rubbery materials is essentially zero.) In elastin, then, there is some chemical process involved in the elastic mechanism, and this process appears to be the unfolding of the hydrophobic region of the globular elastin molecule. Reagents that reduce the strength of the hydrophobic interaction reduce the internal energy change as well. The magnitude of the internal energy change varies with temperature in a manner that corresponds to predictions of NEMETHY and SCHERAGA (1962*b*, *c*) for hydrophobic interactions. It has recently been shown (Gosline, unpublished) that the volume changes of elastin with temperature are directly linked to changes in the strength of the hydrophobic interaction with temperature.

It is difficult to give a complete account of the elastic mechanism because the critical experiments are still in progress, but we can say the following. Elastin can be thought of as a network of globular protein molecules which are compressed or restrained by the tendency of the hydrophobic amino acids to exclude water. The presence of this hydrophobic interaction means that the elastin molecule cannot be described as an unrestrained random-coil molecule. When elastin is stretched, the hydrophobic region must unfold, and this unfolding has two effects: (1) the water exposed to non-polar side groups becomes ordered, and (2) the opening of the hydrophobic region removes some of the restraints on the polymer chain. That is, when elastin is stretched, the water becomes more ordered and the polymer becomes less ordered. This is very different from the mechanism normally associated with rubbery materials, where a random polymer becomes ordered when it is stretched.

The nature of the crosslinks in elastin has been thoroughly investigated; FRANZBLAU (1971) gives a good review of this work. In short, a compound known as desmosine is formed by the condensation of the side chains of four lysine residues. As each of the lysines can come from a different protein molecule, this crosslink can tie together as many as four molecules of elastin. Another compound, lysinonorleucine, which is derived from two lysine residues is also thought to be involved in crosslinking.

It is possible to extract a soluble elastin precursor from the aorta of pigs raised on copper deficient diets (SMITH, BROWN and CARNES, 1972). This soluble elastin, named *tropoelastin*, has an amino acid composition very similar to the one shown in Table 4.1, with the exception that no desmosine is present and lysine is present in much larger quantities. The molecular weight has been estimated at 74 000. Thus, the soluble elastin precursor has much the same molecular weight as the dense protein spheres predicted by PARTRIDGE (1967) in experiments where elastin was used as a gel-filtration medium.

Elastin usually functions as a static elastic element in pliant composites such as skin and the neck ligament of the ungulates. In these static tissues the important properties are the low modulus and the long-range, reversible extensibility; the hysteresis or resilience of these materials is relatively unimportant. However, in dynamic tissues like arterial wall, the resilience is very important. The elastin in arterial wall acts to damp out the pressure pulses generated by the pumping of the heart and to stabilize the flow of blood through the arteries and capillaries. The arteries expand during a pressure pulse from the heart; when the heart is in its recovery phase, the elastic energy stored in the arterial wall can be used to maintain blood flow through the vessel (see ALEXANDER, 1968; BERGEL and SCHULTZ, 1971). This system will work efficiently only if the resilience of the arterial wall is reasonably high.

Recent measurements of the damping of purified elastin samples from beef *ligamentum nuchae* (Gosline and Weis-Fogh, unpublished) are shown in Fig. 4.4. These measurements were carried out at various frequencies at a strain amplitude of 0.11. At low frequencies the resilience is reasonably high (at 1 Hz, $R = 76\%$; $R = 2\pi \tan \delta$) but not nearly as high as resilience of resilin or abductin. The resilience drops sharply as the frequency is increased (at 31 Hz, $R = 31\%$). Elastin must function in arterial wall at frequencies which even in the smallest animals are below about 5 Hz (300 heart beats/min.). However, WEIS-FOGH (1972) has suggested that the low resilience at frequencies of the order of 30 Hz precludes the use of elastin as an energy storing element in the flight system of humming

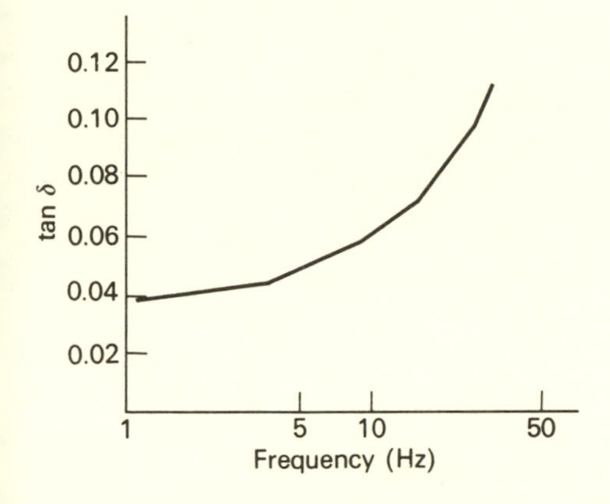

Fig. 4.4 Damping of purified elastin at different frequencies (from unpublished work of Gosline and Weis-Fogh).

birds. Remember that resilin provides an important element in the flight system of insects. Presumably, the inertia of the rapidly moving hummingbird wing is overcome by the powerful wing muscles at considerable metabolic cost to the animal.

4.3 The Mucopolysaccharides

The mechanical properties of the pliant composites and cartilage-like biomaterials depend for a large part on amorphous polymers made up of proteins and polysaccharides in hydrated macromolecular complexes. A variety of names has been used to describe these connective tissue components, but we shall use the presently accepted nomenclature as described by BALAZS (1970). In this system the term mucopolysaccharide has been divided into a number of more specific terms such as *glycosaminoglycan*, a polysaccharide containing amino sugars; *proteoglycan*, a covalently bonded complex of protein and glycosaminoglycan in which the polysaccharide is the dominant feature; and *protein–polysaccharide complex*, a complex held together through non-covalent linkages. The glycosaminoglycans occurring in vertebrate connective tissues have been identified and studied in detail (see BALAZS and JEANLOZ, 1965). They include chondroitin 4-sulphate, chondroitin 6-sulphate, dermatan sulphate, keratin sulphate, and hyaluronic acid (see Fig. 4.5 for the structure of the repeating units of these compounds). These and other similar compounds are found in a number of invertebrate tissues (HUNT, 1970). The glycosaminoglycans are nearly always found in association with protein, either as proteoglycan or as a protein-polysaccharide complex, but our present knowledge of these complexes is quite limited. In fact, the proteoglycans from vertebrate cartilage are the only materials of this kind that have been studied in detail.

The properties of the proteoglycans and protein–polysaccharide complexes depend to a large extent on the poly-anionic character of glycosaminoglycans. The carboxyl groups of the uronic acids and the sulphate groups of the

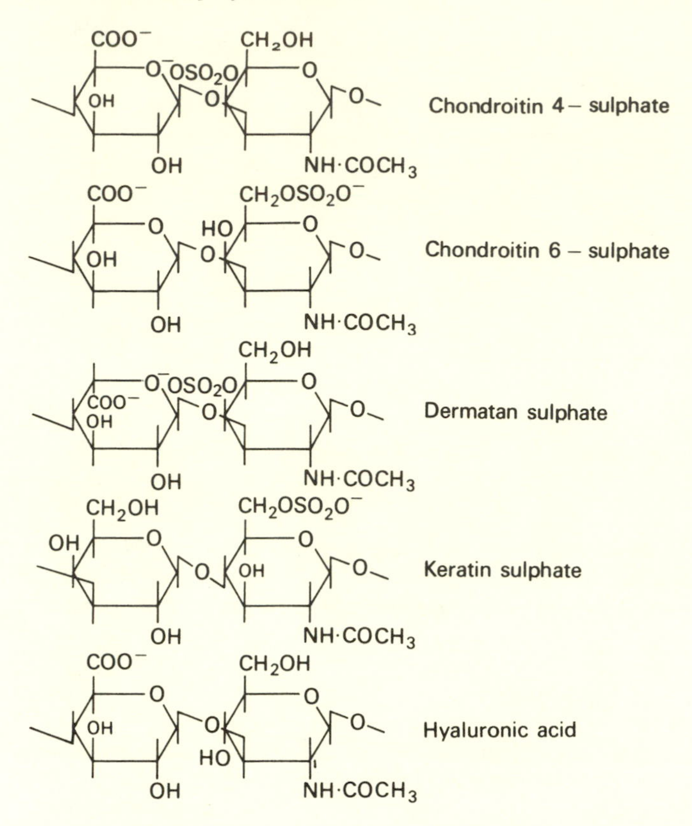

Fig. 4.5 Chemical structure of several glycosaminoglycans.

sulphated amino sugars of glycosaminoglycans in solution at physiological pH will be ionized, giving a polyelectrolyte with up to one negative charge per sugar residue. This high charge density has a marked effect on the shape and properties of these polymer molecules. The repulsive forces between like charges drive polymer segments of this polyelectrolyte chain away from each other, causing the molecule to take on a greatly expanded form (TANFORD, 1961). Because the charges can be masked by ions in the solution, the extent of this expansion is dependent on both the ionic strength and the pH of the solvent medium. Under normal physiological conditions the glycosaminoglycans are dispersed molecules that encompass a much larger volume than would a similar but uncharged polysaccharide molecule. The mechanical importance of this expanded conformation lies in the fact that these polyelectrolyte molecules will begin to interact with each other to form entanglement networks at very low concentrations. This means that very dilute solutions of these materials will have viscoelastic properties similar to those of the high molecular weight, non-crosslinked polymers discussed in Chapter 2B, and if some kind of permanent cross-bridges can form, these materials will become dilute polymer networks or gels with rubber-elastic properties.

The synovial fluid which can be obtained from vertebrate joints provides a well documented example of the viscoelastic behaviour of a protein-glycosaminoglycan complex in dilute solution. This fluid contains hyaluronic acid with a molecular weight of around 2×10^6 (LAURENT, 1966), and protein, some of which may be covalently linked to the hyaluronic acid (OGSTON, 1966). The dynamic mechanical properties of this fluid have been studied by GIBBS *et al.* (1968) in an oscillating Couett rheometer. In this apparatus the fluid is placed between two plates and sheared by sinusoidal oscillations of one of the plates. Stress, strain, and the phase angle between stress and strain are measured and used to calculate storage and loss moduli (G' and G''). The results are plotted in Fig. 4.6 for human synovial fluid with a concentration of 2.38 mg dry mucin per cm^3 of solution (a 0.238% solution). At low frequencies the loss modulus (G'') is larger than the storage modulus (G'), indicating that synovial fluid is essentially

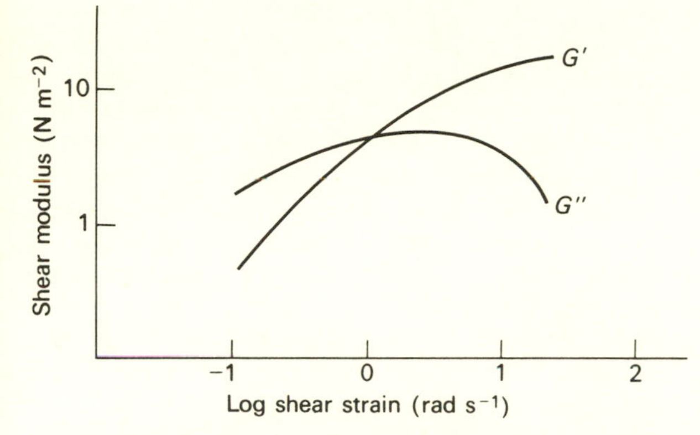

Fig. 4.6 Shear modulus of synovial fluid at different strain rates. G', storage modulus; G'', loss modulus (after GIBBS *et al.*, 1968; courtesy of *Biopolymers*).

a viscous liquid. As the frequency is increased the damping of the system ($\tan \delta = G''/G'$) decreases, and the material begins to show the elastic properties of a high molecular weight, non-crosslinked polymer in the plateau region of its response curve. At the high frequency end of this curve, $\tan \delta$ is down to about 0.1 indicating that synovial fluid is a predominantly elastic material at this frequency. OGSTON and STANIER (1953) demonstrated that synovial fluid could also show long term rigidity when loaded in compression in very thin films. They placed a drop of synovial fluid on an optically flat piece of glass and then compressed the fluid with a convex lens. The thickness of the film between the glass plate and lens was determined by measuring the Newton's rings formed when light was passed through the system. They found that the synovial fluid film did not go to zero thickness even after long periods of time as did water and glycerol films. This suggests that when compressed into thin films, the protein-polysaccharide complexes may interact to form a more permanent network. If synovial fluid is filtered through a material with a 1 μm pore size, the elastic properties described above are not found in the filtrate (OGSTON and STANIER, 1950). This indicates that the protein-hyaluronic acid complex is too large to pass through a 1 μm pore.

Light scattering data suggest that this complex is a solvated sphere with a radius of gyration of about 200 nm (LAURENT, 1970).

We should note here that synovial fluid is believed to play a major role in the lubrication of the joints in which it is found. Early work suggested that synovial fluid acted as a hydrodynamic lubricant. In hydrodynamic lubrication a thin film of liquid is drawn between the joint surfaces by their relative motion, and this fluid film provides a lubricating layer that completely separates the surfaces. OGSTON and STANIER (1953) found that the viscosity of synovial fluid decreased as the shear rate was increased and this non-Newtonian viscous behaviour was thought to indicate that synovial fluid can act as a hydrodynamic lubricating fluid over a very wide range of shear rates. The non-Newtonian viscous behaviour was attributed to changes in the shape and interaction of the hyaluronic acid molecules in the solution. Recent work (LINN, 1968; ROBERTS, 1971) suggests that the viscosity and hence the hydrodynamic properties of the fluid are not as important as the adherence of some compound in synovial fluid to the articulating surfaces to provide boundary lubrication. In boundary lubrication, a compound that is attached to the bearing surfaces alters the surface properties of the material itself. No fluid film is involved. The boundary lubrication in synovial joints is apparently associated with the protein in synovial fluid. The elastic properties and fluid binding capacity of the articular cartilage may also contribute to the lubrication of the joint (McCUTCHEN, 1966). Recent theories of joint lubrication are reviewed by HAMMERMAN (1970). Although the hyaluronic acid in synovial fluid may not play the dominant role in joint lubrication, the viscoelastic properties of synovial fluid described above are certainly associated with the presence of hyaluronic acid, and these elastic properties must contribute to the proper functioning of the joints.

Other materials that contain protein and glycosaminoglycan in hydrated macromolecular complexes show viscoelastic properties like those of synovial fluid. DAVIS and DIPPY (1969) and DAVIS (1970) have shown that the mucus from human lungs and saliva have viscoelastic properties similar to those of synovial fluid shown in Fig. 4.6. In fact, it is a general property of proteoglycans and protein–polysaccharide complexes that they form viscoelastic pastes in low concentrations. These viscoelastic pastes are often referred to as mucous secretions and are used by virtually all animals for functions ranging from feeding nets to slime trails and egg cases. Unfortunately, there is very little else we can say about the mechanical properties of these mucous materials. They are surely very important to the animals which secrete them and deserve further attention.

One other important property of these mucous materials is their tendency to form gels. In a gel these hydrated macromolecules interact to form permanent cross-bridges, probably in the form of crystalline aggregations, which tie the molecules into a polymer network. Such a polymer gel will have rubber-elastic properties, and may provide the elastic component in a pliant composite material. There is very little information available on the nature of the interactions that tie the molecules together in such a gel, but the recent X-ray diffraction study by ANDERSON *et al.* (1969) has shown that double helical regions form between the chains of two gel-forming, sulphated polysaccharides from marine algae. Interactions of this type or interactions between the protein moieties may be important in other mucous gels.

FESSLER (1960) found that he could produce an elastic gel if the precipitated soluble collagen at 37°C in a solution of hyaluronic acid and then centrifuged the mixture. The pellet he retrieved from the centrifuge tube was elastic and had a volume greater than could be accounted for by the collagen alone. When the pellet was treated with hyaluronidase it lost its elastic properties. Fessler calculated from electron micrographs of this material that the collagen fibres formed a lattice-work with a pore size of around 1 μm. This is about the right size to trap the diffuse protein-hyaluronic acid complexes, and Fessler suggested that the trapping of diffuse proteoglycans in fibre networks provided a good model for Wharton's jelly, the hyaluronic acid containing gel-like material from the umbilical cord. As we shall see later, this type of protein–polysaccharide complex provides the basis for a wide range of pliant biomaterials.

4.4 Pliant Composites

In the pliant composites these viscoelastic, gel-forming proteoglycans and protein–polysaccharide complexes and perhaps the protein rubbers as well are combined with high modulus fibres to produce materials that can be stretched and bent as the animal moves or changes shape. Pliant composites are frequently found as thin skins, cuticles or membranes which enclose and support the soft parts of organisms. Most of the pliant composites found in animals contain collagen as the fibrillar component, but materials based on other fibres are found, such as the chitin-containing arthrodial membranes of the arthropod exoskeleton (see VINCENT and WOOD, 1972). Unfortunately very little is known about these other materials, so we will direct our attention to collagenous tissues. In the discussion that follows we will mention only a few of the vast variety of collagenous connective tissues. We restrict ourselves to these few because they are the ones which have been studied in sufficient detail for us to derive some organizational principles, and because we believe that the principles we derive here will apply to those materials we do not mention.

To start, let us assume that the collagen fibres in all these tissues are basically alike and have mechanical properties very similar to those described for collagen tendon in Section 3.3.2. That is, collagen fibres have a Young's modulus of around 10^9 N m^{-2}, and can be extended reversibly 3 to 4%, and break at extensions of 8 to 10%. Although there may be some variation in mechanical properties due to differences in crosslinking, etc., this is probably a reasonable assumption. At least we can be confident that the collagen fibres are orders of magnitude more rigid and very much less extensible than the amorphous polymer components, whatever they may be. Although the thickness of collagen fibres in different tissues varies from 15 to 150 nm, the fibres are generally very much longer than they are wide, and in some cases the mechanical properties suggest that the fibres are continuous throughout the tissue (e.g., tendon). If, as we suppose, the nature of the collagen fibre is constant, we are left with two organizational variables, the distribution and orientation of the collagen fibres in the composite, and the nature of the amorphous component.

4.4.1 Fibre Patterns in Pliant Composites

Several different fibre arrangements are found in pliant composites, each having different mechanical consequences. There are (1) parallel fibre arrays, (2) crossed-fibrillar arrays or, in the case of cylindrical animals, crossed-helical arrays in which the fibres are arranged in discrete layers of parallel fibres, and (3) felt-works in which the fibres are more randomly arranged with some fibres going through the thickness of the material. In each of these arrangements we must distinguish between continuous and discontinuous fibre systems. The term continuous as we use it here does not necessarily imply that individual fibres extend through the entire material. It means that there is mechanical continuity within the fibre system, such that at some point in the extension of the material, the load will be applied directly to the fibre component. This continuity can be achieved either by having long fibres which traverse the full extent of the tissue or by having shorter fibres linked together at junction points by strong crosslinks. In a discontinuous fibre system the fibres may become aligned in the direction of stretch during extension, but the load is transferred from fibre to fibre through the matrix and not applied directly to the fibres alone.

Parallel arrays of continuous collagen fibres have been dealt with previously. These are materials like tendon and are purely tensile in nature. Parallel arrays of discontinuous fibres, however, are important in pliant composites, as this arrangement provides a means of reinforcing the material in one particular direction. In fact, any concentration of parallel or preferentially oriented, discontinuous fibres will result in a mechanically anisotropic material. We shall look at this type of fibre arrangement more closely when we consider the properties of the mesogleal connective tissue of the sea anemone *Metridium*.

Crossed-fibrillar or crossed-helical arrays of *continuous* fibres provide the structural basis for the body wall connective tissues and cuticles of most of the worm-like animals and a number of vertebrates as well (NADOL, GIBBINS and PORTER, 1969; OLSSON, 1964). As shown by PICKEN, PRYOR and SWANN (1947), this kind of fibre system provides a reasonable arrangement for the

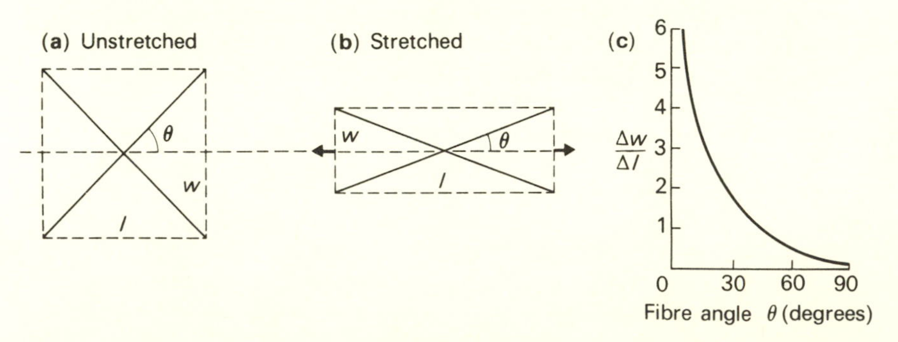

Fig. 4.7 The effects of tensile deformation (Δl) on width (Δw) and on the orientation (θ) of rigid reinforcing fibres in a sheet of material. (a) Before deformation; (b) after deformation in the direction of the arrows; (c) the change in width per unit change in length ($\Delta w/\Delta l$) allowed by the fibre angle (θ) in the material.

extensible covering of a worm-like animal if the fibres are arranged as crossed-helices around the long axis of the animal. We can model the mechanical behaviour of a crossed-fibrillar array with a system of inextensible fibres as shown in Fig. 4.7. Figure 4.7a shows the unstretched form as two crossed fibres with a 45° fibre angle with respect to the long axis of the animal (dashed line). A force applied in the longitudinal direction will extend the model by decreasing the fibre angle (Fig. 4.7b). Note two things about the control of shape change inherent in this model: (1) the extent of the deformation is limited by the initial fibre angle and (2) the rate of change of width (or body diameter) per unit change in length is inversely related to the fibre angle (Fig. 4.7c). When the fibre angle becomes small, the load will be applied almost directly to the continuous fibres, and the material will behave as a tensile material. If these crossed-fibres were embedded in an amorphous polymer network, the load at large fibre angles will be carried primarily by the amorphous component, and as the fibre angle is decreased, the load will be transferred to the fibres. Another important consequence of this type of fibre arrangement is that an extension in the longitudinal direction requires a contraction laterally. Also, the material is essentially inextensible when tested in the direction of the fibres. Thus, the mechanical continuity of the fibre array imposes severe restrictions on extensibility. A fibre arrangement of this type will result in a material that is capable of fairly limited extensions, and the extensions may only occur alternately in perpendicular directions. This, of course, describes the properties found in the body walls of many hydrostatic organisms (see Chapter 7).

Crossed-fibrillar arrays and feltworks of *discontinuous* fibres provide the basis for pliant composites that are capable of extreme extensions in length and width simultaneously. The lack of mechanical continuity means that the fibre system does not impose any limits on the extent of deformation of the tissue. However, if the material as a whole is to be mechanically continuous, the fibres must be embedded in a continuous amorphous phase. The fibres will reinforce the amorphous phase, and depending on the degree of preferred orientation of the fibres, may create a mechanical anisotropy as well, but the amorphous polymer component will play the major role in determining the mechanical properties of the composite.

A feltwork of continuous fibres differs from the crossed-fibrillar system in that the fibres are not neatly arranged into layers of parallel fibres. They are more randomly arranged, and are not restricted to discrete layers but may extend through the thickness of the tissue. As with other continuous fibre systems, the extensibility of a feltwork will be limited by the angle which the fibres make with respect to the direction of extension. The random arrangement will result in an isotropic material. More important, the three-dimensional character of a feltwork, as indicated by the presence of fibres extending through the thickness of the material, provides a resistance to shearing forces that is lacking in the two-dimensional crossed-fibrillar arrays. This type of fibre system provides a suitable arrangement for a flexible, moderately extensible protective covering for an animal's soft tissues. The dermal connective tissues or skins of birds and mammals function in just this manner and are well described as feltworks of continuous collagen fibres.

4.4.2 The Role of the Amorphous Phase

In a pliant composite, a fibre system is embedded in an amorphous polymer phase, and a deformation of such a composite will result in movement of the fibres, as described above, and in the deformation of the amorphous component. The mechanical behaviour of the composite can be analysed in terms of this interaction between the fibre system and the amorphous polymer by considering the force required to move the fibres through the amorphous phase. To do this let us assume that the fibres transmit the applied force through the material but are not themselves deformed, and that the force acts by deforming the amorphous phase. This assumption should be valid for materials with discontinuous fibre systems and for continuous fibre systems when the fibres make a large angle with the direction of the applied force. Of course, we are relying on our educated assumption that the collagen fibres are very much more rigid than the amorphous components. In a continuous fibre system which has been extended to the point where the fibres are aligned with the direction of the applied force, the load will be applied directly to the fibres and the nature of the amorphous component will be of little importance.

If we consider the simple crossed-fibre model of Fig. 4.7a with the added feature of an amorphous phase, we can deduce in qualitative terms the different ways that amorphous polymers can contribute to the mechanical properties of a pliant composite. Consider first the fibre system in a simple Newtonian viscous fluid. The force required to deform the model will act by dragging the fibres through this viscous fluid, and the magnitude of the force required will depend on the viscosity of the fluid and the rate at which the deformation takes place. Thus, we can make a strain-rate dependent material merely by placing the fibre in a viscous matrix. However, all of the energy required to deform this system is lost and there is no stored energy available to return the material to its initial dimensions. Next, consider the fibre system in a viscoelastic proteoglycan paste with mechanical properties similar to those shown for synovial fluid in Fig. 4.6. When tested dynamically at low frequencies, this system will behave like a system of fibres in a simple viscous fluid. However, as the frequency is increased, the elastic properties of this viscoelastic paste will provide a time-dependent elastic restoring force, and a portion of the energy required to deform the material will be available to return it to its initial dimensions. This arrangement of fibres in a viscoelastic paste provides a reasonable description of the crossed-helically fibre reinforced cuticles and basal membranes of most of the worm-like animals. These cuticles are intimately associated with antagonistic circular and longitudinal muscles that deform and restore the external cuticle, so there is no need for an elastic mechanism. The amorphous component provides strain-rate dependence and short term rigidity and may, as we will mention later, provide lubrication for the sliding fibres, but the overall contribution of the amorphous component is relatively small.

Pliant composites that are found in situations where antagonistic muscle systems are not available to restore the material after it is deformed often have rubber-like elastic properties. If we place the fibre system in a proteoglycan gel which is crosslinked to form a permanent network, we shall have such a material with a long range elastic mechanism. However, the presence of this polymer network does not mean that all the energy required to deform the material can be recovered elastically. In fact, in tissues such as the mesoglea of the sea anemone this is definitely not the case. If the network contains a polymer in the transition region of its response curve where segmental viscous interactions are important (see Chapter 2B), the energy required to deform the material will be much larger than the energy stored as a change in conformational entropy of the network polymers during extension. The amount of extra energy required depends on the strain rate, but the magnitude of the elastic restoring force depends on the strain alone. When the applied load is removed, the material will return to its unstrained dimensions, but the recovery will also be retarded by the segmental viscous interactions. Thus, the elastic recovery of such slow elastic systems may take from several seconds to many hours.

Long-range elastic composites based on protein rubbers such as elastin are commonly found in materials such as skin and arterial wall. These materials are usually three-phase systems with collagen fibres and proteoglycan components arranged in parallel with elastin fibres. Because the elastin or other protein rubber can be described as network polymers in the plateau region of their response curves, the strain rate or time-dependent properties of these composites will arise mainly from the interactions of the collagen fibres and the proteoglycan components. If this viscous interaction is not too large, the material as a whole will have elastic properties similar to those of the pure protein rubber, and the composite will be able to function as an energy storing system for relatively high speed cyclic processes. This is certainly the case with vertebrate arterial wall. The proteoglycan will serve as a lubricant to reduce frictional interactions between the sliding collagen fibres. Such lubrication is very important in materials that are subjected to repeated or cyclic deformations, because frictional interactions increase the energy dissipation per cycle, and because abrasion of the fibres may, with time, cause the material to fail.

4.5 Mesoglea

The body wall connective tissue (mesoglea) of the sea anemone *Metridium* provides an excellent example of a pliant composite with a discontinuous fibre system. The mesoglea constitutes the highly extensible container for the variable-volume hydrostatic skeleton which supports this soft-bodied, cylindrical animal. ALEXANDER (1962) found that when mesoglea is loaded in tension with very small stresses it will extend over a period of 12 to 15 hours (5×10^4 s) to more than three times its initial length and then maintain constant length (Fig. 4.8). When the load was removed, Alexander observed nearly complete elastic recovery over a similar time period. The very low equilibrium modulus ($E_{equilib.} =$ *ca.* 10^3 N m^{-2}) suggests that the animal can inflate itself with internal pressures of 1 N m^{-2} or less, generated by ciliary pumping in the syphonoglyph (BATHAM and PANTIN, 1950), but the inflation must take place over a long time. At higher strain rates, mesoglea is much more rigid. This allows the animal to withstand short term environmental stresses such as wave surge.

In the electron microscope, mesoglea appears to be a two-phase composite of collagen fibres in a structureless matrix (GRIMSTONE *et al.*, 1959; GOSLINE,

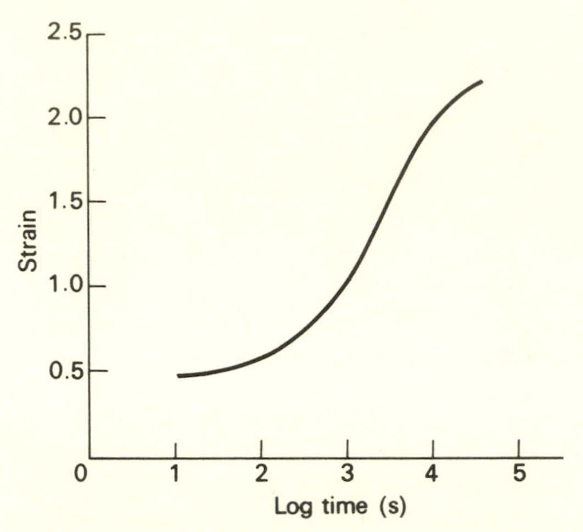

Fig. 4.8 Creep curve for mesoglea from *Metridium* under constant stress (after ALEXANDER, 1962). Courtesy of the *J. exp. Biol.*

1971*a*). With polarized light microscopy the tissue can be seen to consist of two distinct layers. (1) An outer layer making up two thirds of the thickness with collagen fibres arranged in crossed-fibrillar arrays forming right and left-handed helices in the plane of the cylindrical body wall. The fibre angle of this crossed-fibrillar array (the angle between the longitudinal axis of the animal and the fibre direction) in unstressed material is about 45°. (2) An inner layer containing densely packed circumferential fibres (GOSLINE, 1971*a*). If the collagen in this composite is a system of continuous fibres, we would expect that the crossed-helical array in the outer layer would limit longitudinal extension to about 40%, and that an animal which inflates itself by increasing in length will, as a consequence, become thinner. Continuous circumferential fibres in the inner layer should virtually eliminate expansion in width. ALEXANDER (1962) showed that mesoglea could be readily extended more than 200%, and the animal is capable of expanding in a manner that requires simultaneous increase in both length and width. Clearly these are not the properties of a continuous fibre composite.

The collagen in mesoglea has been shown to be structurally similar to the vertebrate collagens described in Chapter 3 (NORWIG and HAYDUK, 1969; GOSLINE, 1971*a*). The equilibrium modulus and the elastic recovery observed by Alexander indicate a long-range elastic mechanism, and as we have seen in Chapter 2B, such long-range elastic properties can be based on networks of random-coil polymer molecules. We can hardly describe the collagen fibre-lattice in these terms, so the elastic properties must arise from some component of the matrix. GOSLINE (1971*a*) reports that the matrix is a highly hydrated protein–polysaccharide complex with a weight concentration of about 2.5%, and GOSLINE (1971*b*) suggests that the matrix contains a dilute polymer network or gel which has extremely long random chains between crosslinks (*i.e.*, M_c is very large). Such a network can account for the observed elastic recovery and low equilibrium modulus. The 'slow elastic' properties have been attributed to a high molecular weight, viscous interaction which retards conformational changes within the polymer network. Thus, many of the viscoelastic properties observed by ALEXANDER (1962) and GOSLINE (1971*b*) can be attributed to this matrix gel. What then is the role of the collagen?

The answer is that the collagen acts to reinforce this flimsy matrix gel by contributing to the high molecular weight, viscous interaction that is responsible for the long time constants of this elastic system. It is possible to observe the effect of this reinforcement by comparing the properties of mesoglea tested in different directions. The outer layer of crossed-helices should respond similarly to extensions in the longitudinal and circumferential directions, as in both cases the collagen fibres will tend to align in the direction of extension (GOSLINE, 1971*a*). The

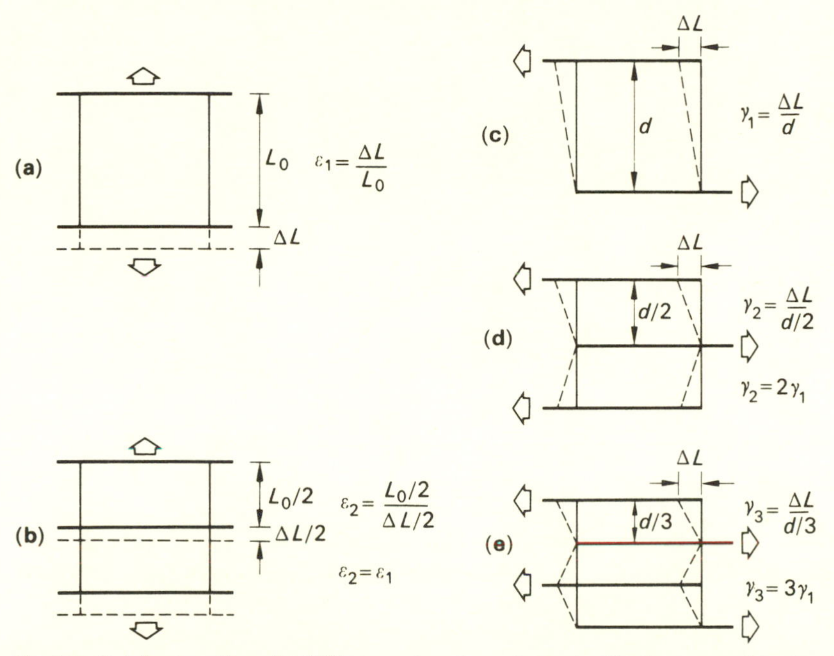

Fig. 4.9 Reinforcing effect of stiff fibres or plates in a soft matrix. This figure is explained in the text.

inner layer of densely packed circumferential fibres will respond differently in these two directions. Longitudinal extension will tend to move the collagen fibres apart laterally and deform the intervening matrix in tension while circumferential extension will tend to slide adjacent fibres along one another shearing the matrix material between them. These two cases are modelled in Fig. 4.9 by considering the deformation of a cube of a soft elastic material between solid plates (heavy lines). In the longitudinal extension the strain ϵ in the elastic material is unaffected by the addition of plates within the original cube. That is, when an additional plate is added (b) the strain in each of the two units formed is the same as the strain in the original cube (a), and the two units combine in series to produce a model with essentially the same properties as that of the original. In circumferential extension, however, the shear strain (γ) in each unit will increase as the thickness d of the unit decreases. Thus, plates added as in (d) and (e) will increase the strain in each unit, and the units will add in parallel to give a system which is considerably more rigid than the original (c). In fact, the force required to deform this series of models (c, d, e, . . .) a distance ΔL will increase as the square of the number of units. The same general arguments applied here to the strain in a soft elastic material between rigid plates can be applied to strain rate in a viscous or viscoelastic material. Again, the spacing will have little effect on the viscous forces that retard longitudinal extension, but close spacing will create substantial velocity gradients in circumferential extension which will increase viscous retardation.

We must be careful in applying these models to mesoglea because the discontinuous collagen fibre system in this tissue cannot be described in terms of rigid plates extending the entire width of the material. Still, we would expect the

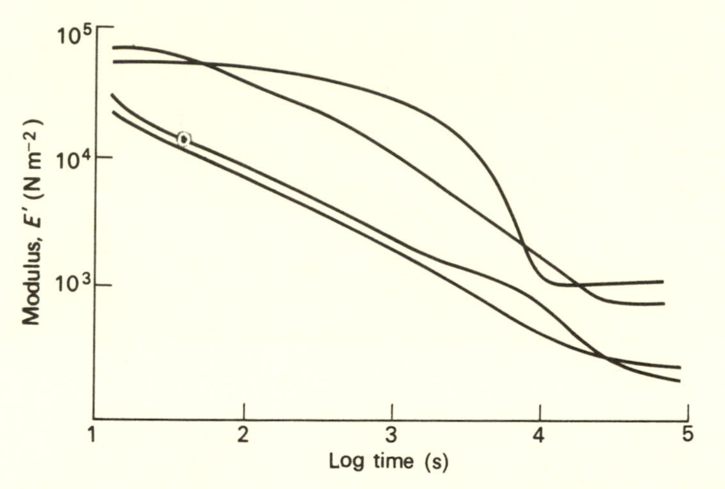

Fig. 4.10 Stress-relaxation curves for circumferentially stretched (upper curves) and longitudinally stretched (lower curves) mesoglea from *Metridium* (after GOSLINE, 1971*b*). Courtesy of the *J. exp. Biol.*

same sort of relationships to govern the deformation of mesogleal matrix between the collagen fibres, and thus we should expect mesoglea to be more rigid when tested circumferentially. Figure 4.10 shows the results of several stress-relaxation tests on circumferentially and longitudinally stretched mesoglea (GOSLINE, 1971*b*). Circumferentially stretched mesoglea is three to four times more rigid over the entire time scale than longitudinally stretched material. These findings are consistent with the hypothesis that the collagen fibres act to reinforce the dilute matrix gel by controlling the extent and the rate at which the matrix material is deformed.

4.6 Uterine Cervix

The uterine cervix provides another interesting but as yet poorly understood example of a pliant composite with a discontinuous fibre system. The resting circumference of the cervix in a non-pregnant rat is about an order of magnitude smaller than that of a mature rat foetus. HARKNESS and HARKNESS (1959; 1961) and HARKNESS and NIGHTINGALE (1962) have studied the changes occurring in this tissue that allow the foetus to pass through the cervix during parturition. Figure 4.11 shows the results of creep tests carried out under constant load on whole cervix from animals at different stages of pregnancy. Before pregnancy and through the first half of pregnancy, which lasts 21 days, the cervix behaves like a continuous-fibre composite in which extensibility is limited by the dimensions of the fibre system (curve A). During the second half of pregnancy there is a major reorganization of the tissue involving substantial tissue growth, an increase in the resting circumference, and the appearance of a viscous mechanism that allows the material to extend over a period of time under relatively small loads

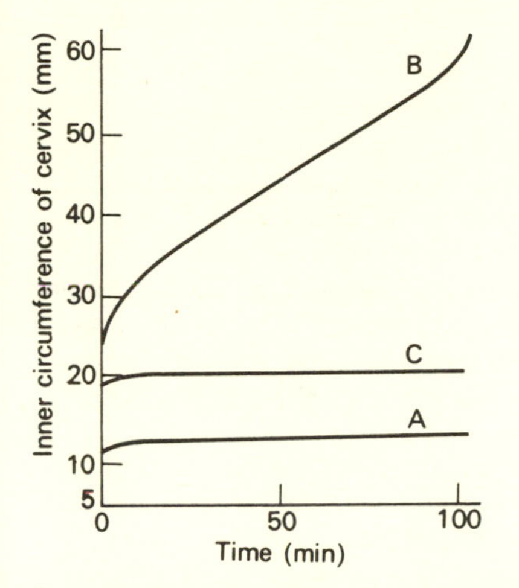

Fig. 4.11 Curves for creep under constant load of whole cervix from rats at (A) 12 days pregnant (load = 600 g), (B) 21 days pregnant (load = 50 g), (C) 24 h post partum (load = 300 g) (after HARKNESS and HARKNESS, 1959).

(curve B). Within hours after parturition, the mechanical properties of the cervix have returned to those of a continuous fibre composite (curve C), and in subsequent days tissue degeneration will return the cervix to its normal state.

The changes in extensibility can be explained in terms of changes in the crosslinks which maintain mechanical continuity within the collagen framework. As yet the nature of these crosslinks is unknown, but it is clear that at parturition the fibre continuity is substantially broken down. The viscous flow probably reflects the shearing of some high molecular weight matrix phase between the collagen fibres as they slide past one another. This interpretation is consistent with the observation that proteolytic enzymes increase the rate of extension under constant load. The enzymes presumably act on some component in the matrix phase and reduce the effective viscosity by lowering the molecular weight of the molecules involved.

Although it is tempting to suggest relatively simple alterations in the organization of this tissue that might account for the observed changes in mechanical properties, we must remember that these changes are associated with tissue growth before and degeneration after parturition. These processes must surely involve substantial reorganization of the tissue. The only phase of this cycle that lends itself to a simple analysis is the decrease in circumference and reduction in extensibility which occurs in the first 24 hours after parturition, as this process takes place before any substantial tissue degeneration takes place. The reduction in circumference has been attributed to the contraction of smooth muscles found in the cervix, but a passive elastic system based on a polymer network similar to that found in mesoglea may contribute as well. The decrease in extensibility must reflect a rapid increase in crosslinking within the collagen framework. Unfortunately the mechanism for this change is unknown.

4.7 Skin

The skin of birds and mammals functions as a boundary to limit the exchange of heat and material between an animal and its environment and to provide a protective barrier to shield the underlying tissues from external mechanical forces. The term skin refers to a complex tissue layer made up of, among other things, a thick collagenous connective tissue (dermis), a basement membrane, and an overlying keratinized epidermal layer. By and large the mechanical properties of interest to us are associated with the dermis (HARKNESS, 1968). However, the zone of weakness between the epidermis and the dermis may provide a mechanism for reducing damage from abrasions by allowing the epithelium to shear away leaving the rest of the skin intact.

The dermis can be described as a three-dimensional feltwork of continuous collagen fibres embedded in a protein–polysaccharide matrix. In addition, elastin fibres are present either distributed throughout the tissue or in some cases concentrated in the deeper layers of the dermis. This arrangement of elastin fibres within a collagen framework results in a material showing rubber-

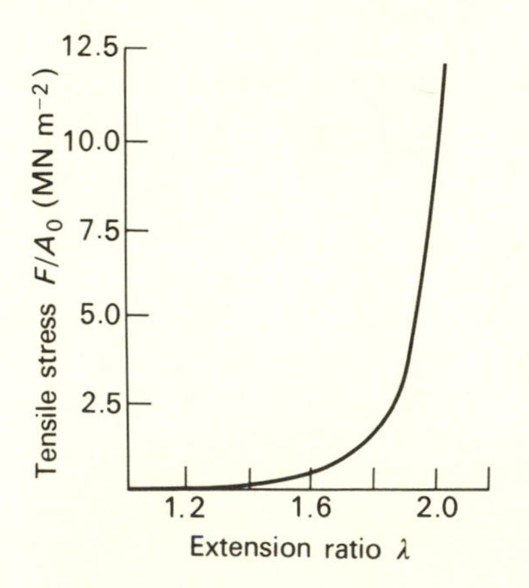

Fig. 4.12 Force-extension curve for cat skin loaded in uniaxial tension to failure (after VERONDA and WESTMAN, 1970). Courtesy of the *J. Biomech.*

elastic properties at relatively small extensions but is limited at longer extensions by the dimensions of the collagen framework (KENEDI *et al.*, 1966; VERONDA and WESTMAN, 1970). This relationship can be seen in Fig. 4.12, which shows a typical force-extension curve for cat skin loaded in tension to failure. The initial low modulus region of the curve (up to about $\lambda = 1.6$) is dominated by the elastin. Although extension in this region of the curve must result in an increase in the preferred orientation of the fibres in the collagen framework, the collagen fibres themselves are not being stretched. Thus, the load is carried primarily by the elastin. As the tissue is extended further, an increasing fraction of the collagen fibres will be brought into line with the direction of extension and be stretched by the applied load. By about $\lambda = 1.9$ the fibre lattice will have a very high degree of preferred orientation parallel to the direction of extension and the skin will become mechanically very similar to tendon.

The matrix phase of skin contains two major polysaccharide fractions, dermatan sulphate and hyaluronic acid (TOOLE and LOWTHER, 1966; HERP and DIGMAN, 1968). The dermatan sulphate is not readily extractable and appears to be tightly bound to the collagen fibre-lattice, while hyaluronic acid can be readily extracted and appears to occupy the spaces between the fibres. The function of dermatan sulphate is uncertain, but it has been observed that dermatan sulphate will cause tropocollagen molecules in solution to precipitate in the form of native collagen fibres (TOOLE and LOWTHER, 1968; TOOLE, 1969). This suggests that the dermatan sulphate may function to orient the collagen fibres during the synthesis of skin. Hyaluronic acid acts as a permeability barrier to large molecules and may act as a lubricant to reduce frictional wear between the collagen fibres (HARKNESS, 1970).

The tensile strength of skin appears to depend primarily on the size and crosslinking of the collagen framework. Evidence for this comes from two sorts of information (HARKNESS, 1970). First, specific chemical reagents and low pH conditions which are known to disrupt the crosslinks in collagen fibres drastically reduce the strength of skin. Second, the extension at which skin fails is independent of the degree to which the collagen crosslinks are broken down. That is, the tissue fails only when the load is applied directly to the fibres. FRY *et al.* (1963) studied the effect of age on the mechanical properties of rat skin. They observed that strength increased with age while extensibility decreased. The increase in strength is partly due to an increase in the collagen content of the skin but primarily to a large increase in crosslinking. The change in extensibility reflects a decrease in the dimensions of the fibre network which results directly from the increase in crosslinking.

The skin of an animal usually exists under tension even when the skin is not being deformed by some movement of the animal. For example, the piece of cat skin which was used to produce the force-extension curve in fig. 4.12 contracted when it was cut from the animal. The resting strain in the animal was found to correspond to an extension of $\lambda = 1.25$ (VERONDA and WESTMAN, 1970), well within the low modulus, elastin-dominated region of the curve. This apparent pre-stressing means that the skin in an animal is continually under tension and that the elastin fibres are laid down while under tension. It is possible to observe this pre-stressing in your own skin by placing two marks on your forearm and then moving them together by pressing on either side of the marks. The skin will remain flat as long as the elastin fibres are under tension but will fold when the tension is removed. The resting strain in human skin varies from about 10 to 30% (HARKNESS, 1968).

Although the fibre system in skin is an apparently random, three-dimensional feltwork, skin is by no means always isotropic or uniform over the surface of an animal. In terms of the force-extension curve of Fig. 4.12, the extent of the low modulus region may vary with the direction of extension at any point on the body or may vary greatly from point to point. In the living animal these kinds of differences must reflect preferred directions of extension which take place during the normal movements and major differences in the mechanical function of skin in different areas of the body. In man, for example, about 60% of the skin shows directional variation in extensibility (GIBSON *et al.*, 1969). This localized anisotropy indicates some degree of preferred orientation within the apparently random feltwork. These directional variations can be demonstrated by drawing a square on your skin and stretching it in different directions.

The skin of the hippopotamus provides an excellent example of the kind of variation occurring in skin from different parts of an animal. The back and side skin of this animal is an enormously thick and rigid layer that functions as a form of armour plate. The belly skin immediately adjacent to this is a much more pliant and extensible tissue that helps to hold in the massive viscera of the animal but that can expand with impact or with slower strain rates such as a large meal (HARKNESS, 1968; 1970). These differences can be explained qualitatively in terms of the limiting dimensions of the collagen framework, i.e. the extension at which an applied force acts directly to stretch the collagen fibres. The rigid side and back skin must be right at this limit as it shows no initial low modulus region. This then is an extreme case of increasing strength at the expense of extensibility by increasing the crosslinking in the fibre-lattice.

4.8 Arterial Wall

The large arteries function in the circulatory system as elastic tubes that expand during a pressure pulse and in so doing store elastic energy that can be used to smooth out the pulsatile flow of blood from the heart. Arterial tissue is mechanically similar to skin. That is, it can be described in terms of a long-range elastic element (elastin) arranged in parallel with a system of continuous collagen fibres that set a limit on extension. However, the actual tissue architecture is rather more complex than that of skin. There are three recognizable layers in an artery. (1) *Tunica interna*—a thin endothelium lining the inside of the artery. (2) *Tunica media*—a thick layer made up of alternating sheets of elastin (elastic laminae) and spaces containing collagenous tissues and smooth muscle cells. In addition to the obvious elastic laminae, there are numerous small elastin fibres forming interconnections between the elastic laminae (PEASE and PAULE, 1960). The *tunica media* makes up the bulk of the thickness of an artery and is primarily responsible for the mechanical properties in the physiological range of blood pressures. (3) *Tunica adventitia*—an outer layer of loosely arranged collagen and elastin fibres. It has been suggested that the adventitial tissues provide additional restraints which prevent bursting at abnormally high pressures (WOLINSKI and GLAGOV, 1964).

The organization of the various components in the *tunica media* varies a great deal in arteries from different regions of an animal. The elastin content is highest in the thoracic aorta and drops off quite rapidly in the abdominal aorta, the carotid and other more peripheral arteries (HARKNESS, 1968). The arrangement and proportions of smooth muscle also varies in different arteries. The large elastic arteries like the thoracic aorta have a relatively small fraction of smooth muscle. The muscle cells run obliquely from one elastic lamina to another and do not form a mechanically continuous ring of muscle around the artery. The muscle cells are also arranged obliquely with respect to the longitudinal axis of the artery and form discontinuous helical spirals in three dimensions (PEASE and PAULE, 1960). Contraction of the smooth muscle in elastic arteries will thicken the artery wall and reduce its internal circumference without increasing the stiffness of the wall (BERGEL, 1966). The distal arteries are considerably more muscular, and the muscle cells form mechanically continuous loops around the artery. Muscle contraction in these muscular arteries will both decrease the circumference and increase the stiffness of an artery (PEASE and MOLNARI, 1960).

Because of the complications imposed by smooth muscle contraction, most of the mechanical studies on artery have been carried out on large elastic arteries in conditions which block smooth muscle contractions. There are two general procedures which can be used to test arteries:

1 The artery can be left free and allowed to expand in length and circumference as pressure is increased, or

2 the artery can be stretched longitudinally and then with length held constant allowed to expand in circumference. As arteries in living animals are stretched longitudinally and do not change significantly in length during a pressure pulse (BERGEL, 1961*a*), the second procedure should be most useful in interpreting the mechanical properties of arteries in terms of their function in the animal.

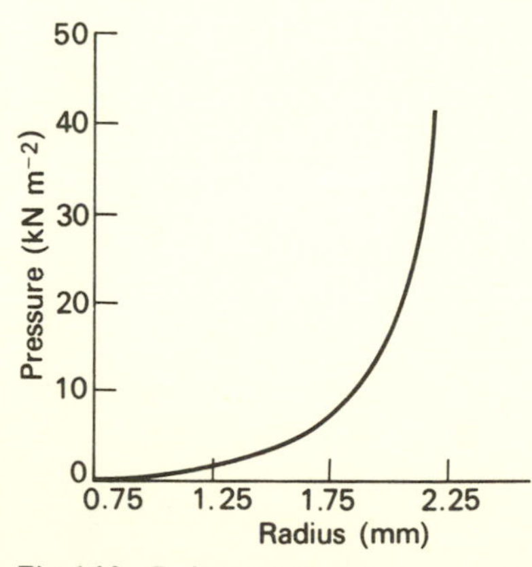

Fig. 4.13 Radius versus pressure curve for canine carotid artery. The artery had been treated with cyanide to inhibit smooth muscle contraction (after DOBRIN and ROVICK, 1969; courtesy of the *American Journal of Physiology*).

If an artery is tested at constant length (procedure 2) the artery will expand in radius with pressure in the manner shown in Fig. 4.13. This curve suggests that the arterial wall becomes more rigid as its radius increases, and that this increase in rigidity imposes a limit on expansion. Because of the particular geometry involved in testing a cylindrical tube with an internal pressure, the slope of the pressure–radius curve does not come close to indicating the rate at which modulus must increase with radius to give such a curve. The Law of Laplace states that the circumferential tension in the wall of a cylinder is equal to the product of the radius and the internal pressure ($T = R \times P$). Thus, even if the pressure were a linear function of radius the tension in the wall must increase as a function of R^2. Further, the stress in the wall material (stress = tension/wall thickness) must increase as a function of R^3, as the wall thickness decreases with increased radius. Thus, if pressure is a linear function of radius, the incremental Young's modulus (see BERGEL, 1961*a*) must increase with the third power of radius. The incremental Young's modulus ($E_{inc.}$) is somewhat different from the Young's modulus we have been using previously. Strain is defined as the change in length divided by the mean length over which the change took place rather than the change in length divided by an initial length. Thus,

$$E_{inc.} = \sigma/\epsilon = \Delta\sigma \Big/ \frac{\Delta R}{R} = \frac{\Delta\sigma}{\Delta R} \cdot R.$$

That is, $E_{inc.}$ equals the slope of a stress-radius plot times the radius at which the slope is taken, and if stress varies with R^3, $E_{inc.}$ will also increase with R^3. Figure 4.13 shows that the pressure in the canine carotid artery increases with the second or higher power of radius, and therefore the incremental modulus must increase with the fourth or higher power of radius. Experimental results (see Fig. 4.14) suggest that this is indeed the case. $E_{inc.}$ for the thoracic aorta increases as a function of R^4, and that for the carotid and several other arteries, increases with even higher powers of radius (BERGEL, 1961*a*). The importance of this rapid increase in modulus with radius can be understood by considering

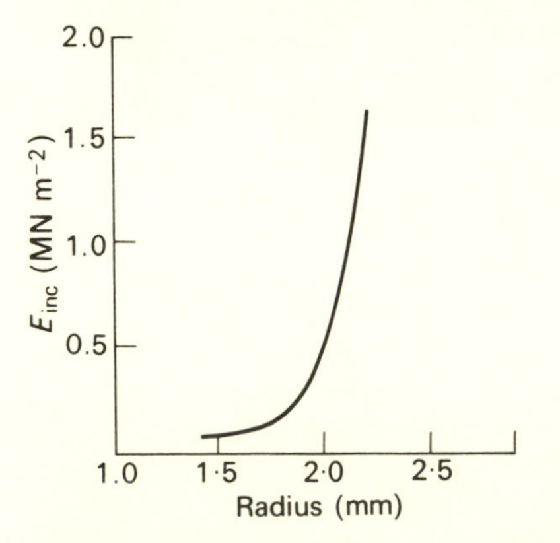

Fig. 4.14 The increase in the incremental modulus with radius of the canine carotid artery (after BERGEL, 1961*a*).

what would happen if modulus did not increase with radius. Under these circumstances some weak point in the artery will expand slightly further than the rest of the artery. Because of the increased radius at this point the stress in the wall will be higher than in adjacent areas and the tissue will continue to expand, further increasing the wall stress. This means that the weakest point will balloon out and burst while the rest of the artery remains unchanged. This does not happen in a real artery because a local hoop strain at a weak point will experience a rapidly increasing resistance to strain at this point. All points on the artery will then equally resist the wall tension and the artery will expand uniformly along its length.

The structural basis for this rapid increase in modulus has been attributed to the parallel arrangement of elastin and the collagen fibre-system. ROACH and BURTON (1957) showed that in artery, as in skin, the initial low modulus region is due to elastin, and the shift to higher modulus with extension reflects a transfer of the load to the fibre system. Figure 4.15 shows how $E_{inc.}$ varies with pressure

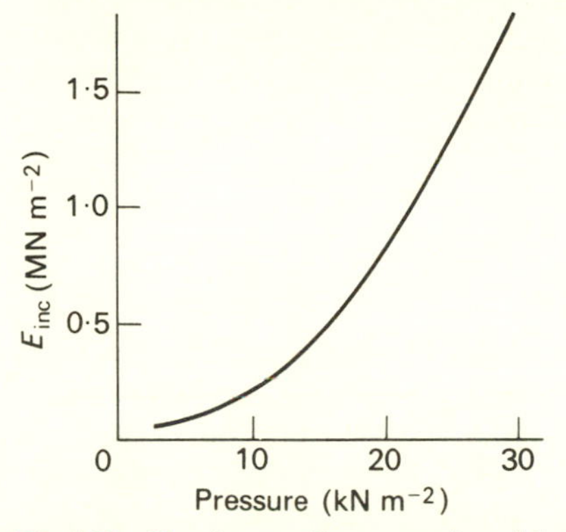

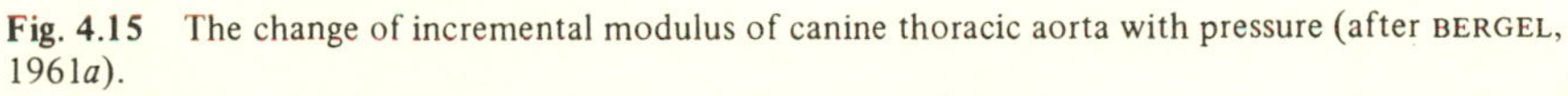

Fig. 4.15 The change of incremental modulus of canine thoracic aorta with pressure (after BERGEL, 1961*a*).

in the thoracic aorta. Over the physiological range of pressures (11 to 21 kN m^{-2}) the modulus increases rapidly with increased pressure. This suggests that under normal physiological conditions the tension in an artery wall is carried by both the elastin and the collagen. More recently WOLINSKI and GLAGOV (1964) studied the structural changes that take place when an artery is allowed to expand in circumference with the length held constant. Transverse sections of artery fixed at low pressures (less than 10 kN m^{-2}) show the elastic laminae to have a wavy appearance. The fine, interlaminar elastin fibres are arranged in a predominantly radial direction. As the internal pressure is increased, the wavy appearance of the laminae disappears, until at a pressure of 10 kN m^{-2} they are straight. At the same time the interlaminar fibres become aligned circumferentially, and at a pressure of about 10 kN m^{-2}, they have a very high degree of preferred orientation in the circumferential direction. At this pressure the collagen fibres also have a high degree of preferred orientation in the circumferential direction, but they are, in fact, arranged in uniform helices of very small pitch. These observations suggest that the fine interlaminar fibres are responsible for the initial low modulus region, and that the rapid increase in modulus reflects the recruitment of the elastic laminae as well as the collagen framework as load-bearing elements.

Studies have also been carried out on the dynamic properties of arterial wall by applying sinusoidally varying pressure pulses to an artery (BERGEL, 1961*b*; LEAROYD and TAYLOR, 1966) or by following changes in the normal pressure pulses in a living animal (GOW and TAYLOR, 1968). In general, the results obtained are as follows. At low frequencies (below about 1 Hz), the dynamic modulus is essentially the same as the static incremental modulus measured at the mean pressure used for the dynamic measurements. Damping is very low. At frequencies around 1 Hz, modulus and damping increase abruptly, but above about 3 Hz both modulus and damping remain nearly independent of frequency. The extent of the increase in the dynamic modulus between 1 and 3 Hz can be directly correlated to the amount of smooth muscle present in the artery. Thus, the thoracic aorta shows little change, and the abdominal aorta, cartoid artery and other peripheral arteries show relatively larger changes. The damping above 3 Hz ranges from about 0.1 to 0.15 and is the same in all types of artery. These results suggest that the hysteresis in arteries at frequencies above about 1 Hz is due to smooth muscle. The results of these dynamic measurements have been used to predict some of the fluid-dynamic parameters of the circulatory system (see BERGEL and SCHULTZ, 1971). However, this is beyond the scope of this book.

4.9 Cartilage

Cartilage could, on the basis of its chemical composition, be classified along with the pliant composites just discussed. That is, cartilage contains a collagen fibre system, a proteoglycan matrix phase, and in some cases elastin fibres as well. However, we regard cartilage as something rather different because the mechanical function and the molecular organization of cartilage are really quite unique. Functionally cartilage acts as a material which maintains the shape of things like ears, the nose, intervertebral disc, etc., and in so doing cartilage must to some extent be able to resist compression and bending forces. In this sense cartilage resembles the rigid materials described in Chapter 5. However, the actual modulus and extensibility of cartilage falls within the range we associate with pliant materials. The molecular organization of cartilage which accounts for the mechanical properties depends to a large extent on the osmotic properties of acidic proteoglycans. Cartilage can be thought of as a hydrostatic system in which the fluid element is provided by the hydration water of the proteoglycan gel and the container provided by a collagen fibre meshwork which immobilizes the molecules of this gel. Thus, the rigidity or turgor of this hydrostatic system arises from the osmotic swelling of the proteoglycan gel against the constraints imposed by the collagen fibre system. This type of fibre-matrix interaction is quite different from that associated with either the pliant composites or the rigid materials.

BARRETT (1968) and the three volume treatise edited by BALAZS (1970) provide access to the extensive literature on the composition, structure, biosynthesis, pathology, etc., of cartilage. We will restrict our discussion here to the macromolecular organization and mechanical properties. As mentioned above, cartilage is very rich in proteoglycans, with chondroitin-4-sulphate and chondroitin-6-sulphate making up the major fraction of the polysaccharides present in the vertebrates. Cartilage is also found in a number of invertebrate animals such as the horseshoe crab *Limulus* and the brain case of cephalopods, and the polysaccharides of these invertebrate tissues are similar but not always identical to those of vertebrate cartilage (PERSON and PHILPOTT, 1969). Although it is possible to break cartilage down into its constituent small molecules, this kind of approach gives very little information about the macromolecular organization of the tissue. It has been found, however, that extraction with neutral salt solutions will yield as much as 50% of the dry weight in the form of soluble macromolecular components if the tissue is thoroughly homogenized. This type of extraction yields a mixture

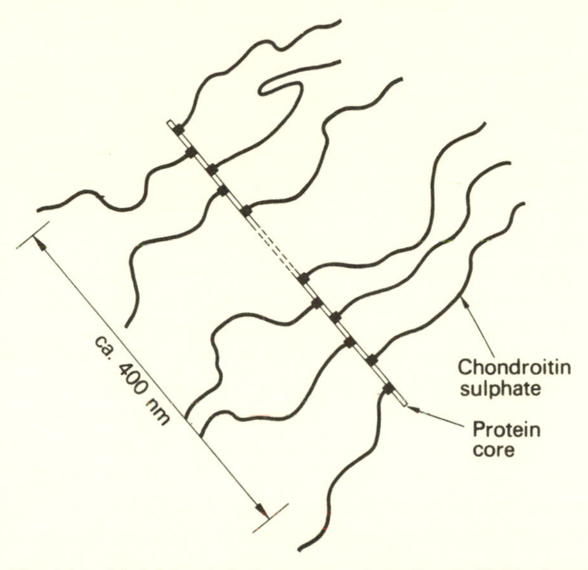

Fig. 4.16 Model for the light protein polysaccharide (PPL) of cartilage proposed by MATHEWS and LOZAITYTE (1958; reproduced from *Arch. Biochem. Biophys.*).

of proteoglycans that fall into two general classes. One is a fraction of relatively small molecules (PPL for light protein polysaccharide) which contains about 15% protein and the rest acid polysaccharide. The other fraction (PPh for heavy protein polysaccharide) appears to contain aggregates of PPL along with a substantial amount of collagen (SCHUBERT, 1966).

The structure of PPL appears to be quite different from that of the proteoglycan found in synovial fluid, where the hyaluronic acid molecules are single linear molecules of extremely high molecular weight. PPL, on the other hand, contains a large number of relatively short polysaccharide chains branching off from a central protein core. Figure 4.16 shows the model proposed by MATHEWS and LOZAITYTE (1958) and MATHEWS (1965) for the structure of PPL. The protein core is seen as having an extended length of about 400 nm with from 60 to 100 chondroitin sulphate chains branching off from the backbone. Each of the chondroitin sulphate chains has a molecular weight of around 50 000 and an extended length of roughly 100 nm. The molecular weight of the complex as a whole is of the order of 2–4 x 10^6. The high negative charge density of the chondroitin sulphate molecules causes these chains to stick out radially from the central core so that the molecule encompasses a huge domain. Light scattering data indicate that PPL in dilute solution has a roughly spherical domain with an average radius of gyration of 140 nm (LUSCOMB and PHELPS, 1967).

The PPH fraction is a complex mixture of particles containing proteoglycan molecules in association with collagen, where the molecular weight of the particles can be of the order of 5 x 10^7. Several studies have shown that the association between PPL and collagen is stabilized by electrostatic interactions between the acid polysaccharides of the proteoglycan and basic groups on collagen (MATHEWS, 1965; ÖBRINK and WATESON, 1971). The exact arrangement of the proteoglycan and collagen is at present uncertain, but the electron microscope study by SERAFINI-FRACASSINI and SMITH (1966), suggests that the interaction is in the form of proteoglycan molecules lying transversely across collagen fibres. Recent evidence (HASCALL, 1972) also indicates that proteoglycan molecules themselves can be linked together through a glycoprotein link molecule to form long linear aggregates. In general it is thought that the proteoglycan gel in the intact tissue is immobilized within a meshwork of collagen fibres, and that this arrangement forms the basis for the structure of cartilage.

The mechanical rigidity of cartilage appears to result from the water binding properties of the proteoglycans immobilized by the fibre mesh. DISALVO and SCHUBERT (1966) made a model cartilage system by precipitating PPL with soluble collagen and obtained elastic pellets which contained up to 100 times more water than an equivalent pellet made from collagen alone. OGSTON (1970) has calculated that the osmotic pressure difference between cartilage proteoglycan and physiological saline is of the order of 150 milliosmolar. Most of the osmotic pressure arises from the Gibbs-Donnan distribution of the counter ions to the numerous fixed charges on the chondroitin sulphate chains. Thus, the osmotic forces cause the proteoglycan gel to expand against the collagen fibre framework or against an applied load with an 'internal' pressure of about 0.35 MN m^{-2}.

If the rigidity of cartilage does indeed arise from the osmotic swelling of the proteoglycan gel within a collagen framework, it should be possible to modify the properties of this system by altering the magnitude of the osmotic forces or by altering the collagen framework. SOKOLOFF (1963) found that the mechanical properties of cartilage could be altered by changing the concentration of the solution with which the cartilage was equilibrated. Free cations can bind tightly to the negatively charged groups on the chondroitin sulphate chains and reduce the osmotic pressure of the proteoglycan gel. Sokoloff found that cartilage became less rigid in strong NaCl solutions. Divalent (Ca^{++}) and trivalent (La^{+++}) cations were more effective in reducing rigidity than monovalent cations, and rigidity was increased when free ions were removed by washing with distilled water. These results are fully compatible with the hypothesis that osmotic forces are responsible for the mechanical rigidity of cartilage (see KATCHALSKY, 1964). REYNOLDS (1966) observed that cartilage grown in tissue culture was more highly hydrated than normal cartilage. This increased hydration was correlated with a reduction in the relative amount of collagen in the cultured tissue. If ascorbate was added to the culture medium, the cartilage produced the normal amount of collagen and there was no excess hydration. This suggests that in normal cartilage the proteoglycan gel is underhydrated and that there is a positive swelling pressure acting against the constraints imposed by the collagen framework.

4.10 Mechanical Properties of Cartilage

The mechanical properties of cartilage reflect the hydrostatic organization of this tissue. Figure 4.17 shows the creep response of human articular (joint) and costal (rib) cartilage under constant compressive load (SOKOLOFF, 1966). This particular experiment is an indentation test where the load is applied over a small region of the sample, and strain is determined from the depth of the

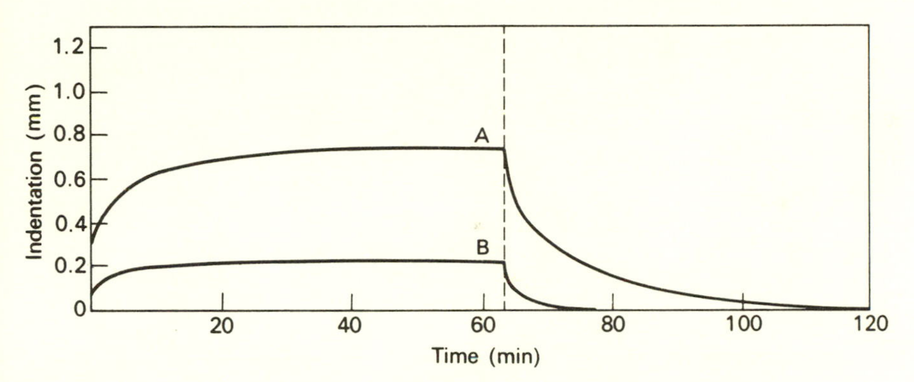

Fig. 4.17 Indentation under constant load (84 kN m^{-2}) of human articular (A) and costal (B) cartilage (after SOKOLOFF, 1966; reprinted from Federation Proceedings, 25, 1089-1095).

indentation formed. The costal cartilage is considerably more rigid than the articular cartilage, but the shape of the creep curves appear to be the same for both. There is an initial rapid deformation followed by a gradual deformation over a period of several minutes until after 30 to 40 minutes an equilibrium value is reached. The initial and equilibrium moduli calculated by Sokoloff are given below:

	Initial Modulus	Equilibrium Modulus
Articular cartilage	2.4 MN m^{-2}	0.7 MN m^{-2}
Costal cartilage	7.8 MN m^{-2}	5 MN m^{-2}

The shape of these curves has been attributed to the fact that in the short term cartilage acts as a constant volume hydrostatic system, but in the long term the volume of the system can change in response to the existing load (SOKOLOFF, 1966). Thus, the initial high modulus reflects the effect of all the fluid bound in the gel acting against the collagen fibre meshwork. However, in localized regions the hydrostatic pressure may exceed the osmotic swelling pressure of the proteoglycan gel, and water will flow away from these regions until the system reaches osmotic equilibrium. McCUTCHEN (1966) suggested that this squeezing out of fluid from the cartilage may act to maintain a lubricating fluid-film between the articular cartilage surfaces in synovial joints. SWANSON and FREEMAN (1970), however, point out that the flow of fluid will be away rather than toward points of contact between articular surfaces and suggest that the remaining thin film of hyaluronic acid acts to lubricate the joint. When the load is removed there is a quick elastic recovery followed by a slower influx of water that rehydrates the proteoglycan gel.

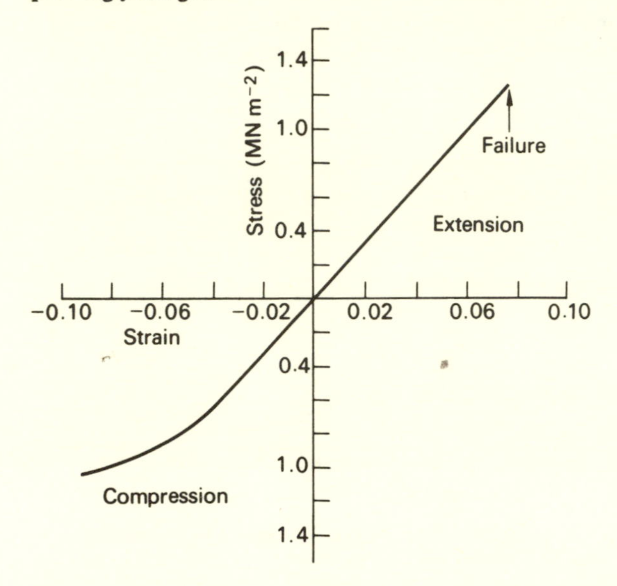

Fig. 4.18 Stress-strain curve for human costal cartilage. Young's modulus in the linear region is 16 MN m^{-2} (after KENEDI *et al.*, 1966; reprinted from Federation Proceedings, **25**, 1084-7).

Figure 4.18 shows a stress-strain plot for human costal cartilage in tension and compression. The values plotted in this figure are equilibrium measurements taken after 60 minutes of stress-relaxation (KENEDI *et al.*, 1966). The Young's modulus in the linear region of this curve is about 16 MN m^{-2}. The tissue fails in tension at a strain of about 0.08. The stress-strain curve in compression is quite linear up to a strain of about 0.06. Although the material does not fail at this strain, the modulus drops off with increased strain beyond this value, and it is probable that this decrease in modulus sets the limit for the useful biological range. This means that cartilage loaded in compression will support a load of the order of 1 MN m^{-2}. It should be remembered that these are equilibrium figures. The instantaneous modulus which would apply for sudden jolts or shocks will be much larger, perhaps by as much as an order of magnitude or more.

One of the interesting consequences of the hydrostatic organization of cartilage is that the collagen fibre system is subjected to a constant tensile stress which arises from the swelling pressure of the proteoglycan gel. Quite frequently the collagen fibre system is concentrated near the surface of cartilage, and if the fibres in this outer layer are cut in an unsymmetrical manner the internal pressure will distort the cartilage sample (GIBSON and DAVIS, 1958; ABRAHANS and DUGGAN, 1964). This distortion is very similar to that observed when a balloon which has tape applied to one side is inflated. The untaped surface will expand more rapidly than the taped side, and the balloon will tend to curve around the taped surface. The importance of stressed cartilage in the integrity of whole animals is discussed in Chapter 7.

Chapter 5
Rigid materials

5.1 Introduction

In Chapters 2A and 2B we discussed, in rather general terms, two classes of materials behaviour: linearly elastic and viscoelastic. Chapters 3 and 4 have dealt, in more detail, with the properties of specific viscoelastic biomaterials. We noted there that, in general, these materials have elastic moduli and strengths which are moderate to low but, at the same time, they possess considerable ductility and time dependent recovery enabling them to accommodate to or to recover from loading conditions that may be applied over long periods of time.

The rigid materials, which form the substance of this chapter tend to show linearly elastic behaviour – at least over the normal working range of load. The basic components of them are, essentially, ceramics (usually oxides, carbonates and/or phosphates) and are covalently bonded. In single crystal form such materials are stiff and strong–that is to say their moduli and strengths are at least one or two orders of magnitude greater than those of the viscoelastic materials. Thus they are, obviously, the right materials to use for supportive and aggressive elements. But, to be offset against high stiffness and strength is the problem of brittle fracture initiated at stress concentrations and controlled, substantially, by the Griffith criterion (Section 2.8).

Now fast unstable fractures, coupled with an inability to absorb much energy, are clearly undesirable in a structural element which is supportive or aggressive. It may come as no surprise to learn that the rigid biomaterials have evolved a strategy to cope with this deficiency. In simple terms this strategy involves no more than combining those linearly elastic and viscoelastic materials as may be available in such a way as to maximize the benefits of both. Such materials are called *composites* and we shall first discuss the theory of simple composites.

5.2 Limiting Behaviour of Composite Materials

The basic theory of composites has been discussed by several authors and comprehensive reviews are available: BIGGS (1966), FORSYTH (1965), HOLLISTER and THOMAS (1966). Here we shall only summarize the main features. We consider two, complementary, approaches.

The first is due to MAXWELL (1873) and considers two limiting cases as shown in Fig. 5.1. In Fig. 5.1a each element of the composite lies parallel to the external stimulus (which may be force, emf, magnetic field, etc.) so that the response of each individual element must be equal in sign and in magnitude if compatibility between layers is to be preserved. Thus, if we consider the elastic modulus of the composite E_c as the property of interest we obtain

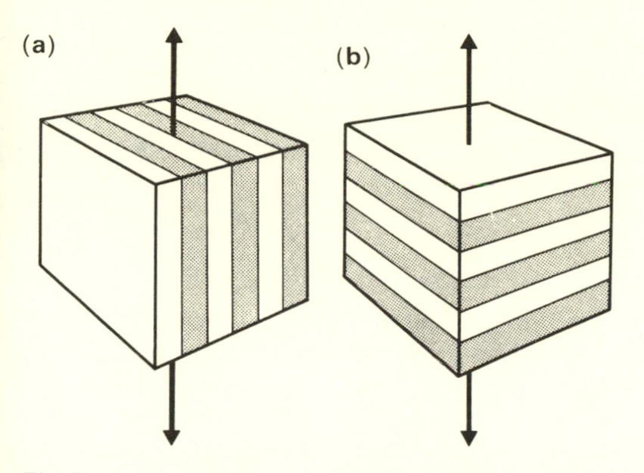

Fig. 5.1 Limiting cases for behaviour of composites (MAXWELL, 1873). These are also called (**a**) equal strain (Voigt) and (**b**) equal stress (Reuss) models (see Section 5.2).

$$E_c = E_1V_1 + E_2V_2 + \cdots + E_rV_r \tag{5.1}$$

where V is the volume fraction and the suffixes 1, 2, ..., r represent the various components. The second limiting case (Fig. 5.1b) occurs when the elements are in 'series' relative to the applied stimulus–each element now carries the same stress but the strains in each may differ depending upon the individual moduli. For this case

$$\frac{1}{E_c} = \frac{V_1}{E_1} + \frac{V_2}{E_2} + \cdots + \frac{V_r}{E_r} \tag{5.2}$$

with the same notation as before. For a simple two phase system ($V_1 + V_2 = 1$) these two equations give upper and lower bounds as shown in Fig. 5.2.

The second approach derives from EINSTEIN'S (1906) expression for the properties (actually, originally, the viscosity) of a liquid containing suspended particles

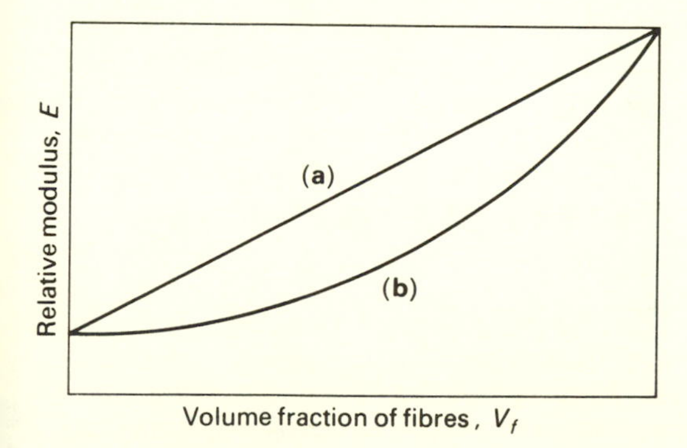

Fig. 5.2 Typical upper and lower bounds for elastic modulus of composites. Lines (**a**) and (**b**) correspond to the limiting cases shown in Fig. 5.1a and b.

$$\eta = \eta_0(1 + kV)$$

where η_0 is the viscosity of the matrix liquid and k is a non-dimensional constant. This expression can, in fact, be applied to many other physical phenomena and, in terms of the elastic modulus, we may rewrite it as

$$E_c = E_0(1 + kV) \tag{5.3}$$

The numerical value of the constant k depends upon the elastic constants of both the matrix and the particle and also upon the shape of the particle—for further details see, JEFFREY (1923), FRÖHLICH and SACK (1946), HASHIN (1955), MACKENZIE (1950).

The two approaches are complementary in the sense that the one can be expressed in the same form as the other. Thus, for a two phase system $V_1 + V_2 = 1$ and, if we let the modular ratio $E_1/E_2 = M$ we obtain

$$\left.\begin{array}{l} E_c = E_2(1 + k_u V_1) \\ \text{and} \\ E_c = E_2(1 + k_l V_1) \end{array}\right\} \tag{5.4}$$

where the 'upper bound constant' $k_u = M - 1$ and the 'lower bound constant' $k_l = 1 - (1/M)[1 - V_1(1 - 1/M)]$ are appropriate for a plot such as that shown in Fig. 5.2.

Most materials fall between these limits and it is often convenient to combine *Eqs. 5.1* and *5.2* (DOUGILL, 1962)

$$\frac{1}{E_c} = x\left(\frac{1}{V_1 E_1 + V_2 E_2}\right) + (1 - x)\left(\frac{V_1}{E_1} + \frac{V_2}{E_2}\right) \tag{5.5}$$

where x represents that proportion of the material which obeys the upper bound solution and $(1 - x)$ that which obeys the lower bound. This expression appears not to have been generally applied to biomaterials though PIEKARSKI (1973) has suggested that, for cortical bone $x = 0.925$. Note, however, that this merely means that 92.5% of bone substance *behaves* as if it were coupled in 'parallel'—it must not be assumed that the components are actually arranged in this way and in these proportions. The dangers of an over-literal interpretation of models have already been pointed out.

Note also that many of these solutions can only be approximate. Thus, rewriting *Eq. 5.5* in the form

$$E_c = x\,[E_2(1 + k_u V_1)] + (1 - x)\,[E_2(1 + k_l V_1)] \tag{5.6}$$

we note that, for a porous material where $E_2 = 0$, *Eq. 5.6* reduces to zero—which is manifestly absurd. Nonetheless the boundary approach described above is a useful one for considering, in a preliminary way, the relative contributions of each of the individual elements in the composite.

5.3 Elastic Fibres in a Matrix

This is, theoretically, the simplest case for it assumes that the fibres are long, oriented parallel to each other and that the fibre material is linearly elastic up to the point of fracture. The matrix, on the other hand, is considered to be linearly elastic up to some value of stress at which point (the yield point) it starts to deform plastically (i.e., non-recoverable deformation). The stress–strain curves of such materials are illustrated schematically in Fig. 5.3.

Within the linear range of both components, the elastic modulus of the composite is given by *Eq. 5.1*

$$E_c = E_f V_f + E_m V_m \quad \text{where } V_m = 1 - V_f \tag{5.7}$$

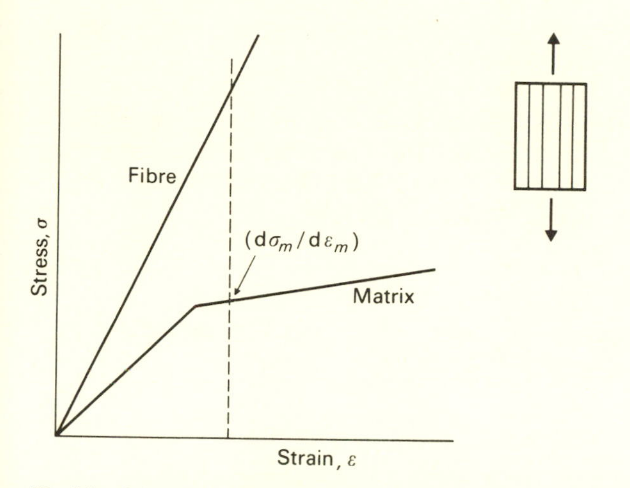

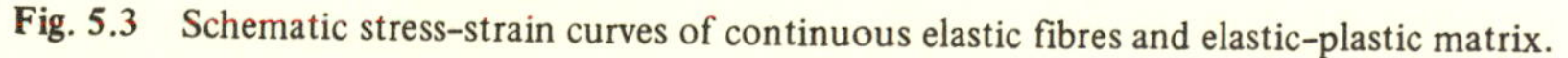
Fig. 5.3 Schematic stress-strain curves of continuous elastic fibres and elastic-plastic matrix.

where the suffixes c, f and m refer to composite, fibre and matrix respectively and V is the volume fraction.

If stressing is continued beyond the yield point of the matrix the modulus is now

$$E_c = E_f V_f + (d\sigma_m/d\epsilon_m)\, V_m \tag{5.8}$$

where $(d\sigma_m/d\epsilon_m)$ is the slope of the stress–strain curve of the matrix at a strain ϵ in the composite. Note that, on unloading from this stage of deformation, there is likely to be some residual deformation. This will be recovered in time if the matrix is viscoelastic but not if the matrix is plastic. For many practical purposes $(d\sigma_m/d\epsilon_m)$ is small compared with E_m so that the contribution of the matrix can be ignored and *Eq. 5.8* reduces, effectively, to $E_c = E_f V_f$.

The strength of the composite follows a similar law–since the fibres will break at a lower strain than the more deformable matrix we find that

$$\sigma_c = \sigma_f V_f + \sigma'_m(1 - V_f) \tag{5.9}$$

where σ_f is the ultimate strength of the fibres and σ'_m is the stress in the matrix when the fibres break.

However, it is important to note that the reinforcing effect of the fibres is critically dependent upon their volume fraction.

Clearly if the strength of the composite is to exceed that of the matrix (i.e., if the fibres are to act as strengthening agents) we must have

$$\sigma_c = \sigma_f V_f + \sigma'_m(1 - V_f) > \sigma_m \tag{5.10}$$

where σ_m is the strength of the matrix. Rewriting

$$V_{crit} = \frac{\sigma_m - \sigma'_m}{\sigma_f - \sigma'_m} \tag{5.11}$$

where $(\sigma_m - \sigma'_m)$ is the difference between the strength of the matrix and the stress in the matrix when the fibres fracture. For volume fractions less than V_{crit}, failure of the composite occurs when failure of all the fibres throws the entire force upon the matrix alone and the strength is given by

$$\sigma_c = \sigma_f V_f + \sigma'_m(1 - V_f) > \sigma_m(1 - V_f) \tag{5.12}$$

so that we obtain a minimum volume fraction which must be exceeded if the strength of the composite is to be given by *Eq. 5.9.*

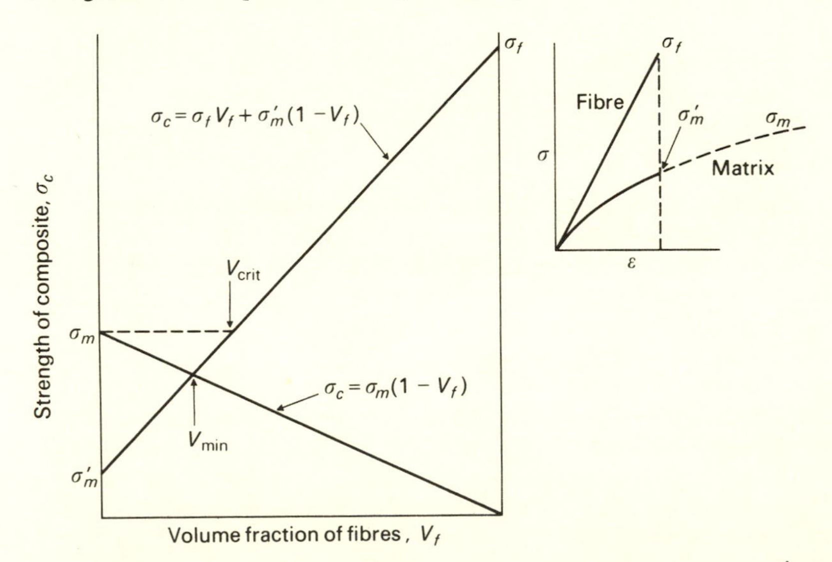

Fig. 5.4 Theoretical variation of strength of composites. σ_m, strength of the matrix; σ'_m, stress in the matrix when fibres fracture; σ_f, strength of the fibres; σ_c, strength of the composite. See explanation in the text.

$$V_{\min} = \frac{\sigma_m - \sigma'_m}{\sigma_f + \sigma_m - \sigma'_m} \qquad (5.13)$$

These relationships are plotted in Fig. 5.4. As an example we may consider bone. There are no useful data for the strength of hydroxyapatite but, assuming it to be perfect (i.e., containing no Griffith cracks) it cannot be far different from 100 MN m^{-2}. Given that the modulus of apatite is ~ 137 GN m^{-2} this suggests a fracture strain of ~ 0.001. From Fig. 3.18 the stress in the collagen at this strain is about 1 MN m^{-2} while the fracture strength of collagen is about 50 MN m^{-2}. Inserting these rough values into *Eq. 5.13* suggests that, in bone, a minimum volume fraction of about 0.35 mineral is needed in order that the mineral should act as a true reinforcement. Since the volume fraction of mineral in bone is close to 0.5 we see that the hydroxyapatite is truly the principal stress bearing element, although not perhaps to the extent one might expect. Collagen in bone appears to be considerably stronger than collagen in, say, rat tail tendon. The reasons for this remain a source of some puzzlement and, as yet, no really satisfactory explanations have been given.

5.4 Discontinuous Fibres

The foregoing discussion was limited to the case of continuous fibres which are long enough to ensure that the force is equally shared between the fibre and matrix. Although a rather special case, this sort of situation is not so uncommon in biomaterials as it is in engineering practice where considerable care (and expense) is needed to achieve an oriented distribution. Obvious examples are the collenchyma in nonwoody stems of higher plants, chitin fibrils in arthropod cuticle, cellulose fibrils in wood—in all of these the fibres are long and well oriented. Indeed arthropod cuticle, as we shall see in Section 5.10, must come close to an engineer's dream of a composite. But many materials such as bone and mollusc shell consist of short fibres and, providing that these are aligned in a parallel array, the differences from the behaviour described above may be summarized quite simply.

(a) In a short fibre the distribution of stress along the fibre is not uniform—it must build up from the end of the fibre as the tensile stress in the matrix is transmitted via a shearing stress at the interface between the fibre and the matrix. Full solutions have been given by various authors (COX, 1952; DOW, 1963) but these are not exact. For the present purpose the distribution of stress shown in Fig. 5.5 will serve.
(b) Figure 5.5 shows that, at each end of the fibre there is a critical length over which the tensile stress in the fibre can build up to a maximum value which is, of course, the fracture strength of the fibre. This is known as the *transfer length* and is given by

$$l_c = r\sigma_f/\tau \quad \text{or} \quad l_c/2r = \sigma_f/2\tau \qquad (5.14)$$

where r is the fibre radius and τ is the shear stress. The ratio $l/2r$ (= l/d) is known as the *aspect ratio* of the fibre and it is clear that, if the aspect ratio is large,

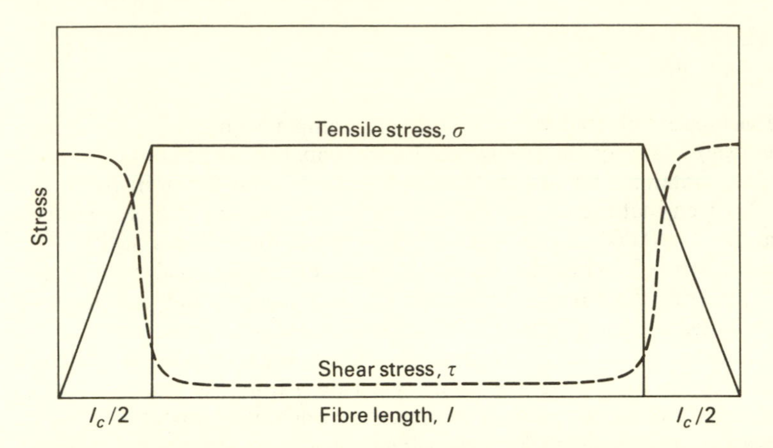

Fig. 5.5 Schematic representation of the distribution of tensile and shear stress in a fibre.

then a fibre of given breaking strength will only need to sustain low shear stresses at the interface.
(c) The average stress in a fibre stretched almost to its breaking point in the central portion is given by

$$\bar{\sigma} = \sigma_f(1 - l_c/2l) \tag{5.15}$$

and, since this is obviously less than σ_f it appears that a composite made of discontinuous fibres must always be weaker than one made of continuous fibres, though the strength will approach that of the continuous composite as the aspect ratio of the fibres increases. KELLY (1964) has shown that in order to achieve 95% of the strength of the continuous fibre composite, the aspect ratio of the discontinuous fibres needs to be about ten times the transfer length for the same volume fraction of fibres.

These observations may well be significant in deciding the influence of spicules in a viscoelastic matrix. We shall discuss this further in Section 5.19.8. but we may note here that, in general, the aspect ratio of spicules is considerably less than the above considerations might suggest, so that it seems likely that, unless they are present in very considerable quantity, spicules do not exert much strengthening effect upon the matrix. However, this is an area where much further research is needed before any firm conclusions can be drawn.

5.5 Effect of Fibre Orientation

The discussion above has been limited to those cases where the fibres, either continuous or discontinuous, are oriented along the direction of the principal tensile stress and it is at once clear that, for tensile loading, such systems possess great advantages. These advantages are displayed in chitinous and collagenous tendon, silk, collenchyma, etc. Similar advantages are obtained when continuous fibres are used as a winding for reinforcing vessels under internal or external pressure–these will be discussed in Chapter 7.

But the disadvantages of unidirectional reinforcement become only too apparent when we consider what happens when a tensile stress is applied at right angles to the fibres. In such a case the fibres contribute little or nothing to the strength and stiffness and the properties of the matrix alone dictate the deformational behaviour of the composite. Thus it is common experience that timber has very different properties across the grain from those in the grain direction and Fig. 5.16 shows that the same is true of bone.

In general, such anisotropy is not easily tolerated as most structural members must sustain a fairly complex loading system and, in fact, simple tension and simple compression are comparatively rare. In any event, as we shall see in Chapter 6, the shape requirements for these types of loading are rather different and it is uneconomical of material to make both short, fat compression members and long thin tension members of the same material. Engineers do this as a matter of convenience, but in critical designs where, say, weight is important, the use of different materials often provides the more effective solution. The mast and stays of a racing dinghy are a familiar example.

The most serious problem, and the one that uses the most non-optimum material, is that of transferring the force smoothly from one member to another. One needs only a small amount of material to make the structure of, say, a rope or liana bridge, but if the knots are to be as strong as the rope, one needs to use about twice as much material. The problem is that, at points where the force must be guided into the member, complex stress patterns are set up and the reinforcement must be arranged in such a way as to cope with these. But apart from this the normal service requirements also impose complex stress patterns. Thus the tibia of an arthropod at rest may be simultaneously under compression (due to self weight), bending (due to the angular relationships between the body, the leg and the ground) and, in many cases, also under internal pressure. Such situations preclude, or at least limit, the use of materials that are stiff and strong in one direction only.

Two strategies are available in the search for isotropy of properties. The first is to use the 'plywood' technique in which a series of unidirectionally reinforced laminae are combined with an angular difference between each. This is admirably exemplified by arthropod cuticle. The second is to arrange the fibres in a two or three dimensionally random orientation so that, for loading from any direction, some fibres are always correctly oriented. Woven bone approximates to this situation. The disadvantages of both are at once obvious as, for a given loading, only a proportion of the fibres are doing any real work–the rest are adding extra weight but are contributing little to the strength or stiffness, which fall off considerably as the structure approaches isotropy.

The problem has been tackled analytically by COX (1952), GORDON (1952), KRENCHEL (1964) and others. The precise results differ slightly according to the initial assumptions. For the present purpose we only discuss one which will serve to show the sort of results which are obtained.

KRENCHEL (1964) defined a term η, the 'efficiency of reinforcement' which is given by

$$\eta = \Sigma\, a_n \cos^4 \phi \qquad (5.16)$$

Table 5.1 Efficiency of reinforcement, η, ignoring Poisson effect (KRENCHEL, 1964).

Fibre orientation	Stress direction	η
1 All fibres parallel	Parallel to fibres	1
	Perpendicular to fibres	0
2 Fibres in two directions, proportions a_1 and a_2 perpendicular to each other	Parallel to a_1 or a_2 fibres	a_1a_2
	Angle $\pi/4$ to fibres	1/4
3 Four equal layers or groups of fibres at $\pi/4$	Parallel to any one group or layer	3/8
	Angle $\pi/8$ to any one group or layer	3/8
4 Fibres uniformly distributed in plane	Any (in plane)	3/8
5 Fibres randomly distributed in 3 directions in space	Any	1/5

where a_n is the proportion of a particular group of fibres lying in a given plane and ϕ is the angle which this group (or groups) makes with the loading axis. The results of such an analysis are given in Table 5.1 where it will be seen that a random distribution in three dimensions is only 20% as efficient as a unidirectional array. Thus the modulus of a composite may be rewritten as

$$E_c = \eta E_f V_f + E_m(1 - V_f)$$

or, if the material is Hookean

$$\sigma_c = [\eta E_f V_f + E_m(1 - V_f)]\, \epsilon$$

whence

$$E_c = \omega E_m$$

where ω is simply a coefficient obtained by collecting up the terms and is given by

$$\omega = 1 + V_f(\eta M - 1)$$

where M is the modular ratio E_f/E_m.

Similar analyses for the case of multiple layers of unidirectional laminates have been made by TSAI (1968). The principle is much the same but the computations are arithmetically tedious and the reader should consult the original papers for details.

5.6 Compression of Composite Materials

It is surprising and, from our point of view disappointing, that the behaviour of composite materials in compression has not received as much attention. The classic study is due to ROSEN (1964) who assumed that failure of the material would be preceded by buckling of the fibres (see also Chapter 6) in one or other of the two modes illustrated in Fig. 5.6. The first is called the 'extensional' mode since here

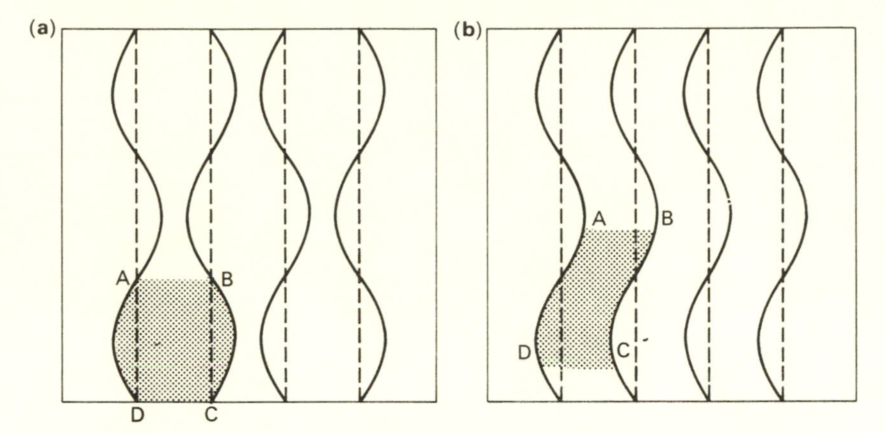

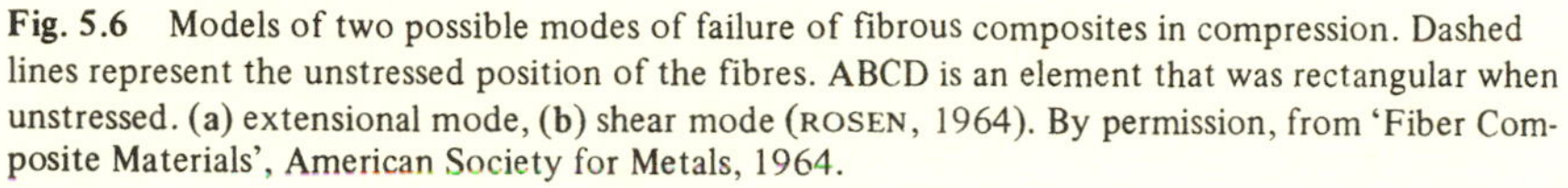

Fig. 5.6 Models of two possible modes of failure of fibrous composites in compression. Dashed lines represent the unstressed position of the fibres. ABCD is an element that was rectangular when unstressed. (a) extensional mode, (b) shear mode (ROSEN, 1964). By permission, from 'Fiber Composite Materials', American Society for Metals, 1964.

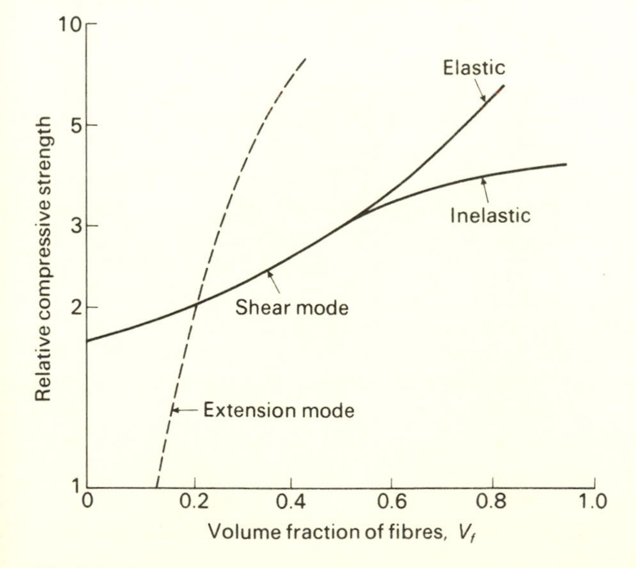

Fig. 5.7 Compressive strength of unidirectional fibrous composites. Dashed line (extension mode) and solid line (shear mode) correspond to the models shown in Fig. 5.6a and b (ROSEN, 1964). By permission from 'Fiber Composite Materials', American Society for Metals, 1964.

the element ABCD must deform laterally in outwards tension in order that the buckling be accommodated in the matrix. The second or 'shear' mode involves a bodily translation of ABCD. The theoretical calculations for the strength achieved by each mode give the results plotted in Fig. 5.7 and it will be noted that for volume fractions less than 0.17, the extension mode is more likely. Above this the shear mode dominates–this is readily equated with the observed mode of failure in wood (see Section 5.18.).

However, the theory predicts strengths which are considerably higher than those obtained by experiment. Rosen explains this discrepancy in terms of non-elastic deformation of the matrix and produces a corrected curve labelled 'inelastic' in Fig. 5.7. Even so the predicted values are high. C. R. Chaplin (private communication) has suggested that, from experiments of glass-resin composites, the surface fibres tend to determine the behaviour since, at the surface, they have a freedom to buckle which is not possessed by the inner fibres which are constrained on all sides. Certainly an outer layer of circumferentially wound fibres considerably increases the strength–this may be of significance in the fracture of wood where the more nearly circumferential S1 fibres may act as a constraint against buckling of the more nearly longitudinal S2 fibres (see Section 5.17) but here, clearly, is an area for considerable investigation.

5.7 Fracture of Composite Materials

One of the important features of fibrous reinforced materials is their comparatively high work of fracture or, to express it in the terms used in Chapter 2A, their resistance to the propagation of fast unstable cracks originating at a sharp crack or notch. This is especially surprising in an engineering material such as 'Fibreglass' where *both* components, the glass and the resin, if tested alone are substantially brittle but which, when combined, show a very high fracture toughness.

The classical theory shows that, in an isotropic Hookean material, very high stresses can exist at the tip of a sharp crack–the absolute magnitude of these has been evaluated by COOK and GORDON (1964). They showed that a surface crack was, in fact, more significant than an internal crack and further that a surface step in a Hookean material was very nearly as effective as a true crack or notch in terms of the local stress concentration. Thus it would appear that in a Hookean material any defect, no matter how apparently insignificant, can provide the necessary conditions for the initiation of brittle, Griffith type fracture. As we saw in Chapter 2A the situation is rather different in a 'ductile' material, i.e., a material in which some plastic flow is possible. Here the tip of the crack or notch is 'rounded off' so that the crack is blunted and the applied stress must be increased in order to make the crack propagate.

COOK and GORDON (1964) have, however, pointed out that the stress configuration around a crack involves a peak value just ahead of the crack tip (Fig. 5.8a) and, moreover, that this stress configuration includes a component operating in the plane of the crack itself. Thus, if a crack were to propagate and to come up against an interface (Fig. 5.8b, c) the transverse stress component will cause the interface to 'open up' and to produce a crack which lies normal to the main crack. This is, in effect, a 'crack stopping' process because the main

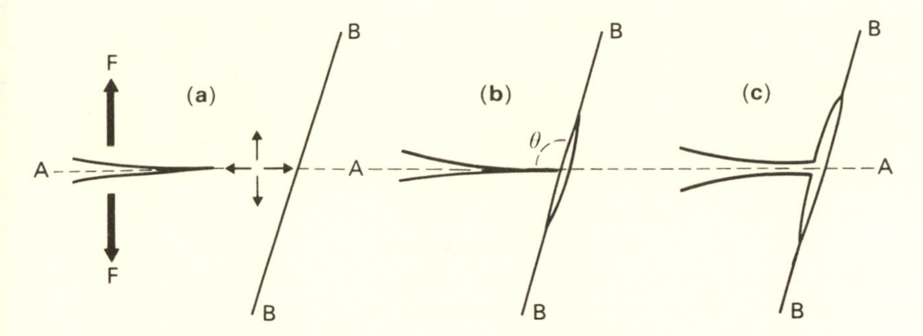

Fig. 5.8 A propagating crack approaching an interface is preceded by stress concentrations (small arrows in **(a)**) that can cause the interface ahead of the crack to widen **(b)** and **(c)** in planes parallel and perpendicular to the crack. A, plane of the propagating crack; B, plane of an interface and of crack after it turns through angle θ (COOK and GORDON, 1964; from *Proc. roy. Soc.* A282, 508).

crack must now be diverted by the vertical crack and more energy must be fed into the system in order to get the main crack to proceed in the horizontal plane.

We may note that the basic principle involved is that of diverting the crack from its usual path so that the stored energy is absorbed by the material. Clearly the whole situation will be further improved if the material on the other side of the interface is, itself, capable of absorbing the elastic strain energy which the system is trying to lose as the main crack propagates. Thus, as a second stage in designing a crack resistant material we would make sure that the material between fibres is soft and ductile and capable of absorbing large amounts of strain energy.

But, even discounting the absorption of energy by a more ductile material, the effect of causing the crack to change its direction frequently is to increase the demand for strain energy to be fed into the system (i.e., to increase the level of the applied load which is needed to produce fracture). This effect has been familiar to materials scientists for many years. All crystalline solids possess one set of crystallographic planes on which fracture by 'cleavage' is relatively easy– these are usually planes of minimum surface energy in the lattice and so, if the surface energy is a minimum, so is the fracture stress (see Chapter 2A, Section 2.8). A polycrystalline aggregate involves areas of mismatch at the grain boundaries so that a fracture must change its direction as it crosses the boundary from grain to grain. But a cleavage fracture in a crystalline material propagates when the stress normal to the cleavage plane reaches a critical value. The more often the crack must change direction, the greater the applied stress which is needed, i.e., the fracture strength of a polycrystalline material is a function of grain size.

Opinions are somewhat divided as to the exact mode or modes of fracture in a fibre reinforced composite–the situation becomes extremely complex if it is necessary to take into account the fact that pre-existing cracks or notches are likely to be distributed in a random way along the length of the fibres. But, in a general way we may recognize four possible modes of failure–these are by no means mutually exclusive and it is likely that, in a real composite, all or any may occur at different locations.

(a) A crack in the matrix advances in a single plane which lies normal to the fibres. The fibres break as soon as their fracture strain is exceeded by deformation of the matrix. This will produce a flat fracture with both matrix and fibre behaving as a single unit. We would expect this to occur if the fracture strain of both the matrix and the fibres were the same and if the interfacial shear strength were the same as the breaking strength of both the fibre and the matrix.

(b) Due to the statistical distribution of flaws, the fibres break in rather random fashion and the matrix cracks in such a way as to join these cracks together. This can arise when the fibres fail at a lower strain than the matrix and when the interface is strong. This mode probably occurs when the matrix is softer and more deformable than the fibres and can thus sustain a high strain prior to fracture. COOPER (1966) has suggested that this type of failure depends largely upon the degree of fibre–fibre coupling. As the crack propagates, the material is held together by matrix ligaments or bridges and it may require considerable plastic work to break these.

(c) Fibre 'pull out'. If the mode of (b) above occurs in a composite where the interface is comparatively weak, the fibres can 'pull out' of the matrix. In the intermediate stage, the composite is effectively held together by the fibres. This mode of failure occurs when the volume fraction is lower than V_{crit} (*Eq. 5.11*) and the fibres have a very large aspect ratio. Failure of long fibres at random locations reduces the system to a set of short fibres which pull out—the work of fracture is, however, generally small because of the comparatively low strength of the interface which is necessary for this mode to occur.

(d) If the interface is very weak, the local stress ahead of the advancing crack causes longitudinal failure at the interface and the crack is blunted as it runs into the separated region. This mechanism, suggested by COOK and GORDON (1964), is not universally accepted but it seems to accord well with many experimental observations. It may, indeed, be a significant feature of the fracture of wood, though the effect here is more likely to be due to a separation along the interface of the S2 and S1 layers caused by differential volume changes under stress. But however caused, this mechanism involves a weak interface and leads to a high work of fracture.

We must, however, remember that all of the above mechanisms are somewhat idealized and bear in mind that none of them is likely to operate exclusively in a real composite. Two other factors may need to be borne in mind when considering the fracture behaviour of biomaterials. The first is that the matrix may, as often as not, be viscoelastic. Thus at low rates of strain we may expect that the matrix is deformable, whereas at high rates of strain it may be linearly elastic—or at least as nearly so as makes no difference. Thus we may expect the fracture mode (and hence the work of fracture) to be highly sensitive to the rate of loading. This is, indeed, found to be so in bone. The second factor is that we really have little information on the degree of perfection of naturally grown fibres, so that we cannot assess—even roughly—the statistical chances of defects. At the same time it seems quite likely that, in biomaterials, the fibres may well be coupled—either physically or chemically—to the matrix so that what appears microscopically to be a fibre of high aspect ratio may behave as if it were a chain of shorter fibres which are 'locked' at various points along their length.

Here again is an area for further study, but it is pretty clear that, as with all other materials problems, the ultimate structure must be a compromise. Strong, stiff fibres in a strong matrix will optimize both strength and stiffness but with penalties of brittleness if the interface is also strong. Oriented fibres will give high strength and stiffness in preferred directions—but will be less efficient in complex loading conditions. A weak interface will give high fracture toughness but may lead to incipient separation of the layers. In what follows we shall discuss some of the strategies adopted by animal and plant materials that have achieved particular types of compromise.

5.8 Voids

A material containing voids represents a special case of a two phase 'composite' since here, the second phase (i.e., the void) has zero strength and stiffness. The problem is, in fact, rather intractable. A formalized analysis has been given by MACKENZIE (1950) but on the whole the effect of voids on the modulus of a linearly elastic matrix is adequately described by empirical statements such as

$$E_c = E_m(1 - 1.9V + 0.9V^2) \qquad (5.17)$$

where E_m is the modulus of the matrix and V is the volume fraction of pores (COBLE and KINGERY, 1956). *Eq. 5.17* shows that the effect of porosity is quite serious—even as little as 10% porosity lowers the absolute value of the modulus by nearly 20%. But it is worth noting here that, providing the pores are empty, the weight of the composite is also decreased so that the specific modulus (modulus/unit weight) does not fall off so rapidly. Nonetheless, the effect on the modulus soon outweighs the advantages of lowered weight and, very broadly, we may expect that porosity in excess of about 15% is undesirable except where weight is at a premium.

The same is very largely true of strength. GOODIER (1933) has made a formal analysis of the effect of porosity on strength. But, here again an empirical formula due to RYSKEWITCH (1953) appears to fit much of the experimental data:

$$\sigma_c = \sigma_0 e^{-nV}$$

where σ_0 is the strength of the nonporous matrix and the exponent n lies between 4 and 7.

In principle at least, the great danger of pores in a linearly elastic material is their potentiality of acting as stress raisers leading to Griffith type fracture. Admittedly the stress concentration around a truly spherical pore is not high (about 3), but only a small degree of ellipticity is needed to increase this factor considerably. Furthermore, as shown in Section 2.7, an accumulation of pores, each lying within the stress field of the next may be tantamount to a crack. The location of the pore is, however, probably more important than its actual size. A pore, lying in a compressive stress field is likely to be unimportant. Similarly in bending, a pore near the neutral axis (Chapter 6) is far less important than a pore lying close to the tension surface. Sandwich bone is, in fact, constructed on this principle: the outer layers are remarkably compact and pore free, the inner region is porous, thereby reducing weight.

5.9 Structure of Arthropod Cuticle

Arthropods are the dominant small- to medium-sized active animals, and their skeletons might be expected to show some splendid adaptations to locomotion, which indeed they do.

The three main components of cuticle are the polysaccharide, chitin, various structural proteins, and calcium carbonate. Calcium carbonate is found mainly in the Crustacea and the Diplopoda. It is virtually absent from the insects. Chitin is found in all true cuticles, though it is absent from a few specialized structures like egg cases. Protein occurs in all cuticles.

We consider first the nature of the protein, that of chitin having been dealt with already in Section 3.5. Then we shall deal with the microstructure of the cuticle, the arrangement of the whole cuticle, and finally with the major variants on the basic arrangement.

Cuticular protein The protein is not well known, as it is difficult to isolate. A review by HACKMAN and GOLDBERG (1971) shows considerable variation among insect proteins (Table 5.2). In general there is much glycine and valine, as might be expected from structural proteins, and much alanine. There is more proline than, for instance, in silk. Since this is a bulky residue it seems that close packing is not important as it is for silk (Section 3.2). There is rather little cysteine.

There is optical, X-ray and chemical evidence that much of the protein is oriented in relation to the chitin, probably parallel to it, and covalently bound to it through aspartate and histidine residues. This is so both in insect and crustacean cuticle (HACKMAN, 1960). HACKMAN (1953) showed that there are different components having different amino acid compositions. PAU *et al.* (1971) found four different proteins in cockroach oötheca that were almost certainly structural. Oöthecal wall is not a cuticle in that it has no chitin, but the proteins of cuticle proper are unlikely to be much less complex.

A most characteristic feature of cuticular protein, which has generated much research, is its tendency to become sclerotized, or tanned. This involves a cross-linking of the protein chains to each other so that the whole protein-chitin complex becomes hard and stiff. It seems that the protein is linked by quinone links. The precise mechanism is still in doubt (BRUNET, 1967; ANDERSEN and BARRETT, 1971). The nature of the linkage will to some extent, of course, control the rigidity of the sclerotin, as the sclerotized protein is called. Short links, such as those suggested by Andersen and Barrett, with only a —C— between residues on different chains, will be less flexible than a tyrosyl-quinone-tyrosyl which, PAU *et al.* suggest, may be common in the cockroach oötheca. However, as shown in Chapter 2B, it is the *frequency* of crosslinks that is of overriding importance in determining stiffness, rather than their nature.

Table 5.2 Amino acid composition of some arthropod cuticles

Amino acid	Residues amino acid/1000 total residues							
	Agrianome spinicollis larvae	*Bombyx mori* larvae	*Lucilia cuprina* Larvae	*Lucilia cuprina* Puparia	*Calliphora augur* puparia	*Xylotrupes gideon* pronotum	*Periplaneta americana* Adult	*Periplaneta americana* Oötheca
Alanine	84.2	123.7	137.2	77.0	78.3	130.2	242.5	24.0
β-Alanine	–	–	–	88.8	114.6	11.2	24.7	–
Arginine	19.2	24.0	16.6	50.2	29.7	18.0	19.0	16.5
Aspartic acid	75.1	88.9	93.9	81.7	97.5	70.9	51.2	16.5
Cysteine†	4.8	6.1	4.6	13.3	14.3	4.0	5.7	0.7
Glutamic acid	98.6	94.3	123.7	94.8	111.4	32.9	51.4	19.3
Glycine	137.7	112.3	140.9	127.1	118.8	299.6	101.2	390.6
Histidine	9.5	13.0	32.1	43.1	37.4	22.8	33.0	11.9
Leucine	52.7	35.0	43.7	39.5	40.5	76.6	40.7	126.8
iso-Leucine	56.8	43.2	23.4	18.1	3.2	51.4	19.9	21.1
Lysine	26.1	34.6	43.9	35.7	28.8	10.4	18.5	19.5
Methionine	1.5	1.5	2.6	2.0	3.3	0.8	–	1.4
Phenylalanine	23.0	24.1	43.4	32.9	26.7	7.6	15.6	24.6
Proline	104.2	98.9	93.6	87.5	66.9	76.7	97.8	74.7
Serine	93.0	74.7	14.3	75.9	80.4	41.9	51.4	14.2
Threonine	54.5	65.2	58.4	39.9	47.9	26.0	32.9	11.9
Tyrosine	57.4	78.9	44.2	36.5	32.4	25.4	76.6	162.4
Valine	101.7	81.6	83.2	55.9	67.8	93.6	118.1	63.8

From HACKMAN and GOLDBERG (1971). Sclerotization of insect cuticle. *J. Insect Physiol.*, 17, pp. 335–347.
† Estimated as cysteic acid.
– None detected.

Though quinone tanning is the common method of crosslinking, HACKMAN and GOLDBERG (1971) point out that many sclerotins contain cysteine, between 0.4% and 1.5% of all residues. These cysteines are probably concerned with making disulphide bridges. They are probably rather unimportant compared with quinone links.

Little is known of the protein of arthropods other than insects. STEVENSON (1969) found that only about half the protein of crayfish was soluble in rather gentle solvents. He supposed that the rest were bound firmly, presumably covalently. DENNELL (1947) found, in the blood and cuticle of various Crustacea, enzymes that are concerned with the tanning reaction in insects. It is likely that the tanning system in Crustacea is broadly similar to that in insects. However, ANDERSEN and BARRETT (1971) found ketocatechols, almost certainly involved in tanning, in all the insects they investigated, but not in a variety of other arthropods. Therefore the tanning system seems to be different to this extent in non-insectan arthropods.

Submicroscopic structure If, as is probable, the protein in arthropod cuticle is oriented in the same direction as the chitin then, if we can visualize the chitin

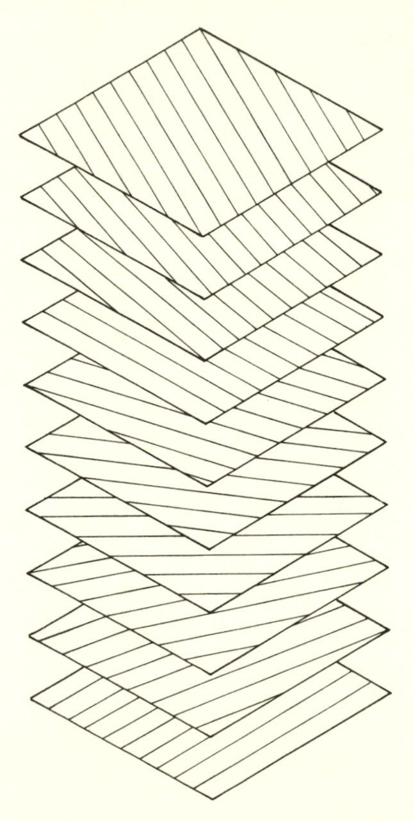

Fig. 5.9 An exploded diagram of the helicoidal model of arthropod cuticle to demonstrate how neighbouring sheets differ in the direction of their fibres. Sheets are in the plane of the cuticle.

fibrils, we can say how the whole cuticle is oriented. This matter has been intensively studied, particularly by BOULIGAND (1965, 1971, 1972); LOCKE (1967); RUDALL (1969); NEVILLE (1967, 1970); DENNELL (1973); and WEIS-FOGH (1970). It is still a matter of some controversy.

NEVILLE and LUKE (1969) suggest that cuticle is arranged in one of two ways. In one, called 'non-lamellate,' the chitin fibrils all have more or less the same preferred orientation, which is in the plane of the cuticle and, in any elongate structure, is along its length. In the other, 'lamellate,' system the cuticle is made of a series of very thin sheets, each consisting of a single layer of chitin fibrils oriented in the same direction. Each sheet is oriented slightly differently from the ones above and below it, and the change is progressive through the thickness of the cuticle. The resulting structure is called 'helicoid' by BOULIGAND (1965) who first proposed it. There will be, for any sheet, other distant sheets oriented in the same direction and closer sheets oriented in other directions (Fig. 5.9). To give some idea of size, in *Hydrocirus*, a water bug, the individual microfibrils are about 4.5 nm in diameter, and the distance between centres of adjacent microfibrils is about 6.5 nm. The sheet direction changes through 180° in about 25 layers, implying a change of 7°-8° between adjacent sheets. Therefore there are about 160 nm between similarly oriented sheets (HACKMAN, 1960). This pattern of sheets gives an extremely characteristic pattern, when sectioned even slightly obliquely, of a series of arches (Fig. 5.10a). The cuticle between any two 'pillars' of an arch is called a lamella. BOULIGAND proposes (1965) that these arches are not, as they appear, continuous microfibrils, but a series of layers producing a moiré pattern (Fig. 5.10b and c). Indeed it is clear that, if Bouligand's helicoid model is correct, these arches cannot be of single fibrils, because fibrils are confined to single sheets. NEVILLE (1970) has published photographs which seem to show the series of layers quite clearly.

Considering insects only, for the moment, we find that the *exocuticle* (the outer part, which becomes tanned) is often entirely lamellate. Therefore the fibres have, overall, no preferred orientation in the plane of the cuticle. There may be many lamellae. Locke, for example, counted more than 400 in larval cuticle of the skipper *Calpodes.* The first lamellae deposited are 0.5 μm thick and take six hours to deposit. Later the lamellae are 0.1 μm thick, and are laid down in 10 minutes. Assuming a value of 3 nm for the centre-to-centre distance of the sheets (which may be a bit low), the thicker lamellae will each consist of about 170 sheets, changing 1° per sheet, each sheet being deposited in two minutes. The inner lamellae will consist of about 33 sheets, each with a 5° shift, being deposited in 20 seconds. It is clear we are dealing with a rapidly formed and very precisely arranged structure.

The *endocuticle*, which is not necessarily sclerotized, has two characteristic arrangements. (a) Non-lamellate cuticle, deposited during the day, alternating with lamellate cuticle deposited in the night. The day and night layers are of roughly equal thickness. (b) Alternating layers of non-lamellate cuticle, are arranged in two directions only, each making an angle of 60° to 90° in its preferred direction to that of its neighbours. This produces a plywood effect. In this latter structure the two oriented layers are connected only by as many sheets of lamellar cuticle as are needed to produce the correct angular difference, so

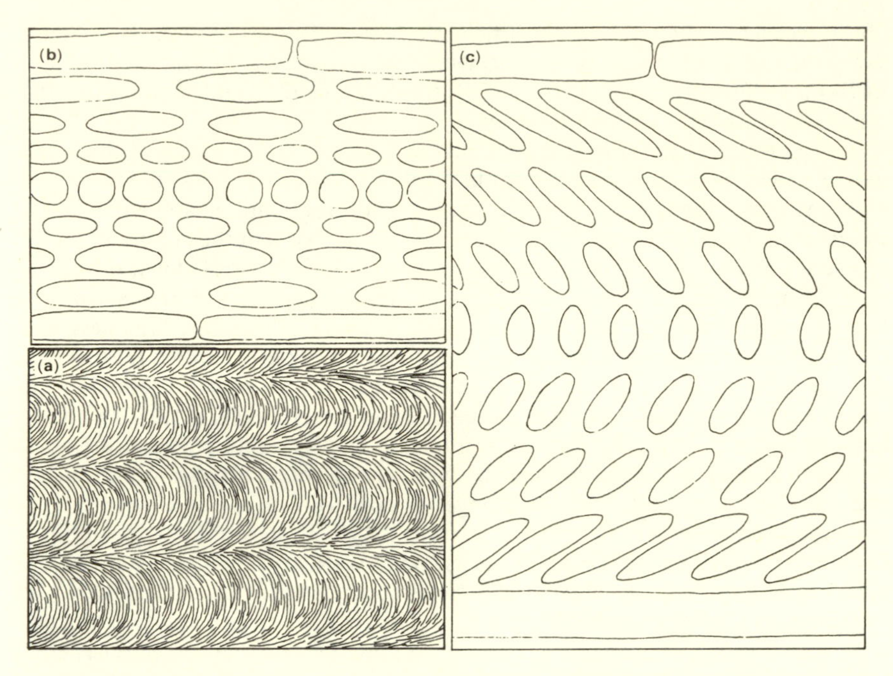

Fig. 5.10 (**a**) The 'Bouligand pattern' seen in electron micrographs of oblique sections of lamellate arthropod cuticle. (**b**) and (**c**) illustrate BOULIGAND'S (1965) helicoid model of lamellate cuticle. Chitin fibrils shown as lines in (**a**) are shown as hollow cylindrical sections in (**b**) and (**c**). (**b**) and (**c**) represent vertical and oblique sections respectively through a single layer (containing 9 sheets of fibres).

there is not an abrupt change of direction from one sheet to the next. The alternate plies are not very thick, often being not much more than the thickness of a normal lamella.

In the Crustacea the organization is similar to that in insects, though the lamellae are thicker, from 2–20 μm in various species, and not as precisely arranged as in insects. However, DENNELL (1973) has produced evidence that in the shore-crab the rather coarse lamellae are indeed entities that can be separated by treatment with KOH. In other arthropods there seems to be the same general arrangement of lamellae as in insects. BARTH (1969) has shown such structures in the spider *Cupiennius salei*, though he is unable to decide whether the lamellae are arranged in the manner suggested by Bouligand, or whether the arches he sees really exist.

Microscopic structure Considering the cuticle as a whole, there is again reasonable uniformity among arthropods, the main variations being in the amount of calcium carbonate and the ratio of tanned to untanned cuticle.

The cuticle of an insect is shown diagrammatically in Fig. 5.11. On the outside is the *epicuticle*. This usually contains several layers with, at least, from the outside in, a waxy layer, a layer of highly oriented lipid and a tanned lipoprotein layer. By definition there is no chitin in the epicuticle. Beneath is the *procuticle*,

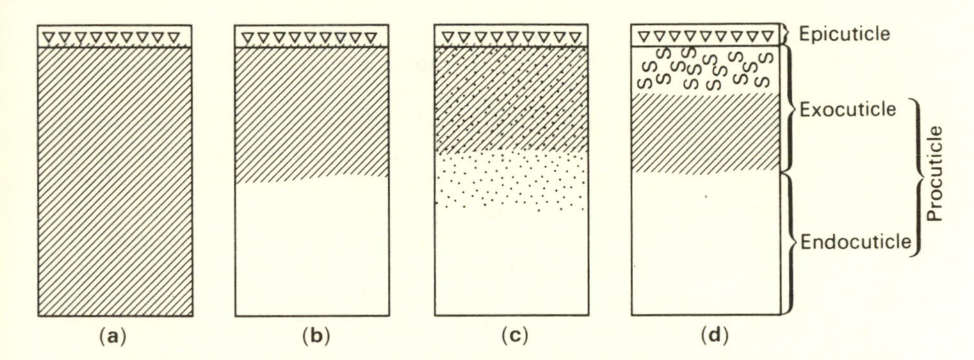

Fig. 5.11 Diagram of the composition of various arthropod cuticles. Triangles, waxy layer; diagonal lines, tanned protein; stipple, mineralization; SS, sulphur-linked proteins. (**a**) Some insects, Ricinulei, some opilionids. (**b**) Most insects, Merostomata, Palpigradi, Solifugae, Aranae, Acari, some opilionids, Chilopoda. (**c**) Crustacea. (**d**) Some scorpions.

forming the bulk of the cuticle, and containing chitin. Its outer, tanned part is called the *exocuticle*; this usually has a completely lamellate structure. Beneath is the *endocuticle*, which is often untanned and has a more complex arrangement of chitin fibrils. Passing right through the procuticle from the underlying epidermis are pore canals; these may end beneath the epicuticle, or pass through it to the surface. In development, the oriented lipid and protein of the epicuticle and the exocuticle are laid down first, before the moult, in an untanned state. Following ecdysis, the exocuticle expands and is then tanned, after which the insect cannot change its overall size. The wax for the waxy layer is squeezed up the pore canals and out onto the surface of the cuticle, and the endocuticle is laid down but may not become tanned.

This is so simplified a description of insect cuticle formation as to be a travesty—for some idea of the variations found see RICHARDS (1967). However, from our mechanical point of view, the really important consideration is perhaps the proportion of tanned and untanned cuticle. In the flightless *Collembola*, for instance, there is virtually no tanned cuticle, whereas in many beetles the cuticle is almost entirely sclerotized. There are also, of course, variations within any animal, most markedly at the arthrodial membrane, which is virtually untanned, and very flexible.

ANDERSEN (1973) has studied the locust. In all animals, the exocuticle is laid down before the animal moults and becomes tanned rapidly post moult. In the immature stages the endocuticle laid down post moult does not become tanned. Andersen claims, however, that in the adult, the whole cuticle becomes tanned with katechols. It seems that sclerotized cuticle cannot be resorbed, and so this strategy would seem adaptive for the juveniles, which are ready to start moulting almost as soon as they have developed a full thickness of body wall.

The ricinulids have remarkably thick and highly tanned cuticles (KENNAUGH, 1968). These strange little beasts have cuticles that are extraordinarily thick relative to their overall size. For instance, an element of diameter 0.26 mm may have a cuticle wall 0.07 mm thick: nearly 80% of the total volume of the leg is occupied by cuticle! The flexible arthrodial membranes are also strange, tending to be thicker than the tanned cuticle and impregnated with phenols though they are, of course, untanned.

KENNAUGH (1959) examined scorpions of two species. In both there was an outer layer of exocuticle that was untanned, but it was instead stiffened by sulphur links, as is keratin. The rest of the cuticle was conventional. BLOWER (1951), investigating both chilopods and diplopods could find, despite careful analysis, no epicuticle. This surprising finding has been contradicted by KRISHNAN and RAJULU (1964) and SEIFERT (1967). However, Krishnan and Rajulu make the interesting observation that two Indian diplopods they worked on had an epicuticle in the summer but in winter, when it is rainy and water loss is not a problem, the animal moults and grows a cuticle with no epicuticle.

Crustacean cuticle, being usually calcified, will be discussed in Section 5.19.5.

5.10 Mechanical Properties of Arthropod Cuticle

The mechanical properties of arthropod cuticle, such as are known, are shown in Table 5.3 (see also Fig. 2.4). It is depressingly short, considering how much work in general has been done on arthropod cuticle. It is clear that insect cuticle is fairly strong in tension, and is also quite stiff, particularly when considered on a stiffness per unit weight basis.

Let us consider the exocuticle of an insect first, specifically the locust. It is made of separate sheets, each containing indefinitely long fibrils of high modulus, high strength chitin, bound together by a matrix that is less stiff. In such lamellate cuticle, each sheet is oriented slightly differently from the ones above and below. The overall effect is of random orientation of chitin fibres in the plane of the cuticle. In a composite the stiffness of a single sheet of fibres varies dramatically with the angle between the direction of the fibres and the line of action of the applied force. What then will be the modulus of a many-plied sheet such as we have here? It will clearly be the same whatever the direction of the force in the plane of the sheet. Unfortunately we cannot say more, accurately, because we do not know the modulus of the protein in either the tanned or the untanned state. However, we can make some rough guesses, and much of what follows does not depend critically on the accuracy of our guesses. NIELSEN and CHEN (1968) show the effect of having a sheet arranged with a two-dimensional random orientation of fibres, instead of in one direction only (Fig. 5.12). The analysis assumes that the fibres are long, which is true in cuticle. It also depends on the volume fraction V_f of the fibres and of the ratio of the elastic moduli of the composite tested along and normally to the fibres. Let us suppose that the matrix, when tanned, is either 5 or 20 times as pliant as chitin. From the observations of LAFON (1943) we can take the volume fraction of chitin to be 0.35.

From standard composite material theory (*Eq. 5.7*) the stiffness of the sheet when tested along the length of the fibres is given by $E_p = E_f V_f + E_m(1 - V_f)$ where E_p refers to the modulus parallel to the fibres, and the subscripts f and m refer to the fibres and the protein matrix respectively. E_n, the stiffness of the sheet when tested normal to the fibrils, is effectively equal to E_m, because in this orientation the fibrils will have hardly any stiffening effect.

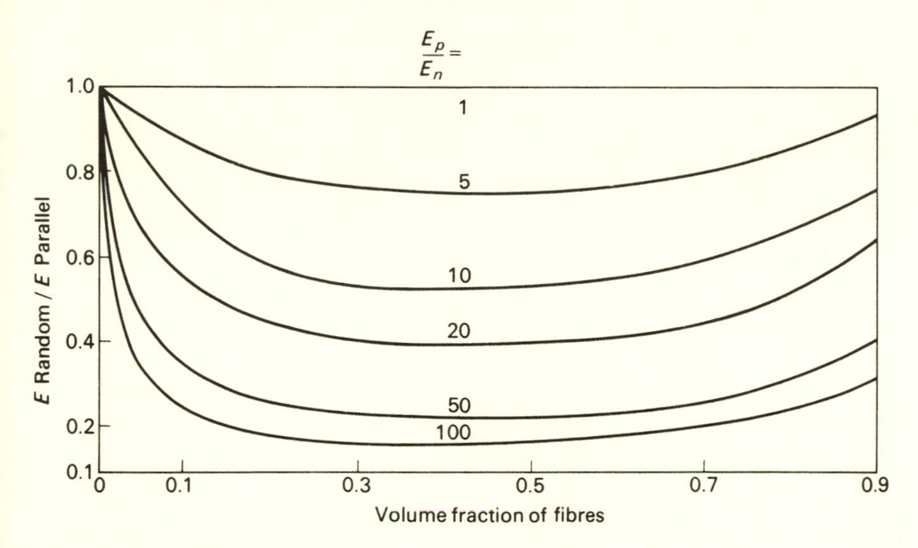

Fig. 5.12 The stiffness of a composite whose fibres make a random two-dimensional net compared with that of a composite with fibres aligned in the direction of applied force (ordinate) as a function of the volume fraction of the fibres (abscissa) and the ratio of the stiffnesses of the parallel-fibred composite loaded parallel to the fibres (E_p) and normal to them (E_n).

Figure 5.12 shows the theoretical relationship between overall stiffness and the stiffness in the two directions E_p and E_n, as determined by Nielsen and Chen. If the differences in the moduli are not very great, say in a ratio of not greater than 10:1, the difference between a uniform sheet loaded in its *stiffest direction*, and a random sheet loaded in *any direction*, is not very great (1.3:1). Even using our second model, in which the difference in modulus between fibre and matrix is 20:1, the difference in modulus of the oriented sheet and the random sheet is only about 1.6:1 (remember, of course, that the oriented sheet would be much stiffer if it were all chitin, but it is not). It follows that the arrangement found in the exocuticle of the locust is probably a rather efficient way of ensuring reasonably high stiffness in all directions in the plane of the cuticle (Fig. 5.13).

The endocuticle of the locust tibia is, according to JENSEN and WEIS-FOGH (1962), almost as stiff as the exocuticle. This is perhaps rather surprising, though ANDERSEN (1971) claims that the endocuticle of adult locust femora does indeed become sclerotized. Let us suppose that the endocuticle protein does become reasonably stiff; what are we to make of the arrangement of the lamellate night layers alternating with the day layers whose fibrils are oriented preferentially along the element? Again, this seems rather good design. The longitudinal layers will be particularly effective in resisting bending forces, while the night layers will provide stiffness against shear forces. Shear will be produced both by bending (in which mode the stresses are not great) and by torsion. The longitudinal fibres are not well adapted to be stiff in torsional loading situations.

We discuss later (Section 5.19.5), the significance of the fact that in many arthropod elements the endocuticle is definitely not as stiff as the exocuticle.

Table 5.3 Mechanical properties of some rigid materials. σ_t, ultimate tensile stress (MN m^{-2}); σ_c, ultimate compressive stress (MN m^{-2}); M_R, modulus of rupture (MN m^{-2}); E, Young's modulus (GN m^{-2}); ρ, density (kg m^{-3}). The last three columns of values are explained in the text. They refer to the weight required to obtain a given strength or modulus. Each value of ρ/σ_t and ρ/M_R has been multiplied by 10^6; values of $\rho/\sqrt{E}$ have been multiplied by 10^4. The values of strength and modulus being expressed in N m^{-2}. $M_R = My/I$ where y is half the depth of the section.

Group	Genus	Part	σ_t	σ_c	M_R	E	ρ	ρ/σ_t	ρ/M_R	$\rho/\sqrt{E}$	Reference
Cnidaria	(stony coral)	Branch	–	–	27	–	2000	–	74	–	1
	Cirripathes	Axis	40	–	–	0.3	–	–	–	–	9
Gastropoda	*Conus*	Shell	36	126	113	68	2700	75	24	103	1
	Cypraea	Shell	–	242	–	–	2700	–	–	–	1
	Lambis	Shell	–	109	–	–	2700	–	–	–	1
	Patella	Shell	33	134	147	60	2700	82	18	110	1
	Strombus	Shell	5	131	58	41	2700	540	47	133	1
	Trochus	Shell	–	269	–	–	2700	–	–	–	1
	Turbo	Shell	121	174	276	54	2700	22	9.7	116	1
Cephalopoda	*Nautilus*	Shell	62	139	207	44	2700	44	13	129	1
Bivalvia	*Anodonta*	Shell	36	–	142	44	2700	75	19	129	1
	Arctica	Shell	–	216	49	66	2700	–	55	105	1
	Atrina (nacre)	Shell	95	198	157	58	2700	28	17	112	1
	Atrina (prisms)	Shell	67	252	129	39	2700	40	21	137	1
	Chama	Shell	–	175	40	82	2700	–	67	94	1
	Egeria	Shell	45	–	90	77	2700	60	30	97	1
	Ensis	Shell	–	–	124	55	2700	–	22	115	1
	Hippopus	Shell	9	–	38	53	2700	300	71	117	1
	Hyria	Shell	72	–	130	67	2700	37	21	104	1
	Ostrea	Shell	–	–	73	47	2700	–	37	125	1
	Mercenaria	Shell	32	182	96	66	2700	84	28	105	1
	Pecten	Shell	42	88	113	30	2700	64	24	154	1
	Pinctada	Shell	52	242	173	48	2700	52	16	123	1
	Pinna	Shell	62	–	–	11	2700	44	–	257	1
	Saccostrea	Shell	31	–	38	29	2700	87	71	158	1

Crustacea	*Homarus*	Cheliped	–	–	7.4	4.2	1900	–	26	293	1
	Cancer	Carapace	–	–	116	11	1900	–	16	181	1
	Carcinus	Carapace	32	–	–	13	1900	59	–	167	1
	Carcinus	Cheliped	–	–	–	18	1900	–	–	142	1
Insecta	*Schistocerca*	Tibia	95	–	–	9.5	1200	13	–	123	2
	Pachynoda	Elytrum	69	–	–	–	1300	19	–	–	3
	Phormia	Wing	–	–	–	6.1	1200	–	–	157	10
Echinoidea	*Centrostephanus*	Spine	–	–	84	74	2000	–	24	232	1
	Cidaris	Spine	–	–	101	–	1800	–	18	–	1
	Echinometra	Spine	–	65	–	–	1800	–	–	–	4
	Echinus	Plate	–	–	23	9.7	1600	–	69	162	1
	Heliocidaris	Spine	–	96	74	–	1700	–	23	–	1
	Heterocentrotus	Spine	–	48	–	–	1100	–	–	–	4
	Stylocidaris	Spine	–	72	–	–	1300	–	–	–	4
Vertebrata	*Bos*	femur	190	–	270	18	2000	11	7	149	5
	Homo	femur	140	210	250	19	2000	14	8	145	5
	Balaenoptera	cancellous	–	4	–	0.3	500	–	–	288	6
		ear bone	–	–	–	35	2400	–	–	128	1
	Homo	enamel	–	–	–	75	2900	–	–	106	7
Gymnospermae	Swedish pine	timber	98.1	–	–	–	500	5.1	–	–	8
	white pine	timber	–	–	–	9.9	370	–	–	37	8
	Douglas fir	timber	–	–	–	16	510	–	–	41	8
	H_2O-sat.	earlywood	32.3	–	–	–	–	–	–	–	11
	H_2O-sat.	latewood	132	–	–	–	–	–	–	–	11
Angiospermae	ash	timber	68.7	–	–	–	500	7.3	–	–	8
	yellow poplar	timber	–	–	–	9.7	380	–	–	38	8
	red oak	timber	–	–	–	6.4	520	–	–	65	12
	yellow birch	timber	–	–	–	9.5	540	–	–	55	12
	sweet gum	timber	–	–	–	7.2	420	–	–	50	12

References: (1) Currey, unpublished; (2) JENSEN and WEIS-FOGH (1962); (3) HEPBURN and BALL (1973); (4) WEBER *et al.* (1969); (5) Burstein and Reilly, unpublished; (6) EVANS and KING (1961); (7) GILMORE *et al.* (1970); (8) KOLLMAN and CÔTÉ (1968); (9) Biggs and Wainwright, unpublished; (10) Rees, unpublished; (11) IFJU (1964); (12) JAMES (1962).

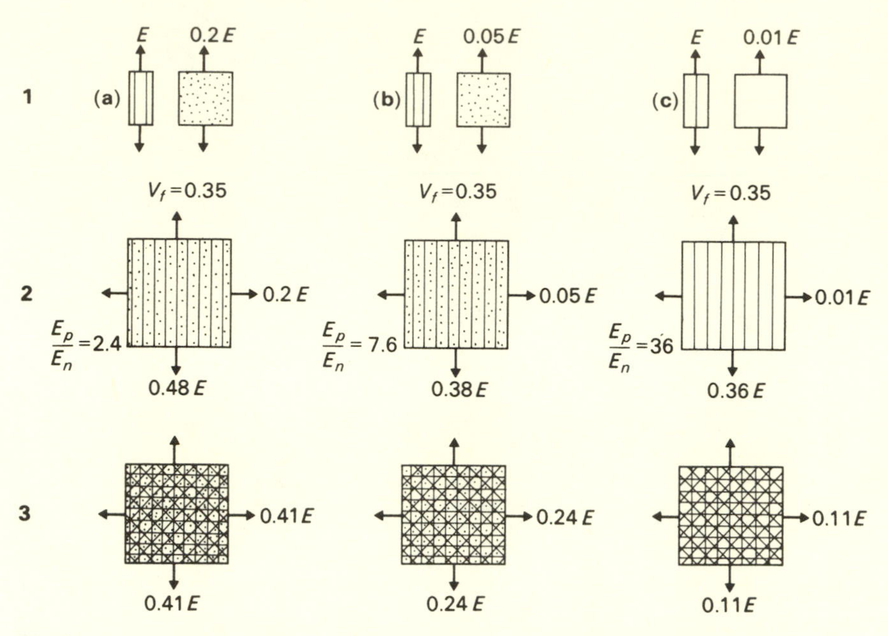

Fig. 5.13 Diagram of calculation of stiffness of cuticle. Row one, assumed values of E for chitin fibres and protein matrix. Row two, resulting stiffness in two directions of cuticle with a chitin volume fraction of 0.35. Row three, resulting stiffness of random network cuticle. The values for the modulus of elasticity (E) are all in relation to the value for the modulus of elasticity of oriented chitin, taken as unity.

Very little is known about the strength of arthropod cuticle. As will be discussed later, it seems that stiffness rather than strength is the important design feature of cuticle, and strength may often be merely of academic importance.

JENSEN and WEIS-FOGH (1962) found that the tensile strength of the cuticle in the locust tibia to be about 95 MN m^{-2}. This value, if generally true, would make the weight for strength (ρ/σ_t) of cuticle and bone roughly the same (see Table 5.3: 11 x 10^{-6} for bone and 13 x 10^{-6} for cuticle). The discussion earlier in this chapter shows that many of the requirements for a strong composite are met by arthropod cuticle:

(a) The chitin fibres have an extremely high aspect ratio, in fact they can be considered to be continuous.

(b) The absolute thickness of the fibres is small, so dangerous cracks cannot form in them.

(c) The volume fraction of the fibres, about 0.35, is sufficiently large for them to act as efficient reinforcers.

(d) The arrangement of the fibres is exceedingly precise, far more so than in man-made composites. As a result cuticle rarely has pockets of matrix or places where several fibres touch each other; such imperfections would tend to reduce the strength of a composite (FRIEDMAN, 1967).

(e) Probably the chitin fibres are almost perfect. It is one of the characteristics of living systems that they are able to synthesize polymers with remarkable perfection and consistency. The replication of DNA molecules is a good example of this. If such near perfection were present in chitin fibres, it would imply that few breaks in the chain are available to produce stress concentrations in the matrix.
(f) In many helically wound structures, there is a danger of delamination between adjacent layers caused by the incompatibility of strain between adjacent fibres at different orientations. Cuticle overcomes this difficulty by having only small angular differences between the fibres in adjacent layers.

It is not possible, yet, to analyse the strength of cuticle further, because effectively nothing is known about the mechanical properties of the protein matrix. All deep analysis of composites requires knowledge of the moduli of the fibres and the matrix, of their strengths and, usually, of their Poisson's ratios. We have some idea of the strength of chitin in bulk, but apart from this we are ignorant.

Clearly, insect cuticle (and probably other arthropod cuticle) is an excellent material for resisting bending. Its specific strength is about the same as that for bone. Its value for $\rho/\sqrt{E}$, which is the important parameter for stiffness in bending and resistance to buckling for a structure of least weight, is better than that for bone (123 as opposed to 149). It achieves this excellence by being a nearly perfectly arranged composite with, presumably, the stiffnesses of the fibres and the matrix having the optimum relationship to each other. However, this mechanical excellence must be paid for. As we shall discuss later (Section 5.20.4), this payment is in the form of a very large metabolic cost in a skeleton that is made principally of protein.

5.11 Structure of Bone

The main constituents of bone are collagen (about $\frac{1}{3}$ of the dry weight, or 50% of the volume), some other proteins, protein-polysaccharides and glycoproteins (HERRING, 1968) and some form of calcium phosphate. This last is some imperfect form of hydroxyapatite, which has a unit cell of $Ca_{10}(PO_4)_6(OH)_2$. Apatite is crystalline, but there is good evidence that at least some of the mineral is amorphous (POSNER, 1969). The amount of amorphous mineral is rather high initially, but decreases as the bone matures (TERMINE and POSNER, 1966). There is also some water in bone.

The structure of bone can be viewed at many levels. We shall start at the lowest (Fig. 5.14). In any small volume of bone, such as may be seen by the electron microscope, the collagen fibrils are more or less parallel and are surrounded by, and filled with, crystals of apatite. There is considerable argument at the moment about the shape of the crystals. One school of thought considers the mineral to be in the form of needles, and another thinks it to be in the form of plates. ASCENZI *et al.* (1965) say that the crystals are needle-shaped, about 4–4.5 nm across, and 100 nm to indefinitely long. BOCCIARELLI (1970) however, by using a somewhat different technique, showed that what were apparently needles were merely plates viewed from the side. He suggests that the crystals

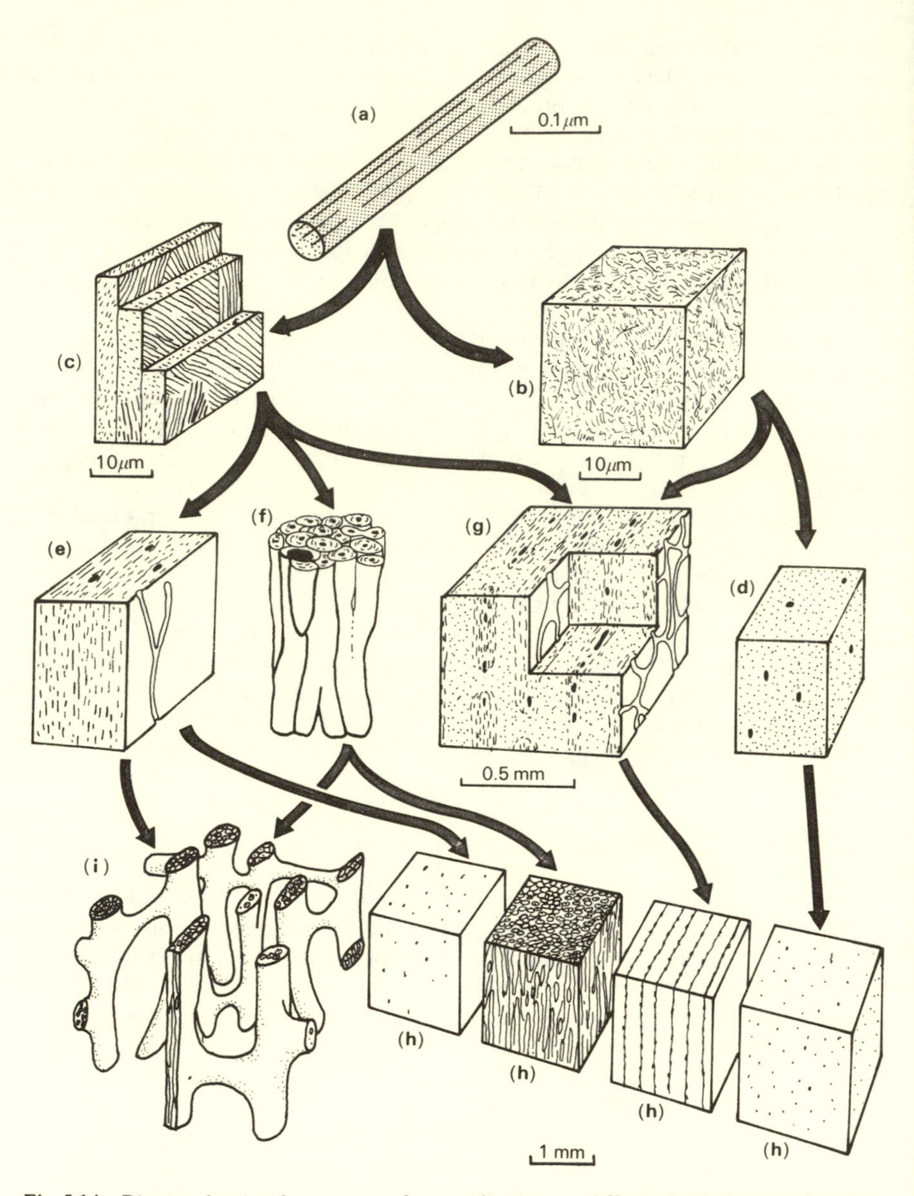

Fig. 5.14 Diagram showing the structure of mammalian bone at different levels. Bone at the same level is drawn at the same magnification. The arrows show what types may contribute to structures at higher levels.

(a) Collagen fibril with associated mineral crystals.

(b) Woven bone. The collagen fibrils are arranged more or less randomly. Osteocytes are not shown.

(c) Lamellar bone. There are separate lamellae, and the collagen fibrils are arranged in 'domains' of preferred fibrillar orientation in each lamella. Osteocytes are not shown.

(d) Woven bone. Blood channels are shown as large black spots. At this level woven bone is indicated by light dotting.

are 4 nm thick and 60 nm or so in the other directions. HÖHLING and SCHÖPFER (1968) investigated bone at the highest resolution then possible. They claim the crystals start as isolated nuclei lying in chains, the centres of the nuclei being about 4–6 nm apart. These nuclei fuse to form needles, but also to a lesser extent join side to side to form plates. HÖHLING *et al.* (1971) report similar findings in dentine which, at the ultrastructural level, is very like bone.

It is unfortunate that these differing views should still be held; no doubt they will be resolved soon. Following Höhling *et al.* it seems to us more probable that the crystals are needle-shaped. Whichever view is right, the crystals are certainly very small, and have one dimension of only 4 nm. Whether the crystals join up to form larger masses is uncertain but, because of the great amount of collagen, unlikely. The needles, or plates, are oriented with one of their long axes parallel to the local direction of the collagen fibril. The crystals are found within and around the collagen fibrils. Some workers claim to have seen large crystals in bone. CHATTERJI and JEFFERY (1968) for example, show scanning electron micrographs with apparent crystal sizes of 1 μm or greater in thickness. However, the collagen was removed from the bone; this would allow small crystals to coalesce, and any scanning electron micrographs of large crystals must be viewed warily for this reason.

It is not known how the crystals and the collagen are bound together, though some form of bonding is almost certainly present.

Mammalian bone is of two basic types: woven and lamellar. In woven bone (Fig. 5.14b), the direction of the collagen fibrils is random over a distance of one μm or so, and the apatite is not as uniformly oriented along the line of the collagen fibrils as is the case in lamellar bone. There is more amorphous calcium phosphate present than in lamellar bone, and there is usually a higher mineral/organic ratio. In lamellar bone (Fig. 5.14c), the collagen fibres are arranged in neat layers called lamellae. Older studies, using polarized light, led to the idea that nearly all the collagen fibrils in any one lamella were oriented in the same way, and the direction changed sharply from lamella to lamella (GEBHARDT, 1906). Were this the case, it would clearly have important mechanical implications. However, recent studies using transmission and scanning electron microscopy show the situation not to be so clear. In some lamellae at least, the fibrils in any particular lamella are in small 'domains,' typically 30–100 μm across. The orientation of the fibrils changes somewhat from domain to domain. Nevertheless, despite our inability to characterize exactly the orientation of the fibrils in a lamella, one can usually make out a generally preferred orientation in any one lamella (BOYDE and HOBDELL, 1969). Furthermore, nearly all the fibrils are oriented in the *plane* of the lamella, though some fibrils pass between adjoining

---

(e) Primary lamellar bone. At this level lamellar bone is indicated by fine dashes.

(f) Haversian bone. A collection of Haversian systems, each with concentric lamellae round a central blood channel. The large black area represents the cavity formed as a cylinder of bone is eroded away. It will be filled in with concentric lamellae and form a new Haversian system.

(g) Laminar bone. Two blood channel networks are exposed. Note how layers of woven and lamellar bone alternate.

(h) Compact bone, of the types shown at the lower levels.

(i) Cancellous bone.

lamellae. The thickness of the lamellae varies, but is typically about 5 μm. It seems to be generally agreed (BOYDE and HOBDELL, 1969; ASCENZI *et al.*, 1965) though not by all (COOPER *et al.*, 1966) that between many of the adjoining lamellae there is a perforated sheet of what has been called 'interlamellar bone,' whose mineral/organic ratio is higher than that of other bone.

In long bones, the lamellae themselves tend to run along the length of the bone, along the length of local blood channels, and to be arranged circumferentially around the whole bone. The blood channels themselves tend to travel along the bones, and do not pass directly from the inner to the outer cortex (BROOKES, 1971). Some of these tendencies are incompatible at times of course, but in general lamellae rarely have their short (5 μm) axis in the same direction as the long axis of a bone.

Most mammalian bone is lamellar, and most has cells within its substance. These are enclosed in small cavities, lacunae, within the tissue. They connect with neighbouring cells, with blood channels, and with the surface of the bone by means of cytoplasmic processes which run through channels, the canaliculi. These canaliculi are about 0.2 μm in diameter. The cell body is subspherical in woven bone, but in lamellar bone it is an oblate spheroid, with the shorter axis aligned with the short axis of the enclosing lamellae.

At a higher order of structure still there are four main types of bone. First woven bone may extend uniformly for distances of many millimetres, this is found only in very young bone (Fig. 5.14d). Primary lamellar bone (Fig. 5.14e) has lamellae oriented parallel to the surface of the whole bone, and disturbance of this pattern round blood channels is slight. In many bones, particularly those of advanced dinosaurs, mammal-like reptiles and mammals, this pattern is modified by the formation of Haversian systems (ENLOW and BROWN, 1957, 1958; CURREY, 1962; DE RIQLÈS, 1969) (Fig. 5.14f). Bone round blood vessels is resorbed, and the resulting elongated cavities are filled with lamellar bone, which is now oriented with relation to the blood channel. The resulting structure of a blood vessel with its associated lamellae is called a Haversian system or secondary osteone. The outer limit of the Haversian system is marked by a 'cement line,' a sheath of calcified mucopolysaccharide from which collagen seems to be absent (ORTNER and VON ENDE, 1971). Metabolically the cement line is important, because virtually no canaliculi cross it, and so cells outside it are virtually isolated from the blood vessel of the Haversian system. This may make the survival of the cells in this position difficult (CURREY, 1964*a*).

Laminar bone is found in the same groups as tend to have Haversian bone. It is a peculiar structure, making sense if its development is known (CURREY, 1960) (Fig. 5.14g). Bovine laminar bone is typical. Here it consists of a series of laminae about 200 μm thick. Each has its thickness in the radial (endosteal–periosteal) direction. The middle of each lamina is an essentially two-dimensional network of blood vessels sandwiched between layers of lamellar bone. This sandwich is itself sandwiched between layers of woven bone; the junctions with adjoining laminae are indicated by a layer of heavily calcified woven bone. Laminae are formed principally in the bones of large animals that have to grow quickly. The bone grows in a series of spurts (Fig. 5.15). On a base of woven bone, a flimsy scaffolding of woven bone is laid down, clear of the surface, and blood vessels

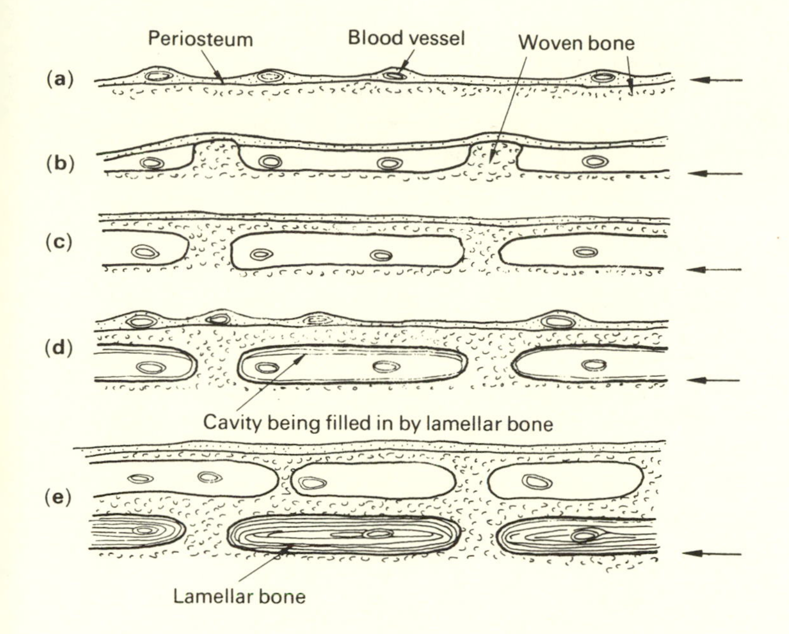

Fig. 5.15 Diagram of the growth of laminar bone. **(a)** Resting surface. The periosteum has blood vessels between it and the woven bone at the surface. **(b)**, **(c)** A scaffolding of woven bone is laid down; blood vessels are trapped in the cavities formed. **(d)** The cavities are filled in with lamellar bone while **(e)** the whole process is repeated. The arrows show the level of the original surface.

are left in the space so formed. The lamellar bone is then laid down more leisurely in the space so formed. While this is going on, one or more further layers of woven bone can be laid down. In this way it is possible for the diameter of the whole bone to increase at a rate faster than that at which bone can be laid down at a single surface.

At a higher order of structure we find the distinction between compact and cancellous bone. Compact bone (Fig. 5.14h), as the name suggests, is bone of any of the types described previously that exists in solid chunks without any spaces except for blood channels and cells. Cancellous bone, on the other hand, is made into slender ties and struts (Fig. 5.14i). Histologically it is like primary lamellar bone or Haversian bone. Mechanically, of course, it is very different from compact bone.

We have only discussed the bone of adult higher vertebrates, and of that, mainly bone occurring in long bones. There is, of course, considerable variation in the histology of bone among animals of different taxa, between different ages in the same species, and between different bones in the same animal. ØRVIG (1967) gives an account of bone and the wide variety of other calcified tissues found in early vertebrates. For some variants of bone we have some idea of the mechanical consequences of the variation; for most we are quite ignorant.

One type of bone is so different from the bone discussed already that we must mention it, though nothing is known of its mechanical properties; this is acellular bone. Bone without cells is found in the primitive, extinct heterostracan fishes, and in most of the more advanced modern teleosts. The bone is completely without cells or blood vessels. Apparently (MOSS, 1961) in the living teleosts the bone cells are incorporated into the bone in the ordinary way, and then they die and the lacunae are filled in with more bone. The physiological consequences of acellularity are obscure, and the mechanical ones even more so.

5.12 Mechanical Properties of Bone

Much research has been performed on the mechanical properties of bone (CURREY, 1970, is a review), but unfortunately much of it has been extremely uncritical. For instance, many tests have used dry or drying bone, whose properties are known to be markedly different from those of wet bone, which is in a more natural state. Nevertheless we have more information about the mechanical properties of bone than for any other stiff skeletal material except wood. Representative values for various properties are given in Table 5.4.

Table 5.4 Some mechanical properties of mammalian bone. 'Longitudinal,' 'tangential' and 'radial' refer to the direction in which the applied forces act. In the fin whale ear bone these directions cannot be specified and in the torsion experiment these directions are inappropriate. Values for modulus of elasticity are in GN m^{-2}, for strength in MN m^{-2}. All tests were on machined specimens except the torsion experiments, which were on whole bones.

Loading mode	Species	Histology	Longitudinal	Tangential	Radial	Reference
MODULUS OF ELASTICITY						
Bending	ox	Laminar	19.3	10.4	—	1
Compression	man	Haversian	18.5	10.0	–	2
Compression	man	Cancellous	0.33	–	–	3
Tension	ox	Laminar	17.2	12.1	15.0	2
Tension	man	Haversian	17.3	11.3	–	2
Ultrasonic	man	Haversian	24.3	–	–	4
Ultrasonic	ox	?	21.3	11.1	–	5
Bending	fin whale	(high mineral)	33			1
ULTIMATE STRENGTH						
Bending	ox	Laminar	264	–	–	6
Bending	ox	Laminar	227	91	–	1
Compression	man	Haversian	220	151	–	2
Compression	man	Cancellous	4	–	–	3
Compression	man	Cancellous	7	–	–	7
Compression	man	Haversian	172	41	–	2
Tension	ox	Laminar	184	69	29	2
Torsion	dog	?	157			8

This table relies heavily on mainly unpublished work of Currey, Burstein and Reilly because their testing procedures were designed to be as uniform as possible in the different modes. Readers wishing to consult others' results may wish to read CURREY (1970) and YAMADA and EVANS (1970) for fairly comprehensive reviews. References: (1) J. D. Currey, unpublished; (2) A. H. Burstein and D. T. Reilly, unpublished; (3) EVANS and KING (1961); (4) ABENDSCHEIN and HYATT (1970); (5) LANG (1969); (6) BURSTEIN *et al.* (1972); (7) WEAVER and CHALMERS (1966); (8) PUHL *et al.* (1972).

5.12.1 Main Features of Behaviour in Relation to Structure

We shall start this section with a brief survey of the main features of bone's mechanical properties, and then attempt to relate these to the structure.

A well controlled, fairly quick tension test, taking about $\frac{1}{2}$ second to failure, of cow femoral cortical bone, gave a load-deformation curve as in Fig. 2.1. For present purposes we can say that the equivalent stress–strain curve would have the same general shape. There is a straight, elastic part, up to a fairly definite yield point. The curve then bends over, and there is a region of plastic flow, the curve rising only slightly all the way to failure. The break, when it occurs, is abrupt and seems to be brittle. The ultimate strain is about 4%, and the elastic strain is about 1%. Bone, therefore, is far from being a brittle material; indeed, in such a test as this, it can absorb about six times as much energy *after* it has yielded as it can in the elastic region. It will be recalled from Chapter 2 that energy absorption is particularly important when structures are loaded dynamically. Dynamic loads are experienced more by active animals than by sessile ones, so it is no surprise that *tough* skeletal materials have been evolved by vertebrates and the arthropods, both of which are active groups.

Bone is a viscoelastic material, but from the point of view of its *function* this viscoelasticity is not nearly so important as it is for the more pliant materials. This is because, except *in extremis*, it is not the function of bone to store or to dissipate energy. Nevertheless, it may be important analytically, for instance the modulus of elasticity and the fracture stress are both quite strongly strain-rate dependent (McELHENEY, 1966) (Table 5.5).

Bone is also more or less anisotropic. Figure 5.16 shows how in different tissues this anisotropy may be more or less. In long bones the highest ultimate tensile strength and the highest value for modulus are roughly those measured along the length of the bone.

The mechanical behaviour of bone must be considered in relation to the obvious fact that it is a composite material. Since CURREY (1964*c*) and MACK (1964) first discussed the mechanical significance of this fact, the theory of composite materials has advanced, and the relationship between theory and the behaviour of bone has strengthened, but there is a long way to go.

Table 5.5 Effects on bone of different strain rates. σ_t, ultimate stress (MN m^{-2}); ϵ_t, ultimate tensile strain; E, Young's modulus (GN m^{-2}).

Strain rate in s^{-1}		Loading mode	σ_t	E	ϵ_t	Reference
0.01		Tension	125	17.0	0.033	1
0.1	(man)	Tension	133	17.4	0.038	1
1		Tension	166	21.4	0.018	1
0.001		Compression	179	18.9	0.018	2
0.01	(ox)	Compression	210	19.6	0.018	2
0.1		Compression	235	24.6	0.018	2
1		Compression	255	28.1	0.012	2

References: (1) Burstein and Reilly, unpublished; (2) McELHENEY (1966).

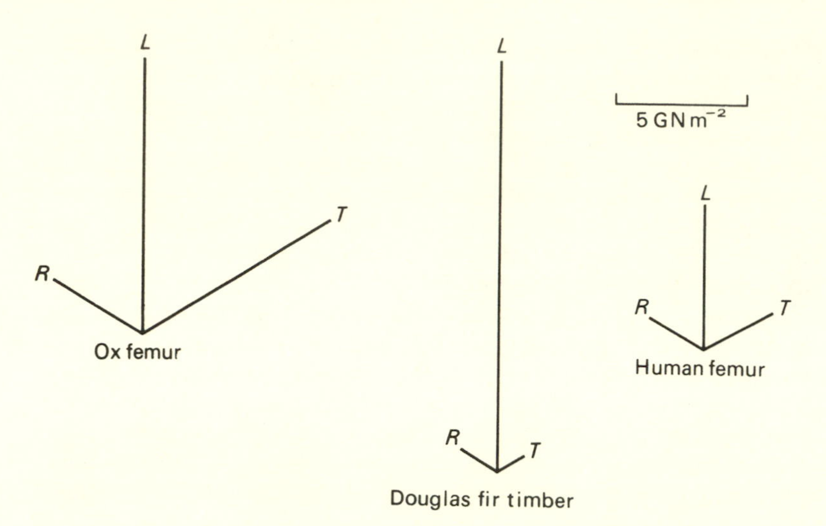

Fig. 5.16 Diagram showing anisotropy of modulus of elasticity in Haversian bone from the femora of ox and man and of timber of Douglas fir. *L*, longitudinal; *R*, radial; *T*, tangential.

Arthropod cuticle is a 'conventional' fibrous composite in that it has very long, thin, stiff fibres in an amorphous matrix. In bone, on the contrary, the obviously fibrous component–the collagen–has a much lower modulus than the mineral. The actual shape of the apatite needles is still a matter of dispute (Section 5.11) but it is clear that bone does consist, in some sense, of rigid needles in a pliant, if anisotropic matrix.

Unfortunately, we are not at the moment in a position to marry theory and experiment rigorously. We shall start by discussing the modulus of elasticity. We saw in Section 5.2, that given the volume fraction and the moduli of the two components, we can use two models to determine the lower and upper theoretical bounds of the modulus. These are the Reuss (equal stress) and Voigt (equal strain) models (Fig. 5.1). KATZ (1971) has applied these models, and refinements of them to bone. He finds that bone indeed falls between the lower and upper limits, but these limits are so far apart that they are not very useful. Katz shows that the Voigt model gives a value for E of about 50 GN m^{-2}, and the Reuss model 2.5 GN m^{-2}. Bone itself is about 18 GN m^{-2}. The great difference between the values given by the two models is a result of the great differences in the values assumed for the two components (1.2 and 114 GN m^{-2} for collagen and apatite respectively). At the moment we must content ourselves by saying that the stiffness of bone is roughly what one might expect from a composite of bone's structure and composition.

The plastic, energy-absorbing behaviour of bone is obviously important for its biological functioning if the bone is in danger of being loaded to fracture–how is this behaviour brought about? In a crystalline material, plastic flow occurs when planes of atoms shear past each other. In a composite like bone, with a fibrous matrix of low modulus, plastic deformation can have several different causes, for instance: plastic deformation of the mineral crystals, with elastic or plastic deformation of the collagen matrix; fracture of the crystals, with plastic deformation of the fibrous matrix; delamination of the matrix from the crystals. The events leading up to and including fracture give us some idea of what plastic flow in bone in fact involves.

When a bone breaks it usually does so with a bang; there is none of the necking down associated with the tensile failure of a very plastic material like steel. The fracture itself is a brittle fracture, but there *is* plastic flow as shown clearly by the load-deformation curve, and bones can be bent slightly and remain bent when the load is removed. So we have plastic flow, absorbing much energy, followed by a rather brittle fracture, with the crack being driven forward by the release of strain energy. The fracture itself does not absorb much energy.

The optical properties of a wet bone specimen loaded in tension will be seen to change when the load-deformation curve bends over, indicating the start of plastic flow. Normally translucent, bone becomes opaque at this point. In a tensile specimen this opacity usually starts in one region and spreads throughout the specimen. We have recently found (CURREY and BREAR (1974)) that the staining characteristics of the bone can also undergo alteration when the load-deformation curve indicates plastic behaviour. The opacity, then, is almost certainly an indication of plastic flow. What is happening? By analogy with plastics that show a similar optical effect, it is most likely that the change in light transmission is caused by the formation of innumerable interfaces (ANDREWS, 1968).

PIEKARSKI (1970), by loading bone in a special way, was able to get a fracture crack that travelled quite slowly. Such a crack required a much greater than normal supply of energy to drive it forward. In these circumstances the fracture surface was very rough indeed, with many Haversian systems being

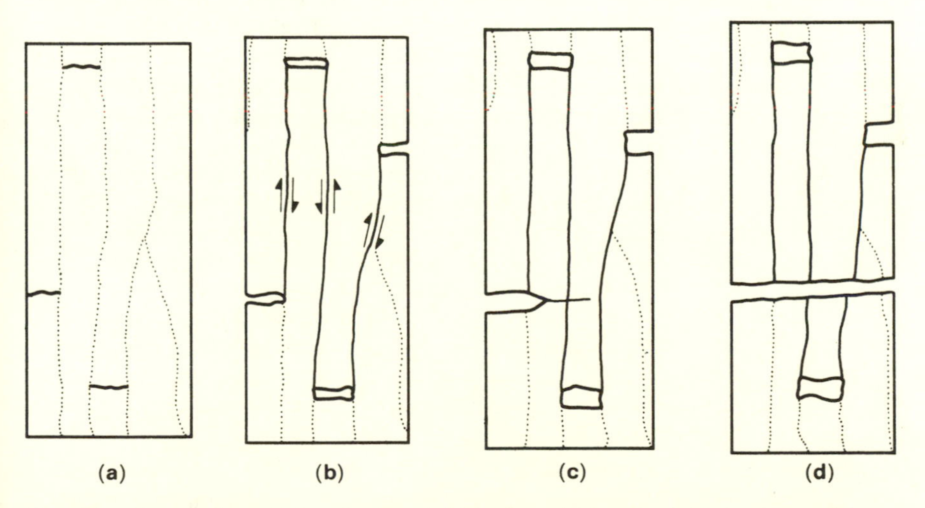

Fig. 5.17 Diagram of suggested mechanism of yield in bone. **(a)** Bone loaded in tension parallel to the long side of the page: transverse cracks appear. These cracks are stopped by discontinuities. Unbroken interfaces are indicated by dots. **(b)** Shear (arrows) allows relative movement of parts of the tissue and the cracks open. **(c)** A crack starts to travel across old interfaces. **(d)** The bone breaks, apparently by brittle fracture.

pulled out from one surface and appearing as little spikes on the other, and vice versa. When, as often happened in these experiments, the crack started slowly but speeded up, the surface reverted to the normal, 'brittle,' appearance, and the energy absorbed fell to the usual value.

What we suggest is happening when a normal bone is fractured in tension is that the bone material yields and plastic flow takes place. This plastic flow is of a kind involving the production of new internal surfaces throughout the yielded region. There is probably delamination of Haversian systems, laminae and lamellae from their neighbours, and the newly formed surfaces then shear past each other. For this shear to occur, each sub-unit must break *across* in some place, but the crack does not travel far, being interrupted by the planes of weakness between the sub-units (Fig. 5.17). These processes require energy. However, the process of shearing cannot continue indefinitely, and sooner or later one of the cracks oriented more or less at right angles to the internal faults and surfaces caused by the yielding manages to achieve a reasonable length and speed, and then travelling very quickly spreads and produces a brittle fracture (Fig. 5.17). It travels quickly because it is able to use much of the strain energy in the bone as it drives forward.

Many composite materials, such as Fibreglass, are tough because it is difficult to get a crack to start travelling. This is also true of bone. However, in Fibreglass the crack remains extremely reluctant to travel fast; this is not what happens in bone. Nevertheless bone material in tension will undergo some plastic flow before a crack can start to run, and this property gives it a reasonable amount of toughness.

The way bone fails in compression is quite different from the mode in tension. Instead of a general rather diffuse yielding with, at first, no obvious histological changes, failure in compression is indicated by lines that run rather straight, often at an angle of about 30° to the applied load. Polarized light studies by TSCHANTZ and RUTISHAUSER (1967) show that these lines are places where the lamellae of the bone seem to have buckled (Fig. 5.18). This again is a fairly standard, though not unique, way for fibrous composites to fail in compression and is shown particularly clearly in wood (Fig. 5.18c). Whether this is the sole mode of failure that is occurring in compression is unknown. Certainly when bone begins to break up in compression, actual cracks and splits occur along these 30° lines.

Although the investigation of the compressive failure properties of bone might tell us more about bone as a material, we shall not consider the matter further here because, in life, bone fails in compression much less frequently than it fails in tension. For instance, the most common mode of failure in long bones is probably brought about by bending. In bending the stress on the tensile side is about the same as that on the compressive side, so that the bone, having a lower yield stress in tension than in compression, yields on the tension side. As the bending continues, some cracks may appear on the compression side, but the crack that travels and breaks the bone in two initiates on the tension side (BURSTEIN *et al.*, 1972). Even when a long bone is loaded in compression the failure is often a tensile one because the bending moment produced is more important than the net compression. Finally, bones, particularly their cancellous ends, and cancellous vertebrae, are often crushed in compression. However, the actual mode of failure of the individual parts of the bone is extremely complex in

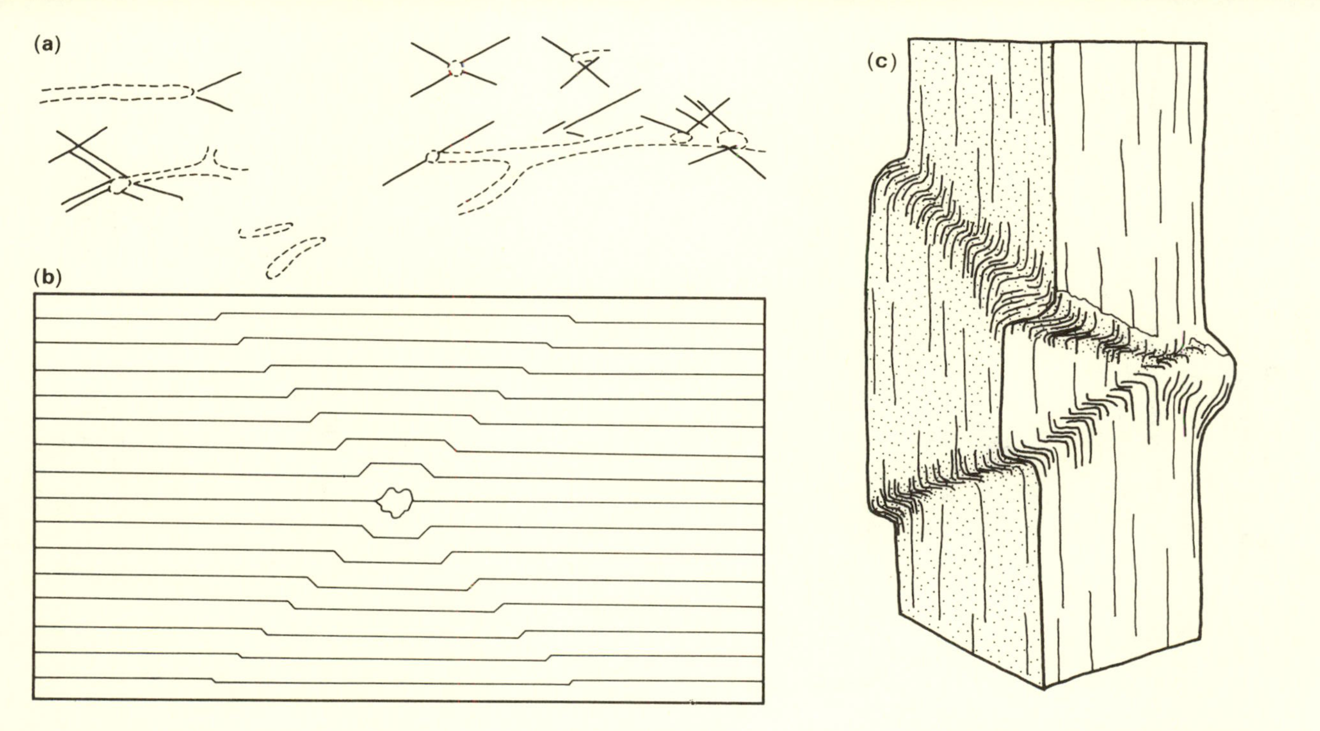

Fig. 5.18 Compression failure in bone and wood, **(a)** Camera lucida drawing of a section of a bone specimen that was compressed from right and left. Dashed lines represent blood channels; solid lines are cracks in the bone. Note how cracks tend to initiate at blood channels, especially where the channel is at a large angle to the loading direction (shown by complete rings). **(b)** Schematic diagram showing how cracks in **(a)** are produced. Solid lines delimit bone lamellae. Buckling has produced two discontinuities that cross. **(c)** Failure in a block of timber compressed from the ends.

these situations, involving a great deal of buckling. The failure is quite unlike the failure of a cubical block of compact bone loaded between two flat platens, which is what is investigated in a conventional compression test.

5.12.2 Anisotropic Behaviour of Bone

Anisotropy in bone has some interesting consequences. That bone is anisotropic is to be expected, of course, from its structure. Because the apatite needles, collagen fibres, lamellae, laminae, Haversian systems and blood vessels show a clear tendency to be oriented along the length of a long bone, it is intuitively clear that the tensile strength and stiffness of the bone will be greater along its length than at a large angle to it (Fig. 5.16).

The different structures of laminar and Haversian bone may be important here: Haversian bone is effectively the same, structurally, in any direction at right

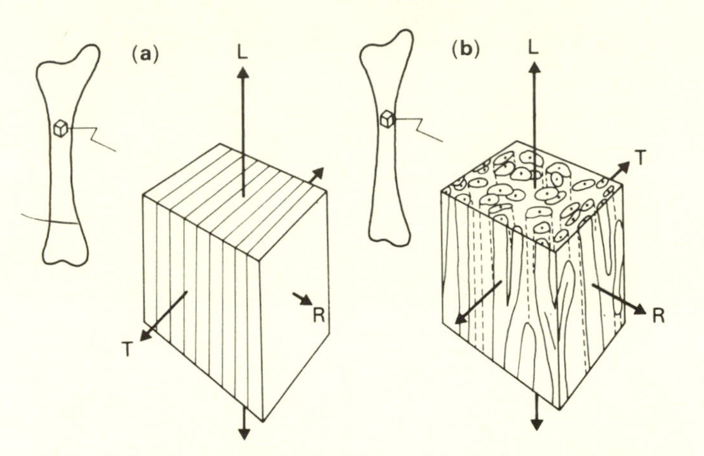

Fig. 5.19 Diagram showing how laminar bone (**a**) differs more between radial and tangential directions (*R* and *T*) than does Haversian bone (**b**). The arrows are vectors representing tensile strength in various directions.

angles to the long axis of the Haversian systems, whereas laminar bone shows great differences between the radial and tangential directions (Fig. 5.19). Little is known about the mechanical consequences of these differences because reports of tests on bone not carried out in the longitudinal direction have not sufficiently characterized the different histological types. However, in laminar bone the tensile strength is less in the radial direction than in the tangential, because the interlaminar blood channels will be a great source of weakness. Again, the strength in the tangential direction is less than that in the longitudinal direction because the general grain of the bone is longitudinal. Haversian bone, on the other hand, is likely to have similar behaviour in the radial and tangential directions.

In general, most of the dangerous loads in bone are likely to be those acting along its length. This is true for bending and compression. Tension is rarely applied to a bone *as a whole*. Torsion, another dangerous mode of loading, will produce maximum tensile forces at roughly 45° to the long axis. Combined bending and torsion will incline the maximum tensile forces on the tensile side towards the long axis. In none of these fairly simple cases will any important loads be directed radially. The situation is as in arthropod cuticle, and structure and function are again seen to be closely related: the material is strong and stiff in the plane of the sheet (if the cortex of a bone can be considered as a sheet) and less strong and stiff in the direction normal to this. Haversian bone is probably strong only in the long axis of the bone.

In general, the great advantage of an anisotropic composite material is that it can be so arranged that in at least one direction the material is much stronger and stiffer than an isotropic material of the same composition could be. This excellence is paid for, of course, by weakness in at least one other direction. This does not matter if it can further be arranged, as in bone and cuticle, that this other direction is one in which large forces do not act. When this cannot be arranged, as in the insertions of muscles acting at a considerable angle to the long axis of the bone, the bone histology alters locally so that its grain is in line with the muscle pull, smoothly altering so as to align itself with the long axis a short distance away from the insertion.

Woven bone, found in immature animals, and in layers in mature laminar bone, has no obvious grain, and it is reasonable to assume that it is fairly isotropic both for strength and for modulus. If so, this must mean that in laminar bone there will be layers that are weak (with reference to loads imposed along the long axis of the bone) sandwiched between strong layers (the layers of lamellar bone). Will this mean that the weak layers will break before the strong layers? Not necessarily for, as the *strains* through the thickness of the bone will be the same in simple tension or compression (the layers will act in parallel) and, since $\sigma = E\epsilon$, reduction in the modulus of the woven bone compared with that of the lamellar bone will correspondingly reduce the stress in it.

The isotropy of woven bone makes it well suited to rapidly growing bones. It is in such bones particularly, in which very active reconstruction is taking place, that the orientation of any particular piece of bone is likely to be considerably altered in time. Therefore, a highly anisotropic bit of bone might be found to lie in a very unsuitable orientation at a later stage of development. Nevertheless, it is probably the great speed with which woven bone can be laid down that makes it particularly suitable for embryos, and for the fossil-scaffolding position it has in laminar bone.

5.12.3 Stress Concentrations in Bone

Although bone is quite tough, it is quite sensitive to stress concentrations. That is, the plastic flow round the tip of a notch is not sufficiently great to prevent the notch acting as a stress-concentrator. In simple theory, relating to brittle materials, notches have the effect of reducing the load a specimen can bear. This particular effect is insignificant in bone, but the effect of notches on the *toughness* of bone is most marked. A specimen of bone with a small notch can absorb only about 15% of the energy absorbed by a similar, but unnotched specimen (Currey, unpublished observation). It can be seen from the load-deformation curves in Fig. 5.20 of beef tibial specimens with and without a fine cylindrical hole that this reduction in toughness is brought about by the inhibition of plastic flow in

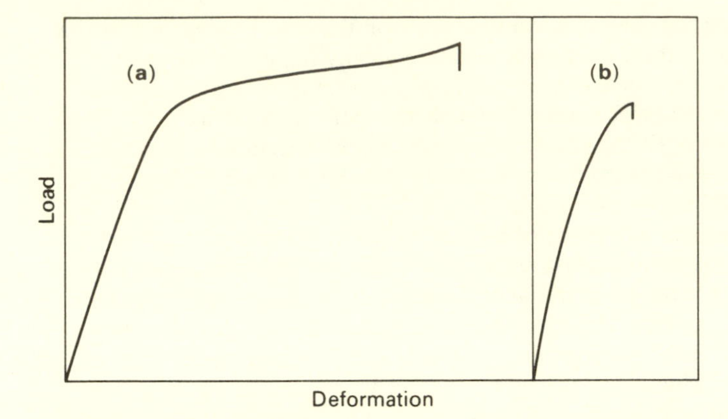

Fig. 5.20 Load-deformation curves for beef tibial specimens. (**a**) Intact bone tissue. (**b**) Bone with a fine cylindrical hole drilled through it. Stresses at failure were about the same but the strain in (**a**) was 4 times the strain in (**b**).

the specimen. Some flow certainly occurs, right at the root of the notch, but much less than occurs in a smooth specimen.

There is, presumably, some lower limit to the size at which stress concentrations are important, though we do not know what this is. However, it is clear that bone is designed to reduce the stress-concentrating effects of all the potentially dangerous discontinuities in it. For instance, if we consider a long bone, we can see that the dangerous forces imposed on it are likely to be along its length. This is particularly so of bending forces, which will produce compression on one side and tension on the other. The main potential stress concentrators in bone are sharp notches on the surface, and blood vessels within the cortex. The outer surface of bones is smooth, so there are no dangerous notches. The blood channels in the cortex are arranged in general along the length of bones with rather few branches travelling across the cortex. One large vessel in long bones, the nutrient artery, must traverse the cortex, and it is surely significant that it is oriented at a very shallow angle. This minimizes the stress-concentrating effect of its channel.

The osteocyte lacunae are themselves possible stress concentrators. In lamellar bone, the flattened lacunae are oriented with one long axis along the length of the bone. This will minimize their stress-concentrating effect. The more or less spherical osteocyte lacunae of woven bone are another manifestation of the isotropy of this type of bone. These devices, and others, are discussed by CURREY (1964*b*).

Of course, if the various substructures in bone are oriented favourably for loads acting in one direction, it is likely that they will be badly placed for loads acting in other directions. Some recent work by Burstein and Reilly (personal communication) illustrates this. They investigated the effects of small (0.33 mm diameter) holes drilled in specimens of laminar bone. Specimens were tested in the longitudinal and tangential directions. Intact specimens are considerably stronger when tested longitudinally than when tested tangentially (184 as opposed to 69 MN m^{-2}). Furthermore, the strain at failure was much greater (0.041 as opposed to 0.008). This might be expected from the general difference in grain direction. Drilling a hole in the specimens had little effect on the static strength and, as we have said, this is a general finding with bone. However, the effect of the stress concentrator on final strain, and hence on energy absorption, is quite different in the two directions. The hole reduced the mean final strain (in the longitudinal direction) from 0.0406 to 0.0121, a reduction to 30% of the unflawed value. In the specimens loaded tangentially the reduction is from 0.0083 to 0.0050, a reduction to only 60%. The laminar bone loaded in the tangential direction appears, in fact, to behave as if it were already subject to stress-concentrators (in the form of badly orientated blood channels and so on) and so the addition of a fierce stress-concentrator has, proportionally, much less of an effect than it has on the specimens loaded longitudinally.

5.12.4 The Effect of Mineralization on Bone

Being long and fairly slender, most long bones must be made of a stiff material to function properly. It is known (ABENDSCHEIN and HYATT, 1970; CURREY, 1969*a, b*) that the stiffness of bone depends critically upon its mineral content: the greater the mineralization, the greater the stiffness. Furthermore VOSE and KUBALA (1959), and CURREY (1969*a*), showed that in the normal range (about 65–70% by weight of mineral or about 50% by volume) the static bending strength also increases with mineral content. Why then, are not bones more mineralized than in fact they are? It is certainly possible, because the tympanic bulla of the fin whale *Balaenoptera physalus* has a very high mineral content (*ca.* 85%) and is very stiff. It is also, however, extremely brittle. This latter fact is an indication of what prevents the long bones of mammals from becoming more highly mineralized. CURREY (1969*a*) has shown that although stiffness and static

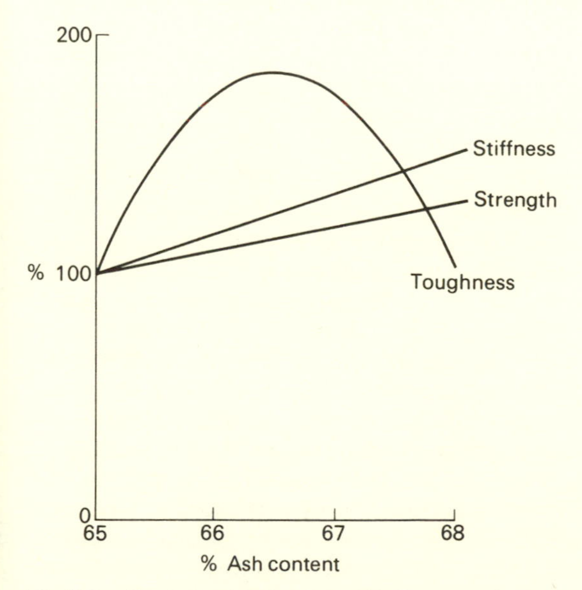

Fig. 5.21 Graph showing the relationship between ash content of bone and its stiffness, strength and toughness. Ordinate: each value of a quality is compared with its value at 65% ash.

strength increase with mineralization, the amount of energy absorbed in static loading and in impact rises, and then falls. The fact that the median amount of mineralization (in rabbits' metatarsals) corresponds quite closely to the highest point on the energy absorption curve indicates that toughness, or energy absorption, is an important variable, and it is this that natural selection acts upon, rather than upon static strength (Fig. 5.21).

The tympanic bullae of whales, and the auditory ossicles of many mammals have a high mineral content. It is satisfying that this is the case, for these bones are in a position where they do not normally have to bear large loads and, their normal mechanical function being removed, they can become modified for another function.

5.12.5 Fatigue in Bone

When a structure is loaded repeatedly it may break at a load that would not cause it to fail if it were loaded once only. This type of failure is known as fatigue failure. Fatigue failure in bone has been analysed *in vitro* by KING and EVANS (1967) and FREEMAN *et al.* (1971) but in a living tissue which can repair defects, at least to some extent, it is dubious to what degree *in vitro* studies are relevant to the life situation.

Fatigue fractures are rather charmingly called 'stress fractures' by orthopaedic surgeons. It has recently been suggested that the fatigue fracture, as seen clinically, in fact comprises two very different sequences of events (Burstein and Frankel, personal communication).

(a) The bone is subjected to quite high loads for a large number of cycles, and eventually a crack appears. This kind of fatigue failure is probably the 'normal' one, initiating from a stress-concentration of some kind. The bone is probably put into a state of high loading by the fatigue of muscles that would normally act to prevent this loading from occurring. It is seen frequently in army recruits who are made to march a great deal when they are not used to it. It also occurs in racehorses. The crack spreads faster than the bone can heal itself, and it is characterized by intense pain on walking.

(b) The bone is subjected to a rather small number of very high loads, and suddenly breaks in two. Such failure is seen in racehorses who have not been lame at all (in contrast to the situation in (a)). The metapodial bone will quite suddenly break into two or more pieces. Again, this type of fracture often occurs in army recruits who have extremely strong, sergeant-major induced, motivation to perform a task that will load the leg bones excessively. Such a task is running fast while carrying a man. There is no premonitory pain, the tibia just snaps. Frankel and Burstein suggest that this type of failure occurs when the normal inhibitions against overloading are overridden, and in consequence the bone is loaded into the plastic region. It does not require many repetitions of such overload for a sufficiently large yielded region to appear for a brittle crack to propagate from it. (Our continual harping on army recruits and racehorses is not coincidental, these are two groups of organisms whose mechanical behaviour is fairly closely monitored.)

At the moment we do not know whether such analysis is correct; if it is, perhaps the important point about it is that it is the muscles surrounding the bones that play the crucial role in determining whether the bones suffer fatigue failure or not. The muscles are strong enough to break many of the bones in the body outright, and it is only the nicely tuned inhibitory system of the musculo-skeletal apparatus that prevents this from happening more often.

Fatigue fracture in bone is extremely difficult to investigate, because the fracture is so much a function of the body's ability to repair incipient fractures before they become dangerously long. We have effectively no information about the body's capabilities in this respect.

5.12.6 Adaptive Growth and Reconstruction in Bone

If a bone is broken and set badly, then with the passage of time it may undergo a process of reconstruction so that it is better able to bear the loads put on it. For instance, a femur set at an angle may, with luck, become straighter by means of bone being added to concave surfaces and taken away from convex ones. If the angulation is too severe, this may not happen, but the excessively stressed cortexes become thick, whereas the less stressed ones remain the normal thickness (PAUWELS, 1968). Similarly, if for some reason a bone is exposed to a greater than normal load, it will, in time, become more massive so that the stresses caused by such a load are reduced. Such adaptive remodelling can occur throughout life, but is most efficient in young animals (LIŠKOVÁ and HEŘT, 1971). The opposite effect is well shown in the wasting of the skeleton of someone confined to bed. Astronauts, unaffected by any gravitational force, start to lose bone.

Bones, then, can adapt their shape, at least to some extent, to the loads imposed upon them. We must consider briefly how this adaptation is brought about. The subject has recently been reviewed by BASSETT (1971). In general, of course, it is clear that the bone responds to strains in the tissue by adaptive reconstruction. (We assume that it is strain that is important, rather than stress, because there is no system that can measure stress directly.) However, there are two problems that must be clearly separated:

(a) How does the bone, or do the bone cells, receive information about the state of strain in the bone?
(b) Why is the reconstruction arising from this information adaptive and appropriate?

It is interesting how much the literature concentrates on the former question, with the implicit assumption that the reconstruction will be adaptive, as long as the local state of strain is 'known'. This is far from being the case, and we shall deal with this latter aspect first.

Take a fairly simple case. A hollow bone is broken, and ill set, and heals (Fig. 5.22a). On being loaded, there will be strain at the four surfaces as indicated. Now, in order to remodel adaptively, surfaces A and C must have bone added, while surfaces B and D must have bone removed. Yet many people writing about this problem (e.g., BASSETT, 1965; JUSTUS and LUFT, 1970) assume that if bone is added where strain is compressive, and removed where it is tensile, then

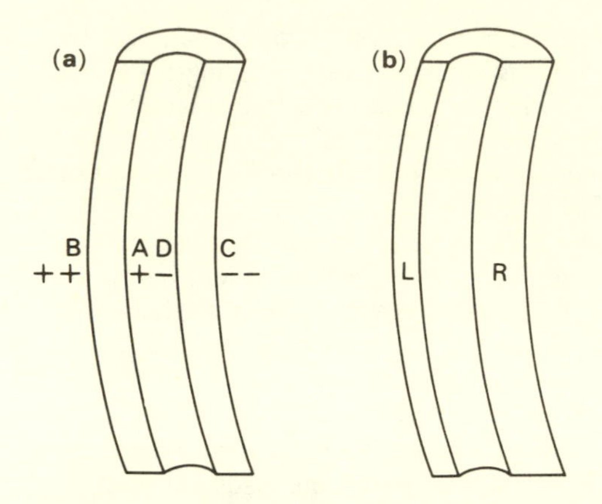

Fig. 5.22 A broken bone, ill-set, heals so that it is curved when not loaded. The strains on being loaded axially are indicated in (**a**) which shows a segment taken from the middle of the bone. The reconstruction that would occur if bone is deposited under compression (−) and resorbed under tension (+) is shown in (**b**).

adaptive remodelling will occur. In the simple example discussed here this will have the alarming result that the right-hand cortex in Fig. 5.22b will increase in thickness and the left-hand cortex will wither. Clearly, such a simple approach will not suffice as an explanation. FROST (1964) suggested a more sophisticated scheme. In this the bone cells measure the *change in curvature* of the surface when load is applied. When a surface becomes more concave, it should be built up;

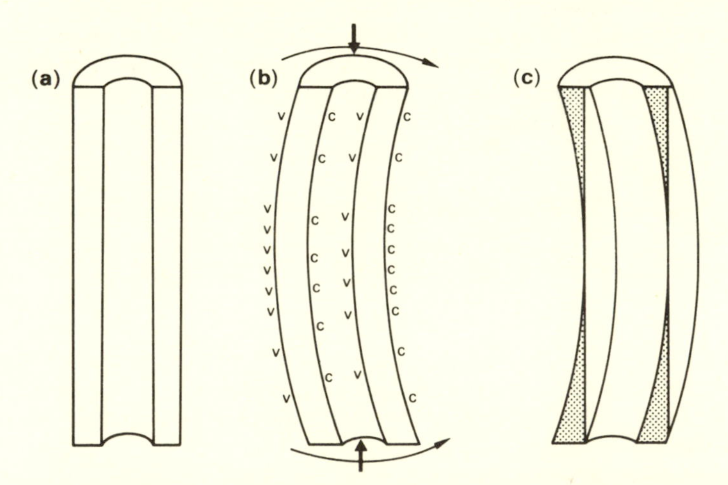

Fig. 5.23 Frost's model of adaptive bone reconstruction. (**a**) A piece of an unloaded bone. (**b**) The bone is subjected to an axial load and to a bending moment. The bending moment bends the bone, and so the 'axial' force also has a bending moment, which is more severe towards the middle. v = surface becomes more convex; c = surface becomes more concave. (**c**) The bone is remodelled. Dotted areas—old bone. When loaded as in (**b**) it will straighten under the influence of the bending moment, so the axial force will act axially.

when it becomes more convex it should be broken down. Reference to Fig. 5.23 shows that this scheme will indeed work. CURREY (1968) pointed out that changes in curvature would be extremely difficult to measure, and that it was much more likely that *changes in strain* with distance from the surface could be measured. This would give the same information. Osteocytes, with their canaliculi penetrating deep into the bone, seem ideally placed to measure such changes. If the strain becomes more tensile, or less compressive, with depth, then the surface should be built up, and vice versa. Currey also pointed out that if a bone were loaded in net tension, even this scheme would produce the wrong kind of reconstruction. He suggested that it was possible that some bones might be loaded in net tension, and that the scheme needed to be still further complicated to take account of this. OXNARD (1971) has produced anatomical evidence that they may not be so loaded. The details of the argument are complex, but what remains agreed is that *no* very simple signalling system will produce adaptive remodelling. This agreement is confirmed by the obvious fact that many structures, such as the coronoid process of the jaw, the zygomatic arch, the neural crest of the vertebrae, the calcaneus and the olecranon process, must be able to ignore any signalling system, otherwise they would alter their shape so that they would experience smaller bending stresses, but would also become useless for their function. This is true, in fact, of virtually all parts of bone that have to apply turning moments about joints. See Currey and Oxnard for a fuller discussion of these points.

Having demonstrated the complexity of the signalling necessary to produce adaptive remodelling (or to inhibit it in many cases!) we can now cast a wary eye over the suggestions as to the nature of the signals themselves. The most favoured signal is that of electric potential probably produced by piezoelectric effects (BASSETT, 1971; GJELSVIK, 1973*a, b*). Bone, like many other organic materials, exhibits the piezoelectric effect–an electric potential that develops when the material is strained. It is suggested that the bone cells could measure this potential. JUSTUS and LUFT (1970) suggest that bone cells respond to the calcium concentration in the vicinity, and that this is determined by the altered solubility of apatite when loaded. Another possibility is that the bone cells respond directly to strains in their own membranes. Experimental evidence for the various suggestions is scant and equivocal. BASSETT (1971) has reviewed the whole question comprehensively.

At the moment we can say little more. The whole subject of adaptive ontogenetic changes in bone is of very great interest, and has attracted a thick fluffy coating of woolly thinking. This is unfortunate, because not only is the matter of considerable clinical and intellectual interest, it is also of great importance to the theme of this book. What we are saying, in effect, is that vertebrates can change, in a single lifetime, the design of their skeletons in response to changed circumstances, but we really don't know how.

5.13 Keratin

Although keratin is frequently referred to as an α-helical fibrous protein along with fibrin and muscle myosin, the term 'fibrous protein' is rather misleading because it suggests that keratin is a fibrous, tensile material like silk or collagen (see Chapter 3). Actually keratin is a rigid material with a structural organization similar to that of insect cuticle and other rigid composites in which a crystalline polymeric fibre is embedded in a highly crosslinked amorphous polymer matrix. The fibrous component of keratin is made up of an assembly of α-helical protein chains or, in the case of bird feather keratin, a complex helical structure derived from β-pleated sheet proteins. The amorphous polymer phase is also a protein and is crosslinked to itself and to the fibres through disulphide bonds between cystine residues. A good example to demonstrate that keratin does indeed function as a rigid material is provided by the insulating down feathers of birds. The insulating properties of down arise from the air which is trapped and essentially immobilized within the dense mass of keratinous barbules. Down is one of the best known insulating materials because a small mass of keratin can trap an enormous volume of air. This high volume to mass ratio is of course important to a flying animal and is possible only because the keratin itself is very rigid. That is, the barbules on the down feather can be extremely thin and yet stiff enough so that the down will not collapse under its own weight or even reasonable compressive loads. A similar argument applies to mammalian hair, and of course keratinous structures such as horn, reptilian scales, hooves, etc., clearly reflect the rigidity of keratin. Keratin is a rather unusual biological material in that it is laid down intracellularly within the membrane of a living cell. This is quite different from the extracellular deposition of bone, insect cuticle and most other biological materials. It means that keratinous structures are in fact made up of dead cells that are packed full of the complex structural material we call keratin.

Many aspects of the biology and chemistry of keratin have been reviewed by MERCER (1961), CREWTHER *et al.* (1965) and RUDALL (1968). We will limit our discussion to its structure and mechanical properties. Early X-ray diffraction studies revealed the presence of two basic forms of crystalline organization in wool fibres; one (the α-form) which existed in the unstretched fibre, and another (the β-form) which was created when the fibre was stretched in steam and held in the extended state. ASTBURY (1933) deduced from the strong similarity between the X-ray diffraction pattern of the β-form and that of silk fibroin that the protein in the β-form was present as extended polypeptide chains, and that the protein in the α-form was folded or coiled in some manner. However, the true nature of the coiled protein in α-keratin was not worked out until the α-helix was proposed as a stable conformation for proteins by PAULING and COREY (1953*a*). The α-helix alone can account for most but not all of the features on the α-keratin diffraction pattern. For example, the 0.54 nm pitch of the α-helix does not match with the 0.515 nm meridional spacing observed for α-keratin. This difference could be accounted for if the α-helix is distorted to form a super-helix, and several coiled-coil structures based on aggregates of α-helices have been proposed including a three-stranded rope (CRICK, 1953) and a seven-stranded cable (PAULING and COREY, 1953*b*). More recent X-ray evidence tends to support a structure similar to the three-stranded rope proposed by Crick (FRASER *et al.*, 1962). Electron microscope studies clearly reveal the two-phase organization of keratin. Stained sections show closely packed, rod-like microfibrils about 7.5 to 8 nm in diameter surrounded by a densely staining, amorphous matrix. High resolution micrographs (FILSHIE and ROGERS, 1961) have revealed a protofibrillar substructure within these microfibrils. It appears that the microfibril is made up of a ring of nine protofibrils with the possibility of an additional two protofibrils within the circle of nine, and it has been suggested by FRASER *et al.* (1962) that the protofibril is a three-stranded rope of α-helices (see Fig. 5.24).

The structure of β-keratin was analysed recently by FRASER *et al.* (1969). They obtained β-keratin by stretching porcupine quill in steam and observed an X-ray diffraction pattern indicating that the crystalline protein in β-keratin exists in the form of anti-parallel, β-pleated sheets. They concluded that the crystallites are on the average 2.6 sheets thick with 10 chains per sheet. This means that each crystallite contains about the same number of protein chains as does the microfibril of the α-form, and it is suggested that each β-crystallite is formed from a single α-microfibril when keratin is stretched. Also, the anti-parallel arrangement

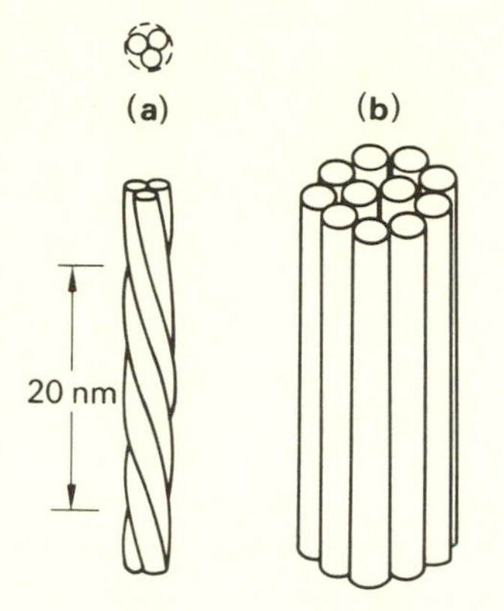

Fig. 5.24 The structure of the α-keratin microfibril as proposed by FRASER *et al.* (1962). (**a**) The protofibril is a 3-stranded rope of α-helices. (**b**) The microfibril contains a ring of 9 protofibrils and possibly 2 central protofibrils.

of the β-crystallite implies that the α-microfibril contains an equal number of oppositely directed α-helices.

The fibre component, of course, accounts for only about half of the keratin structure. Of equal importance is the crosslinked matrix that provides the mechanical continuity between the individual microfibrils. It has been known for a long time that keratin is very rich in the amino acid cystine, which makes up from 8 to 9% of the total amino acids present. Further, keratin is insoluble in most aqueous solvents and can be readily dissolved in reagents that break disulphide bonds. Many workers have isolated and analysed the soluble proteins from dissolved keratin and found that the proteins fall into two general classes. One, the low-sulphur fraction, is presumably derived from the microfibrillar fraction, and the other, a high-sulphur fraction, is derived from the amorphous matrix. For example, ALEXANDER and EARLAND (1950) used peracetic acid to oxidize the disulphide bonds of wool. They isolated a low-sulphur fraction accounting for 60% of the total protein and containing 1.9% sulphur, and a high-sulphur fraction that accounted for 30% of the total and contained 5.8% sulphur. In general, the proteins in the low-sulphur fractions can be shown to form α-helices spontaneously in solution while the proteins of the high-sulphur fractions have a random-coil conformation (see CREWTHER *et al.*, 1965). These observations are consistent with the idea that keratin is a rigid composite in which α-helical protein fibres are stabilized by a highly crosslinked amorphous protein matrix through disulphide bonds.

The literature on the mechanical properties of keratin is dominated by the extensive work carried out by the textile industry on wool. Although much of this work is useful in understanding the relationship between the structure and mechanical properties of keratin, very little of it applies to the biological function of keratin. For example, Figs. 5.25a and b show a series of stress-strain curves for wool under differing conditions of temperature and water content (PETERS and WOODS, 1955). In each curve there is an initially high modulus or elastic region

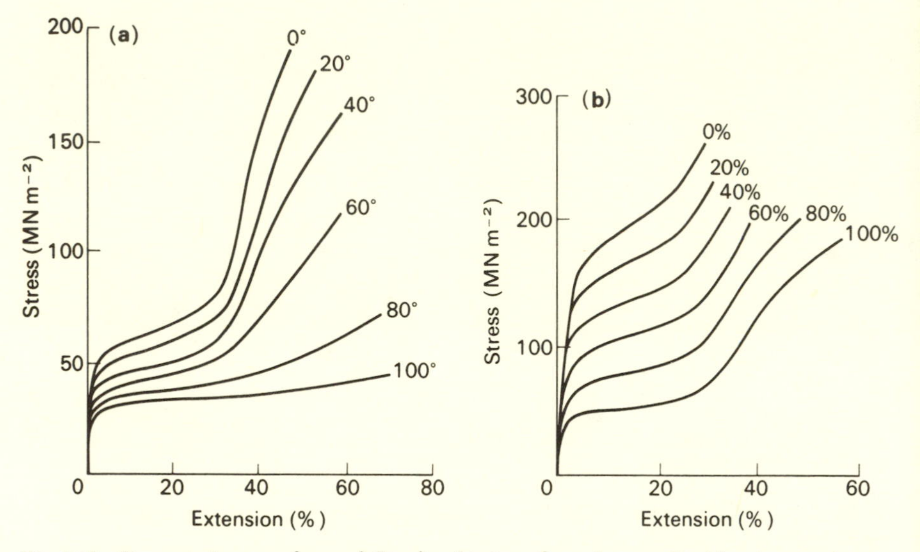

Fig. 5.25 Stress-strain curves for wool. Results of tests performed on wool in (a) water at various temperatures and (b) air at various relative humidities (PETERS and WOODS, 1955).

where the fibre shows typical Hookean elasticity. This is followed by a shoulder or yield region and then by a region of increased modulus or post-yield region. Quite clearly, the biological range must lie within the initial elastic region of these curves because beyond this region the modulus drops by a factor of ten or more. This abrupt drop in rigidity must surely mean that a structure made of keratin would cease to function properly at this point. The modulus in the elastic region is about 4×10^9 N m^{-2}, and this value appears to be the same for all the test conditions shown. The extent of the elastic region, however, varies a great deal with both temperature and water content.

It is generally accepted that the transition from the elastic region to the yield region occurs when the α-helical structure of the proteins in the microfibrils begins to break down and when the extended β-keratin is first formed. X-ray diffraction studies of stretched wool fibres clearly show a decrease in the intensity of the α-diffraction pattern and an increase in the β-pattern with extension beyond the elastic region (ASTBURY and WOODS, 1933; BENDIT, 1957). The β-pattern, however, does not form immediately when the fibre is stretched, but only after the fibre has been held in the extended state for a considerable period of time. There is an extensive literature on all aspects of this $\alpha \rightarrow \beta$ transition and on the significance of the post yield region (see CREWTHER *et al.*, 1965) which is not directly relevant to this discussion. One interesting point, however, is that keratin stretched beyond the elastic region shows considerable creep or stress-relaxation, and the mechanism associated with these long term changes involves an exchange of the disulphide bonds within the amorphous matrix phase. This reflects the importance of the contribution that the matrix makes to the overall properties of keratin, and it suggests that the variations in the extent of the elastic region (Fig. 5.25) may be due to a loosening of the matrix structure at higher temperature or with increased water content. Of course, both temperature and water content may directly affect the stability of the α-helical microfibrils.

The keratinous structures found in mammals are all of the α-form and have mechanical properties as described above. Another form of keratin, however, is found in the scales of reptiles and in the feathers of birds. This keratin appears to have a crystalline organization in the unstretched state that is similar to the β-form created when α-keratin is stretched. The organization of feather keratin as observed with the electron microscope appears to be very similar to that of α-keratin from wool. Feather keratin contains closely packed microfibrils embedded in an amorphous matrix, but the microfibrils are only 3 nm in diameter rather than the 8 nm diameter of the α-microfibrils (FILSHIE and ROGERS, 1962). In a recent X-ray diffraction study of feather keratin FRASER *et al.* (1971) concluded that the microfibrils contain β-pleated sheet structures twisted in a helical manner around a central axis. The basic crystalline unit contains two sheets symmetrically disposed around the central axis, and each sheet contains four polypeptide chains. FRASER *et al.* (1971) suggest that this complex β-structure is probably found in the β-keratin of reptiles as well. Because the protein chains in the feather keratin microfibril exist in an extended rather than an α-helical form, there should be no yield region associated with the $\alpha \rightarrow \beta$ transition. This should make feather keratin and other β-keratins more useful as rigid structural materials because they will probably have a much higher strain at 'failure'. In fact, RUDALL (1947) observed that the keratinous epidermal thickenings which make up the protective scales on lizards contain β-keratin while the softer inter-scale regions of the lizard skin contain α-keratin.

5.14 Gorgonin and Antipathin

Gorgonin is the name given to the dark brown, organic, flexible material that either constitutes or is the major constituent of the axial skeleton of the 'soft' corals in the octocorallian order Gorgonacea, suborder Holaxonia. The X-ray diffraction pattern of gorgonin has the characteristics of a pattern of poorly aligned collagen (MARKS *et al.*, 1949) and the presence of 40 to 60 imino acids per 1000 residues (ROCHE *et al.*, 1963; GOLDBERG, 1973) also indicates collagen. GOLDBERG (1973) has found an axially repeated pattern of 33–35 nm that is subdivided into 3 equal segments in transmission electron micrographs. There is also a high content of iodo- and diiodotyrosine found in gorgonin that is not found in other collagens. The structural relationship between the amino acids and the 10 pentoses and hexoses per 1000 residues is unknown. In some genera notably *Plexaurella*, gorgonin is intimately associated with polycrystalline bodies of calcite (Fig. 8.5). Calcified octocoral axes will be discussed in Section 5.19.8.

Gorgonin is deposited extracellularly in concentric layers around a central hollow, chambered canal that seldom exceeds 100 μm diameter. GOLDBERG'S (1973) scanning electron micrographs of fracture surfaces show these layers to be made of fibres. The fibres appear to be lath-shaped (rectangular in cross-section) rather than round (Wainwright, unpublished). The material has a strong axially positive birefringence, but polarized light microscopy of thin tangential sections indicates that molecular orientation probably varies at very low angles to the axis.

Each concentric layer seen in cross section consists of a thick sublayer with tangential birefringence and a very thin sublayer with weak radial birefringence. The 3-dimensional fibrous texture of gorgonin is complex, but does show a high degree of preferred axial orientation.

Gorgonin is flexible, allowing axes of less than 5 mm diameter to be tied in knots without breaking. The tensile strength and modulus are what one would expect for an oriented polymer with a fair degree of crosslinking. Thicker axes are quite rigid. When dry, the material is brittle.

Antipathin is a material that superficially resembles gorgonin: it is a dark brown (it is, in fact, the semiprecious 'black coral'), fibrous (in SEM: GOLDBERG, 1973) polymeric material comprising the axial skeleton of certain soft corals in which it shows concentric layering and axial birefringence. Antipathin differs from gorgonin in that it is known only in the Antipatharia, an order of poorly known soft corals in the subclass Zoantharia (gorgonin occurs in subclass Octocorallia). Antipathin contains about 10 imino acids per 1000 residues but has from 254 to 277 aromatic amino acids per 1000 residues (gorgonin has 40 to 60 per 1000)

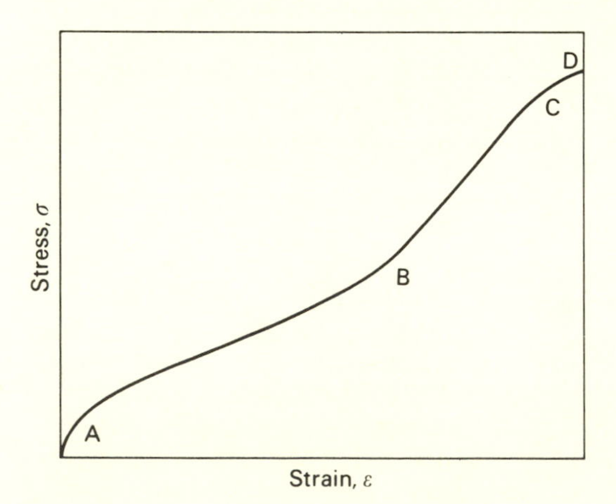

Fig. 5.26 Stress-strain curve for the axial skeleton of *Cirripathes* sp. The figure is explained in the text (Biggs and Wainwright, unpublished).

(GOLDBERG, 1973). Although the macromolecular organization is unknown, a favourite hypothesis is that the aromatic amino acids may be involved in cross-linking the structural proteins: this function of aromatic amino acids has been established in tanned insect cuticle (Section 5.9).

The tensile stress–strain curve for the axial skeleton of *Cirripathes* sp. is shown in Fig. 5.26 (Biggs and Wainwright, unpublished). Ultimate tensile stress and strained tensile modulus are given in Table 5.3. Slopes of the regions AB and BC are linear and fracture occurs beyond C. The ratio of the moduli of the two regions (slope AB/slope BC) and the strain ($\epsilon_c = ca.\ 0.1$) at which the changeover occurs are very nearly the same for all specimens. The increase in modulus with increased strain is characteristic of a strain-crystallizing polymer, wherein the change in mechanical behaviour depends on a critical state of strain.

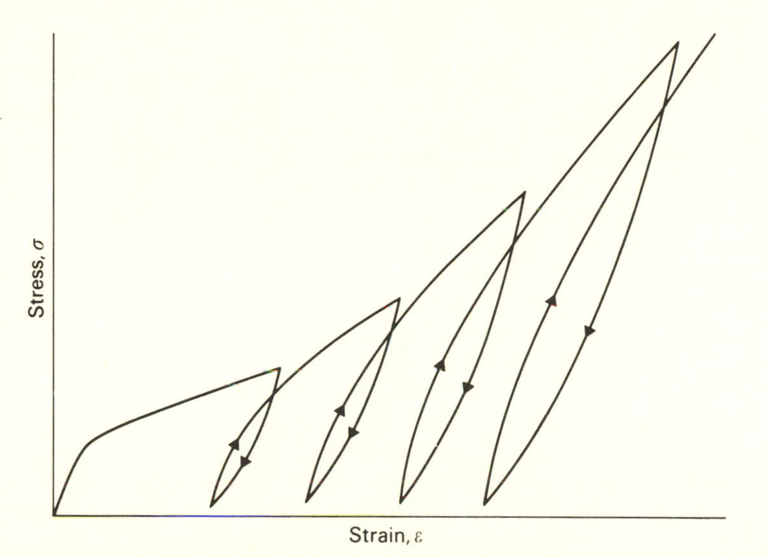

Fig. 5.27 Stress–strain curve for the axial skeleton of *Cirripathes* sp. subjected to cyclic loading and unloading (Biggs and Wainwright, unpublished).

Cyclic loading and unloading of the material confirms this interpretation. Figure 5.27 shows that the modulus increases on successive loading in the same way that it does in a single test. Further, the higher modulus is retained by an unloaded specimen for several hours at 18°C, but after 24 hours a tensile test shows that the modulus has reverted to that of the nonstressed state. Thus, structural changes that cause an increase in modulus are apparently reversible but in a very time-dependent way.

Stress–relaxation curves of antipathin are exponential for about 3 hours (10^3 s) and then tend to flatten, implying a relaxation spectrum. Relaxation times of non-prestressed specimens (10^5 to 10^6 s) and of prestressed specimens (10^{11} to 10^{12} s) are further evidence of structural changes wrought by stressing.

The mechanical behaviour of antipathin is similar to that of keratin, some other crystalline polymers and partially crosslinked rubbers. The various regions of the stress–strain curve (Fig. 5.26) are generally attributed to:

OA: Molecular extension of both fibrous and amorphous constituents.
AB: Plastic deformation of amorphous component; elastic deformation of fibrous component.
BC: Transition of coiled molecules into aligned, crystalline regions, followed by further elastic extension of fibrous component.
CD: Further deformation of the amorphous component leading to fracture.

That the more highly crystalline state remains for 10^3 s but finally reverts in 10^5 s implies a high viscosity for the amorphous component.

The importance to the coral of these interesting properties is not yet known, and again we may ask if this arrangement has been selected for its mechanical function or for, say, the energetic cost of synthesis.

5.15 Structure of the Plant Cell Wall

In botanical parlance, a 'fibre' is a type of wood or secondary xylem in angiosperms; it is also used to indicate a single cell of specific type, e.g., the cotton fibre. The unit of cellulose that is most often seen in electron micrographs is called a 'microfibril' and is known to be an aggregate of cellulose molecules, presumably in parallel array. We shall follow this nomenclature when speaking of plant structures.

The essence of plant cell wall is that it is composed of a composite material of cellulose microfibrils in an amorphous polymeric matrix. The woody supportive cells of all higher plants, the cells of fungal hyphae and algal filaments and the fibres of the cotton seed pod are all more or less cylindrical in shape. Since we have some knowledge of mechanical properties of these cell walls and very little of cell walls having other shapes, we shall make the simplifying assumption that plant cells are microfibril-reinforced cylinders. A modern treatment of the plant cell wall as a cellular component is that by FREY-WYSSLING and MÜHLETHALER (1965), while ROELOFSEN (1959) gives a more comprehensive account of the diversity in cell wall structure.

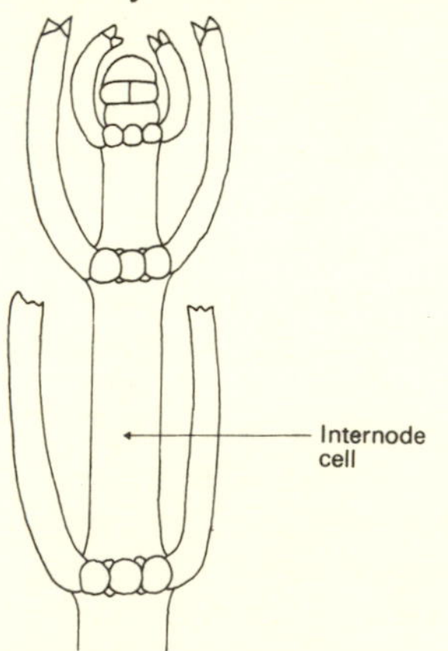

Fig. 5.28 The growing tip of a *Nitella* shoot. Internodal cells attain dimensions of 100 mm long and 0.9 mm diameter.

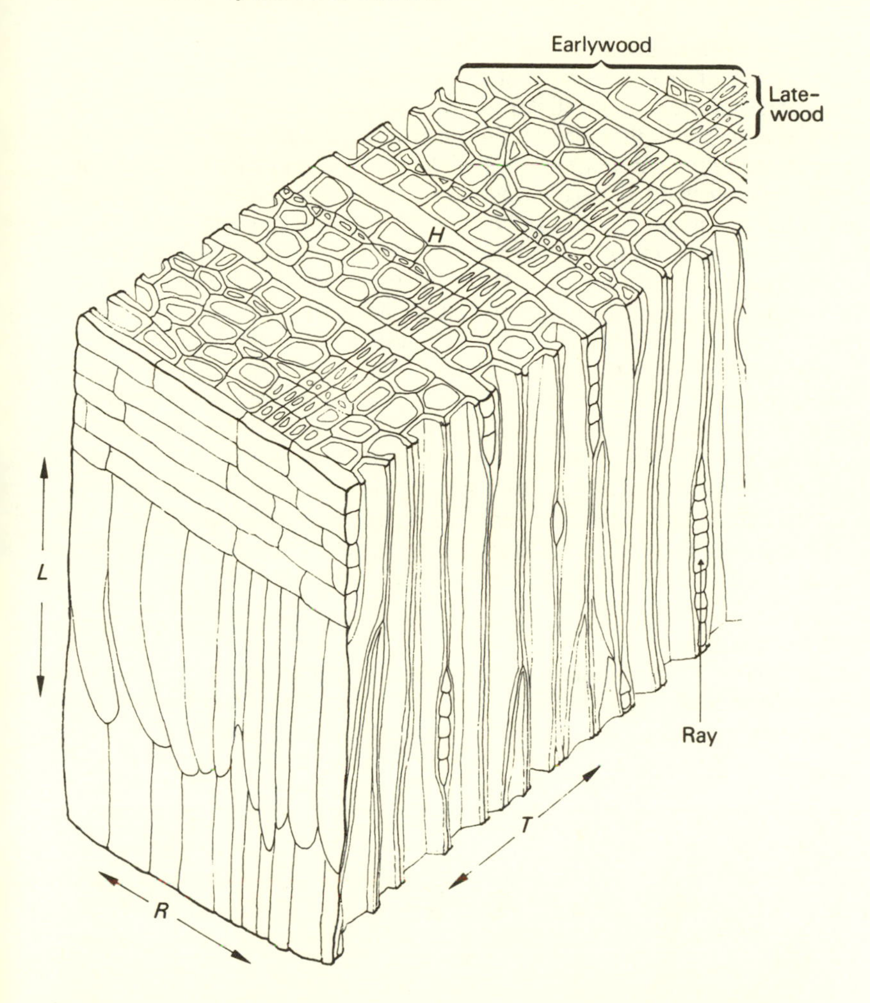

Fig. 5.29 3-dimensional view of woody tissue. *H*, horizontal section; *R*, radial section; *T*, tangential section. *L*, longitudinal section. All vertical cells shown are thick-walled secondary xylem fibres and vessel members. Horizontal thin-walled cells are ray cells that are involved in radial translocation of solutions in the stem.

The cell wall has a concentrically layered structure. The outer layers are the first formed in ontogeny. The cellulose microfibrils within a layer have a particular degree of preferred orientation: both the direction and the degree of preferred orientation of cellulose varies from layer to layer, as does the volume fraction. The chief structural differences between the walls of cells of an isolated filament, such as *Nitella* (Fig. 5.28) and of the fibres in woody tissue (Fig. 5.29) are in the nature of the outermost layer, the degree and angle of preferred microfibrillar orientation and the ratio of outer cell diameter to wall thickness D/t. The latter is very high ($\sim 10^4$) in the cells of algal filaments.

5.15 Structure of the Plant Cell Wall

5.15.1 Cell Wall Structure in Nitella

Nitella is a genus of demersal plants of still fresh water. The shoot is upright and is a linear series of short multicellular nodes separated by long internodes. Each internode is a single cell that may attain dimensions of 100 mm long x 0.9 mm diameter at maturity. This large size is the chief reason that *Nitella* internodal cells have been selected for mechanical studies. The cell wall has a thin external cuticle, of unknown composition and properties (GREEN and CHAPMAN, 1955). Inside the cuticle, the rest of the wall is a series of layers of microfibrils and matrix. The microfibrils of the outer layer are dispersed and have preferred axial orientation; those of the innermost layers are more close-packed and have preferred circumferential orientation; those in intermediate layers have intermediate orientation (PROBINE and PRESTON, 1962). The total distribution is predominantly circumferential: the cell wall as a whole shows negative axial birefringence. At maturity, the innermost layer is suggested by Probine and Preston to be of crossed fibrillar structure. PROBINE and BARBER (1966) published a transmission electron micrograph of a section of the innermost wall layers of *Nitella*. The micrograph shows a pattern of microfibrils that is indistinguishable from the helicoidal array of chitin fibrils in the cuticle of insects and crustaceans. This is discussed as the 'Bouligand pattern' in Section 5.9.

5.15.2 The Tracheid

'Tracheid' refers to the predominant cell type of gymnosperm wood. Tracheids and fibres differ in that tracheids have wide imperforate ends and conduct fluid, whereas fibres are nonconducting and have tapered, imperforate ends. A wood

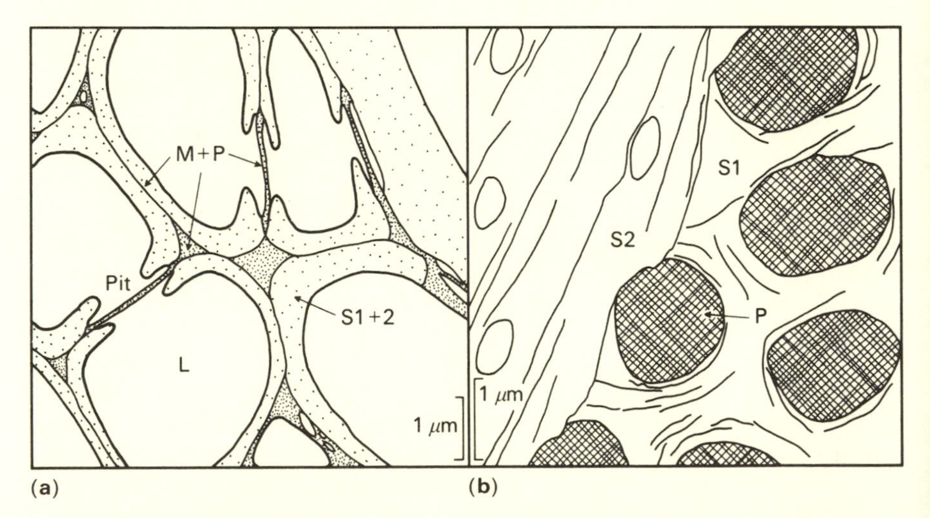

Fig. 5.30 Fine structure of wood cells. (**a**) Transverse section through several tracheids (*Quercus rubra*) showing their close-packing, pits and wall layers. (**b**) View of vessel wall (*Tilia americana*) from inside. S2, shown on the left has been peeled off on the right revealing S1 and the full diameter of the pits. L, lumen of tracheid; M, middle lamella; P, primary layer; S1, S2, secondary layers. (Drawn from electron micrographs in KOLLMAN and CÔTÉ, 1968.)

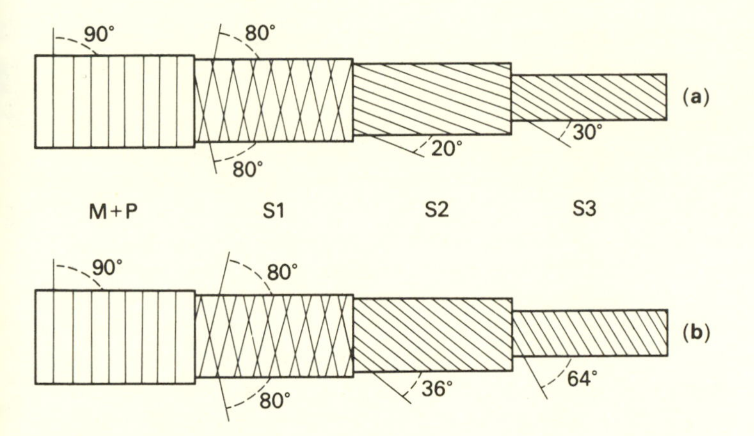

Fig. 5.31 (a) Tangential and (b) radial walls of juniper tracheid showing mean fibre angles. M, middle lamella; P. primary layer; S1, S2, S3, secondary layers (after MARK, 1967; courtesy of Yale University Press, New Haven).

fibre or a tracheid at its functional maturity is a cell wall whose protoplast has disappeared and is replaced by 'sap', air, or resin. This cell wall is essentially a shell with a hollow, attenuated cylindrical shape and a polygonal cross-section (Fig. 5.29). It is often many times longer than wide.

Much of the information on structure and mechanical properties that follows is taken from MARK's (1967) uniquely thorough analysis of the mechanical features of tracheids of *Juniperus virginiana*.

Tracheids are bonded together in woody tissue by amorphous matrix materials in a layer called the middle lamella that is continuous around the tracheids throughout the tissue (Fig. 5.30). The first-formed and outermost layer of the wall is the thin primary layer (P) containing a dispersed network of cellulose microfibrils that have a preferred circumferential orientation. The next layer inward is the secondary layer in which cellulose microfibrils are in close-packed array and make up the bulk of the wall material (Fig. 5.30). This secondary layer is separable into three component layers: (1) A thin outer (S1) layer has sublayers containing microfibrils lying in closely parallel helices around the cell. The *fibre angle*, the angle between the microfibrils and the long axis of the cell, is greater than 55° (Fig. 5.31). Helices in alternating sublayers in S1 are right- and left-handed. (2) In S2, making up 50–60% of the thickness of the entire cell wall, the microfibrils all lie in helices of a single direction and their fibre angle is less than 45°. (3) S3 is a thin innermost layer whose microfibrils describe helices in a single direction and whose fibre angle is somewhat greater than that of S2.

Tracheids are close-packed and, due to their origins as radial series of cells—each series arising from the division of a single cambial cell—most approach a rectangular shape with radial and tangential sides (Fig. 5.31). There is an abrupt change in the fibre angle in S2 and S3 at the corners such that fibre angles on radial walls are 15° to 25° greater than those on tangential walls.

Secondary walls are perforated by holes, called pits (Fig. 5.30) that allow the fluid cellular contents to move laterally from cell to cell across a membrane that is supported by the diaphanous primary walls of the two cells. Mechanically, pits are interruptions in wall structure and therefore may be stress raisers, crack stoppers and sites where shear between microfibrils may occur. Pits may be simple, slit-like gaps between microfibrils of secondary walls, or they may have a border of circumferentially arranged S2 microfibrils. The pit border may extend as a dome over the pit membrane so that the hole in the S2 is a third the diameter of the pit membrane (Fig. 5.30).

The proportions of chemical constituents vary throughout the thickness of the woody cell wall. Figure 5.36 shows the approximate percentages on a dry weight basis of cellulose, lignin and hemicelluloses across a woody cell wall. Cellulose contributes half the dry weight of wall material in S2 and is present in substantially smaller proportion in S1 and S3. It is obviously highly dispersed indeed in the primary wall. In order to understand mechanical properties of a composite, we need to know the volume fractions of the constituents. MARK (1967) finds that for late springwood of *Juniperus virginiana*, cellulose is 50% of the tracheid weight, thus accounting for 44% of the volume. Hemicellulose is present as 4% of the volume. We have little such information, but we can predict that the amorphous component will contain more water per unit weight of dry matter than will the crystalline cellulose fraction.

5.16 Mechanical Properties of Cell Walls

PROBINE and PRESTON (1962) cut narrow rectangular strips from walls of *Nitella opaca* and determined a number of tensile properties in directions perpendicular and parallel to the cell's long axis, and thus respectively parallel and perpendicular to the direction of preferred orientation of cellulose. Under cyclic loading, they observed an extension versus time relationship as shown in Fig. 5.32. Upon loading a longitudinal strip, an initial extension increased rapidly for a few seconds and then levelled off to a slow creep rate. Extension upon subsequent loading was greater. On unloading, less than a third of the extension was recovered

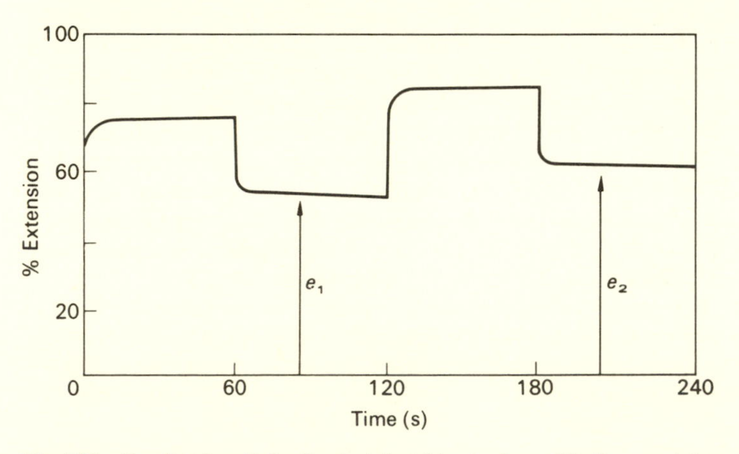

Fig. 5.32 Results of cyclic loading tests-(axial tension) on a *Nitella opaca* internodal cell (PROBINE and PRESTON, 1962). Courtesy of the *J. exp. Bot.*

instantaneously. In creep tests of 6000 s duration, strain rates remained the same or slightly increased throughout the last decade of time. Initial longitudinal extensions (1 minute after loading) were of the order of 10 to 20% according to the stress. Maximum extension after 6000 s was 30%. No figures are given for strains and strain rates in the transverse strips of cell wall material. Transverse creep was within the limits of error of measurement for loads that were great enough to break longitudinal strips.

Nitella cells grow in length much more rapidly than in girth. Young cells may increase in length by 20% per 24 hour period. Throughout growth, turgor pressure is maintained at a constant level. It is noteworthy that the longitudinal modulus (E_L) is initially a fifth of the transverse modulus (E_T), whereas, at maturity E_L has increased to half the value of E_T. E_T increases slightly during this growth period. There is a decrease whose nature is not understood in water-soluble material of the wall as growth rate decreases. The chemical basis for the increased longitudinal rigidity is not indicated.

Stress relaxation experiments of whole *Nitella opaca* cell walls stressed longitudinally were performed by HAUGHTON *et al.* (1968) over a temperature range of 0° to 50° C and time intervals 1 to 300 s. They record an approximately 13% decay in tension over 300 s at every temperature; at any time after setting the strain, they found a 14% reduction of tension on raising the temperature from 0° to 50° C. The relaxation spectrum of *Nitella* cell walls is flat over 10^3 s, insensitive to temperature and has a high value: 100 MN m^{-2}. This behaviour is characteristic of highly crystalline polymers and of amorphous polymers below their glass transition temperatures.

Cellulose is a highly crystalline polymer (TRELOAR, 1960) that has a Young's modulus of 100 GN m^{-2}—a hundred times that of *Nitella* wall (1.0 GN m^{-2}: PROBINE and PRESTON, 1962). It therefore appears that the measured tensile modulus of *Nitella* wall is primarily that of the amorphous component and that this component is a rather highly crosslinked one—much more so than, for example, the matrix of mesoglea discussed in Section 4.5. That E_T, parallel to the preferred orientation of cellulose in *Nitella* wall, is two to five times higher than that of E_L indicates—in common with mesoglea—that the crystalline fibrillar component is acting as a reinforcing filler. Therefore, we may expect there to be no direct continuity among cellulose microfibrils.

Plant cells increase in volume during ontogeny. At maturity they are more-or-less rigid bodies whose viscoelastic properties allow them to sustain and to recover from deformations due to externally applied forces. The bodies of animals, e.g., sea anemones, are highly flexible and sustain large deformations. Such animals have a 'resting shape' that is achieved by slow elastic recovery of the mesoglea. This slow elastic recovery is possible because there is a three-dimensional network arrangement joining the crystalline collagen molecules to the amorphous matrix molecules. Dynamic tests of *Nitella* and wood cells (see below) do not reveal such a network relationship between cellulose and its amorphous matrix. Once plant cells have sustained a 10% strain, most of it remains as a permanent deformation.

Woody cells, such as tracheids of *Juniperus* to be discussed here, also may be expected to have different properties during growth than at maturity. However, we have information only on the mature tracheids. Again the information of MARK (1967) is selected for use here because of its breadth, depth and internal consistency.

The primary load-bearing component in the tracheid is cellulose, in close association with amorphous hemicelluloses. Xylans act as intermicrofibrillar adhesives in the wall, while other glycans perform this function in the middle lamella between walls. Finally, the role of lignin appears to be to act as a bulking agent and to control the hydration of hydrophilic moieties in the wall.

Mark's mathematical analysis predicted the greatest stresses to be shear stresses in the S1 and that this layer should be the first layer to fail in tension. Also, since the shear stresses of S1 and S2 are of opposite sign, he predicted that separation of these layers would be a second point of fracture. Both these phenomena were confirmed by experimentation.

For reasons that will be made clear in Section 7.1.2 we can predict that changes of diameter in opposite directions will occur in S1 and S2 when the tracheid is put in tension. Fibre-wound cylinders in tension will become longer and thinner.

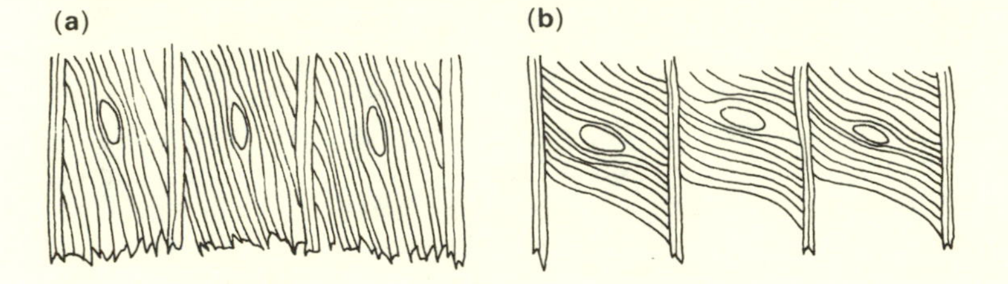

Fig. 5.33 Tough (a) and brash (b) fracture in wood as it is related to the fibre angle in the S2 layer.

The decrease in diameter per unit increase in length varies with the initial fibre angle as shown in Fig. 4.7c. Since the fibre angle of S1 is 80° and that of S2 is 20-35°, we would expect fracture to arise at this interface as S2 shrinks away from S1.

Mark reports this separation and a helical splitting between microfibrils of S2 as two zones where fracture occurred first. Failure was never observed to initiate in the middle lamella. He attributes the concomitant fracture in S1 to the high shear modulus and high shear stresses there.

Redistribution of stresses following failure in S1 may lead to two types of fracture of the tracheid–and of wood–depending on the fibre angle. Figure 5.33 shows brash and tough fracture of wood samples. In brash fracture, secondary bonds are broken, allowing separation of microfibrils in S2. In tough fracture, the primary covalent bonds of cellulose chains must break. Brash fractures follow suddenly after failure of S1, whereas tough fracture will proceed more slowly, splintering the wood and requiring much more energy.

Many of Mark's specimens included cell wall fragments as well as the whole cell walls being tested. He made a most interesting discovery about the different mechanical behaviour of an entire cylinder and a fragment of cylinder wall material. Figure 5.34 shows that such an entire cylinder shows linearly elastic behaviour over a large range of stress. The half-cylinder shows curvilinear stress–

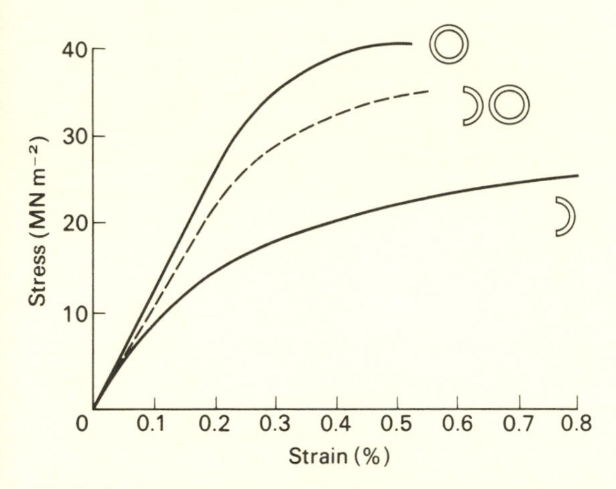

Fig. 5.34 Stress–strain curves for fibre-wound plastic cylinders and cylinder fragments. Cross-sections of the specimens are shown by their respective curves. The dashed curve represents the total load on both specimens divided by their combined cross-sectional area (from MARK, 1967; courtesy of Yale University Press, New Haven).

strain behaviour from the first. A plastic cylinder with its elastic winding fibres intact or an intact tracheid will behave as the Voigt equal strain model in Fig. 5.1a. In the fragment of plastic cylinder or in the tracheid fragment, the fibres are not continuous and the behaviour is that of the plastic matrix in series with fibres acting as fillers as is represented by the Reuss equal stress model in Fig. 5.1b.

This means that an important difference between non-lignified *Nitella* cell walls and tracheids is that the cellulose is neither continuous nor tightly crosslinked in

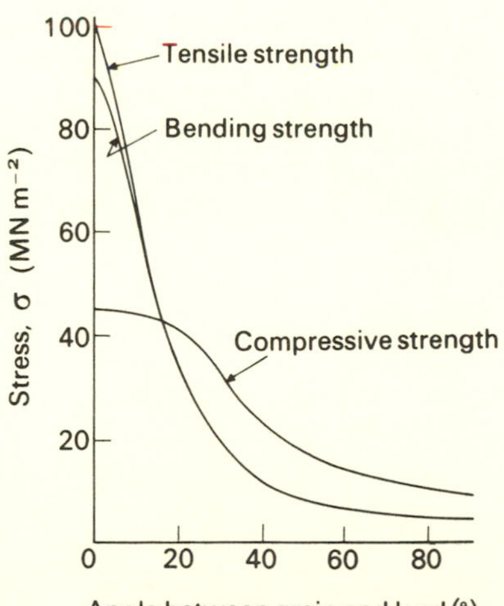

Fig. 5.35 Graph showing the strength of ash timber at various angles to the grain (after KOLLMAN and CÔTÉ, 1968).

the former but is so tightly crosslinked in the lignified tracheid as to convey to it linearly elastic behaviour of continuous cellulose. Figure 5.35 shows values for the strength of ash timber according to the angle between the applied force and the grain.

The viscoelastic properties of individual cell walls have been studied only in *Nitella*. The stress relaxation curve shown in Fig. 3.28 has already been discussed at the end of Section 3.4.2.

5.17 Structure of Wood

Wood is complex structurally and functionally. In our obsession with mechanical function, we must not forget that while wood does support the tree, it also is the plumbing system that transports water and minerals to all parts of the tree up the stem from the roots. This makes it easier to understand why tracheids are hollow and perforated by pits and are not as solid as cotton fibres.

As a material, wood is structurally complex in four basic ways. First, it is a hydrated composite of the high modulus, crystalline polysaccharide, cellulose, and an amorphous moiety of hemicelluloses, lignin and other compounds. Second, wood is an aggregate of microscopic cylinders (cell walls) of the material just described, and these cylinders lie parallel to the long axis of the stem or root. Third, the cellulose in the cell walls has preferred orientation that varies according to its position and hence to its age in the wall (Fig. 5.36). Fourth, the wood of many plant species occurs in concentric growth layers of cell walls having large lumens (earlywood) alternating with layers of cells having small lumens (latewood). These structural aspects tell us that wood is anisotropic and that it should be

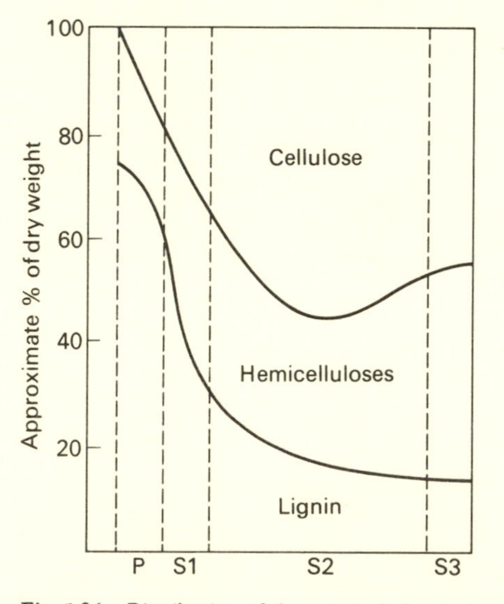

Fig. 5.36 Distribution of the principal chemical constituents within the various layers of the cell wall in conifers. P, primary layer; S1, S2, S3, secondary layers.

viscoelastic. In some measure, we know how each of these four aspects affects the mechanical properties of materials in general. We should be able to predict how each may vary in any particular biological context and how this variation in structure will affect the properties.

Wood is anisotropic because most of its constituent cylindrical cells are oriented parallel to the axis of the stem or other organ in which they lie. Figure 5.29 is an oversimplified diagram of wood that shows only cell wall thickness, lumen extent and the shape and orientation of cells in the tree trunk. Rays are small parcels of cells lying radially throughout the wood. There are no cylindrical cells whose long axes lie in a circumferential direction.

The chief difference between a single tracheid and the wood in the trunk of a living tree is that wood is an aggregate of tracheids bonded together by the amorphous substance of the middle lamella. The importance of the middle lamella is emphasized by the fact that Mark, throughout his extensive study, never saw a fracture initiated in the middle lamella. Unfortunately, the theory of aggregates has not yet been developed to include aggregates of filament-wound cylinders in an amorphous matrix. Nor has wood been studied from this point of view.

Once again, we caution the reader against extrapolating from published values of mechanical properties of dry timber (including Mark's otherwise admirable study) to wood as it functions in living trees. How rigid is a tree? What wind force can it withstand? What is the allowable strain a tree may take (a) briefly, in a storm, or (b) for several days or weeks under a load of snow? We must learn the mechanical properties of wood in the living tree.

5.18 Mechanical Properties of Wood

Timber is a much studied commodity, and the literature of mechanical relevance is vast. Timber, as fashioned into houses, violins and telephone poles, is wood that has been dried to about 12% of its saturated moisture content. It is known that the mechanical properties of timber vary with moisture content. Nevertheless, the literature on wood as it functions, water-soaked, in a tree is small and does not allow a satisfactory account to be given at this time. Therefore, we will discuss those properties of timber that are known or can be predicted to be the important tree-supporting properties. A more complete display of data on physical and mechanical properties of timber is given by KOLLMAN and CÔTÉ (1968).

The structural anisotropy of wood is faithfully reflected in its mechanical properties. Tensile and compressive properties (E, J, K, etc.) have greater values in the longitudinal direction than in radial and tangential directions (Figs. 5.16 and 5.37). The stress-resisting tensile properties are slightly greater in radial than in tangential directions. One microstructural feature that contributes to this latter anisotropy is the presence of rays (Fig. 5.29). Rays are evenly dispersed throughout the wood and are the conducting pathways by which fluids are translocated radially in woody tissues.

Earlywood is formed each year in the first part of the growing season. It has less cell wall material per unit volume than has the latewood that is formed in the remainder of the growing season. Strength and modulus of timber vary directly

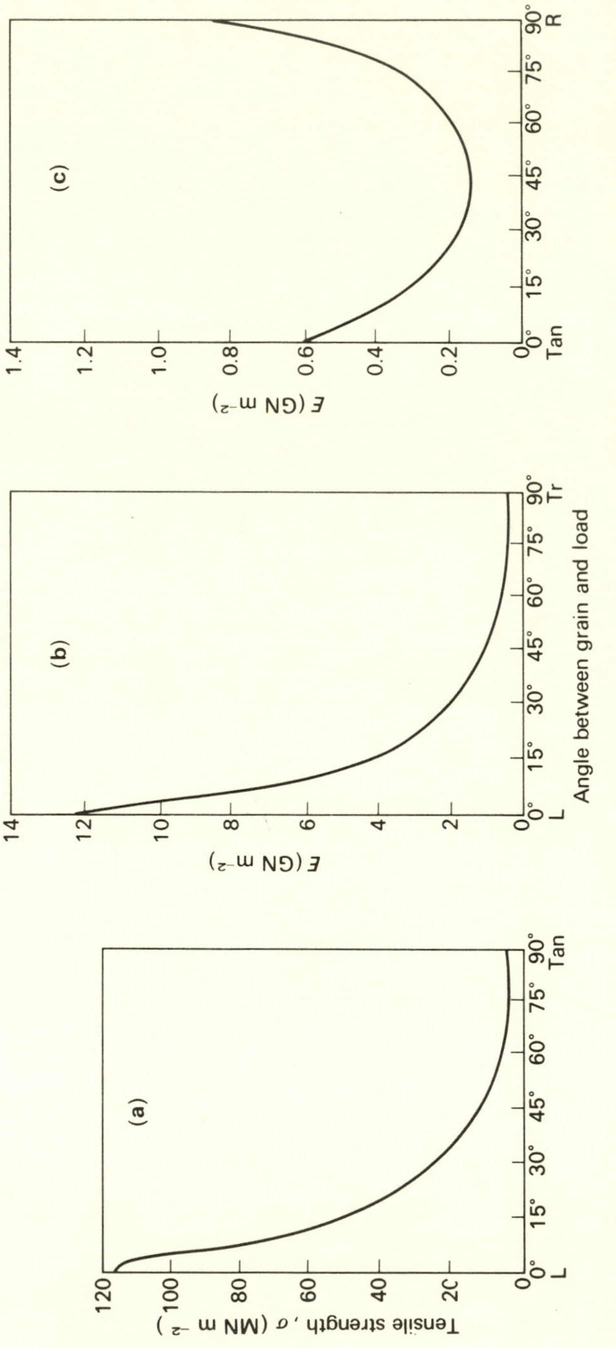

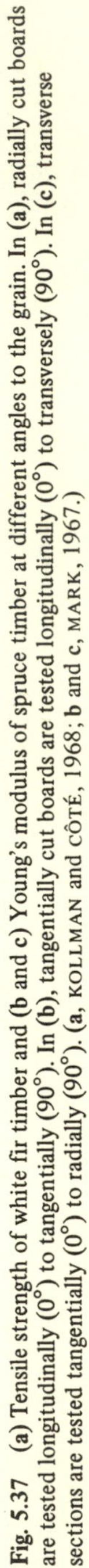

Fig. 5.37 **(a)** Tensile strength of white fir timber and **(b** and **c)** Young's modulus of spruce timber at different angles to the grain. In **(a)**, radially cut boards are tested longitudinally (0°) to tangentially (90°). In **(b)**, tangentially cut boards are tested longitudinally (0°) to transversely (90°). In **(c)**, transverse sections are tested tangentially (0°) to radially (90°). (**a**, KOLLMAN and CÔTÉ, 1968; **b** and **c**, MARK, 1967.)

with density (Table 5.3), but specific strengths of early- and latewood are nearly the same.

Strength and modulus along all orthogonal axes of timber vary inversely with the moisture content as it drops below saturation levels. As timber dries it shrinks anisotropically along orthogonal axes. Shrinkage is inversely proportional to tensile strength and modulus. It is believed that, as water is lost from the cell wall material, progressively more cellulose microfibrils become close-packed and form reversible interactions that effect the rise in strength and modulus.

The effects of drying on the mechanical properties of timber are particularly clear in a study of creep in red beech timber (KOLLMAN, 1962). He measured compressive strain in samples that were cyclically stressed and unstressed. Figure 5.38a shows that samples with 12% moisture content showed progressively more strain and progressively higher strain rates in successive loading cycles. Permanent deformation also increased at each loading. This behaviour is expected from a

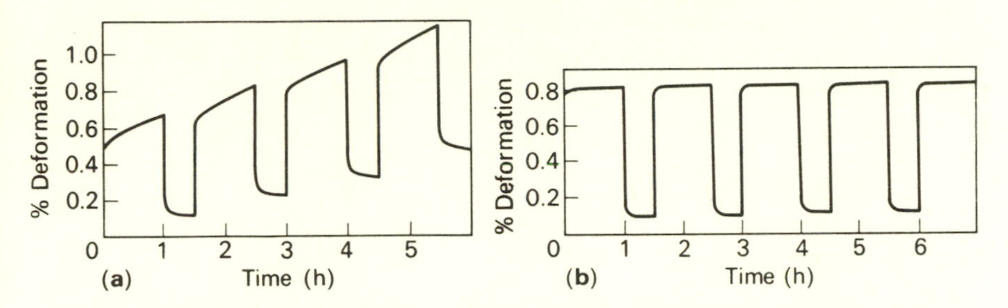

Fig. 5.38 Results of tests in which red beech timber is compressed parallel to the grain. (**a**) Timber (moisture content 12% of saturation) loaded to $\sigma = 49.0$ MN m^{-2}. (**b**) Oven-dry timber loaded to $\sigma = 68.6$ MN m^{-2} (KOLLMAN, 1962).

fibre-and-matrix composite in which the matrix is more viscous than elastic. As individual fibres break and the matrix gives, strain increases; as more fibres break, the greater is the stress on remaining fibres and the greater is rate at which they will break. Figure 5.38b shows results of a similar experiment with oven-dried timber. After six hours of repeated load bearing, both instantaneous creep and permanent deformation increased by very small amounts. Drying appears to have eliminated the viscous component and left the material rigid.

The cellulose fraction of Douglas fir timber was randomly depolymerized with various doses of gamma radiation by IFJU (1964). The tensile strength of samples having a low degree of polymerization of cellulose is lower, especially in the more hydrated state, than is the case in highly polymerized samples. This information is consistent with the explanation given above for Kollman's data on the behaviour of red beech.

The coefficient of thermal expansion,

$$\alpha = \frac{dl}{l_0 dT}$$

of timber varies linearly with temperature from −60° to +60°C (WEATHERWAX and STAMM, 1946). Values of α are an order of magnitude greater in radial and tangential directions than they are parallel to the grain of the timber. For example, values for white pine (*Pinus strobus*) of 4% moisture content and density 0.39 mg mm^{-3} are $\alpha_L = 4.00 \times 10^{-6}$ and $\alpha_T = 72.7 \times 10^{-6}$ (HENDERSHOT, 1924). Raising the temperature of cellulose whose crystalline regions have a length-to-width ratio of 10 will cause increased thermal oscillations perpendicular to the molecular chain that are 10 times greater than oscillations along the chains. The importance to trees of the dependence of α on temperature is that peripheral splits occasionally occur in hardwood trees exposed to severe frosts. During a sudden drastic drop in temperature, outer layers will shrink before inner wood is chilled, resulting in longitudinal-radial splits in the bark and outer wood (KOLLMAN and CÔTÉ, 1968).

It is said that, above saturation, water content of timber is not known to affect its mechanical properties (KOLLMAN and CÔTÉ, 1968). However, STEVENS and TURNER (1948) conclude that the failure of very fresh wood in compression is partly due to wood cells bursting from increased hydrostatic pressure. They note also that freshly cut wood bends easily. Timber is commonly steamed or boiled to achieve larger strains. Air-dried beech timber can take strains of 0.75 to 1.0% and can be bent to

$$\frac{a}{R} = \frac{1}{67} \text{ to } \frac{1}{50},$$

where a = sample thickness and R = radius of curvature, without failure. After steaming, strains of 1.5 to 2.0% and curvatures of $a/R = \frac{1}{33}$ to $\frac{1}{25}$ are obtainable.

KOLLMAN (1951) reports values of compressive strengths for timber from 20 hardwood and 10 softwood species. They vary, of course, but they are, in general, about 50% of tensile strengths for the same species (Fig. 5.35).

Although we have already discussed failure in wood at the level of the cell wall (Section 5.17), it is important to the tree that wood has very low resistance to shear forces. A graphic example of compression failure in timber is shown in Fig. 5.18c.

When compared with other rigid materials, wood is unremarkable in its various strengths and moduli. However, the low density of timber, varying from 250 to 1250 kg m^{-3}(pick an average value of 600 kg m^{-3}) makes wood a very favourable material for man to use in situations where it is important to save weight in the total structure (see the last column of Table 5.3). The superior specific properties of timber are difficult to bring into man-made structures: the weakness in shear causes timber to fail in shear around most fastenings and joints. Trees solve this problem by growing branches that are smoothly continuous with the main stem.

Cellulose is a tensile material without peer in terms of its specific properties (Fig. 2.4): it is ideal for use in minimum weight structures. Wood is a tough, energy-absorbing material that incorporates the advantageous tensile properties of cellulose into a low density composite material that effectively supports all higher plant structures against bending.

Despite the large number of tests which have been made involving viscoelastic properties (creep, delayed elasticity), hard experimental data are lacking. MORIIZUMI *et al.* (1973*a, b*) have reported the results of stress relaxation tests on wood of *Cryptomeria japonica* and *Chamaecyparis obtusa* and have derived values for the relaxation spectra—these are given in Fig. 2.24 where it can be seen that, by comparison with *Nitella* (Fig. 2.23), the elements having the longer relaxation times contribute more to the total modulus than do those with short relaxation times. This seems reasonable since elements with long relaxation times are the more nearly linearly elastic components and we would expect that, in wood, cellulose is the most linearly elastic component.

Unfortunately quantitative comparison is made difficult by the fact that different investigators do not adopt consistent units. In Fig. 2.23 the data for *Nitella* are given conventionally (and properly in our opinion) as $\log H(\tau)$ versus $\log \tau$. The data for wood are derived from the specific modulus $E(t)/\rho$ and are presented as $\log H(\tau)/\rho$ versus $\log \tau$ in Fig. 2.24. The increased interest and expanding work into the viscoelastic properties of biological materials requires that consistent methods of presenting data be early adopted.

5.19 Stony Materials

We here distinguish three types of mineralized skeletal materials. The first is that of materials consisting almost entirely of mineral, with a small (less than 5% by weight or 10% by volume) amount of organic matter. Within this group we make a distinction between the vast majority, in which the mineral is produced by the animals themselves, and those that are arenaceous, the material being picked up from the environment.

The second type has more than roughly 10% by volume of organic matter. Included here is an awkward group, whose members have skeletons lying on the borderline between *materials* and *elements*. These materials have marked differences in composition over distances of the order of one tenth of a millimetre or more. An example is the axial skeleton of certain Cnidaria (e.g., *Isis*), in which there are almost entirely organic nodes separated by long internodes in which there is little organic material. In this group are spicular skeletons. Mechanically, spicules are isolated stiff blocks, often pointed, embedded in a matrix which is much more pliant. Unfortunately for classifications, spicules tend to fuse to form rigid structures. The third type is that of teeth.

We describe these types using a classification by organism. Apart from incidental remarks, however, we do not describe the mechanical significance of what we describe until Section 5.20.1. However, we shall be emphasizing the absolute size of the grains or crystals in the materials because, as will be shown, this size strongly affects the mechanical properties. Table 5.6 shows the sizes of some structures in stony skeletons.

5.19.1 Porifera

Sponges typically have isolated spicules, but VACELET (1970) and HARTMAN and GOREAU (1970) have described sponges that are quite massive. In these

Table 5.6 Size of crystallites in rigid skeletal materials. 'Lowest,' 'mid-' and 'high level' refers to structures of different size in the same material: high level structures are composed of structures from lower levels. At each level, 'small,' 'medium' and 'large' refer to dimensions in different directions. All sizes are in μm. 'Organic' refers to the thickness of the organic layer between blocks of mineral. Values in this table have been garnered from many sources and some involved the measurement of diagrams and so on when sizes were not stated. In general, a range of values is seen, and here the larger values are given. The table gives a general idea of the sizes involved and is no doubt inaccurate in many places.

	Lowest Level					Mid-Level					High Level					
Group	Shape	Small	Medium	Large	Organic	Shape	Small	Medium	Large	Organic	Shape	Small	Medium	Large	Organic	
Cnidaria																
Scleractinia	Needle	0.3	0.3	5	Little	Needle	10	10	Hundreds	Little	Spherulites	←	Indefinite	→	Little	
Tubipora	Needle	1	1	20	Little						Spicules	*ca.* 200	← Irregular →		Little	
Brachiopoda																
Articulata 1° layer	Irregular	←	Very Small	→	Little											
2° layer	Needle	3	20	mm	0.2											
Crania 1° layer	Needle	0.2	0.4	Tens	Little											
2° layer	Plate	0.3	50	50	0.1											
Lingula, calcified layer	Needle	0.02	0.02	0.1	Much	Needle	(indefinite, not large)				Sheets	50	← Indefinite →		Much	alternates with organic
Bryozoa																
Schizoporella 1° layer	Brick	1	1	2	0.005											
2° layer	Plate	0.5	0.5	2	0.005											
3° layer	Needle	0.5	0.5	2	0.005											
Mollusca nacre	Plate	< 3	10+	10+	0.3						Sheets	<3	← Indefinite →		0.3	
foliate	Laths	0.5	4	20	0.02						Sheets	0.5	← Indefinite →		0.02	
simple prisms	Needle	3	3	8	0.3						Prisms	10+	10+	Indef.	0.5+	Spherulites
composite prisms	Needle	5	5	30+	0.2						Prisms	300	300	Indef.	Little	Spherulites
crossed lamellar	Needle	2	2	8+	Thin	Lath	8+	500	Indef.	Little	Lenses	500	← mm →		Little	
complex crossed lamellar	Needle	1	1	4+	Thin	Lath	8+	(indefinite)		Little	Blocks	←	1 mm or less →		Little	
homogeneous	Lump	3	3	3	Thin											
Echinodermata, normal plate	Any	←	Indefinite	→	V. Little											
Vertebrata, eggshell	Needle	3	3	300	Little						Prisms	50	50	300	1.0	Spherulites
enamel	Needle	0.04	0.04	0.15	V. Little						Prisms	4	6	mm	0.3	

sponges the calcium carbonate skeleton is in the form of innumerable fine spicules that are laid down in a diaphanous organic matrix, as yet uncharacterized chemically.

Most sponges use isolated spicules, and these may have two functions: to support the body, keeping all its orifices and passages open, and also to act as a deterrent to predators. Most sponges are encrusting, not erect, and have no need for solid mineral skeletons. Their spicules are either calcium carbonate, usually calcite with 5-15% magnesium carbonate, or amorphous silica.

The shapes of the calcite spicules may be complex (Fig. 5.39a). However, JONES (1970) has shown that each spicule is a single crystal in the sense that the lattice orientation is uniform. In this respect the spicules are like the elements of echinoderm skeletons. There seems (*pace* TRAVIS, 1970) to be no organic

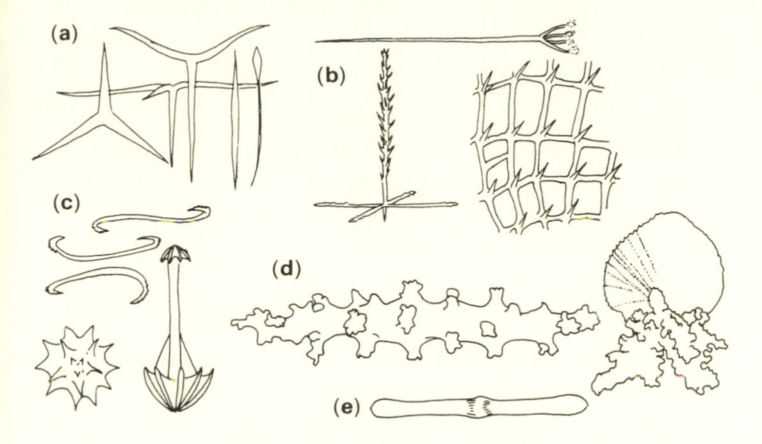

Fig. 5.39 Some spicules from (**a**) calcareous sponges and (**b**, **c**) silicious sponges and (**d**, **e**) gorgonacean corals. (**b**) Hexactinellid sponge, *Euplectella*. (**c**) Various demosponges. (**d**) Cortical spicules from gorgonaceans. (**e**) Axial internodal spicule from *Melithaea ocracea*. Not to scale, but all are less than 0.5 mm long.

material within the spicules (JONES, 1970). The complicated external shape of many spicules and the uniform lattice of the calcite show there is no *necessary* relationship between the two.

Siliceous spicules of amorphous SiO_2 are superficially rather like calcareous spicules. However, internally there is a definite organic thread, probably protein, but certainly not collagen.

FJERDINGSTAD (1970) describes quartz crystallites arranged with their long axes at right angles to the long axis of the spicule. However, the specimens were prepared by ultrathin sectioning of undecalcified specimens and GARONNE (1969) reports similar effects from knife-induced artifacts. TOWE and HAMILTON's demonstration (1968) of the striking crystallite-like artifacts that can be introduced in mineralized tissues by knives makes Garonne's analysis the more credible.

Although most spicules in sponges remain separate, there is a tendency for them to become connected by a cement of calcium carbonate or silica as appropriate. This leads to a much more rigid skeleton. A fine example of this is Venus' Flowerbasket *Euplectella* (Fig. 5.39b).

The Demospongia use silica or spongin or both. In forms with spicules, these lie within the spongin trabeculae (Fig. 5.39c) and of course considerably increase the stiffness of the whole sponge.

5.19.2 Cnidaria

There are several groups in the Cnidaria with members having very stony skeletons. The scleractinians are the stony corals, and are easily the most important skeleton builders in the Cnidaria. Their skeleton grows by spheritic calcification; that is, there is a centre of calcification and polycrystalline fibres of aragonite radiate from it (BRYAN and HILL, 1941). These fibres keep more or less the same diameter, so when growth has gone a little way new fibres can form in the angles between the original fibres. On any surface, crystals will grow if there is sufficient supersaturation of the fluid bathing it. The aragonite will usually grow on pre-existing crystals so that the crystal is continuous. Such growth is called 'syntaxial' or 'epitaxial.' Initially crystals may grow at all angles, but those growing more sideways will be shortlived because they will grow into the side of crystals growing straight ahead. When a sclerodermite starts off with a burst of activity, spaces will be left behind unfilled by material. In spheritic growth these are usually not filled by matrix, and remain as very small voids.

The polycrystalline fibres are about 10 μm across, though there is much variation. It has been shown in one scleractinian (*Pocillopora damicornis*) that each fibre is in fact a bundle of very fine needle-shaped crystals, all with their longitudinal axes (aragonitic *b* axis) in the direction of the fibre, whereas the aragonitic *a* and *c* axes are randomly arranged in the plane perpendicular to the *b* axis. The smaller crystallites are about 0.3 μm across, and considerably longer (WAINWRIGHT, 1964). WAINWRIGHT (1963) found the matrix to consist of chitin fibrils in a network whose enclosed size and shape were identical to that of the whole skeleton. The proportion of organic material was very small, about 0.02–0.2% by volume. YOUNG (1971) found amino acids in hydrolysates of skeletons of this and other species, but we have no idea how such proteins might be disposed in the skeleton.

The shapes of skeletons resulting from the type of growth described above are marvellously varied. They may be massive or delicate, solid or trabeculate or with a solid cortex and a trabeculated medulla. They may be flat, mushroom-shaped, fan-shaped or tree-like. The particular habit adopted is a function both of the species and of environmental conditions (BARNES, 1970).

The octocorallian, *Tubipora*, has a most characteristic skeleton. This is the 'organ pipe' coral and its skeleton consists of long, separate, infrequently branching tubes, with an occasional transverse platform between the tubes stabilizing the whole mass. The skeleton is made of cemented separate spicules, with an apparently weak glue between the spicules. SPIRO (1971), shows that the spicules themselves are made of needles about 0.4–1 μm in diameter. Organic membranes occur periodically around the needles.

Of all the cnidarian classes, the Anthozoa has the most skeleton-bearing species. As mentioned above, these may be of massive aragonite, or they may be of gorgonin (Section 5.14). However, many species make use of spicules of aragonite, and some of calcite. The relationship between the spicules and the organic skeleton in which they occur is very varied. (It is also extremely difficult to analyse mechanically.) Figure 5.46 shows some of the spicule-matrix relationships. There is a strong tendency for the spicules to become cemented together by mineral, forming concretions within the horny organic material. Alternatively they may become so numerous as to make the skeleton quite stiff, though they do not fuse. Spicules vary greatly in size, from about 50 μm to 5 mm or more in length. They tend to be rather blunt and knobbly, and not to be the single-minded prickers so characteristic of the sponges (Fig. 5.39d). The range of combinations of spicules plus matrix (calcareous, organic or both) is great and our knowledge of it is being augmented by work in progress (Muzik and Wainwright, in press).

5.19.3 Mollusca

The shells of most molluscs are of calcium carbonate with a small amount of organic material. The organic material is mainly protein, though occasionally there is chitin. The disposition of the shell materials produces rather characteristic microstructures, which we shall briefly describe, as some have different mechanical properties from others. (Figure 5.40 and Table 5.6 summarize and fill out these descriptions.) In general we follow TAYLOR *et al.* (1969) and KENNEDY *et al.* (1969) in these descriptions. The organic matrix is called *conchiolin.*

Nacreous structure Mother of pearl. The crystallites are flat plates of aragonite, arranged in layers, as in a brick wall (sheet nacre) or in columns (columnar nacre). There are sheets of organic matrix, not more than 0.3 μm thick, cementing the plates together.

Simple prismatic structure Polygonal columnar crystals, of calcite or aragonite, run normally to the surface through the shell thickness. They are large, 10–200 μm across, and may be millimetres long. The organic sheet surrounding the columns is 0.5–8 μm thick. There is a lower order of structure, with finer crystals 1–8 μm long, 0.2–3 μm across, surrounded by correspondingly thinner organic sheets.

Composite prismatic structure The large prisms are indefinitely long, about 300 μm across, are directed along the length of the shell, and are always the outermost layer of any shell in which they are present. Again, there are smaller crystallites, more than 30 μm long, and about 5 μm wide.

Crossed-lamellar structure This has a lamellate arrangement rather like plywood, in which each lamella is built of long aragonite needles all oriented the same way. In adjacent lamellae the needles are oriented in different directions.

Complex crossed-lamellar structure Rather like crossed-lamellar structure, but the lamellae interpenetrate in a more complex fashion.

Foliated structure Long, lath-like calcite crystals, about 2–4 μm wide and 0.2–0.5 μm deep are arranged in overlapping sheets, like tiles on a roof. Each crystal is wrapped in an organic sheath about 0.01–0.02 μm thick.

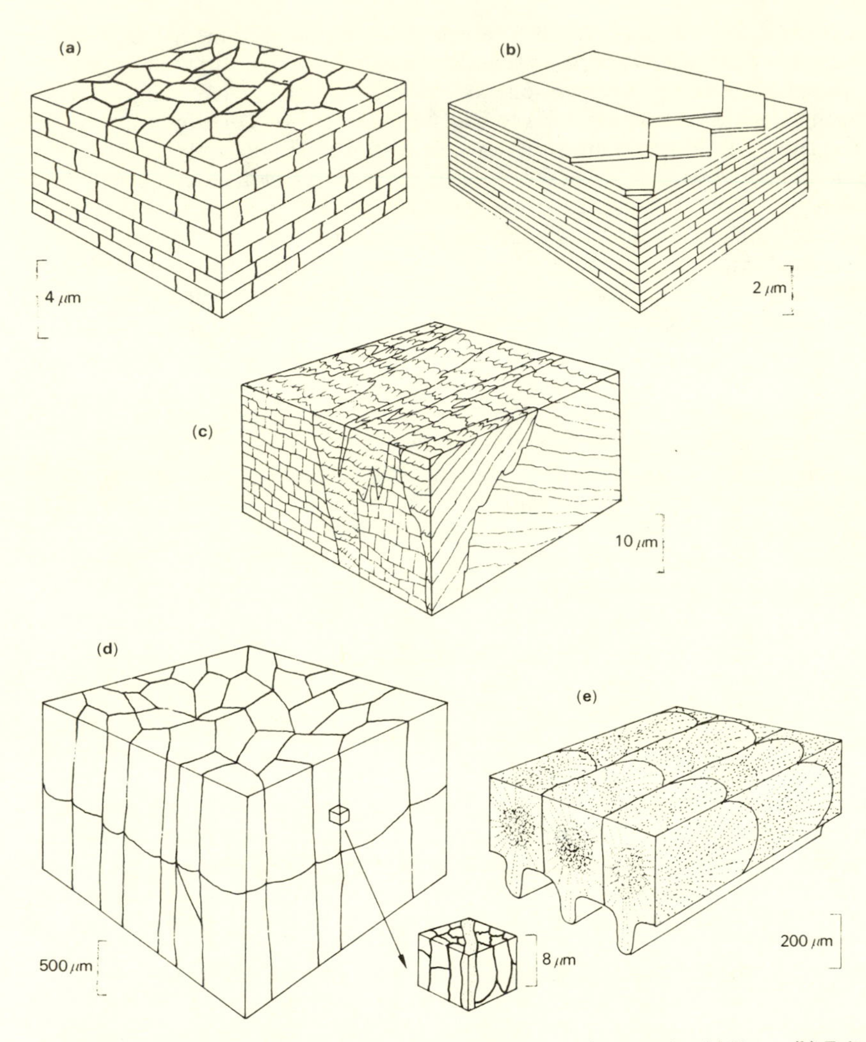

Fig. 5.40 Diagrams of molluscan shell structures. Note the different scales. (**a**) Nacre. (**b**) Foliated structure. (**c**) Crossed-lamellar structure. (**d**) Simple prisms. A small part is enlarged to show the very thick conchiolin sheath between the major prisms. (**e**) Composite prisms.

Cross-foliated structure This consists of calcite laths arranged into lamellae, rather like the lamellae in crossed lamellar structure

Homogeneous structure This is made of very small granules of aragonite, about 0.5-3 μm in diameter, in a very fine organic matrix.

Underneath muscle attachments a special structural type, *myostracum*, appears. This consists of extremely irregular prisms, between 10-50 μm in diameter, with a thin conchiolin sheath around them. The long axes of the prisms are oriented in the direction of muscle pull. Because the muscle attachments move over the inner surface of the shell during growth, myostracum forms as a continuous sheet from the dorsally located site of the muscles' original attachment to the present attachment site, sandwiched between two other layers, and outcropping only where the muscles attach at the moment (Fig. 5.41).

The nature of the organic material has been reviewed by WILBUR and SIMKISS (1968). There seems to be considerable variation in the primary structure of the proteins and in any one shell and in any one structural type there may be several different fractions. It seems that there is often a high proportion of water-soluble protein, and this is present actually within the mineral, while water-insoluble protein forms sheets round the crystals. In general, hydroxyproline is absent. There is little cysteine, and glycine has usually about 100–300 residues per thousand. Therefore one can say, without surprise, that the protein is neither collagen,

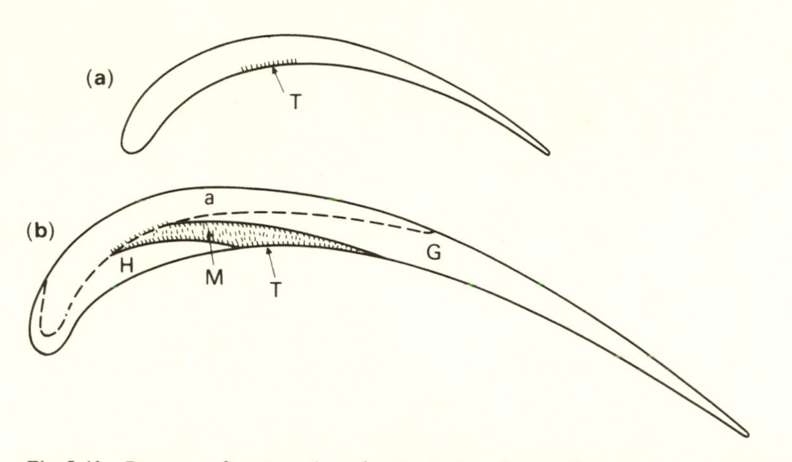

Fig. 5.41 Diagrams of sections through a bivalved mollusc shell at two stages in its growth showing how the young shell (a) is retained in the mature shell, (b) and how the myostracum (M) runs as a seam through the shell. New shell material has been added (H) between muscle and hinge and (G) between muscle and margin. T, site of adductor muscle attachment.

nor keratin, nor silk. There are varying amounts of glucosamine, though always much less than there is of protein. Chitin is certainly present in the shells of some molluscs, but not all.

The amount of conchiolin varies over several orders of magnitude. In general, crossed-lamellar structures and foliated structures have little, 0.02–0.8% by volume, while nacreous and prismatic structures have more, 2–8%. The shells of steno-glossan gastropods have the least organic material, and the shells of cephalopods tend to have the most (HARE and ABELSON, 1965).

The literature gives the impression that the structure of mollusc shells is rather uniform. This is far from the case, as reference to BØGGILD (1930) or KENNEDY *et al.* (1969) will show. A few generalizations do emerge if all groups are surveyed. In the bivalves and gastropods there is a tendency for there to be a layer of prisms on the outside and nacre on the inside. This arrangement is often considered 'typical.' Prisms are never the only shell structure. However, shells often have varying mixtures of crossed-lamellar, complex crossed-lamellar and homogeneous structures. This is true, for instance, of nearly all heterodont bivalves.

5.19.4 Brachiopoda

The brachiopods are almost unknown mechanically, but they have many structures analogous to those of molluscs, about which more is known, and so it seems worth describing the brachiopods briefly.

There are two groups of brachiopod shell, the chitinophosphatic and the protein–calcitic. The calcitic shell has evolved independently at least twice. For an excellent functional account of the brachiopods read RUDWICK (1970).

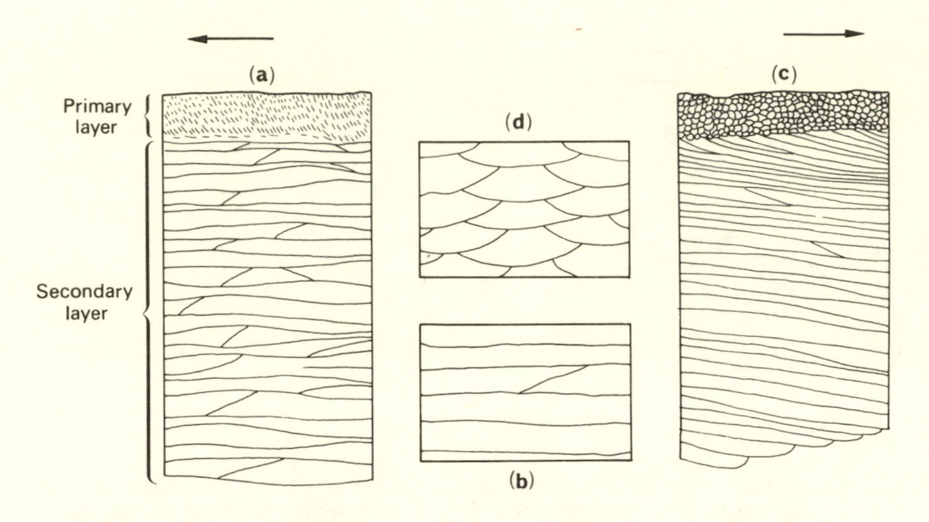

Fig. 5.42 Structure of brachiopod shell. (a) Section of the shell of *Crania.* (b) A section of the laminae in the secondary layer of (a). Note how the units of the secondary layer are sheets, extending in two directions. (c) Section of the shell of a typical articulate brachiopod. (d) A section of the secondary layer, at right angles to (c). Note the secondary layer is made of fibres, not plates. Arrows indicate the direction of the free margin of the shell.

The shell structure of many brachiopods (in the Articulata) is rather similar (Fig. 5.42). On the outside, beneath the periostracum, is a thin primary layer, occupying about 20% of the thickness, of very finely divided crystalline calcite. The layer has a delicate protein matrix. The deeper, fibrous, secondary layer is made of elongated calcite crystals, each surrounded by a protein sheath. The characteristically shaped crystals are 20 μm by 3 μm in cross section, and may be several millimetres long. They lie nearly parallel to the plane of the shell itself (WILLIAMS, 1968; 1970). Sometimes, when the shell becomes very thick, as in the Spiriferids, and mechanical considerations are presumably less important, the neat division of the secondary fibres breaks down, and rather larger chunks of calcite are formed.

5.19.4 Brachiopoda

The other calcitic brachiopods, among the inarticulates, evolved the calcite mineralogy independently. *Crania* and its allies sit firmly attached by one valve to a rock. This valve is made of small crystals about 0.4 μm across and fairly long, inclined nearly normally to the surface. There is some organic matrix. The upper valve has this pattern on the outer surface for about 50 μm, but inside there is a secondary layer of overlapping laminae, each about 0.3 μm deep but extending many tens of micrometres in the plane of the shell (Fig. 5.42). There is a substantial organic matrix. Under the muscle attachments of the upper valve the crystals are oriented differently, being aligned with the muscle pull. The analogy with the prismatic myostracum of bivalve molluscs is clear (Section 5.19.3) (WILLIAMS and WRIGHT, 1970).

So, the inarticulate brachiopods have evolved the calcitic shell independently of the articulates, and they have solved what must be a basically similar mechanical

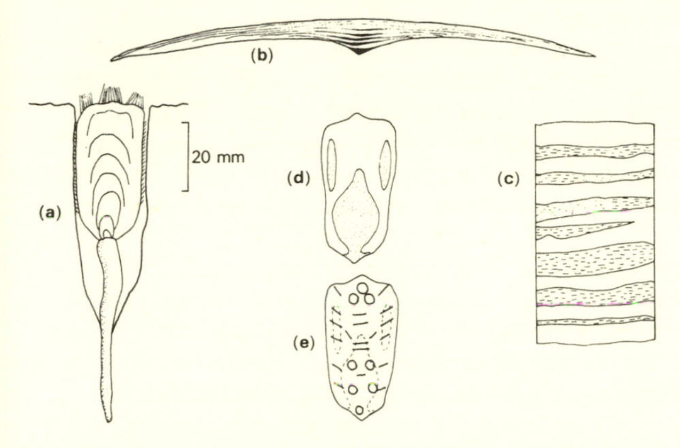

Fig. 5.43 The shell of the inarticulate brachiopod ***Lingula unguis.*** (**a**) General view of the animal in its tube. (**b**) Cross-section of a single valve showing alternating layers of heavily calcified material (black) and lightly calcified layers. (**c**) Enlarged cross-section. Mineral layers stippled. (**d**) plan view of shell, showing more heavily calcified regions (stippled). (**e**) Orientation of crystallites (calcitic c-axis) shown by lines. Disorientated crystalline regions shown by open circles.

problem in almost the same way, with a thin outer layer of small crystals and a thicker inner layer, the long axis of whose crystals is nearly in the plane of the valve.

The chitinophosphatic Lingulacea, as exemplified by *Lingula*, have a shell with alternating layers of more or less highly mineralized material. Different regions of the shell have different proportions of the two kinds of layers (Fig. 5.43). The heavily calcified layers are composed of small crystallites of calcium phosphate (said to be a calcium fluorapatite by McCONNELL, 1963), about 100 nm long and 5–20 nm across. Their long axes are in the plane of the shell. In the most heavily calcified regions the needles may become fused into larger blocks. Where the shell is more mineralized, there is less protein relative to the chitin. This is just like the situation in the Crustacea (KELLY *et al.*, 1965). The organic sheets are themselves layered, with regions of high protein/chitin and low, alternating about every 2.5 μm (JOPE, 1971). As a result of the alternation of the thin layers of mineral with organic sheets, the shell of *Lingula* is extremely flexible.

5.19.5 Arthropoda

The cuticle of arthropods has, in the main, been dealt with already, but some arthropods have impregnated the cuticle with calcite, and these we shall deal with briefly.

The cuticle of crustaceans has been reviewed by DENNELL (1947, 1960). The presence of calcium carbonate is not universal, but the mineral is often an importan constituent of the skeleton. The epicuticle is normal. The procuticle has an outer 'pigmented layer' (which may be colourless) which is always tanned and nearly always calcified, a 'calcified layer,' which is calcified but not tanned, and finally an inner uncalcified, untanned layer. The layers vary considerably in their relative thicknesses.

It seems that in general the more the procuticle is calcified, the lower is the protein-chitin ratio. This is found, for instance, in different parts of the same animal, in the same parts of different animals, and in the same part of one animal at different times post-moult (LAFON, 1943).

DUDICH (1931) made an extensive survey of calcification in Crustacea. He described very thin calcareous plates between 0.5 and 5 μm thick in the thickness of the cuticle, considerably greater in other dimensions. It seems that centres of crystallization are present in the cuticle, and crystal growth starts almost simultaneously at the various centres. The tiny platelets grow until they meet, at which point they do not seem to fuse. DIGBY (1968) reports similar findings in *Carcinus maenas*, the shore crab.

The ostracods have heavily calcified shells, often with about 90% $CaCO_3$ by weight. It is probable that the organic component is mainly chitin, but good analysis is difficult. Sometimes the organic and inorganic components are laid down in alternating layers, as in *Lingula*; more often they are intermingled.

In Crustacea the mineral is often amorphous; this is unusual in calcium carbonate skeletons. It seems to be associated with the presence of more than a few percent of phosphate ions, and these disrupt the calcitic lattice.

Skeletally, of course, the barnacles are atypical Crustacea. The exoskeleton is quite massive and heavily calcified. Nothing seems to be known about its ultrastructure. However, in light microscopic examination of whole scutes and thin sections, NEWMAN *et al.* (1967) observed increased size and preferred orientation of calcite crystals with increased distance from centres of calcification. This feature is very like that in stony corals, which have planar structures formed adjacent to an epithelium and, like the stony corals, the morphological and crystallographic axes of the oriented crystals are normal to the formative epithelium. Barnacle scutes however, contain a much higher proportion of organic material than do coral skelet

The diplopods (millipedes) are the other major group of arthropods to have a calcified cuticle, though not all species have calcified cuticle. CLOUDSLEY-THOMPSO (1950) reports that in *Glomeris* the order of layers from outside inward is: a thin epicuticle, a tanned uncalcified layer, a calcified untanned layer, and an inner untanned uncalcified layer. It is curious that the tanning and the calcification should be mutually exclusive, unlike the situation in the Crustacea.

5.19.6 Echinodermata

The echinoderms have a skeletal structure unlike any other. In all other heavily mineralized skeletons, except in the tiny protozoa, the mineral is divided into small blocks, usually by an organic matrix. Even where the evidence for organic material is slim, the mineral is broken into blocks, visible in surface replicas with the electron microscope. In general, in the echinoderms, this is not true. Each skeletal element in an echinoderm, whether it be a spine, a plate of the test, or a spicule, behaves optically as if it were a single crystal of calcite. In spines this may mean that the crystals are several centimetres long. Now, the fact that the skeletal elements behave *optically* like single crystals does not preclude their being made of many crystals all oriented in the same direction. But if the crystals were of different orientations this would show up in the scanning and conventional electron micrographs of broken surfaces. These instruments do not show any irregularities, at least over a greater part of the skeletons. The broken surface appears smooth, with only occasional sharp steps, as are also produced on the fractured surface of a native crystal of calcite (NICHOLS and CURREY, 1968). In most elements the only evidence of a clear substructure is seen on etching smooth, broken surfaces with acid. Surface peels then show some sort of polygonal patterning. The same kind of appearance is seen on similarly treated native crystals. Sponge spicules show a similar patterning, which in that case at least, is not related to the crystal substructure.

MÄRKEL *et al.* (1971) have demonstrated that a few echinoid elements are in fact polycrystalline, but these are such special cases that they rather prove the rule. These authors found polycrystalline material in the cortex of the primitive cidarids, and in some tooth structures.

It has been asserted by TRAVIS (1970) that there is a collagenous matrix to the skeleton, but it seems to us that the undoubted collagen fibrils appearing in her preparations come from the fibres binding the skeletal elements together (KLEIN and CURREY, 1970). It is, however, possible that there is some matrix. Klein and Currey report enough proline in the skeleton to be equivalent to 0.1–0.3% by weight protein, possibly therefore approaching 1% by volume, though even this is probably an overestimate caused by contamination.

It is interesting that HENISCH (1970) shows that an apparently perfect crystal of calcite can be grown in silica gel, and that parts of the gel become incorporated,

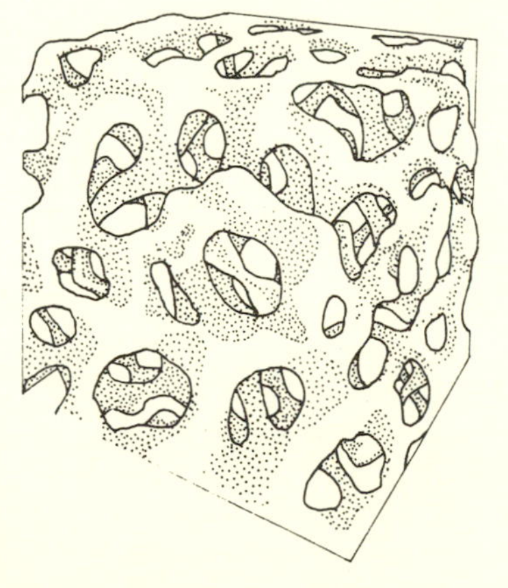

Fig. 5.44 Stereogram of typical echinoderm skeletal material, full of interconnecting holes.

making the crystal turbid, and reappearing as a ghost if the crystal is dissolved. It is possible that there may be such an extremely diaphanous matrix in echinoderms, even though echinoderm crystals seem quite clear. But there is virtually no evidence that the calcite is divided up into small crystals.

We have so far been talking about the organization of the skeleton at the order of size of one μm or so. The uniqueness of the echinoderms is also apparent at higher levels. The skeleton is full of interconnecting pores or canals. These canals are from about 10 μm to 500 μm across, and make any fragment of echinoderm skeleton instantly recognizable to the practised eye. In life they are filled with living tissue that functions, among other things, to keep the surface very smooth (NICHOLS and CURREY, 1968). The porosity of the skeleton varies, but is characteristically about 50%. Figure 5.44 shows the typical appearance of echinoderm skeleton.

5.19.7 Birds' Eggshells

These consist mainly of calcite, unlike the majority of vertebrate hard tissues, which are phosphatic. Their structure is reviewed by WILBUR and SIMKISS (1968). The structure of the domestic fowl eggshell is typical (Fig. 5.45, TYLER, 1969). Beneath an outer protein cuticle is a 'palisade' layer, a series of prisms arranged normal to the surface, interdigitating in surface view. On the inner side is the 'cone' layer. The cones insert into an outer shell membrane. Eggshell structure can be understood as spheritic calcification starting in the middle of the cones, and proceeding towards the outer surface. This outward growth would be preferred because as calcium and carbonate ions are precipitated, they are replaced from the outside. As the crystals grow they abut and so form elongate prisms. The situation is very reminiscent of that in stony corals. The prisms are about 300 μm long and 50 μm across. However, fracture surfaces show much smaller crystallites within the prisms. There is an organic matrix, about 2–4% by volume, which is certainly not confined to the boundaries between the major crystals,

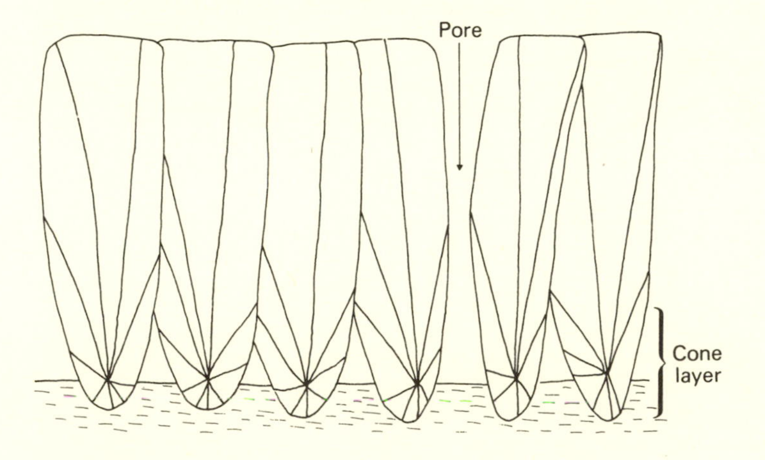

Fig. 5.45 Diagram of a section through a piece of bird's egg shell.

because on decalcification the organic matrix is seen to form a spongy and unstructured mass, unlike that seen in molluscs on decalcification. The matrix is about 70% protein, containing little cysteine and no hydroxyproline. There are some polysaccharides and smaller sugar molecules.

Avian eggshells are pierced by innumerable narrow pores connecting the inside to the outside. Gaseous exchange takes place through these stress-concentrating pores.

5.19.8 Spicules: Mechanical Considerations

Spicules have two main functions, one is to make the animal unattractive as food, and the other is to increase the stiffness or strength of the material in which they are embedded.

Spicules may often transform soft, unprotected prey into a most unpalatable meal. In such animals the spicules are often rather sharp, and can only be eaten by animals with hardened teeth. In spicules performing this function, strength is not a particularly valuable property, in fact the more the spicule tends to splinter as the animal is being eaten, the more unpleasant it may be. Spicules will not be much of a deterrent if the animal is gulped down whole, though glassy spicules, not being soluble in ordinary stomach acids would be unpleasantly indigestible. It is, therefore, a moot point to what extent the spines and spicules of the Protozoa protect them from being eaten. Sedentary animals, and large ones, are those that will have most advantage in having spicules.

The mechanical effects of spicules are little understood. They are sometimes so dense in a tissue as to make an appreciable difference to its stiffness—one sees all levels from the chitons and holothurians, in which the density of spicules is low, and the stiffness of the tissue cannot be very different from similar tissue without spicules, to animals like *Tubipora* in which the spicules are so dense that they fuse. However, this last class is less interesting than the case in which the spicules are very dense but do not fuse.

Unfortunately the analysis of the effects of blunt spicules in tissues has not been carried out. Most spicules are so short (that is, they have such a low aspect ratio) that they cannot be considered as ordinary composite, fibre-reinforced materials. A better analogy would be elastomers with fillers, but the interesting cases, when the spicules reach high volume fractions, are of scant interest to elastomer technologists, and so have been little analysed. Probably all we can say is that the effect on stiffness in a very pliant matrix will not be increased markedly until the volume fraction is very high. In a more rigid matrix the stiffness increased rapidly after 20% volume fraction (THINIUS and HÖSSELBARTH, 1970).

Similar vagueness characterises our knowledge of the effect of spicules on strength. In general, again, they can have little effect until the volume fraction is high. In the case of tough materials like gorgonin it may well be that the spicules actually reduce the strength, due to the introduction of potential Griffith cracks in the brittle spicules. The spicules must be introduced for their effect on stiffness, and some concomitant effect on strength is acceptable.

So far we have considered the spicules as isolated from each other. When they are joined up they produce a much stiffer structure. For instance, if we have a structure such as in Fig. 5.46a and d, it will, in shear, be hardly any stiffer than

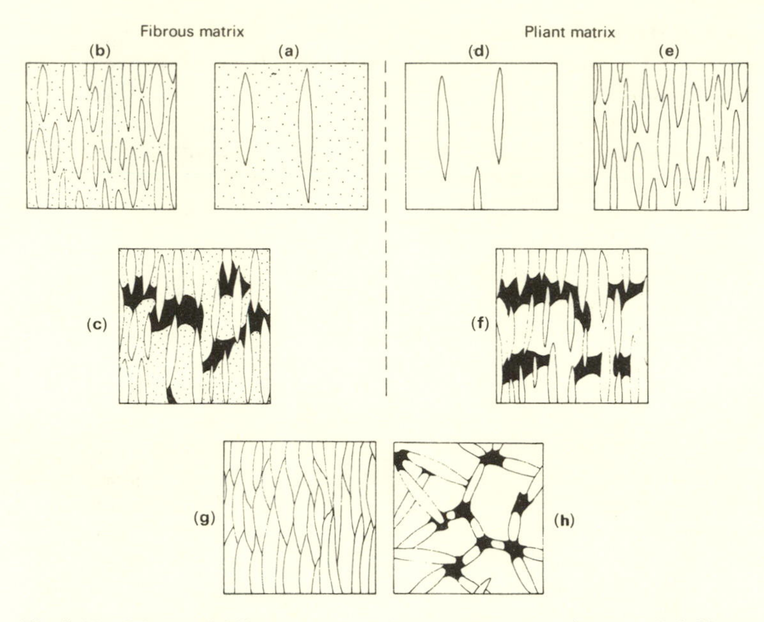

Fig. 5.46 Scheme of different ways spicules can exist in animal tissues. With fibrous matrix: (**a**) Few isolated spicules: low stiffness but high extensibility; (**b**) many isolated spicules: high strength and stiffness; (**c**) spicules joined by cement: stiff, strong and tough. With pliant matrix: (**d**) few isolated spicules: extensible and discourages predators; (**e**) many isolated spicules: stiff; (**f**) spicules joined by cement: stiff; (**g**) spicules fused: very stiff and brittle; (**h**) spicules arrayed in a framework, joined only at ends: flexible.

the matrix itself. However, if the spicules are joined end to end (Fig. 5.46c and f), the structure will be much stiffer. This is because in the first case the stiff spicules are not distorted when the whole structure is distorted, but in the second case they are distorted, and so their stiffness is brought into play.

The skeleton of the fan-shaped scleraxonian, *Melithaea ocracea* is 'jointed': long, narrow internodes alternating with short, wide nodes. MUZIK (1973) has shown that both nodes and internodes contain the same simple-shaped spicule (Fig. 5.39e). The arrangement of spicules in nodes and internodes is different and, when lateral force is applied to the axis, it bends only at the nodes. In the rigid internodes, the spicules are aligned parallel to the axis and are tightly fused (as in Fig. 5.46g); in the flexible nodes, the spicules are in an open 3-dimensional framework and are joined only at their ends (as in Fig. 5.46h).

This is a good example of the adaptation of a composite material to different functions by altering the angle of orientation and the magnitude of tight junctions of the stiffer component. It is also an example of how the same materials in units of the same size and shape can be organized into composite materials having quite different mechanical properties. We have now given theoretical reasons and examples of how any one of the four major morphological variables may affect mechanical properties of composite materials. These variables assume no change in chemical constitution. They are (1) size, (2) shape, (3) volume fraction and (4) orientation of component particles.

5.19.9 Teeth

Many groups of animals have teeth or spines for gripping, piercing or mashing. Often this is the only stiff part of the whole skeleton. Groups that have teeth or spines but no other stiff skeleton include the Trematoda, Cestoda, Nemertina, Echinoderida, Acanthocephala, Rotifera, Priapulida, Chaetognatha, Hirudinea, Tardigrada, Onychophora and Pentastomida.

The major requirement of teeth and spines is that they should be stiff and sharp. Very often, if they are used for mastication, they need to be hard as well, and sometimes do not need to be sharp. Unfortunately, for the majority of teeth and spines, there is little information about the details of structure and chemical composition. Frequently they are made of chitin, frequently of sclerotized protein, and frequently of both.

The mouthparts of arthropods have been investigated by BAILEY (1954). The mechanical test he used was the standard Mohs scratching test, determining what minerals would, and what would not, be scratched by the mouthparts. In the arthropods he studied four insects, a crustacean, a centipede, a scorpion and a spider, and found that calcite could be scratched with difficulty, implying a hardness of 3 on Mohs' scale. It is interesting that, as the sclerotized mouthparts could scratch calcite, the addition of calcite to the skeleton, as in the Crustacea, might not have made the mouthparts much, if at all, harder. Arthropod mouthparts tend to be dark, usually a sign of particularly heavily sclerotized cuticle.

The radula of molluscs is a most effective structure. It consists of a flexible strap to which are attached rows of teeth. The strap is pulled to and fro over a cartilaginous pad, and the food is rasped, grated or pierced according to the type of tooth on the radula (Fig. 5.47). The teeth are usually of sclerotized protein, and as we have already seen, this is probably enough for most purposes. However, molluscs browsing on organisms encrusting rocks have special problems for, unless they have quite hard teeth, they will be worn too quickly by the rock.

TOWE and LOWENSTAM (1967) examined a chiton, *Cryptochiton*, and showed that the mature radular teeth consist of magnetite ($Fe_2O_3 \cdot FeO$) in a regular meshwork of organic material, probably protein. Magnetite is hard, 7.5 on Mohs' scale. The blocks of magnetite are about 0.1–0.2 μm across. The iron seems to be deposited as amorphous ferric oxide, and is later partially reduced and changed to a crystalline structure.

Runham and his colleagues (RUNHAM and THORNTON, 1967; RUNHAM *et al.*, 1969) have studied the limpet *Patella*. The tooth is complex, with different regions having different proportions of iron (probably in the form of goethite and $FeO \cdot OH$) and silica (Fig. 5.47c and d). There is an iron-rich and an iron-poor region. In life the iron-poor region is less resistant, and wears down more rapidly, so maintaining a chisel edge. This differential wear is assisted by the orientation of the fibres in the teeth. The fibres are about 0.8 μm across, and probably contain most of the silica, the iron being in a matrix. The fibres are oriented as shown in Fig. 5.47e. The iron-rich part has fibres nearly normal to the surface being rasped, the silica-rich region has fibres more nearly parallel to the surface. It is clear that this arrangement will assist the more rapid wear of the silica-rich part.

The teeth of echinoderms are more complex than the rest of their skeletons. The teeth of echinoids are the most complex, and have been fully described by

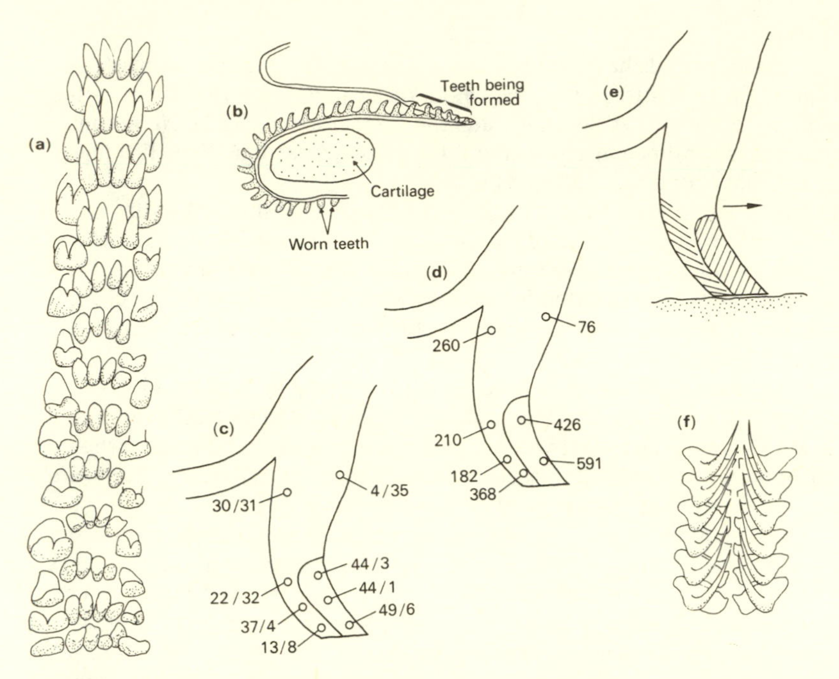

Fig. 5.47 Teeth from radulae of gastropod molluscs. (**a**) the browsing limpet *Patella vulgata*. Older teeth at the bottom of the figure have become worn. (**b**) Diagram of a radula from the side (anterior to left) showing how it is pulled over a cartilaginous bar. (**c**) Sagittal section of a lightly worn tooth of *Patella*. Numbers for each dot represent the amount of Fe and Si respectively. (**d**) Same as (**c**). Numbers represent Vickers hardness values. (**e**) The way a tooth is pulled over the substrate in browsing. (**f**) Teeth of a carnivorous gastropod (**c** and **d**, RUNHAM *et al.*, 1969).

Märkel (MÄRKEL and TITSCHAK, 1969; MÄRKEL, 1969, 1970*a*, 1970*b*, 1970*c*). The detailed structure varies, but usually the teeth are long and narrow, gently curved with a keel on the concave side (Fig. 5.48). There is a flexible proximal growing region, covered by an organic sheath, a long middle region that is attached to and grows through parts of Aristotle's lantern, and the business end, which is sharp.

The tooth is made of a series of three-sided, baseless, hollow, inverted pyramids, stacked one inside the next. The pyramids have two flat walls, each a sheet of calcite; the third wall is made of a tuft of calcite prisms and needles, all attached delicately to the apex of the pyramid (Fig. 5.48c). As the pyramids leave the growing region, the various parts are cemented together by other blocks of calcite The tuft of needles grows and fills the central part of the tooth. The degree of cementing varies, being almost complete in the region where the prisms and needles are attached to the apex of the pyramids but being progressively less in other regions. The central region, called the stony region, is the part that does the rasping; the more peripheral parts break off as they near the tip of the tooth.

The stony region is remarkable, for echinoderm skeleton, in having a very high proportion of magnesium. Most echinoderm 'calcite' has about 8–15% $MgCO_3$ in solid solution with the $CaCO_3$. But SCHROEDER *et al.* (1969) have shown by

5.19.9 Teeth

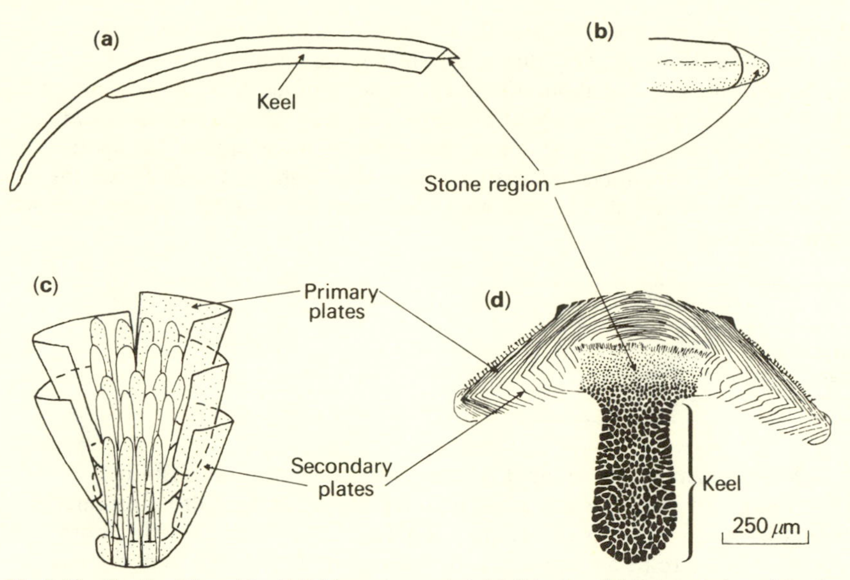

Fig. 5.48 Tooth of the echinoid *Echinometra mathei.* (a) Side view. (b) Enlarged view of the tip, seen from convex surface. (c) Tooth consists of pyramids nesting into each other. 5 are shown; alternate ones are stippled (after MÄRKEL, 1970*c*). (d) Cross-section of a tooth of *Stomopneustes* near the rasping tip (after MÄRKEL, 1969).

electron-probe analysis that in the stony region there is about 43% $MgCO_3$, making the mineral effectively dolomite, $CaMg(CO_3)_2$. The high proportion of magnesium will have the effect of making this part of the tooth harder than the rest: calcite has a Mohs hardness of 3, whereas dolomite has one of 3.5–4. The other remarkable thing about the stony part is that it has an organic matrix, unlike normal echinoderm skeleton. This matrix is trapped as the calcite cement is formed. MÄRKEL (1970*c*) remarks 'The stone part has a considerable amount of organic matrices, and that is why the stone part is not as brittle as the remainder of the tooth skeleton, which is nearly free from organic matrix.' This seems a perfectly satisfactory suggestion. However, recently MÄRKEL and GORNY (1973) have found stone regions with little organic material. In these teeth the stone region bears a remarkable resemblance to glass fibre reinforced plastic. There are many fibres of calcite set in a matrix of much more disorganized calcium carbonate.

The teeth of vertebrates are nearly always in the form of a capping of enamel on a base of dentine. For a full account of the structure of teeth see MILES (1967). Dentine is an interesting variant of bone. At the ultrastructural level it is very like bone. However, it does not contain cells, instead it has 'dentinal tubules' running in from the surface. These tubules contain the cytoplasmic extensions of odontocytes that line the inner surface of the dentine.

Enamel is the hardest and stiffest of the vertebrate skeletal materials. The hardness of enamel is roughly that of apatite–5 on the Mohs scale. This is harder than calcite (Mohs hardness 3) but softer than quartz (Mohs hardness 7). The structure and chemical composition of enamel are reviewed by EASTOE (1971).

Enamel consists, like bone, of crystals of apatite in a protein matrix, but apart from this these tissues are very different. The apatite crystals are much larger than those of bone, being about 40 nm across, as opposed to 4 nm in bone, and so have a cross-sectional area about 100 times as great. They are about 150 nm long, which may or may not be longer than the long dimension of the apatite crystals in bone. The inorganic/organic ratio is very high. The table shows the proportions by weight of the three main constituents of enamel in immature and mature enamel of pigs.

	Just formed	Mature
Inorganic	37.0	94.0
Organic	19.0	1.8
Water	44.0	4.3

As the enamel matures the amount of organic material declines absolutely.

The protein of enamel is not collagen–there is very little hydroxyproline; nor is it keratin–amongst other things there is hardly any cysteine. Furthermore, the protein seems to be amorphous; early ideas of a fine fibrillar network have been superseded by the concept of an amorphous gel being compressed into particular regions as the crystals grow during maturation.

At the microscopic level enamel is composed of a series of rods or 'prisms', each extending from the dentino-enamel junction to the outer edge of the tooth. The enamel is deposited extracellularly by ameloblasts, and at first contains much protein and water. The apatite crystals grow, and in doing so displace the water and the protein, which are forced along the length of the crystallites towards the free border.

The apatite crystals are in a gentle fanning arrangement. Where the adjacent prisms abut there is, necessarily, a discontinuity in crystal orientation. The protein squeezed out by the maturing crystallites tends to follow the direction of the crystallites and so it accumulates at the prism boundaries. In this way there develops a relatively organic-rich 'prism sheath' round each prism. But these sheaths are in no way comparable in development to the sheaths round the prisms in the secondary shell layer of articulate brachiopods, which they superficially resemble, for these latter sheaths are payed out, as such, by cells. Mechanically, of course, they may well have similar functions.

The enamel prisms run fairly straight to the surface, though they do tend to wind helically round each other.

5.20 Mechanical Properties of Stony Materials

In considering the properties of stony materials found in skeletons, we must amplify some statements made about fracture in Chapter 2A.

Many things determine whether a particular ceramic material is strong or not, and there are some ceramics that are stronger than the minerals used by animals. (For our purposes 'stony' and 'ceramic' can be considered synonymous.) Diamond, alumina, silicon carbide and aluminium nitride are all quite strong in bending, for instance. We assume there are various biological reasons why such materials are not used by animals; probably the most important is that they require a high temperature for their production. However, given that a skeleton is made of a particular material, there are two things that will have a most striking effect on its strength and stiffness: grain size and porosity.

5.20.1 Grain Size

The effects of grain size in the present context are far from well understood. This is because shell structures, though bearing superficial resemblances to, are quite different from those of most of the materials in which the effects of grain size on mechanical properties have been studied. We begin by considering two extremes of behaviour, that of polycrystalline ceramics, and that of fibre-reinforced polymers.

In general polycrystalline ceramics have what we shall call 'coherent' grain boundaries—that is, although the orientation of the atoms may change across the boundary region, the type of atom in this region is the same as elsewhere. In effecting a change of orientation, certain atoms have been displaced from their normal lattice positions and occupy positions intermediate between the positions appropriate to the grains on either side.

When a single crystal fractures we find that one particular plane of separation is preferred. This accounts, among other things, for the ability of the diamond cutter to split a gemstone along certain planes without wrecking the stone. In a disoriented aggregate of many crystals, the preferred fracture plane must, then, change its direction at most grain boundaries. The force tending to keep the crack moving is the force acting normal to the plane of the crack and if the crack, proceeding under a force F in plane A, is made to change direction through angle θ, the force component opening the crack travelling along plane B must be larger than that necessary for plane A. As the surface area of the fracture surface is increased by all these changes in direction of the crack, more surface energy is required to fracture the material than if it were a single crystal. We showed in Chapter 2A that the fracture strength depends upon the *total* work that must be done, so we may expect that the more often a change in direction is needed, the greater the resistance to fracture under a given stress. This leads to the well established relationship of PETCH (1953),

$$\sigma_f = \sigma_i + K \cdot d^{-1/2} \qquad (5.18)$$

where d is the grain diameter and K and σ_i are material constants, the latter approximating to the strength of a single crystal considered alone.

This relationship was developed for metals, but is found also to apply to ceramics (e.g., CARNIGLIA, 1965, 1966). Data for polycrystalline BeO are shown in Fig. 5.49. Similarly ILER (1963) showed that translucent flint, having a mean grain size of rather less than 1 μm, had a modulus of rupture of 190 MN m^{-2}, while opaque flint, with a grain size of 2–10 μm, had a strength of 120 MN m^{-2}. Flint is a remarkably strong naturally occurring rock. (Granite, for example, has a modulus of rupture of at most 40 MN m^{-2}.) Iler attributes the high strength of flint to its fine grain structure.

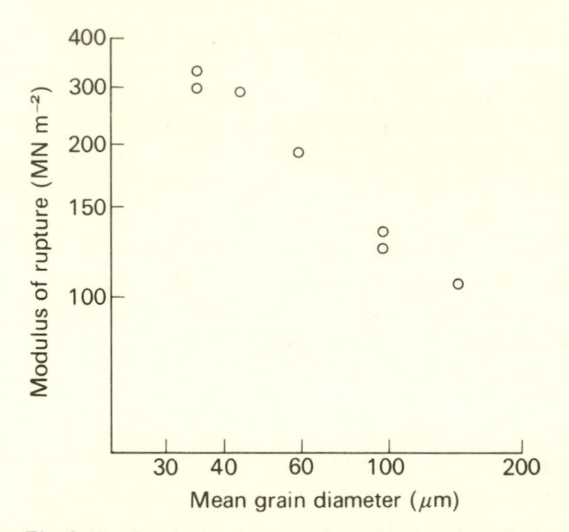

Fig. 5.49 Graph showing the effect of grain size on the modulus of rupture of beryllium oxide. Note log scales.

Equation 5.18 predicts that the fracture strength should increase without limit as grain size decreases, and it will be seen that the second term in *Eq. 5.18* has much the same form as the Griffith equation

$$\sigma = \left(\frac{2ES}{\pi}\right)^{1/2} \cdot a^{-1/2} \tag{5.19}$$

but with the grain size d taking the place of the crack length a. The reason for this correspondence, which at first sight seems reasonable, is still the subject of considerable discussion. However, the similarity of Petch's relationship to the Griffith equation (*Eq. 2.21*) suggests that, at this stage, we should consider some absolute values.

The critical crack length for a brittle material is given by *Eq. 5.19*. All the important skeletal minerals are covalently bonded, and in single crystal form are brittle. The Griffith equation can be expected, therefore, to apply to them as much as to any other material. For calcite the surface energy has been reported as 0.23 J m^{-2} (GILMAN, 1960), and E as 137 GN m^{-2} (BHIMASENACHAR, 1945). For the tensile stress at fracture we can take two values, 50 MN m^{-2}, which falls rather high in the range of tensile strengths of calcite-containing materials, and 100 MN m^{-2}, which is above the range except for *Turbo* shell (see below). For these two cases the critical crack lengths are approximately 8 μm and 2 μm. The surface energy calculation was made in experiments involving very cold calcite. Possibly at a higher temperature there might have been some plastic flow, in which case the critical crack lengths would have been greater. In general (Table 5.6) most stony skeletal materials have grains with at least one dimension smaller than about 3 μm, but usually not a great deal less, so that we may conclude that grain size plays some part in determining the fracture strength.

At the other extreme we have the mode of crack propagation in fibre-reinforced materials which was described in Section 5.7. The essential feature here is that of a propagating crack which is deflected at a matrix–fibre interface, and energy is thereby absorbed. If the fibre and the matrix have equal fracture strengths and fracture strains and the matrix is rigidly bonded to the fibres, there will be no extra energy absorbed as the crack goes from fibre to matrix or vice versa. Conversely, in a material consisting of stiff brittle fibres in a pliant or even a plastic matrix, cracks in the fibres will be unable to propagate because the stress concentrations at their tips will be blunted by the pliant matrix.

We suggest that the behaviour of shell materials falls between these extremes and depends upon (a) the size and shape of the mineral phase and (b) the amount and properties of the cementing phase. Given that the cementing material is less brittle than the mineral, it is fairly easy to see that the actual amount of cement is important. A thin layer will offer little or no possibility of plastic flow as the crack traverses the boundary and we might expect the behaviour to be determined primarily by misorientation effects between grains, i.e., that *Eq. 5.18* might describe the behaviour. A thicker layer of a more ductile cement would permit plastic flow and, while *Eq. 5.18* might still apply, it would need to be corrected–in much the same way as described in Chapter 2A for a plastic work term additional to a surface energy term.

For a thin layer of a hard, brittle cement, we may expect separation along the boundary as the crack advances through the crystalline phase, giving behaviour similar to that suggested by COOK and GORDON (1964) for failure in fibre reinforced solids. But a thick layer of brittle cement would offer no barriers and may indeed provide a source of additional cracks. Such a case would, surely, reduce to little more than the Griffith criterion.

In the absence of hard experimental data concerning the strength and modulus of the cement–which is of course the matrix–we can only speculate. However, the matrix may be quite strong. *Atrina* shell whose tensile strength was found to be about 70 MN m^{-2} (Table 5.3) has a prismatic structure, and the fracture line ran between the prisms, and therefore was through the matrix, or through the mineral-matrix interface. (Normally prismatic structures would probably not be exposed to tensile stress, so its high strength is even more creditable.)

Not knowing the modulus of the matrix we can perform no useful estimates though, as a guess, the modulus of tanned arthropod cuticle (10 GN m^{-2}) is unlikely to be far different from the modulus of the organic matrix of ceramic skeletons. The organic phase is likely to show plastic deformation, and to be tough. Even so, the fact that the crystallites are generally small indicates that the Griffith criterion is also important. Furthermore, many shells have very little organic material indeed (less than 0.05% by volume). In these the mechanical effects of the organic phase must be small. Indeed, the fact that stony skeletons have a characteristically small crystal size seems, in hindsight, inevitable, and it would be disturbing if they did not.

5.20.2 Porosity

The amount of space within a ceramic material, its porosity, is a very imporant feature. If the porosity is large, the strength of the ceramic will be considerably less than if it had no voids within it. A small part of this effect is caused by the reduction in the cross-sectional area subjected to load. However, a more important effect is that voids produce stress concentrations and, if there are connections between the voids, there may in effect be long cracks present. The effect of porosity has not been much analysed theoretically, because the effect of pores is to turn a solid block into a structure of awesome complexity as far as stress analysis is concerned. However, an empirical formula due to RYSKEWITCH (1953) seems to work reasonably well. This predicts $\sigma = \sigma_0 e^{-nV}$ where σ is the actual strength, σ_0 is the strength of the same material with no porosity, V is the porosity (the fraction of the total bulk occupied by void), and n is a number lying somewhere between 4 and 7. Most ceramic skeletal materials have a volume fraction of organic material of between about 0.2 and 8%. There is usually very little actual void at all. If we suppose that the sole function of the organic matrix were to fill in holes that would otherwise exist, would it have an important effect upon the strength? Probably not. Taking a value of 8% for porosity, and 7 for n, the value of strength from Ryskewitch's formula is 0.57 of the strength of the same material without voids. This difference, though fairly large, is obtained only if both parameters in the equation are at an extreme value, when the effect will be most marked.

There is another consideration. The matrix may be filling up the voids, yet, because it has a much lower modulus than the stony material, it should not be particularly effective in reducing stress concentrations. Therefore, the ceramic whose voids are filled with matrix will not, according to this argument, be as strong as one that has no voids at all. Even so, it is possible that the matrix may have other good effects than those indicated by the formula of Ryskewitch. AUSKERN and HORN (1971) performed some experiments in which they compared concrete with about 12% porosity, and concrete of the same initial porosity in which most of the pores had been replaced by polymethylmethacrylate (PMMA). This plastic has, of course, a much lower modulus of elasticity than the cement or the stones making the aggregate. The following table shows some of their results:

	Standard concrete	Concrete with PMMA	Increase
Compressive strength	36	155	x 4.3
Tensile strength	2.9	11	x 3.8
Modulus of rupture	5.1	18	x 3.5
E	24 000	43 000	x 1.8

(all values in MN m^{-2})

There is a considerable increase in strength and, remarkably, in stiffness. GEBAUER *et al.* (1972) have confirmed these remarkable results. Auskern and Horn suggest that, apart from filling up the voids, the PMMA improved the bonding between the interfaces. It would be rash, of course, to analogise concrete with its large porosity, and ceramic skeletons very closely. There are, however, obvious similarities between the two materials.

A remarkable feature of most animal ceramic skeletons is that they are almost without voids caused by deficiencies in packing (even spheritic structures have only very small voids). Usually, if there are voids, they occur quite clearly as part of the designed structure. For example, the scleractinian corals have a complex, porous structure. The shells of many brachiopods are pierced by punctae, which possibly allow noxious fluid to pass through the shell onto the outer surface, discouraging would-be settling larvae. The eggshells of birds are pierced by pores which allow gases to pass through. In each case we suggest that the advantages of having these voids outweigh their possible disadvantageous stress-concentrating effect. Finally, of course, there are the echinoderms, of which LIBBY HYMAN (1955) said so aptly "I here also salute the echinoderms as a noble group especially designed to puzzle the zoologist". Their skeleton is so unlike any others that it will be treated separately.

5.20.3 The Function of the Organic Matrix

A great deal of this chapter has gone towards showing that rigid materials are either classical fibrous composites or are ceramics with very small grain size in an organic matrix. We have argued that this latter structure is the result of mechanical design, because materials made in this way are much stronger than they would be if they consisted of larger blocks.

This idea has not, as far as we know, been put forward with such force before, nor with so many data showing how universal are the structural facts on which it is based. (A shorter version has been put forward by CURREY (1970*a*) and other authors have considered the composite nature of skeletal materials, and discussed the possible mechanical implications of such structures, e.g. JENSEN and WEIS-FOGH (1962), GORDON (1970) and MARK (1967).)

There is one rather obvious criticism of this idea that we shall discuss. This concerns the necessity for *nucleating sites.* It has frequently been suggested that the organic matrix of calcified tissues has the function of acting as an energetically favourable site for the nucleation of appropriate crystals. This has been suggested, for instance, for bone (SOBEL *et al.*, 1960), mollusc shells (TRAVIS, 1970) and Foraminifera (TOWE and CIFELLI, 1967). It could be argued that, since organic matrices are needed for the nucleation of crystals, the principle of parsimony would not require us to seek other functions.

Even accepting that an organic surface is necessary for nucleation, would this account for the general distribution of the organic matrices in organic skeletons? Surely not, for it is the large number and small size of the crystallites, and their tendency to be *wrapped up* in organic sheets that is impressive, not the mere presence of an organic sheet somewhere in the system.

Recently TOWE (1972) has produced strong arguments against the idea that the organic matrices of mineralized tissues are serving merely, or even principally, as templates or nucleating sites. For instance, if the crystallites are bound up in polygonal chambers, only one wall of the chamber can act as a nucleating site, otherwise the crystals will grow in from all faces, and where they meet will not be crystallographically compatible, yet it is an observed fact that the crystallites in such chambers are usually optically single crystals. Furthermore, evidence is appearing that, in these compartments, the mineral does not always nucleate in contact with the limiting membrane. There are other difficulties.

Indeed it is clear, particularly in the case of calcite, that once a crystal is nucleated in a saturated solution, the difficulty is to prevent it growing large, but ill-formed, very rapidly. HENISCH (1970) discussing crystal growth in silica gels, points out that the function of the gel is to *inhibit* nucleation, so that reasonably well-formed crystals can form.

It is also quite clear that the organic sheets do often have the function of organizing the crystallites. BEVELANDER and NAKAHARA (1969) and NAKAHARA and BEVELANDER (1971) demonstrate this for molluscan nacreous and prismatic structures. In both these structures organic sheets grow and form virtual spaces within which the crystallites themselves later appear. The final size of the crystallites is determined, therefore, before they exist. A similar situation is seen in the prismatic (fibrous) layer of articulate brachiopods, in which the mineral prisms lie in tubes of matrix paid out by the mantle cells.

The question of the initiation, growth and compartmentalization of stony material is too complex to pursue here. But the fact that the strongest materials have the most organic material, and that some structures have very little organic material, so that very little is necessary to initiate calcification, indicate to us most strongly that there is an important mechanical role for the organic matrix of rigid skeletal materials.

The extremely close packing of animal ceramic skeletons, is, of course, important mechanically, and is brought about by the matrix, which ensures the orderly deposition of the crystallites.

There is another important function that the matrix serves, and that is to produce a great *uniformity* of crystallite size. The inverse relationship between crystallite size and strength that has been discussed above implies that if there are occasional large crystals these, on their boundaries, could act as 'weak links', causing the material to fail at a low stress (RICE, 1972). But what is so striking about many biological ceramic structures is their uniformity, particularly the lack of crystallites or substructures much above the median size. This is particularly clearly seen in nacre, in prismatic structure, in the fibrous layer of brachiopods, and in enamel.

The basic mechanical properties of stony skeletal materials that we and others have been able to determine are shown in Table 5.3. There is a fairly large series for mollusc shells, and we shall concentrate on this. The main fact that stands out is the great variability of the median values for strength. It is difficult to assert that these differences are all real, but some certainly are. Figure 5.50 shows the individual results classified according to shell structure. It is clear that these differences in properties accord, to a large extent, with differences in structure. In general nacre is strong compared with foliated, crossed-lamellar, complex crossed-lamellar and homogeneous structures. The shell of *Anodonta* is the only rather weak nacre in tension. Calcite prisms seem to occupy an intermediate position between nacre and the rest. In elastic modulus there is little difference between the various types, except that foliated structure seems rather low, and calcite prisms have a very low value in *Pinna*. *Pinna* shell was unusual in that of all the shells we tested it was the only one to show a considerable amount of stress relaxation (CURREY and TAYLOR, 1974).

The load-deformation curves for the weaker types are effectively linear till fracture, and the fracture spread immediately across the specimen. The load-deformation curve of most nacreous structures in bending showed considerable

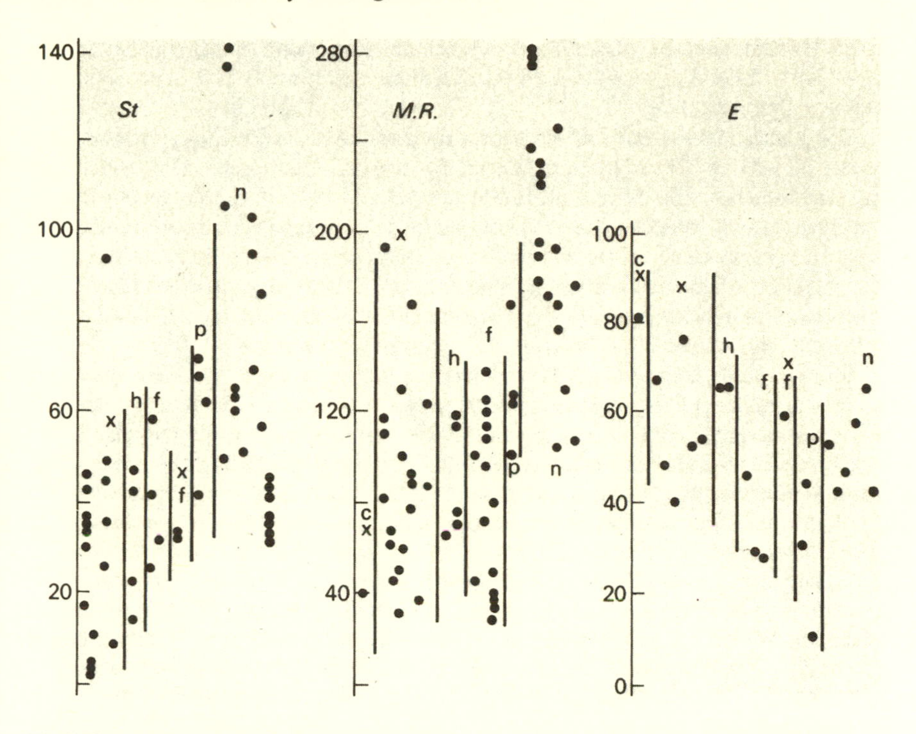

Fig. 5.50 Diagram of relationship between different structural types of mollusc shell and various mechanical properties. The dots in a single column refer to tests on different specimens from a single species. Different structural types are separated by thin vertical lines. cx, complex crossed-lamellar; x, crossed-lamellar; h, homogeneous; f, foliated; xf, cross-foliated; p, calcite prisms; n, nacre. *St*, ultimate tensile strength in MN m^{-2}; *M.R.*, modulus of rupture in MN m^{-2}; *E*, modulus of elasticity in GN m^{-2} (CURREY and TAYLOR, 1974; courtesy of the *J. Zool., Lond.*).

plastic deformation. Interestingly the shell of *Anodonta*, the weakest of the nacreous structures, sometimes shows brittle fracture. Calcite prisms and cross-foliated structures also showed a little plastic deformation (Currey, unpublished).

It is clear that nacre is superior in some of its mechanical properties to crossed-lamellar and homogeneous structures. The greater strength of nacre is not associated with any marked difference in modulus or in density.

Nacre and calcite prisms are different from the other structures in having a high organic content (about 1–4% by weight, which will be equivalent to 2–8% by volume approximately). Crossed-lamellar shell has 0.01–0.4% (0.02–0.8% by volume) and foliated shell 0.1–0.3% (0.2–0.6%). The amount of organic material in homogeneous shell seems to be unknown, but it is certainly very low (HARE and ABELSON, 1965). The prisms, and the blocks in nacre, when seen under the scanning electron microscope, are much more clearly separate from each other than are the various substructures in the other types. Referring to our discussion earlier in this section, it seems that although small-sized crystallites are in general advantageous, it is also important for strength to have some separation of the blocks by organic material. Even so the blocks in nacre are small, relative to likely

Griffith flaws, and the prisms, though large, are composed of still smaller crystallites (Table 5.6). As mentioned above, prismatic structure in fact tends to cleave *between* the prisms.

The plastic deformation of the whole specimen before fracture is presumably caused largely by plastic deformation of the organic component. This will allow increased energy absorption before fracture, compared with a completely brittle material, though whether this is a biologically important effect is unknown. The most important effect of the organic component is, no doubt, truly to isolate the crystallites from each other, so that fracture of any block is not followed by fracture of neighbouring ones. For calcite prisms the fracture is in the main interprismatic, and in nacre the fracture runs mainly between the blocks.

Mollusc shells seem, therefore, to have two different strategies. One is to have a fairly granular polycrystalline material, with small crystals, so that the size of any pre-existing Griffith cracks is limited. The amount of organic material is small, and is probably important mechanically only in that when the shell is laid down it prevents the formation of large crystals. Fracture probably occurs because some intergranular boundaries are near enough to the direction of maximum tensile stress to act as fracture-initiating cracks. When the crack has started to run it seems to travel indifferently between or through the grains. The other strategy is to have similarly small crystals, but also to have a much greater amount of strong organic material which can deform elastically and, as the stress increases, plastically. The strength of the crystals is probably high enough so that when sufficient fracture initiates in the matrix, the strength of the matrix is the limiting factor.

Why is it that some molluscs have shells made of material apparently worse mechanically than others? First, of course, this may simply not be true, because we do not know the compressive strengths of many shell structures. Even so, it seems improbable that the compressive strength of, say, crossed-lamellar structure could be so much greater than that of nacre to outweigh a nearly 50% disadvantage in tensile strength. TAYLOR and LAYMAN (1972) have tested the *hardness* of various bivalve shell types and have found that crossed-lamellar, complex crossed-lamellar and homogeneous structures are harder than the rest. The structures found on the outside of burrowers' shells are just these. Taylor and Layman suggest, reasonably enough, that hardness is probably associated with abrasion resistance. Those burrowers that use nacre or prisms nearly all live in very fine sediments, and are not very active, so the tendency for wear will be small. This may be part or, conceivably, the whole of the explanation. We certainly cannot adopt the idea that it is the most primitive groups of molluscs that have the weaker shells. Quite the reverse is the case. The primitive groups had, in general, a shell with prisms on the outside and nacre on the inside. Shells with weaker structures have evolved from this condition. This trend has been clearly documented for the bivalves by TAYLOR (1973). In general, the trend has been towards reduction of the organic content of the shell. Unless the abrasion resistance of the structures is sufficient to account for their evolution, we must seek a non-mechanical reason. One that suggests itself is that the less neatly packaged shells can be built much more quickly. There seem to be no studies on this point.

Finally, we should mention the fact that stony skeletons of calcium carbonate exist in two forms, either calcite or aragonite. (Some other crystal types occur to an insignificant extent.) The subject is reviewed for bivalves by KENNEDY *et al.* (1969). The type of mineral produced is almost entirely genetically determined, though the proportion of the different shell layers found may depend marginally on the ambient temperature. The mechanical significance of the calcite *versus* aragonite dichotomy, if any, remains obscure.

5.20.4 Stony Skeletons with many Holes

We have mentioned that, other things being equal, a skeleton with many voids is weaker than a solid one. The corals, the echinoderms and to rather a lesser extent the bryozoans may have skeletons that are anything but solid. The corals, particularly, often have a fairly delicate tracery of stony material. If it is so much weaker, why do the animals build it?

Ryskewitch's empirical formula, $\sigma = \sigma_0 e^{-nV}$ implies a considerable drop in strength with more than a small amount of porosity. We did some rough calculations to see whether it applied to coral. Three specimens of an unidentified scleractinian from Eilat, Israel, were tested in three-point bending, and gave values for modulus of rupture of 27, 27, and 29 MN m^{-2}. The sums were done as if the skeletons were solid but the isolated, large protuberances on the surface were ignored, as they would contribute virtually nothing to the strength. The calculations were rather rough and ready, because the shape is irregular. However, the agreement of the results with each other is good, and they are probably not far from the truth. Strong mollusc shells have a modulus of rupture in the region of 200 MN m^{-2} (except for *Turbo* which is exceptionally stronger than the rest), though many are not as strong as this. The coral is, therefore, about one eighth as strong as strong mollusc shell. The porosity of the coral, ignoring a small central canal, was about 40%. By Ryskewitch's formula, using the extreme values of 4 and 7 for n, the strength with 40% porosity would be about 0.2 to 0.06 times the strength of the solid equivalent. We are, of course, comparing apples and pears here, particularly as the modulus of rupture is not the same as the tensile strength, but it seems that the coral is in the range one might expect.

Having shown, therefore, that the production of voids in the skeleton does produce a reduction in strength over and above that that might be attributed to the reduction in volume, we can enquire why it is that the coral builds skeletons weaker than need be. To some extent the animals live in the voids, and are protected. But, beyond this, the skeletons hold the polyps in position in space, and this is probably their primary function. Coral colonies can grow only if there is sufficient space between the polyps. There will be sufficient space only if the total surface area of the skeleton is sufficient. Therefore growth involves laying down a supporting skeleton, but this is expensive in energy and time. Because of surface area/volume relationships, the polyps would need to add on ever-increasing amounts of skeletal material for each new polyp for whom space would be made thereby. It seems, therefore, that the porosity of the skeleton is an adaptation for the need for area, rather than very great strength. Similar arguments will apply to those bryozoans that have porous and rather insubstantial stony skeletons.

The echinoderms are a more baffling problem—since, among other things, they seem to have holes in even the smallest pieces of skeleton. The only exception to this seems to be in the teeth of echinoids, which have been mentioned above. Values of the mechanical properties of echinoderm skeleton, insofar as they are known, are high, considering that they are full of holes (Table 5.3). The porosity was found to be about 50% for those we tested in bending. Ryskewitch's formula gives values of between 0.135 and 0.03 of the solid equivalent, that implies values of 290–1130 MN m^{-2} for the solid equivalent. However, if one looks at the echinoderm skeleton under the microscope it becomes apparent that it is unrealistic to apply, too slavishly, a formula which was found to work for substances with large irregular closed pores. The interconnecting voids of echinoderm skeletal plates are obviously neatly arranged. Suffice it to say that, weight for weight, echinoderm skeleton is as strong as reasonably strong mollusc shells, even if the living tissues in the interstices, which will contribute nothing to the strength, are taken into account in determining specific gravity. WEBER *et al.* (1969) came to the same conclusion after testing echinoid spines and gastropod shells in compression. However their gastropod values may be too low (CUMEY 1975).

How are we to explain the functional significance of the very great porosity of echinoderm skeleton and the fact that, alone in the metazoa, its stony skeleton is not made of polycrystals? CURREY and NICHOLS (1967) and NICHOLS and CURREY (1968) suggested that the trabecular structure would prevent cracks from running far, because any crack would soon run out of the trabecula in which it started. This idea was taken up by WEBER *et al.* (1969). If the system works like this, it would be somewhat analogous to a polycrystalline ceramic. Nichols and Currey also suggested that since the trabecular surfaces are covered by living tissue, this tissue could function to keep the stony surfaces, where the dangerous flaws form, free from such flaws. They showed that indeed the surface was very smooth where covered by tissue, but not elsewhere. There is no mechanical information to test this idea.

Unfortunately the theory of trabecular ceramics has not attracted much attention from engineers or materials scientists. Presumably this is because only small animals, mostly rather obscure (to an engineer) use such an unpromising material. The echinoderms are unique in possessing a matrix-free skeleton in the form of trabeculae. Also unique is the stone region of their teeth (MÄRKEL and GORNY, 1973). As mentioned before, this region of the teeth of some Echinoids consists of calcite fibres in a calcium carbonate matrix. The mechanical consequences of such an arrangement are not known, but it would seem probable that such a structure will act as a classical fibre-reinforced composite.

5.21 Rigid Skeletal Materials: Some Final Remarks

In this review of the main types of rigid skeletal materials we have shown that the material is so arranged as to make the greatest mechanical advantage of the structure—ceramic materials are made small-grained, with some matrix and without voids, composite materials have the fibres arranged in the most advantageous directions and so on. However, apart from a few passing remarks, we have not discussed why particular animals have particular kinds of skeletal material in the first place. Why are not corals made of bone, or insect cuticles of aragonite, for instance?

We cannot, at the moment, do much more than try to apply biological common sense in answering such questions, for our knowledge is extremely scant. However, we shall state a principle that should guide us in our discussion. This is that we shall not consider the idea that skeletons are what they are because the animal is incompetent to make anything else. We shall not do this for two reasons: we believe such an idea to be false and, more importantly, to act as if it were true would be sterile (CAIN, 1964). There seem to be no remotely convincing cases where one can show that groups have remained unchanged for a long time, even though it would have been advantageous for them to change. Of course, the great majority of species have become extinct, usually because they did not change in response to altered selective circumstances quickly enough. But these were crisis situations, over almost instantaneously in geological time. *Lingula*, which has remained effectively unchanged in its shell characters since the lower Palaeozoic, is sometimes quoted as an animal that cannot change. But, as we have shown, its shell is specialized, adapted to a rather unusual life habit and, in a quiet sort of way, *Lingula* is successful. SIMPSON (1953) discusses the idea of rates of evolutionary change in organisms. Natural selection has had hundreds of thousands of generations in which to act and therefore all skeletons, except for those of just a few species that are in the process of evolving rapidly, are nearly optimal for the conditions in which the organisms find themselves.

We should have to consider many factors in order to account fully for the features we see in skeletons. However, for most purposes, four are enough (and usually more than we can say sensible things about, unfortunately). These are: (a) mechanical function, (b) availability of raw materials in the environment, (c) ease of fabrication, and, (d) suitability of the ontogeny of the skeleton. We shall briefly exemplify the four factors.

(*a*) *Mechanical Function* The significance of this is by now familiar to the reader, we hope. It is perhaps worth pointing out, however, that the lack of skeleton with particular mechanical properties is sometimes an absolute bar to the existence of any organisms in a particular habitat or having a particular way of life. For example, shingle beaches are notoriously poor in animal life because any animal trying to settle on the pebbles would get smashed as the tide line passed over it. Yet if a barnacle-like creature, with a covering of steel, could establish itself during one high tide it would find itself, at high tides, covered by water from which, because of lack of competition, very little food had been removed. Since the skeletons of animals cannot be made in this way, shingle remains barren.

(*b*) *Availability in the Environment* Organisms on the whole do not make use of very bizarre materials for their skeletons. Strontium sulphate may be used by the actypilinid protozoans, but if all species that use calcium carbonate as their skeletal material used this instead, the biomass of shelled animals would be sharply reduced. Organisms usually use materials that are locally abundant. Only modern man is exempt from this, but the cost of transport remains a limiting factor.

(*c*) *Ease of Construction* Animals must extract materials from the environment and incorporate it into the skeleton at body temperature. This of course rules out any material that is prepared by melting. We have discovered no useful studies of the metabolic cost of producing a skeleton. Such cost would include the actual calorific value of the skeleton, the work that has to be done to concentrate the raw materials in one place against a concentration gradient, and the work that has to be done in organizing the material once it is there. This last cost will be particularly difficult to quantify.

(*d*) *Ontogeny* By this we mean the difficulties associated with the growth, as such, of the skeleton. We may give as examples the difficulties of ecdysis, which is itself necessitated by the inability of the arthropod cuticle to grow by intercalation; the weakness of the epiphyseal plates in mammals, caused by the need to have a growing zone away from the articular surface; and the awkward and wasteful shape of snails' shells, caused by the snail's inability to remove material once it is laid down.

We shall now discuss some of these factors, showing how they may interact. Table 5.3 shows the mechanical properties of many of the stiff skeletal materials of which we have knowledge. The three columns on the right show the weight per unit strength or stiffness. The lower the value the more efficient the material. We take, as an example, the arthropods. These animals have, essentially, three characteristic types of skeletal material. One kind is chitin and protein. This is found in insects, some crustacea, and all other arthropods except the diplopods. Another kind is made of protein, chitin and calcite, with the calcite more or less replacing the protein. This is found in the diplopods and most Crustacea. Finally, there is skeleton made of calcite with very little organic matter. This is found in the barnacles.

Can one give convincing reasons for the existence of these three separate types? In our opinion we can. First compare the insects with the ordinary crustaceans. The specific strength and stiffness of the few crustacean cuticles that have so far been examined are lower than those of insect cuticle. It would seem, then, that crustacean cuticle would be mechanically better if it were like insect cuticle. But of course, the metabolic costs of the skeletons are different. In the sea, which is often supersaturated relative to calcite (SIMKISS, 1964), the metabolic cost of precipitating calcite is probably not great. Even in fresh water calcium carbonate is usually reasonably easy to obtain (MACAN, 1963). The metabolic cost of producing the protein, which in the Crustacea is to a large extent replaced by mineral, is certain to be considerably greater. The crustacean skeleton, therefore, may be able to be made of a somewhat mechanically inferior material, because it is easier to make, and therefore imposes less of a strain on the animal. (See also the discussion on safety factors in Section 7.13.) For the crustaceans, most of whom live in water, a slight increase in the mass of the skeleton needed to make up for its mechanical deficiencies is probably no great matter. For insects, most of which fly, it could be crippling. Most other land arthropods also have no mineral in the skeleton, and though the cost of a heavy skeleton will not be so great for a ground-living animal as for a flying one, it could still be considerable, particularly if the animal is very active.

Nevertheless it is interesting that crustacean cuticle seems to be inferior to many mollusc shells both in weight per unit strength and weight per unit stiffness. Therefore, in terms of both metabolic cost and mechanical properties, the crustaceans are inferior to, say, the gastropod *Turbo*. We can only suggest that the crustacean cuticle is superior, from the crustaceans' point of view, because it can be laid down so quickly. A newly moulted lobster is a helpless animal (as anyone will know who has seen the carnage when a lobster moults in a pen with other lobsters present) and the rapidity with which a functional skeleton can be made is obviously important. A lobster can go from one fully hardened exoskeleton to a different one in a few days. It is most unlikely that a mollusc shell could be laid down, with the same precision of form, in the same time.

The diplopods are terrestrial and do have calcium carbonate in their skeleton. However, as has been shown by MANTON (1954), much of the structure of diplopods can be explained as a result of their habits of walking and of burrowing their way through soil or litter. In these latter circumstances, where there are hollow, thin-walled structures pushing hard, the stiffness of the material making the structures will be very important, for otherwise they will buckle. The resulting stiffness may be achieved at some cost, but we are not now able to assess this cost. For complicated structure like the exoskeleton of millipedes, it may well be that an increase in thickness is not a very acceptable way of increasing stiffness, and absolute, rather than specific stiffness is required.

The barnacles, on the other hand, have typical ceramic skeletons with a very small proportion of organic material. Why do they not have the same kind of skeleton as the rest of the Crustacea? For animals that are sedentary, weight, or mass, becomes rather unimportant. Therefore, if the mechanical advantages in having a skeleton with much organic material are only marginal, or anyhow not large, then the metabolic disadvantages of having much organic material may be important. For sedentary animals that produce the necessary stiffness and strength with only a small amount of extra material, the good mechanical qualities of bone, or conventional arthropod cuticle, become nugatory. Their relative metabolic disadvantage, however, remains. Furthermore, uniquely among the arthropods, the calcareous barnacle shell is not moulted, so a skeleton that can grow only slowly is satisfactory.

We have proposed that the structure of the skeleton of sessile organisms is much less constrained by the necessity for high specific stiffness and strength than is that of active animals. However, there must still be limits on the weakness of these materials. If a material is weak, a greater *volume* of it may be needed to produce a structure of the right properties. Great volume will be a disadvantage for at least four reasons. First, there are structures that, because of their nature, cannot be large. For instance, in the hinges of bivalves the interconnecting teeth must be reasonably small, and therefore strong. Secondly, large surface area can be a disadvantage. Many animals, bivalve molluscs particularly, but also some gastropods and echinoids, have shells or hard skeletons and are also burrowers. The resistance to burrowing is roughly proportional to surface area, and therefore, for an animal of a given size, there will be an advantage in having a small skeleton. Again, for animals that must hold on to surfaces when buffeted by waves or currents, the large surface area of a massive shell will be a disadvantage, because it will experience a force from the waves or current proportional to its surface area. Only really massive shells, such as those of the large tridacnids, will have a sufficient mass for their surface area to be mechanically unimportant.

Thirdly, although the metabolic cost of making a stony skeleton may, in water, be less than that of making an organic skeleton, there is still some cost. So, the less the strength of the material per unit volume or mass, the greater the cost of making a structure with the desired qualities. Fourthly, the larger the volume of the skeleton, the longer, in general, it will take to make it. There must in many species be strong selection for members that can achieve a strong skeleton quickly.

We have produced arguments for the advantage of stony skeletons in sessile or almost sessile animals, and we have shown that even in sessile organisms, strong stony material will be better than weak material. It may seem that the advantages of having a stony skeleton have been stated in a rather negative, and certainly non-mechanical way. One of the great mechanical advantages of the stony skeleton is that it is usually more resistant to abrasion, and to boring, than the organic one. For animals that live inside shells which are themselves constantly being attacked by wave-borne sand or pebbles, or which are used for burrowing through a sandy substrate, or which are being radulated by a hungry whelk, such resistance will be important. Great mass may be important on occasions. The tridacnids have been mentioned as being possibly unaffected by watercurrents. There are many bivalves, and many brachiopods, that have thick shells and which need such shells to act as ballast. The same is true of mushroom corals that float on sandy substrata.

The Bryozoa are an example in which the habit of particular animals is imposed by the limitations of the mechanical properties of the skeleton. *Bugula* is bushy and, living in the littoral zone, cannot possibly have a brittle skeleton, and so its skeleton has a high organic content. The Crisiidae have blocks of organic material between the calcified regions, allowing considerable flexibility overall. These differences in habit have been investigated by SCHOPF (1969), who draws data from a large number of bryozoan species off New England. He found that, at different stations, the percentage of erect species as opposed to encrusting species increased with depth. At 10 metres less than a quarter of the species were erect, while deeper than 100 metres more than half were. Furthermore, among the erect species, there is a marked change with depth in the proportion that are flexible rather than stiff. Above 35 metres, all the erect species are flexible, while beneath this depth only 50–75% are flexible. Such a distribution is, of course, what one might expect, with stiff erect forms found only below the level of wave disturbance.

The Bryozoa have hardly been mentioned in this book because, as they are so tiny, it is difficult to have much idea about the forces that their skeleton must withstand. However, KAUFMANN (1971), in a charming paper, has analysed the likely forces on the avicularia of *Bugula* as it closes its jaw on would-be settlers. This analysis is possible because the loading system is particularly simple. The skeleton consists of chitin, protein, and calcium carbonate. Kaufmann finds, first, that the avicularia is designed to reduce to a minimum bending and shear forces. Second, that it is possible to characterize those parts of the avicularia that are subject to tension, and which to compression. The compression-resisting regions have much calcium carbonate, whereas the regions that are subjected to tension have little calcite, and considerable amounts of the tensile material chitin. All this satisfactory mechanical design, it should be added, is in a complex structure little over one millimetre long.

We shall now briefly consider another major division of animals—between those that have sclerotized cuticle and those that have bone. This is a much more difficult difference to explain than that between the various groups within the arthropods. The arthropods and the vertebrates are both, in the main, active animals. Bone and cuticle are not markedly different in their specific strength and elastic modulus. At the moment we have no idea which is the more expensive material to produce. Collagen is probably a rather cheap protein, compared to sclerotin, as it has so much glycine and alanine. On the other hand phosphate is often difficult to win from the environment.

We can suggest a possible reason for the retention of different skeletal materials by these two groups. Probably, as we get more knowledge, it will be shown that it is inadequate.

We shall show, in Section 6.3, that for fairly large active animals like vertebrates an exoskeleton is not suitable. Therefore vertebrates must have an endoskeleton as their main supportive structure. But sclerotized cuticle has a great disadvantage as an endoskeleton—it is extremely resistant to enzymes. An exoskeleton must be capable of being eroded, otherwise a very heavy skeleton, with no space for many internal organs, will result during growth. The arthropods overcome this problem by shedding their sclerotized cuticle occasionally as they grow. Such a course is not really open to an animal with an endoskeleton, although it is quite remarkable how the arthropods manage to shed their internal apodemes with the rest of their exoskeleton. Bone, on the other hand, although stiff and seemingly immutable, is in a dynamic state of erosion and deposition the whole time, and therefore the remodelling necessitated by growth is easily brought about.

We are not suggesting that this is the whole story or even that, if bone were an impossible tissue for some reason, vertebrates would not develop enzymes capable of breaking down tanned protein (though the fact that arthropods have not done so suggests that this might be very difficult). We do suggest, however, that if the mechanical advantages of one kind of skeletal material are not overwhelming, other factors may decide the issue.

Clearly, in all these discussions we are critically hampered by our lack of knowledge of the metabolic cost of building different kinds of skeleton. Compared with determining their cost, determining the mechanical properties of skeletons is child's play—though difficult enough.

Part II
Structural elements and systems

Chapter 6
Elements of structural systems

6.1 Introduction

For an element subjected to simple loading in tension, compression, bending or shear, the engineering analysis is elementary and the optimum shape can easily be derived. But real structures are not like this. The loading conditions may produce complex stress patterns, the load may be applied over a long or a short period, it may be applied suddenly or slowly and it may be once only or repeatedly. And to complicate matters still further many elements, such as the limbs of active animals, must be able to move relative to each other whilst still being able to transmit a force–often a very considerable one. Thus the simple problem of designing an element to resist a given force applied in a given direction becomes complicated by the concurrent necessity of providing actuating and restricting muscles, tendons, etc., which must insert into the element at those points where they can offer the maximum mechanical advantages.

Therefore, in this and the succeeding chapters we must, of necessity, start to simplify and idealize. However, we can at the outset state the broad requirements. Since any structure must equilibrate stresses and since all the members of the structure must be compatible (this merely says that there can be no resultant force displacing the structure and that all the bits must fit–and remain fitted–together) it is fairly easy to write down the requirements of a particular element.

(a) It must transmit or carry a force over a certain distance.
(b) In doing so it must neither break nor deform excessively.

In addition it is desirable (though not essential)

(c) That it should have sufficient reserve of strength to cope with overloads due to accident, etc.
(d) It should use the least amount of material–for reasons of metabolic cost of synthesis, maintenance or transport.

Also, in certain cases, relative movement must be possible with least effort.

These considerations then dictate the course of the present chapter. We first consider the idealized situations for beams in bending, ties in tension, and struts or columns in compression, and show the methods of simplified stress analysis and the effects of cross-sectional size and shape upon the stresses and deflections generated by externally applied forces. We then consider the material requirements–i.e., the ways in which materials fail, and the sort of design strategy that minimizes the chances of failure and that uses the least amount of material in doing so. Finally we discuss some of the factors (especially the need for relative movement) that require adaptations both in the shape of the element and in the material of which it is composed—this is restricted, mainly, to a discussion of the joints primarily because these have been more fully studied.

6.2 Bending

Although the discussion of properties in the previous chapters has been, very largely, in terms of the tensile properties, pure tension is, in fact, a comparatively uncommon state of stress. A spider hanging from a single thread is perhaps an example, but if a wind blows the thread is in bending as well as in tension. In fact, a state of pure tension, i.e., a force transmitted along a line is rare except in limiting cases such as a single strand of muscle fibre or a single element of plant cell wall, etc. By far the commonest situation is that of bending; for example the branch of a tree under self weight or the forearm with a load on the hand. In engineering practice this also is true and leads to a whole subsection of engineering analysis known as 'beam theory'. We discuss it first here because it introduces a number of important fundamental concepts.

We must first state the assumptions. Simple beam theory—and indeed the whole of simple engineering analysis is based upon two assumptions:

(a) The material is linearly elastic and isotropic and remains so throughout the loading range of interest.
(b) That sections of the element which may be considered as plane initially remain plane throughout.

These assumptions are perfectly adequate for a stiff, linearly elastic material such as a metal, shell, hard coral, etc., but are clearly unwarrantable if the material deforms inelastically or viscoelastically as do bone, wood and arthropod cuticle. To meet the first case, the technique of finite element analysis is used—here each small element is considered on its own and the conditions which govern the requirements of equilibrium and compatibility are considered at the junctions between the elements. In the case of viscoelastic (i.e., time dependent)

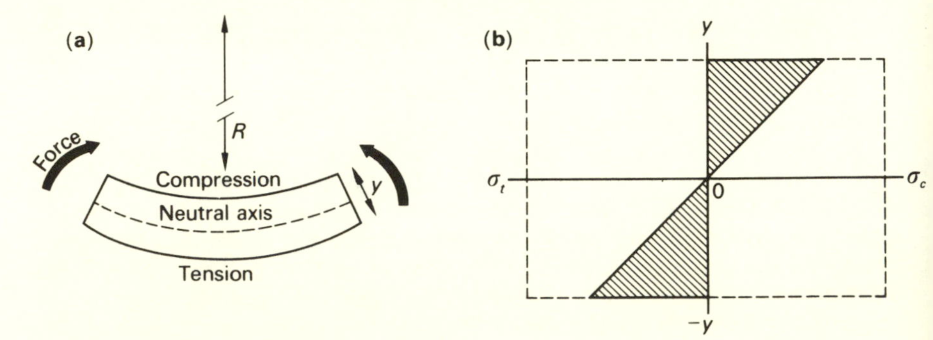

Fig. 6.1 (a) Bending of a linearly elastic rectangular beam. (b) Stress distribution about the neutral axis. Dashed lines show the cross-section.

deformations, the above assumptions can still be used providing that, in the analysis, each of the material parameters is replaced by the appropriate time-dependent parameter (see, for instance FAUPEL 1964). For our present purpose we accept the assumptions above and consider a prismatic bar of material bent into a radius R as shown in Fig. 6.1a. It is intuitively clear that the top surface has shortened and that the bottom has lengthened so that, since $\epsilon \propto \sigma$, the stress distribution across the section must be as shown in Fig. 6.1b. At the mid-point there is no stress—the material here neither extends nor contracts—this is known as the *neutral axis* (*NA*). Now if the bar of Fig. 6.1 is long enough to be bent into a circle, the length of the neutral axis will be $2\pi R$, but any layer of material above or below this axis will suffer a change in length given by

$$2\pi(R \pm y) - 2\pi R = \pm 2\pi y$$

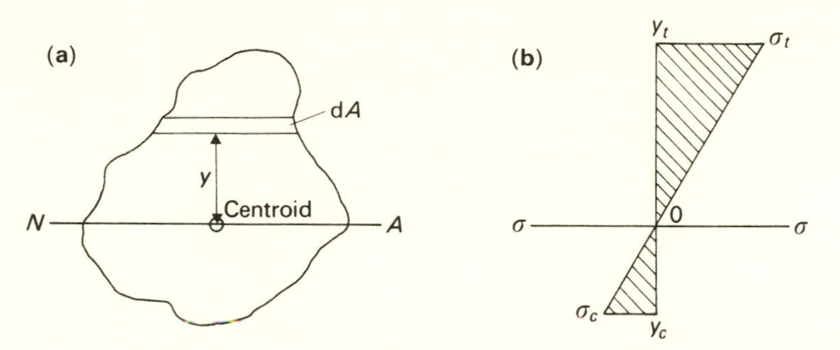

Fig. 6.2 (a) An irregular cross-section. (b) Stress distribution across the section in bending about the neutral axis (*NA*).

The strain now is

$$\epsilon = 2\pi y/2\pi R = y/R$$

and, since, for a Hookean material $E = \sigma/\epsilon$ it follows that

$$y/R = \sigma/E \quad \text{or} \quad \sigma/y = E/R \qquad (6.1)$$

This is clearly true at any *point*, even if R varies locally along the length, and it follows that, at any given section σ/y is *constant* for a material which obeys Hooke's law.

The neutral axis *must* pass through the centroid of the section. This is easily shown if we consider any irregular cross-section (Fig. 6.2) where the maximum stresses are σ_t at y_t on the tension side and σ_c at y_c on the compression side. In the general case, note that $\sigma_t \neq \sigma_c$ and $y_t \neq y_c$ but that $y_t + y_c = y$ = the full thickness. If plane sections are to remain plane there can be no resultant force along the bar so that the sum of all the σ_t must be equal and opposite to the sum of all the σ_c.

Consider an elemental strip of area dA distant y from the neutral axis and parallel to it. It carries a stress σ so that the total force in the strip is $\sigma\,dA$. But σ/y is constant so that

$$\sigma/y = \sigma_t/y_t = \sigma_c/y_c$$

or

$$\sigma = \sigma_t y/y_t$$

and the tensile force in the element is therefore

$$\sigma\,dA = \frac{\sigma_t}{y_t} y\,dA$$

The total tensile force is then

$$\int_0^y \sigma\,dA = \int_0^{y_t} \frac{\sigma_t}{y_t} y\,dA = \frac{\sigma_t}{y_t}\int_0^{y_t} y\,dA$$

and since

$$\sigma_t/y_t = \sigma_c/y_c$$

then

$$\int_0^{y_t} y\,dA = \int_0^{y_c} y\,dA \qquad (6.2)$$

Now $\int y\,dA$ is defined as the *first moment of area* of an element about the neutral axis and represents the analytical method of finding the position of the centroid of the section. The only condition which allows the equality of *Eq. 6.2* is that the neutral axis passes through the centroid (centre of gravity) of the section and that it will do this whatever the plane of the bending moment.

To calculate the bending moment and the stress in a given section:

Force in elemental strip = $\sigma\,dA$
Moment of force about $NA = y\sigma\,dA$
But since

$$\sigma = \frac{\sigma_t}{y_t} y$$

the moment of the strip

$$\frac{\sigma_t}{y_t} y^2\,dA = \frac{\sigma_c}{y_c} y^2\,dA$$

and, over the whole section, the moment

$$M = \frac{\sigma}{y} \int_{y_t}^{y_c} y^2 \, dA$$

The term $\int y^2 \, dA$ is called the *second moment of area I* or, more often and less precisely, the *moment of inertia* of the section, so that we can rewrite *Eq. 6.1* as

$$M = \frac{\sigma_t I}{y_t} = \frac{\sigma_c I}{y_c} = \frac{EI}{R}$$

or combining

$$\frac{\sigma}{y} = \frac{M}{I} = \frac{E}{R} \tag{6.3}$$

which is the basic formula of all beam theory written, in its most generally useful form, as

$$\sigma = My/I \tag{6.4}$$

We shall return to a discussion of I in Section 6.5, meanwhile it is not necessary to understand its physical significance and it may be taken simply as a term which describes the geometry of the cross-section and which may be varied in equations such as *Eq. 6.4.*

We should notice two things about *Eqs. 6.3* and *6.4*:

(a) Suppose that we wish to minimize the maximum stress that a given material must withstand in bending. Since, for a particular loading situation (force and distance) the bending moment M is fixed, we can only minimize σ in *Eq. 6.4* by maximizing I or by minimizing y or by maximizing the ratio I/y. This is then a matter of mere geometry—the properties of the material are not relevant (except insofar as they set an acceptable value for σ).

(b) Suppose now that we want to minimize the deflection—this is equivalent to making the radius of curvature very large for a given bending moment M. Since $R = EI/y$ it is clear that we must maximize EI and minimize y or, since the modulus is fixed for a given material, we must again maximize I/y.

Notice that, in both cases, for a given material the desired effect can be achieved by changing the geometry—specifically, the ratio I/y which is often known as Z, the '*section modulus*,' a bad name and one which you must *not* confuse with the elastic modulus which is a material property. The other parameter is the *flexural stiffness EI.* It is not always possible to use an alternative material of different modulus but, where this can be done, it is clear that the same resistance to deflection can be achieved by a large cross-section of a floppy material or by a small cross-section of a stiff material. This much, at least, is common experience.

ALEXANDER (1971) applies simple beam theory to the trunk of a tree of radius r sustaining a drag D due to wind at a height h above ground. It is a familiar example but it is also difficult to find a better one so, with due acknowledgement, we restate it here. Figure 6.3 shows that the tree is a simple cantilever whose bending moment M is given by

$$M = Dh$$

and this is a maximum at the fixed end (i.e., the ground).

From *Eq. 6.4* the maximum stress in the trunk

$$\sigma = Dhr/I$$

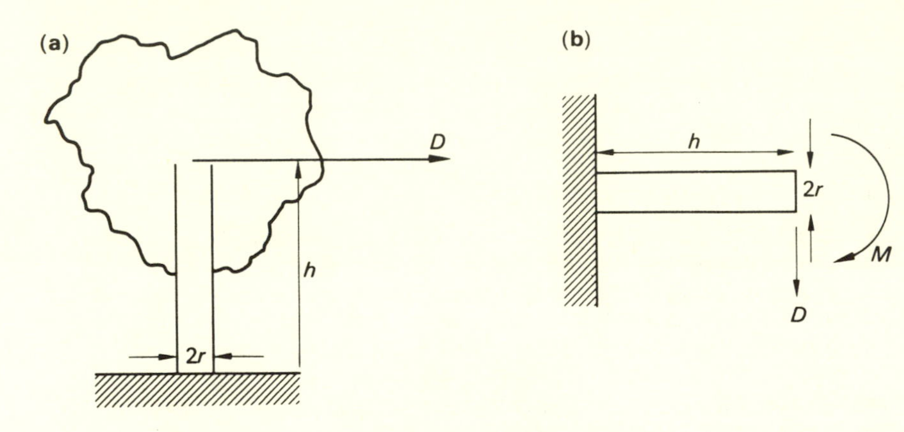

Fig. 6.3 (a) Wind forces acting on a tree trunk above ground. (b) Treatment as a simple cantilevered beam. Symbols are explained in the text.

and, anticipating Section 6.5, the second moment of area for a circular section of radius r

$$I = \frac{\pi r^4}{4}$$

so that we obtain

$$\sigma = \frac{4Dh}{\pi r^3} \tag{6.5}$$

Now to assess the limiting conditions we let σ be the maximum strength of the material, when the minimum radius which will ensure that failure of the tree will not occur by fracture in the trunk is

$$r_{\min} = \left(\frac{4Dh}{\pi\sigma_{\cdot}}\right)^{1/3} \qquad (6.6)$$

Alexander, quite rightly, points out that we should not be too ready to apply *Eq. 6.6* to real trees because there are wide variations in drag caused by such factors as wind speed, area actually presented to the wind, alignment of branches in wind, etc. A real tree is not merely trying to achieve the best design to meet a wind–at least one other factor influencing the height/diameter ratio is competition for light. A tall spindly tree will be more easily destroyed by wind whereas a short, rigid tree may suffer from malnutrition in a dense forest. The tree has more problems than that of stress to contend with. Nonetheless *Eq. 6.6* is informative, for it shows that for the most efficient use of material the trunk should taper, being thickest at the base where the bending moment Dh is the greatest.

We must, at this point, make one other reservation. *Equation 6.6* (and indeed many of those which follow) are to be used *only* where strains are small and elastic (i.e., the material may be regarded as Hookean). When deflections get very large (even though they be still elastic) a more elaborate treatment may be needed. In this connection see, for instance, CHARTERS *et al.* (1969) for the use of MITCHELL'S (1957) large deflection formulae in a study of the behaviour of the kelp stipe. This is discussed further in Chapter 8.

6.3 Compression and Buckling

Simple compression of a short fat specimen is simply the reverse of tension, though in calculating stresses we must note that in the case of deformable materials, the area *increases* under load, whereas in tension it decreases. Thus, on such materials, the stress-strain behaviour in tension and compression will only be strictly comparable if the data are expressed in terms of true stress and true strain (*Eqs. 2.3* and *2.4*).

An axial compressive force applied at the ends of a long, slender member (such as the long bones of the leg of a vertebrate, the tibia of a locust or the erect stem of a plant) will produce buckling in which the member bows laterally under load. Providing the deformation is wholly elastic, the original shape is recovered when the force is removed and we therefore call this *elastic buckling* or, sometimes, *primary buckling.* The latter term should be used with caution however as it strictly describes a particular buckling mode as shown in Fig. 6.4–other modes are 'secondary', 'tertiary' etc. as shown.

The elementary theory of elastic buckling is normally credited to Euler and a full analytical treatment may be found in standard engineering texts (e.g., FAUPEL, 1964; SINGER, 1962; SHANLEY, 1957). The derivation of the so-called Euler buckling formula is not difficult and we give it here without proof

$$F_E = \frac{n\pi^2 EI}{L^2} \qquad (6.7)$$

where F_E is the critical (Euler) buckling force, L is the length of the column and

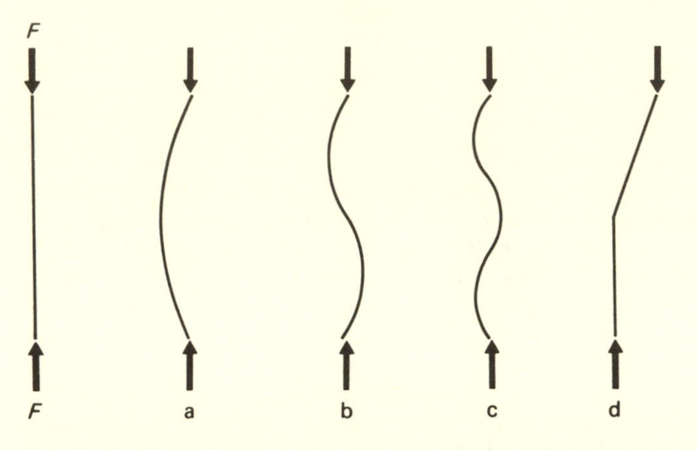

Fig. 6.4 Buckling modes of simple columns under force F. (**a** to **c**) Euler buckling in primary, secondary and tertiary modes. (**d**) Local buckling.

n is a coefficient whose value depends upon the nature of the restraint at the ends of the column–it is unity for a pin-ended strut (i.e., free to rotate in its own plane), 1/4 for one pin-end and the other firmly fixed (a useful approximation for plant stems), and 4 if both ends are firmly fixed.

We notice a number of things about the Euler formula:

(a) For a column of given length F_E is a function only of the flexural stiffness EI. Since we have already seen that, in simple cases at least, E is fixed for a given material, F_E can only be increased by increasing I.
(b) F_E is a function of length–it is common experience that a long slender cane buckles more easily than a short one. This is generally expressed by using the equation

$$I = Ar^2 \tag{6.8}$$

where A is the area of the cross-section and r is the *least radius of gyration* of the section. This is a distance such that, if the whole area of the section were to be concentrated at a point which is distance r from the neutral axis, the second moment of area relative to the axis would be the same. Substituting *Eq. 6.8* into *Eq. 6.7* we obtain

$$F_E = \frac{n\pi^2 EA}{(L/r)^2} \tag{6.9}$$

or, in terms of stress,

$$\sigma_E = \frac{F_E}{A} = \frac{n\pi^2 E}{(L/r)^2} \tag{6.10}$$

where L/r is called the 'slenderness ratio.' Engineers customarily use the plot of

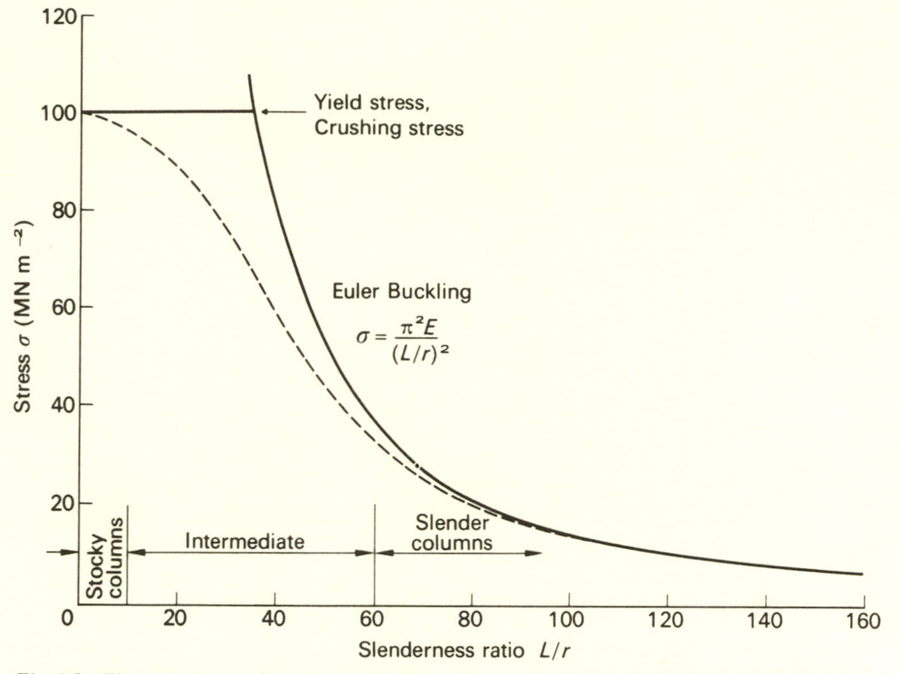

Fig. 6.5 Theoretical curve for pin-ended timber columns of unit area. Dashed line shows typical experimental curve.

slenderness ratio against F_E or σ_E to give the so called 'column curve' which is illustrated for timber in Fig. 6.5.
(c) The behaviour expressed by *Eq. 6.7* does not involve any departures from Hookean elasticity–when F_E is exceeded the strut will bow elastically and when the force is reduced the original shape is restored. The fact that the *total* deflection is large–in most cases quite obviously visible–is a consequence of the summation of many, very small, Hookean deformations occurring along the whole length of the strut.

We should note here that any slight initial deflection which may exist in the unloaded strut or any eccentricity of loading which may tend to deflect the strut initially in a given direction will result in a drastic reduction in F_E. This occurs in the human femur which, when considered in front view (Fig. 6.6) is both bent and eccentrically loaded so that the adductor muscles must exert forces of considerable magnitude in order to counteract the bending stresses which might otherwise lead to buckling failure.

Figure 6.5 shows that, at low values of the slenderness ratio L/r the curve becomes asymptotic to a single value. This corresponds to the mode of failure in short, fat columns which fail by crushing–either by yield or by fracture depending upon the capacity of the material to undergo plastic flow. At high values of L/r Euler buckling occurs at very low stresses indeed. Most useful columns lie somewhere in between and the failure mode in these is a combination of buckling and crushing.

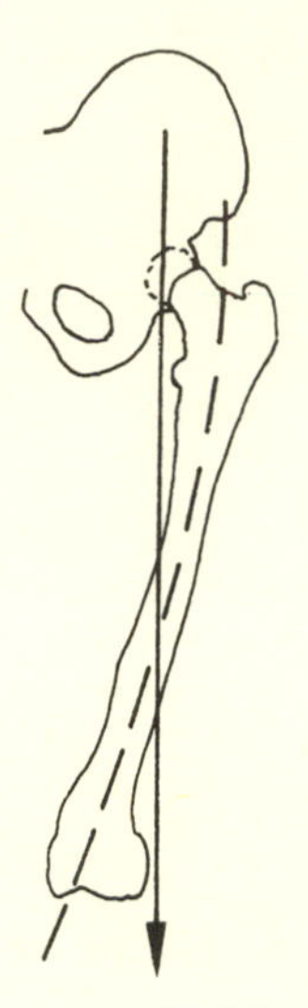

Fig. 6.6 Front view of left human femur showing eccentricity of loading and initial curvature.

We return now to conclusion (a) above and we again anticipate Section 6.5 where it will be shown that, for equal cross-sectional area (which is the same thing as equal weight of material) a tube is much more effective than a solid bar because the second moment of area I is much greater. In those situations where compressive forces must be transmitted over some distance and where weight should be kept low, a tube is very much better than a solid bar. Tubular furniture and bicycle frames are common domestic examples. But common sense tells us that we cannot go on indefinitely increasing the diameter of the tube and, at the same time reducing its thickness lest we reach a stage where the wall thickness is so small that a permanent kink forms and the tube crumples. Conversely if we start with a tube of fixed dimensions (like a soda straw) and apply an increasing load we shall first of all go into elastic buckling and, eventually, into a disastrous and damaging kink. This is known as *local* buckling and this mode of failure sets a limit both to the diameter/thickness ratio of the tube (D/t) and to the greatest permissible applied force.

It is not easy to derive analytical expressions for local buckling, and a number of rather empirical factors must be used in order that the experimental data shall accord with the calculated stresses. For most purposes local buckling occurs at a stress which is adequately given by

$$\sigma_L = \frac{kEt}{D} \qquad (6.11)$$

where k is a constant which is theoretically 1.2 but which, in practice, usually lies between 0.5 and 0.8—the discrepancy being generally ascribed to local imperfections and irregularities in the cross-section.

6.4 Torsion

Figure 6.7 shows a solid rod, of radius R and length l fixed at one end and subjected to a torque T about the z–z axis at the other end—this torque produces a twist θ in the bar.

It is intuitively clear that the twist varies from zero at the centre line to a maximum at the surface—i.e., the distribution of stress is analogous to that in pure bending, i.e., that

$$\tau = Cr \tag{6.12}$$

where C is a constant.

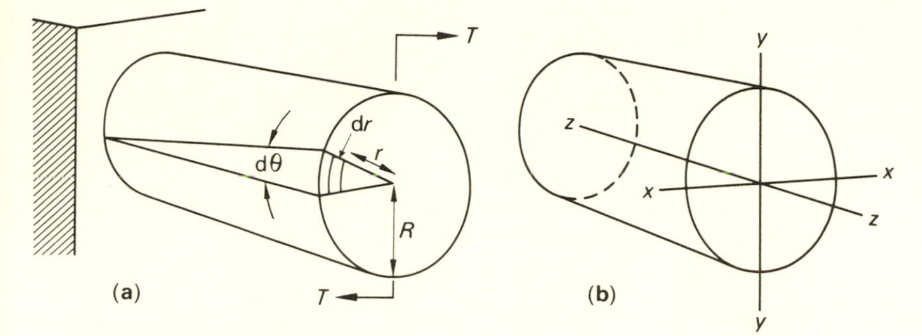

Fig. 6.7 Solid rod in torsion.

In a Hookean material, the shear strain γ is related to the shear stress τ by the shear modulus G

$$\gamma = \tau/G$$

whence, substituting into *Eq. 6.12* we obtain $\gamma = Cr/G$.

To evaluate the constant C we note that the net force on the cross-section must balance the applied torque—i.e., that

$$T = \int \tau(2\pi r)(\mathrm{d}r)r$$

or, substituting from *Eq. 6.12*

$$T = \int Cr(2\pi r)(\mathrm{d}r)r$$

whence

$$T = 2\pi C \int_0^R r^3 \, \mathrm{d}r = \frac{\pi C R^4}{2}$$

and therefore

$$C = \frac{2T}{\pi R^4} \tag{6.13}$$

Substituting this into *Eq. 6.12* we obtain a value for the shear stress τ in terms of the variable radius r thus

$$\tau = \frac{2Tr}{\pi R^4} \tag{6.14}$$

Now the quantity $\pi R^4/2$ is called J, the *polar second moment of area* (polar moment of inertia) for a circular cross-section about the z–z axis. It is analogous to the second moment of area I. For a circular section

$$I_{xx} = I_{yy} = \pi R^4/4$$

whereas

$$J = \int r^2 \, dA = \int_0^R \int_0^{2\pi} (r^2 \cdot r \, d\theta) \, dr = \frac{\pi R^4}{2} \tag{6.15}$$

i.e., it is *twice* as great as the second moment of area I and is simply the sum of the separate moments I_{xx} and I_{yy}. Thus we may rewrite *Eq. 6.14* as

$$\tau = \frac{Tr}{J} \tag{6.16}$$

which may be compared with the bending formula (*Eq. 6.4*)

$$\sigma = \frac{My}{I}$$

Note that the polar moment of area J varies as the fourth power of distance R—so that a small increase in R causes a considerable increase in J.

6.5 Cross-Sectional Shape

The second moment of area $I = \int (y^2 \, dA)$ is essentially a measure of the way in which the material is distributed about a given axis. Taken on its own, it has little significance—its importance only becomes clear when it is considered in the context of a bending formula such as *Eq. 6.4.*

It is, unfortunately, generally called the moment of inertia of the section. The reason for this is clear if we recall that, in dynamics, to move a mass (inertia) we express the force as

$$F = ma$$

i.e., force equals inertia times acceleration. For rotating bodies with angular acceleration ω the force is given by $F \cdot d = \int (y^2 \, dM)\omega$ which may be stated as *moment* of force (Fd) equals *moment* of inertia times acceleration, i.e., the term $\int (y^2 \, dM)$ is the moment of inertia of a *mass*. By analogy then, the expression $\int (y^2 \, dA)$ is known as the moment of inertia of an *area.* But moment of inertia is a thoroughly bad and confusing name, though it is now so beloved by engineers that it is unlikely to be changed.

We can best try to describe I in terms of simple bending. Figure 6.1 shows that the maximum stresses occur in the upper (compression) and lower (tension) surfaces—this immediately suggests that most material should be concentrated here rather than near the neutral axis where the stresses are lower. Thus, the most efficient distribution of the material would be that shown in Fig. 6.8a. The

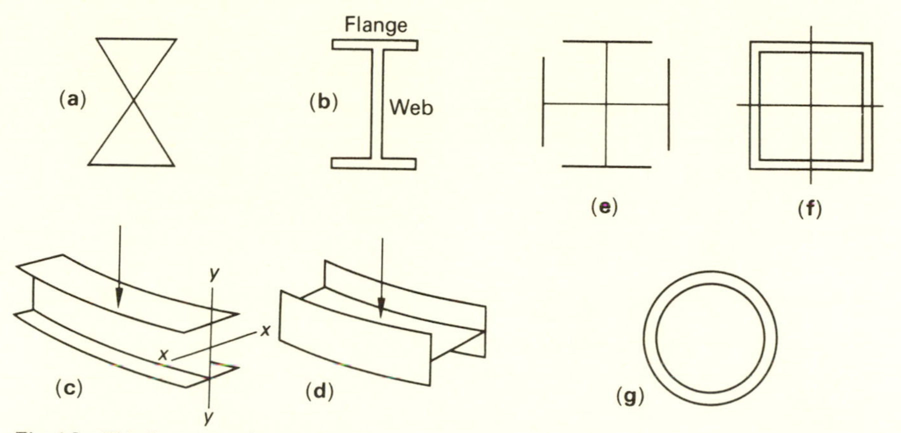

Fig. 6.8 This figure is explained in the text.

engineer achieves this in the familiar I-shaped section (Fig. 6.8b) where the maximum amount of material is concentrated at the tension and compression surfaces and the web is, essentially, a device for holding these surfaces apart. This is fine for situations where the bending is known to occur about the x–x axis (Fig. 6.8c). But, if the same section is bent about the y–y axis, the situation is less happy (Fig. 6.8d). The second moment of area for (c) is twenty times that for (d) so that, in the latter case, the extreme surface layers are carrying an abnormally high stress for the same bending moment. This is obvious if we recall that $I \propto y^2$, so that by placing the material as far as possible from the neutral axis, we maximize I and hence minimize stress. Notice that, in this case, it does not matter whether the force comes from above or below, since the section is symmetrical about both axes.

I sections are not common in biological structures (though it is interesting to note that the stiff sclerenchyma of some plants has shapes not unlike the I section (see NACHTIGALL (1971), Fig. 33). The reason is fairly obvious. An I section is the most efficient solution when the direction of the bending moment is fixed, but for a situation in which bending may occur about either axis the material placement strategy leads to (e) and to the box section (f). And so, when bending may be about any axis we come to a circular tube (g) as the most efficient section.

Some useful formulae for calculating I are given in Table 6.1; standard engineering texts such as FAUPEL (1964) and SINGER (1962) contain others. More complex sections are usually divided into a set of simpler elements whose individual moments relative to the axis of interest are summed (see for instance, Singer, p. 536). Note that since the second moment of area is the product of an area and a length squared, it has dimensions of length to the fourth power.

Table 6.1 Formulae for second moment of area for simple sections. c shows the position of the centroid.

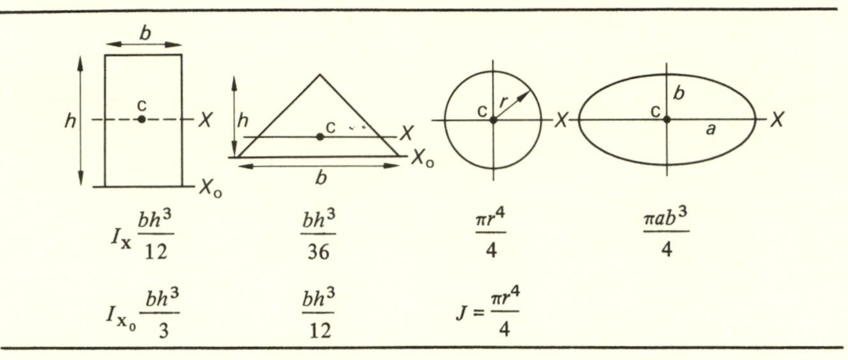

For a circular tube of radii R and r

$$I = \frac{\pi}{4}(R^4 - r^4) \tag{6.17}$$

and, if the wall thickness $t(= R - r)$ is small there is no great loss of accuracy if *Eq. 6.17* is written

$$I = \frac{\pi \bar{D}^3 t}{8} \tag{6.18}$$

where $\bar{D}$ is now the mean diameter $(2R - t)$.
The area of a thin walled tube is, approximately $\pi \bar{D} t$ so that

$$I = \frac{A\bar{D}^2}{8} \tag{6.19}$$

showing that, for a series of tubes of equal area (and hence equal weight) the second moment of area increases as $\bar{D}^2$.

Table 6.2 gives some relative values of I for different sections of equal area—these are listed by comparison with a solid circular rod whose second moment of area I and section modulus Z are taken as unity.

So far we have discussed the influence of I on strength, but it is clear that

Table 6.2 Relative moments of area for simple sections of equal size. In hollow sections, thickness (t) is constant. In asymmetrical sections, the long axis is arbitrarily set at 1.5x the short axis.

	I_{xx}	I_{yy}	Z_{xx}	Z_{yy}	J
	1	1	1	1	1
	17.4	17.4	5.8	5.8	17.4
	1.1	1.1	0.6	0.6	1.1
	13.8	13.8	5.9	5.9	13.8
	12.4	6.6	4.3	3.4	9.5
	18.0	9.5	6.4	5.1	13.8
	17.0	5.8	7.3	2.5	11.4

similar remarks apply also to stiffness. All deflection formulae contain the term EI, the flexural stiffness, thus for the tree considered as a cantilever in Section 6.2 the maximum deflection at the top is given by

$$\delta_{max} = \frac{Dh^3}{3EI}$$

Clearly, δ will be minimized if EI is increased—since the modulus of the material is generally fixed this is tantamount to requiring only an increase in I.

Similar comments apply also to the polar moment of area J—since the torsion formulae are, analytically at least, similar to the bending formulae, the same considerations apply and the tube is seen to be the most efficient way of utilizing the available material in bending, elastic buckling and in torsion.

To summarize then—the strength is proportional to I/R, the stiffness to I. CURREY (1970) presents the following table for tubes of constant area such that I is unity when R is unity.

R	1	2	4	10
I	1	7	31	199
I/R	1	3.5	7.8	19.9

This table shows dramatically the advantages of using thin walled tubes where both I and I/R increase rapidly and without limit.

Why then are all load bearing elements not composed of large tubes of minimal wall thickness? Most plant stems are and arthropods have developed their limbs in this way. The limit is, of course, set by the possibility of local buckling when the wall of the tube becomes very thin.

We recognize two important cases (CURREY, 1967). The first arises when a long slender tube may fail either by Euler buckling or by local buckling and the limiting situation will be reached when $\sigma_E = \sigma_L$. Thus, from *Eqs. 6.7* and *6.11* we may write

$$\sigma_E = \frac{F_E}{A} = \frac{n\pi^2 EI}{AL^2} = \frac{kEt}{D} = \sigma_L$$

where the symbols have the same meaning as before. Rearranging

$$\frac{D}{t} = \frac{k}{n} \cdot \frac{AL^2}{\pi I} \qquad (6.20)$$

we see that the limiting dimensions for a given length are dictated entirely by the geometry and not by the material.

Using the approximations for thin walled tubes (*Eq. 6.18*), we may substitute for A and I in *Eq. 6.20* to obtain

$$\frac{L}{D} = \left(\frac{n\pi}{8k} \cdot \frac{\bar{D}}{t}\right)^{1/2} \qquad (6.21)$$

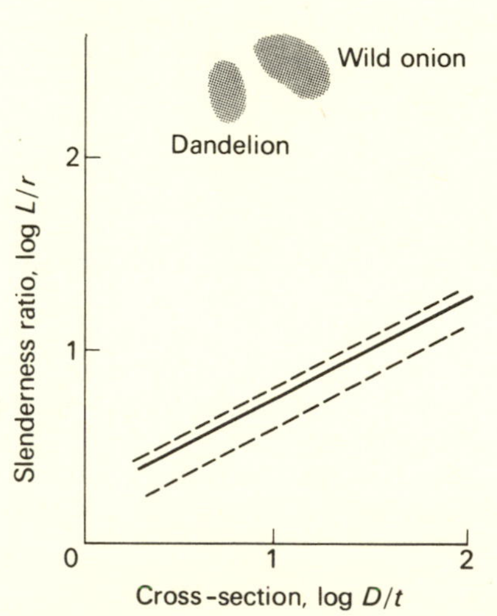

Fig. 6.9 Slenderness ratio, L/r, vs D/t from *Eq. 6.21* showing critical dimension for hollow columns in bending. Solid line assumes $\sqrt{n/k} = 1$; dashed curves show extreme limits. Shaded areas enclose observed values for plant stems.

i.e., the maximum height/diameter ratio is determined by $(\bar{D}/t)^{1/2}$. This is plotted in Fig. 6.9 together with some observations made on several plant stems and it is clear that, in general, there is ample reserve against local buckling (failure) of the stem and that tolerable Euler buckling is a more likely occurrence. This may account in some measure for the tendency of tall hollow stems to develop a noticeable curvature during growth, this being a primary buckling induced by the self-weight of the stem, their self-weight is sufficient to exceed the Euler condition if the stem grows very tall. Indeed slenderness ratios of 150 or so are not uncommon in hollow-stemmed plants such as wild onion (*Allium* sp.), dandelion (*Taraxacum* sp.) and many cereal grasses so that these are, by engineering terms, slender columns (see Fig. 6.5).

The second case arises when failure can either take place by Euler buckling or by local crushing. If σ_c is the compressive strength of the material, the condition for local failure is now

$$\sigma_c = \frac{kEt}{D}$$

or

$$\frac{D}{t} = \frac{kE}{\sigma_c} \tag{6.22}$$

i.e., D/t is determined by the ratio of compressive strength to elastic modulus.

Currey has applied this to a consideration of arthropod legs—for which $\sigma_c \approx E/100$. Thus taking a realistic value of $k = 0.5$, the limiting D/t ratio here is about 50 whereas, for most arthropods, D/t is generally closer to 10 or 15. Thus we assume that local buckling is not an important factor except possibly in certain spider legs (*Heteropoda*, *Grammostola*, *Trochosa*). These however tend to be stiffened hydrostatically, thus compensating for a potentially weak design. The same considerations show that, in general, the long bones of vertebrates have an effective 'safety factor' against local buckling of about 5.

One other important aspect of local buckling needs to be mentioned here. The same mode of failure can apply in the bending and torsion of thin tubes and, indeed, of any thin section. The main reason for using a thick central web in the engineer's I beam is to preclude the possibility of local buckling here. Thin, unsupported sections are rarely found in compressive situations.

We conclude this section with an example of an adaptation of shape in the frond of the coconut palm (*Cocos nucifera*). Here the petiole springing from the trunk adopts the shape of a bent cantilever as it grows (Fig. 6.10a). The leaflets form a flexible blade as shown in Fig. 6.10b and, in a wind, the petiole is twisted to allow equalization of wind forces on the two sides of the petiole. A typical petiole, 4.5 m long had, over the first 200 cm or so from the trunk, the cross-section shown in Fig. 6.10c. We can consider this as a compromise between two requirements. For maximum bending resistance which involves a principal force downwards, the most appropriate section would involve a flattened upper face—as in the I beam. Torsion, however, requires that J be maximized—i.e., that the second moment of area be as high as possible about *both* the x–x and y–y axes—

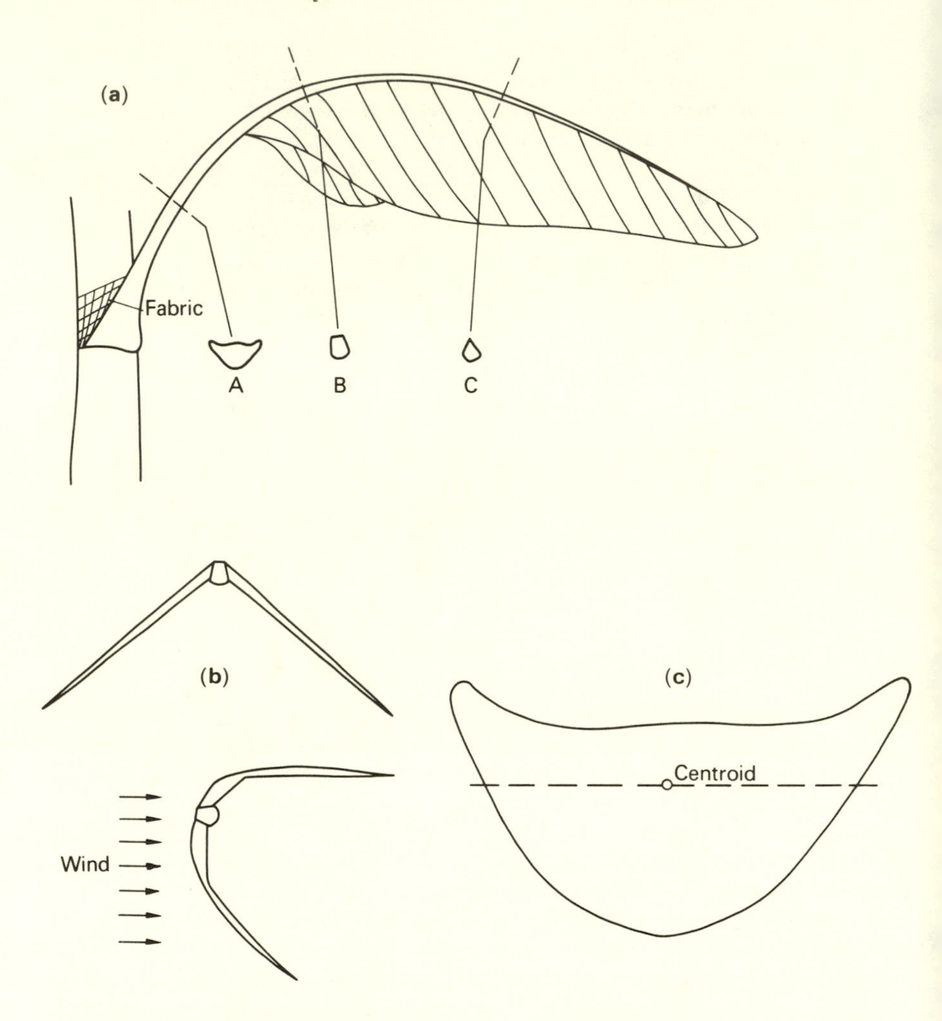

Fig. 6.10 Petiole of the coconut palm. (**a**) Diagram of the entire leaf attached to the tree. A, B, C are cross-sections of the petiole at the points indicated. (**b**) Diagrammatic cross-section of the frond showing two leaflets suspended from the petiole. Upper figure: in still air; lower figure: with wind from the left. (**c**) Cross-section of petiole at point A in (**a**).

so that a circular section would be more appropriate. The most efficient section therefore would be that which maximizes both I and J, this can be done by maximizing the geometric mean $\sqrt{IJ}$. As an exercise we consider four possible sections—all of equal area—which might have been adopted. Table 6.3 compares these, again using the circular rod as unity and assuming that in the semicircular and triangular sections the upper face is in tension. We see that the semicircular section offers a good compromise. It is not the most efficient for simple bending (though the asymmetry of the section increases Z_x and therefore helps to keep tension stresses low), but it maximizes I_y and J. We suggest therefore that lateral rigidity and resistance to torque are important—this would seem to agree with the

Table 6.3 Comparison of moments of area of some possible cross-sections of palm petiole

	I_x	I_y	Z_x	Z_y	J	$\sqrt{I_x J}$	$\sqrt{I_y J}$
	1	1	1	1	1	1	1
	0.6	2	1.2	1.5	1.27	0.87	1.6
	0.5	2	0.7	1.4	1.24	0.8	1.6
	1.2	1.2	1.6	0.9	1.21	1.2	1.2

sort of flexure imposed by a wind in which the aerodynamic lift of the leaflets would tend to produce lateral and torsional forces as suggested in Fig. 6.10b. Why not a triangular petiole? Primarily because of the thin section at each corner which would tend to buckle locally as would an elliptical section (at the minor circumference). The semicircle also has sharp, unsupported corners and we suggest that the two triangular ribs at the edges of the upper surface act as 'stiffeners': if a large lateral force should produce buckling, it will occur at a point which is remote from the axis of the petiole.

6.6 Shells

When one thinks of a 'skeleton', one usually imagines that of a bony vertebrate. However, practically all the skeletons one actually sees, in the garden, in the quarry, on the seashore, are shells. Shells are fascinating, and have been collected for their beauty and for their use as money since prehistoric times. They are the subject of intense research by many biologists and have been for centuries, and yet their mechanical properties are almost completely unknown. There are three reasons for this apparently rather remarkable state of affairs.

First, it is exceedingly difficult to describe their morphology in quantitative terms. RAUP (1966) has recently clarified this matter to some extent, showing how most variants in shell shape can be described in terms of four parameters. Even so, such a description refers only to shells with no alteration in their mode of growth during ontogeny, and with no spines or other irregular ornamentations.

Second, it is exceedingly difficult to analyse the mechanical properties of such complicated shapes as shells. They are obviously not solid, and so cannot be analysed as such. On the other hand their walls are too thick to be considered as variously shaped membranes. Even if they were membranes, analysis would be formidably difficult.

Third, it is exceedingly difficult to discover the actual mechanical function of shells. Are they designed to resist hydrostatic compression, localized chipping, crushing, or merely abrasion? Usually one has little idea and, frequently, one has the thought that the answer is: 'none of these'. STANLEY (1970) has recently written a most lucid account of the natural history of bivalve shell function which is an excellent source where we can find many untested hypotheses about their function. The main thrust of Stanley's analysis is that most of the obvious features of bivalve shells are not related to mechanical (in the sense of load bearing) characteristics at all.

We shall consider three characteristics of shells—thickness, folding and ornamentation—as examples of our ignorance.

Thickness The thicker a shell, the longer it will take to bore through it, and the greater the loads it will be able to bear without breaking, and so the advantages of a thick shell seem clear. Yet to Stanley, thick shells are overdesigned for load bearing, and their function is to provide for the need for density and mass, for stability in currents and waves. Stanley adduces the mussel *Mytilus edulis* as evidence for overdesign in the shells of other species. The mussel lives on wave-battered and stormy shores, and yet it has a fairly thin shell. Stanley argues that, if *they* are capable of surviving, bivalves in general cannot need thick shells for durability (he allows a few exceptions). One could argue, by analogy, that the thickness of a tortoise's carapace is unrelated to mechanical properties because a lizard does not have a carapace at all, yet survives. The argument is faulty, but the conclusion remains convincing: probably most thick-shelled bivalves would have thinner shells if there were no need for stability. CHIA's (1973) observation that some young echinoids (*Dendraster excentricus*) ingest grains of sand that are on average considerably denser than the average for the substrate shows that the need for stability is probably not confined to the molluscs. Of course, there are shells that are almost certainly thick for mechanical reasons. Winkles and whelks that live above the low tide line are inevitably occasionally dislodged and thrown around by the sea, and they have extremely thick shells compared with many of those living beneath the tide line, and most land snails. Furthermore, it is known that thickness of shell will prevent certain predators from eating an animal, and many predators will attack young forms, but leave adults alone, because the shell is presumably too thick (ORTON, 1926, and discussion of safety factors, Section 7.13).

Folding By folding we mean the presence of folds or wrinkles in a shell that extend right to the interior. Folding is a method of increasing the stiffness and strength of a shell without greatly increasing its mass. A fine example of this is seen in the scallop *Chlamys opercularis.* This animal swims quite frequently, and so needs a light, and therefore thin-walled, shell. Also it needs a shell whose overall shape is streamlined, not globular. Therefore, the valves cannot be strongly arched, and the folds presumably make the shell stronger than it would be without them. An interesting contrast to this is the edible sea scallop of the northwestern Atlantic, *Placopecten magellanicus*. It is an active swimmer whose shell commonly attains a length and depth of 18 to 20 cm but is exceedingly thin (3 mm) and bears no hint of folding.

Another example of the functioning of folding is seen in the septa of ammonoid and nautiloid cephalopods. These have the problem of withstanding considerable hydrostatic pressure. The cavity inside the coiled shell of these animals is interrupted periodically by septa. In most nautiloids the septa are strongly domed, concave to the outside world. This is well shown in the extant *Nautilus*. This shape is very efficient as far as the septa themselves are concerned—this is the favoured shape for the ends of pressure vessels. However, it is not a good shape for supporting the walls of the shell itself; any tendency for the walls to collapse will be aggravated by the septa. One wonders why the outwardly convex septum that is a design feature resisting implosion and that is found in some ammonoids is not also found in *Nautilus*. However, the nautiloids had thick shells, and were and are capable of withstanding the pressure without much assistance from the septa. The ammonoids, however, had much thinner shells. In these animals the septal shape is quite different. The septum runs more or less straight across, but it is heavily fluted. This fluting has two effects. One is that the septum is itself much stronger than a simple flat sheet of the same thickness or weight. The other, and more important, effect is that these septal walls supported the walls of the shell in a way the nautiloid septa do not. This is especially the case because the septa are quite close together, and usually interdigitated (RAUP and STANLEY, 1971).

A rigorous mechanical analysis of these two systems has not been carried out, and would in fact be ferociously difficult to do, so the balance of advantage between the nautiloid and ammonoid systems is not known.

Ornamentation Often the shell is not folded, but is nevertheless ornamented, that is, it has spines, ridges, or other features on the outside. Ridges tend, in the bivalves, to run either from the dorsal part of the shell to the margins (radial) or to be parallel to the margins (concentric). Both these designs may strengthen and stiffen the shell somewhat, but it is noticeable that these ridges are found principally on thick and globular shells, and are infrequent on thin and flat shells which, were the ridges for stiffening, would seem to require them more. It is argued by STANLEY (1970) that these ridges have functions unconnected with stiffness or strength. Radial ridges may help a rather unstreamlined animal 'saw' its way into the sand as it rocks back and forth when burrowing. Both concentric and radial ridges will help to anchor an animal in the substrate.

Spines may have an effect on the mechanical properties of some shells. Some bivalves produce broad, thin spines. One of the giant clam species, *Tridacna squamosa*, has these 'spines' and lives on the open reef tops where it may be struck by storm-swept coral debris. LaBarbera (unpublished) reports that impact blows that will break the shell of a smooth-shelled species will only break a few spines in *T. squamosa*. Spines will increase friction between shell and substrate. Active burrowers rarely have spines. Spines may also act as a first line of defence against predators, as they have to be removed before the shell itself can be reached. This is seen most clearly in sea urchins. The 20 cm long spines of the urchin *Diadema* appear to be effective protection against smaller wrasses (family Labridae) and surgeon fishes (family Acanthuridae). However, large parrot fishes (family Scaridae) will readily attack *Diadema*, chew the spines down in the same way a cartoon rabbit eats a carrot, and then bite through the test to eat the soft parts.

This description of some features of shells has dealt almost entirely with bivalves, and has left out many features that *may* be of mechanical importance. These features include hinge design in bivalves and brachiopods, the serrated margins of these two groups; the thickened lips of many gastropods; and the fluting of brachiopods which, since brachiopods are attached to the sea floor, cannot be for swimming. We have said essentially nothing about the overall shape of shells. There is much that we could say, but we are not convinced that such descriptions would help the reader towards a deeper understanding of the 'mechanical design in organisms'. The state of analysis of shell form is still pathetically primitive, and is a most fruitful field of research for well-endowed research teams.

6.7 Materials for Minimum Weight

In the preceding sections we have considered only the shape of the cross-section acting through the moment of inertia I or the section modulus Z. We have also noted that the flexural stiffness EI controls the deformation in all cases except that of simple, uniform tension or failure by compressive crushing. We now consider the importance of weight in the design of elements.

There is probably no reason to suppose that minimum weight is a necessary factor in all designs. Clearly it is less important in aquatic situations than in airborne ones. For terrestrial situations, the need for minimum weight varies, but it seems fair to assume that, wherever weight can be saved, it will be metabolically advantageous to do so. Most minimum weight engineering structures are designed upon the basis of a given strength or stiffness per unit weight of material–thus materials are often compared on the basis of the specific strength (σ/ρ) or specific modulus (E/ρ) where ρ is the density of the material. We believe that our approach must be the reverse of this–our interest surely lies in understanding the properties which determine how much strength (or stiffness) we can achieve from a structure of given weight. In other words the weight is the critical factor and the one which must be generally minimized. SHANLEY (1957) has developed this theme at some length and his original reference should be consulted for more details. However, the present argument will try to develop the important material parameters in terms of weight/unit strength rather than in terms of the more familiar strength/unit weight.

We may assume that, in any structure, two factors are 'fixed' in advance–one is the greatest load to be carried and the other is the distance over which it must be carried or transmitted. In addition, we have the requirement that the material we use must only be stressed to a certain level or failure of some sort will occur–this may be by yielding, by fracture or by the development of excessively large strains.

There are several complementary ways of expressing these conditions. The most obvious is to use a 'weight/strength' parameter W/P where P is the load to be carried and W is the weight of material needed to carry it. The other is to use a parameter which is often called the 'structure loading index'. For the case of simple tension this is merely P/L–the ratio of the greatest load to the distance over which it must be carried. It is not, in fact, a dimensionless coefficient (though it can easily be made so) but, like a dimensionless factor, it can be used to characterize the relationship between strength and size which is common to an entire family of structures, i.e., two structures which are geometrically similar will behave in a manner which is determined by a simple scaling factor. We shall use both parameters in the following discussion.

Consider first simple tension (or compression, provided that failure is by crushing and not by either Euler or local buckling). The allowable stress (which may be the yield stress or the fracture stress) is given by

$$\sigma = P/A$$

where P is the load to be supported and A is the area of the cross-section.

Now the weight is given by

$$W = \rho A L$$

where ρ is the density so that the weight/strength ratio is

$$W/P = L\rho/\sigma \qquad (6.23)$$

Rewriting this in terms of the structure loading index P/L we obtain

$$\frac{P}{L} = \frac{W}{L^2} \cdot \frac{\sigma}{\rho} \qquad (6.24)$$

These equations show that minimum weight/strength for a given length is obtained by minimizing the term ρ/σ or, alternatively, for given ratio of load/length, by maximizing the specific strength σ/ρ. In both cases the desired result is obtained by using a material of higher strength and lower density.

For simple bending, the same argument can be used. *Equation 6.4* gives the allowable stress as

$$\sigma = My/I$$

where M is the bending moment.

For a given bending moment the weight/strength ratio is

$$\frac{W}{M} = \frac{yLA}{I} \cdot \frac{\rho}{\sigma} = L \cdot \frac{A}{Z} \cdot \frac{\rho}{\sigma} \qquad (6.25)$$

or, in terms of the structure loading index

$$\frac{M}{L} = \frac{W}{L^2} \frac{Z}{A} \frac{\sigma}{\rho} \qquad (6.26)$$

In both of these expressions we note that the term which governs the material properties is the same as before, namely σ/ρ. But we should also note that in this case the geometry of the cross-section enters into the optimization via the term A/Z which must also be minimized if W/M is to be as small as possible. For a given area of cross-section (and hence a given weight), this means that we must maximize the section modulus Z or, for a given shape of fixed section modulus, we must minimize A. This agrees with our previous findings in Section 6.2.

However, it is the column under Euler buckling conditions that provides us with the most illustrative example. Here, as we have already seen, length, area and I are involved and to assess properly the influence of the material parameters, we must adopt a slightly more complicated approach. We first try to express the critical stress for buckling in terms of both the geometrical and material parameters.

The critical load (*Eq. 6.7*) is given by

$$P = \frac{\pi^2 EI}{L^2}$$

whence the weight/strength parameter is

$$\frac{W}{P} = \frac{\rho A L^3}{\pi^2 EI} \tag{6.27}$$

Now the structure loading index for columns (see SHANLEY (1957) for derivation and examples) is P/L^2 and, since the critical buckling stress is

$$\sigma = \frac{P}{A} = \frac{\pi^2 EI}{AL^2}$$

we obtain

$$\frac{P}{L^2} = \frac{W}{L^3}\frac{\sigma}{\rho} \tag{6.28}$$

Now we can collect all the geometrical terms into one parameter if we write

$$\sigma^2 = \frac{\pi^2 EI}{AL^2} \cdot \frac{P}{A}$$

whence

$$\sigma = \left(\frac{\pi^2 I}{A^2}\right)^{1/2} \left(\frac{P}{L^2}\right)^{1/2} E^{1/2} \tag{6.29}$$

or

$$\sigma = \beta_c E^{1/2} \tag{6.30}$$

where β_c is called the 'column buckling parameter' containing both the geometrical terms I/A^2 and the structure loading index P/L^2. Thus, from *Eq. 6.28*

$$\frac{W/L^3}{P/L^2} = \frac{\rho}{E^{1/2}} \cdot \frac{1}{\beta_c} \qquad (6.31)$$

The term $(W/L^3)/(P/L^2)$ obviously reduces to W/PL, which is often called the 'column efficiency' i.e., the weight W per unit length L and unit of load P.

But *Eq. 6.31* shows that, for a given geometry, the column efficiency for failure by Euler buckling varies as $\rho/E^{1/2}$—engineering texts often write this in the form E/ρ^2 and, although this is not the correct form for the weight/strength argument used here, it does, in this form, show the tremendous advantages to be

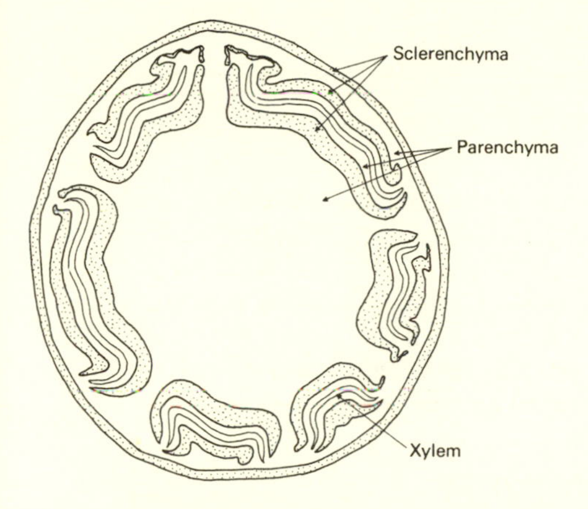

Fig. 6.11 Transverse section of the trunk of the tree fern, *Sphaeropteris lunulata*, showing the distribution of rigid sclerenchyma and spongy parenchyma. The diameter of the section is 105 mm.

gained from decreasing the density of the material. Even greater advantages arise in the case of plate columns (thick plates loaded along one edge) where the relevant parameter can be shown to be $\rho/E^{1/3}$ (or E/ρ^3). This helps explain the advantages which aircraft constructors gain from the use of lightweight honeycomb sandwich materials. Analogous biological structures that function in this way may well include the patterns of distribution of sclerenchyma (Figs. 6.11 and 6.19) and collenchyma (Fig. 7.18) in some plant stems. More relevant here is the case of the flat panel stiffened with ribs—a first approximation to, say, the wing of an insect. GERARD (1956) shows that the material efficiency parameter for the entire structure here is $\rho/E^{3/5}$ (E^3/ρ^5).

Some typical data for biomaterials and for some engineering materials are given in Table 6.4. It is interesting, though not unexpected, to note that the tensile materials, cellulose, collagen, silk and chitin all have low values of ρ/σ and further, with the exception of collagen, they also show good resistance to elastic buckling (low $\rho/E^{1/2}$). The fact that they appear, in fact, to be superior to the harder composite materials such as bone, cuticle, etc., should not surprise us

Table 6.4 Material efficiency parameters for minimum weight/strength

Materials	Strength $\rho/\sigma \times 10^6$ m^{-2} s^2	Column buckling $\rho/E^{1/2} \times 10^3$	Edge buckling $\rho/E^{1/3} \times 10^2$
Aluminium alloy	5.4	9.9	66
Fibreglass	1.3	8.9	53
Timber	7.3†	6.5‡	28‡
Cellulose (ramie)	1.5	7.2	41
Collagen	1.8	44.4	140
Silk	2.2	13.0	60
Chitin	2.7	7.5	45
Bone	10.4	14.8	76
Locust cuticle	12.5	12.2	57
Sea urchin spine	23.7	24.1	103
Shell (various)*	62.0	10.7	72

† ash, ‡ red oak, * median values

greatly, since their higher efficiency here is a consequence of their lower density. Of the rigid materials, locust cuticle is, as expected, very efficient while the value for shell is surprisingly good. But it should be noted that this is an average value for various shells and the spread is such as to include some very inferior examples indeed. The apparently poor value for ρ/σ for shell is probably due to the fact that most shells tested in the laboratory contain defects, some of which may be due to uncontrolled drying, which tend to give low values of strength.

6.8 Principles of Structural Optimization

With the foregoing comments in mind, we are now in a better position to start thinking about the design of individual members and about the distribution of material. This area is ill-served by experimental data in engineering as well as in biology, so that we can do little more at this stage than try to formulate some general principles which are as follows:

(1) In cases of simple tension, the maximum allowable stress per unit weight of material is governed by the specific strength and the maximum allowable elastic deformation per unit weight of material by the specific modulus. The shape of the cross-section is immaterial and will, in general, be the simplest. In practice it will usually be the one that minimizes free surface area—so that in principle all simple tension members can be rods or wires.

(2) For all other types of static loading the maximum allowable stress is a function of the cross-sectional shape as well as of the material properties. The maximum allowable deflection is governed by the flexural modulus EI which must be maximized in those cases where either elastic deformations must themselves be minimized, or where excessive elastic deformation may lead to non-recoverable plastic deformation, e.g., in the local buckling of tubes.

(3) Maximizing I must not, however, be carried to the stage where local instabilities develop. Thus although, on the whole, tubes are preferred to solid members, the wall thickness must always be great enough to ensure that localized plastic deformation does not occur. This is especially important in bending (and hence in Euler buckling) where high stresses can be reached in those fibres which are furthest from the axis. Alternatively it may be possible to use stronger material as noted in (5) below.
(4) For members where energy absorption is the predominating need, the major requirement is that of a uniform cross-section with the elimination of local stress raisers (see Chapter 2A). For tension and simple compression, the greatest energy absorption is given by the solid round rod which in the most extreme form may be a string or fibre under axial load. All other types of loading are less efficient but may be maximized by the use of tubular members so that the stored energy per unit volume is as high as possible.
(5) For minimum weight/strength it is evident that density should be minimized. This becomes especially important in members subject to compressive buckling, where density operates via a power term, and small changes in density are much more effective than increases in modulus in enhancing the structural efficiency. Since we must accept the fact that modulus is generally fixed by the material available, density becomes the most significant parameter in determining the structural efficiency, except where modulus is variable, as in cuticle.

6.9 The Failure of Elements (and Shells)

The second requirement which we laid down was that the element should not fail under the working conditions imposed on it. As we have already said, this does not necessarily imply that failure occurs by fracture–an element has failed when it can no longer perform its function properly. Euler and local buckling are two cases where failure does not necessarily involve fracture. Indeed, what may be failure in one situation, may not matter in another, and may even be part of the function of the structure in another. For instance, noticeable wear on a synovial joint would cause immobilizing pain. In a hoof such wear would be unimportant. In the incisors of a rodent the dentine *must* wear considerably in order that the chisel edge of the enamel should remain sharp.

There are various ways we can classify types of failure, it is particularly instructive to classify in a twofold way: by the duration of time taken for failure to occur, and by the amount of deformation that occurs before failure.

The time taken may be very short, in other words failure takes place as a result of impact loading, or it may take a second or so, which is effectively the result of static loading, or it may take place as a result of loads applied (often intermittently) over a matter of hours, days or weeks. This is usually known as *fatigue failure*. It is convenient also to consider here abrasion and the boring of holes (the latter is rarely a cause of failure in engineering structures).

The deformation may be small before failure, and not contribute to it. This is the case, for instance, in the breaking of most bones and the smashing of bivalve shells. Or, the deformation may be small, but contribute to the failure. This is seen, for instance, in the local buckling of the femur of a locust. A wrinkle appears which is at first very small, but if the load is maintained, failure is inevitable once the wrinkle has appeared. Alternatively, the deformation may be considerable. The tibia of a gibbon, if loaded in compression, would deform into a noticeable bow before it became unstable and failed by primary buckling. Large deformation, even if stabilized, may in itself be a failure; a rickety tibia, bent into a bow, would not necessarily break, but in life someone with such a deformity would be at a dreadful selective disadvantage. In some cases, a very small deformation may seriously reduce the effectiveness of a structure, and may cause failure. This would often be true, for instance, in the sliding joints of arthropods, which are often very precisely related to each other geometrically. The distortion of the shell of a bivalve,

Table 6.5 Qualities which resist failure in elements and shells

	DEFORMATION BEFORE FAILURE		
Loading	Very small	Small plastic (local buckling)	Large elastic (Euler buckling)
High Strain Rate ('Impact')	*Material* (a) High strength (b) Low modulus (c) Long plastic region (d) No cracks *Shape* (a) Uniform section (b) Tube (c) Large volume	*Material* (a) High compressive strength (yield strength) *Shape* (a) Uniform section (b) Tube with low D/t	*Material* (a) High modulus (b) High strength *Shape* (a) Large I/A and/or low L/r
Low Strain Rate ('Static')	*Material* (a) High strength (b) Plastic region (c) No cracks *Shape* (a) High Z	*Material* (a) High strength (b) Plastic region *Shape* (a) High Z (b) Tube with low D/t (c) No local imperfections	*Material* (a) High modulus (b) Plastic region desirable *Shape* (a) Large I/A^2 (b) Low L/r (c) Straight
Repeated ('Fatigue')	*Material* (a) High strength (b) Plastic region (c) No cracks *Shape* (a) Uniform section (b) No stress concentrations	*Material* (a) High yield strength (b) Plastic region *Shape* (a) Uniform section	*Material* (a) High modulus (b) Plastic region desirable *Shape* (a) Large I/A^2 (b) Low L/r (c) No stress concentrations
Abrasive	*Material* (a) High hardness (b) Plastic region *Shape* (a) Thick	N/A	N/A

so that a minute gap was left somewhere between the valves would constitute failure, since it would allow the penetration of the proboscis of a predatory gastropod, which otherwise would have to grind its way laboriously through.

Table 6.5 shows the desirable qualities of *materials* in rigid elements and shells prone to failure under various loading conditions but especially in bending. The table shows that the requirements are not the same for all types of failure, indeed, some requirements are contradictory. For instance, if deformation must be small, then great stiffness is required, but if the rate of loading is high, and the amount of deformation before failure is not very important, then stiffness should be low in order to maximize energy absorption. Furthermore, if the material is likely to be attacked by boring, then hardness is good, but hardness almost always goes with high stiffness. For static loading when deformations are small, stiffness is unimportant. For fatigue failure, the ability to remove or to be insensitive to stress concentrations is important, as is the ability to rapidly repair incipient cracks. Occasionally creep may be important. For instance, as *Nautilus* spends its time beneath the sea surface, it will be subjected to continuously high hydrostatic pressure, and it is important that the structure of the shell should not be greatly altered by such pressure.

It is possible for the 'build' (that is the shape *and* size) of the elements and shells themselves to overcome partially any deficiencies in the skeletal material and vice versa. Table 6.5 also shows the desirable build for elements. In general, of course, large size is good (from the point of view of failure, though not for

Table 6.6 Probable modes of failure of different kinds of rigid materials and elements

	DEFORMATION BEFORE FAILURE		
Type of loading	Very small	Small plastic (local buckling)	Large Elastic (Euler buckling)
High Strain Rate ('Impact')	Bones Shells and eggshells Enamel Calcareous tubes Stiff arthropod elements Stony corals Echinoderms (rare)	Thin walled arthropod limb elements	Slender arthropod elements Plate-like arthropod elements Slender bones (rare)
Low Strain Rate ('Static')	Bones Shells and eggshells Enamel Calcareous tubes Stiff arthropod elements Stony corals Echinoderms	Thin-walled arthropod limb elements Terrestrial plants	Plate-like arthropod elements Terrestrial plants
Repeated ('Fatigue')	Bones	Terrestrial plants	Terrestrial plants
Abrasion	Shells, teeth, corals		

hauling it around), and the table shows how the mass may best be distributed. In general, a large value for the second moment of area is good. In impact loading, the total volume should be as large as possible. Also important is that the stress distribution should be as uniform as possible. Usually again this implies the need for a high moment of area for a given volume, since this means that most of the material will be a long way from the neutral axis, and will therefore be more uniformly loaded. Otherwise, the only point to note is that for boring or abrasion, it is clear that only thickness can prevent failure.

Table 6.6 shows, *very* roughly, the kinds of ways that different elements and shells may fail during life. This table is, of course, merely the product of our thinking about how things *ought* to fail, hard evidence being almost completely lacking.

One fact that emerges from this table is that shells are often loaded to failure in impact, but they tend to be very stiff, so they are not well adapted for that

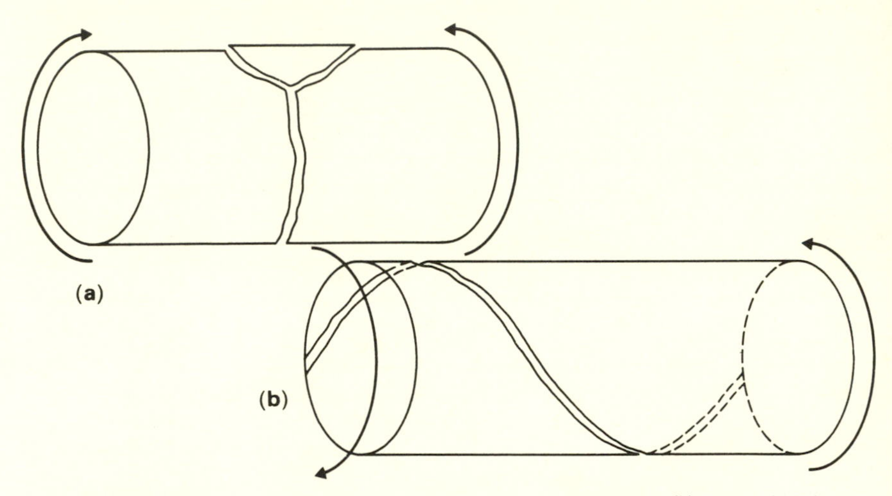

Fig. 6.12 Patterns of bone fracture resulting from (**a**) 3-point loading and (**b**) torsional loading.

particular loading system. Bones also usually break in impact, though, as a material, bone is much tougher than shell and so will survive loads that would smash shells. Failure can occur from direct or indirect loading. In direct loading, the bone is subjected directly to force (with the intervention of skin and muscle of course). This type of loading will often produce conditions like the classical three-point loading test, and long bones will break across transversely (Fig. 6.12a). Indirect loading occurs when the load is transmitted through a joint to the bone. A common type of injury caused in this way is the so-called spiral fracture of the tibia caused by a fall in skiing (Fig. 6.12b). The tibia is broken by torsional forces transmitted through the ankle joint.

The fact that it is impact loading that usually causes failure in bones means that energy absorption is more important than static strength, and indeed bones are quite good at absorbing energy. CURREY (1969*a*) investigated the relationship between different amounts of mineralization of rabbits' metatarsals and strength. He found that the bones with the greatest mineralization have the greatest static bending strength. However, the bones that absorbed the greatest amount of energy, both in static and impact loading, were those with a medium amount of mineralization. This is what one would expect, because if static strength were the most important factor, then the bones should be more mineralized than they in fact are. This study would have indicated, therefore, even if we did not know that bones usually fail in impact, that energy absorption is more important than static strength.

Failure of human bones through buckling is known only in children, but buckling presumably occurs when the slender limbs of a gibbon are loaded in compression after a fall. Bones are also prone to fatigue failure.

Shells are known to fail both by impact and static loading and, of course, usually rather little deformation takes place. Oyster-catchers, *Haematopus ostralegus*, open mussel shells by prodding at the junction of the two valves with their beaks (NORTON-GRIFFITHS, 1967). When the shell chips under these circumstances it is an impact failure, and it is clear that if the shell were more leathery it would be safer from *this* kind of attack. The same is true of land snails that are hammered to pieces on stone anvils by the thrush *Turdus philomelos.* But these shells are also loaded statically by small rodents using their teeth. Crabs and fishes also usually break shells by crushing them with their claws or teeth, respectively. Some whelks chip the edges of bivalves with the thick lip of their own shell (CARRIKER, 1961). It is clear that shells must sometimes fail by impact, and sometimes by static loading.

The boring of shells, both by predators and by animals and plants merely gaining protection, is commonplace. A comprehensive review is by CARRIKER *et al.* (1969). The following groups have some members who bore into living animals' shells: bacteria, fungi, algae, lichens, sponges, sipunculids, polychaetes, barnacles, bivalves, gastropods, cephalopods, bryozoa, turbellaria and phoronids. Organisms merely gaining protection, like sponges and algae, do not need to bore right through the thickness of the shell; predators, however, must reach the soft animal to gain any return for their boring activity, and this may take time. A whelk may take 70–100 hours to bore through an oyster shell, using a mixture of mechanical and chemical attack (ORTON, 1926), and many similar times have been reported (FRETTER and GRAHAM, 1962; CARRIKER, 1969). It is clear from these long times that if a shell is thick, it must considerably increase the chances of its possessor escaping death from predation. Octopuses are voracious predators of bivalves and rasp the valve with their radula; they achieve much higher boring rates, up to 1 mm per hour (ARNOLD and ARNOLD, 1969).

What characteristics make a shell resistant to abrasion and boring are unknown. However, subjective experience gained by filing and drilling test specimens of shells, and experiments involving tumbling (CHAVE, 1964) show that different types of shell vary markedly in their resistance. TAYLOR and LAYMAN (1972) report that in bivalves composite prisms, crossed-lamellar and complex crossed-lamellar structures are considerably harder than foliated and simple prismatic structures. Hardness is strongly correlated with resistance to wear. It is interesting that crossed-lamellar and composite prisms include the least and the most organic contents among bivalve shell structures, and yet they are the hardest. This is a subject of great interest that should rapidly repay study.

Arthropod elements have been little studied in regard to their failure. It is probable that most uncalcified arthropod elements do not rupture, but buckle. If one roughly manipulates a freshly killed locust, it is quite difficult to find places that will snap in two. The distal end of the femur will, for instance, snap, but the proximal end merely buckles, as does the tibia and virtually all the other elements in the skeleton. The same seems to be true of many other insects, spiders and other arthropods. Beetles have a heavily sclerotized cuticle, and their elements indeed often snap, as will parts of the calcified skeleton of crabs and lobsters, though in Crustacea there is often considerable deformation before failure finally occurs.

It would be extremely interesting, though extremely difficult, to make a survey of the way elements in arthropods fail in life. It seems, from our crude 'prodding' survey, that very often they will buckle, because they are not stiff enough, rather than snap because they are not strong enough. We shall discuss later the possibility that the function of the unsclerotized endocuticle is to prevent the exocuticle snapping when bent into a very small radius of curvature (Section 6.12).

Since many arthropod elements collapse by buckling, one might expect to see many examples of stiffening ribs. These certainly do exist. The wings of insects have thickened veins. The thorax of flying insects is stiffened by great internal ridges, the apophyses and pleural ridges. The edges of the carapaces of crabs and ostracods have internal buttresses. The elytra of beetles often have a sandwich construction. Even so, one cannot escape the feeling, in surveying the arthropods, that more stiffening would often be advantageous. Presumably there are factors we are overlooking.

One of these oversights may be engendered by the fact that in our heavily engineered environment, complete disruption of the shape of structural members usually implies failure. But this, of course, need not necessarily be the case. The cartilage of our ears normally holds the ears out so that they help match the impedances of the outside world and the cochlea. But, in a rugby scrum or elsewhere the ear can become crumpled without damage, as the elastic cartilage allows the ear to spring out again. In the same way, the abdomen of many insects is armoured by cuticular plates. These plates maintain the shape of the abdomen, preventing it from dragging on the ground, and protect the contents from abrasion and desiccation. Yet the plates are flimsy, and can bear little load without buckling. But, when buckled, the plates are so thin that the material undergoes little strain, and so remains elastic. The cuticle springs back on removal of the load and the squashy abdominal contents move, unharmed, back to their original position. It is probably the case in many arthropods that, as long as the elements are capable of withstanding the load produced by the muscular activity of the animal, other loads are best dealt with by elastically collapsing, rather than by stiffly resisting, with the attendant dangers of rupture.

In our discussion of failure, we are greatly hampered by often having little idea of what happens in life. However, what we have said does show that simple 'strength' is by no means always the most important factor in determining whether a skeletal element will fulfil its function or not.

6.10 Joints

In Section 6.3 we considered the various structural elements–struts, ties and beams as if they were simple bars or tubes with idealized end fixings–pin joints or rigid joints.

This sort of approach is commonly the first stage of analysis of an engineering structure–the engineer then finds that his real problems begin when he considers the way in which a force in one member may be transferred to the next. Invariably he finds it necessary to alter the geometry–and quite often to increase the amount of load-bearing material–at the joint. Even for static structures the problems of load transfer at joints are quite complex (even though the solutions are pretty arbitrary and generally traditional) but, for a joint which must allow relative movement of the members and still transfer a force, the problems are formidable. It comes as no surprise then to learn that, in articulate skeletal systems, the principal factor which causes modifications to the basic shape of a limb element is the necessity for movement at a joint.

We do not propose, here, to discuss all the many variations and specific solutions which have been found by both vertebrate and invertebrate systems to this problem. Rather it is our intention to try to formulate some very general principles, even though these idealized situations may be exemplified by only a few particular cases.

There are, in general, three questions which need to be asked–how much load is to be carried, how much relative movement is required and in how many directions must the movement go? This leads rapidly into a complex area of design which goes far beyond our present topic. But in biological systems, some of the complexities may be reduced if we make one initial assumption. This is that the number of possible movements required is not a factor in determining the design–though, as we shall see, it may determine the application. We shall now try to justify this.

6.10.1 Degrees of Freedom

Consider two elements AB and CD with a joint at B-C (Fig. 6.13). The freedom of movement of the joint can be described in terms of six degrees of freedom–three displacements parallel to the coordinate axes and three rotations (i.e., in the plane of each pair of axes).

The simplest movable joint to consider is a joint having only one degree of freedom–a familiar example is the hinge that can rotate in one plane only. This is exemplified by the badger's jaw joint–and once we have specified how far the lower jaw has moved with respect to the upper jaw, we have said all that is possible about it.

The most complex joint has six degrees of freedom–this is exemplified by the cow's jaw though here it is important to note that the amount of movement in any one mode may be small, and also that the range of movement in one direction may be restricted by a simultaneous movement in another. Thus if you open your lower jaw wide, you will find it difficult to swing it to the side. In other words the mere fact of availability of some movement does not mean that all possible modes can be used, to their full extent, at the same time.

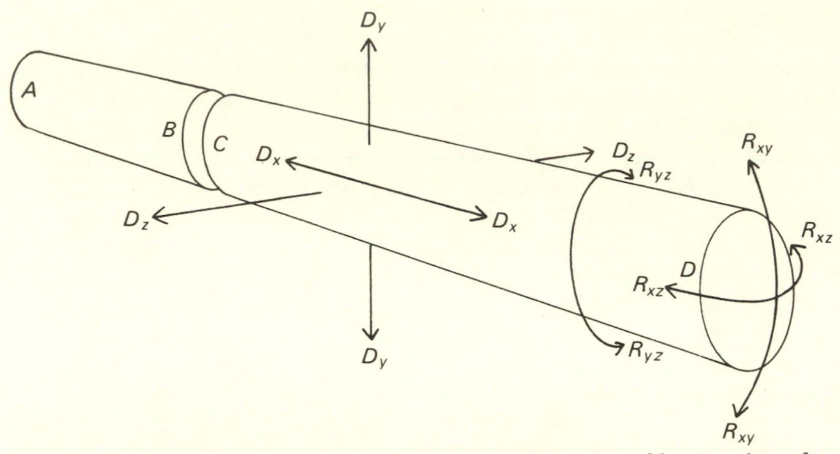

Fig. 6.13 Degrees of freedom of a joint. There are six possible degrees of freedom; three of rotation (R_{xy}, R_{xz}, R_{yz}) and three of displacement (D_x, D_y, D_z).

However, despite the fact that there are, in principle, seven possible types of joint (one rigid and up to six with varying degrees of freedom) we suggest that most biological systems end up with only a choice of three–rigid, one degree and six degrees (or as nearly six as makes little difference). The likely reason is obvious if we think about it in purely engineering terms. A simple hinge has only one degree of freedom–but now try to conceive of a hinge which can translate, say along the hinge line–as well as rotate. Or a hinge with two modes of rotation. There are, of course, perfectly good solutions to these problems–but as they generally involve placing one type of joint in series with another, they are not only clumsy but require a complex set of control systems in order that the relative motions may be properly directed. We propose then that, for any joint requiring more than one degree of freedom, it is as easy to substitute a system which has all six and then to limit, by mechanical or other means, those motions which are not needed. This is exemplified, for instance by your fingers–these have two notable displacements (forwards and sideways) and one small backwards displacement.

Forward, backward and sideways displacement are under control, extension (though generally accompanied by a hideous cracking noise) and rotation through a very small arc are possible under excessive load. The joint itself has six degrees of freedom–three of which are not used, and we may suggest therefore that the number of degrees of freedom was not a factor in determining the design of the joint.

But the number of degrees of freedom does determine one thing–namely the number and location of both the actuators and stops. Each degree of freedom requires at least two muscles in order to move it in each direction in that mode and additional muscles are often required to prevent disarticulation. The one degree jaw of the badger needs no muscles to prevent disarticulation (as attested by the fact that the jaws remain attached to the dried skull) but four muscles are needed to actuate it–one to open and three to shut. At the other extreme, the human shoulder joint has considerable freedom in all three rotational modes, so that the joint requires a minimum of six muscles. In fact there are, broadly speaking, fifteen of which at least four are used simply to keep the head of the humerus in place and to prevent disarticulation. Thus freedom of movement must be offset against complexity of control system and economic use of space, so that we may expect to find that, where many degrees of freedom are needed, animals with exoskeletons have had to find different solutions from animals with endoskeletons.

6.10.2 Forces and Directions

If we assume that the degrees of freedom are less important as a design parameter, we are left with the magnitude of the load and the amount of excursion. Of these the magnitude and direction of the load is the most important. Limbs are, in general, made of stiff materials so that they can resist bending, whereas bending is what many joints are designed to do. But while limbs can also resist tension, certain types of joint cannot, so that, once again, we see the need for

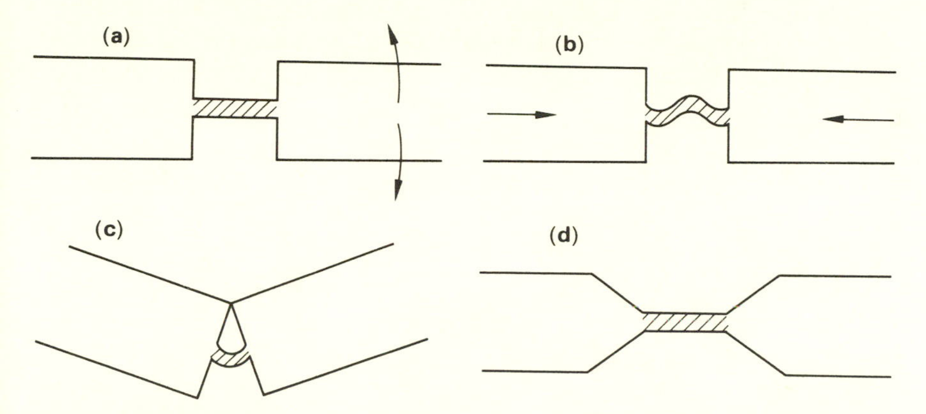

Fig. 6.14 (a) Simple hinge joint composed of a flexible strip between two rigid elements. **(b)** The same joint under compression. **(c)** Joint movement restricted by abutment. **(d)** Increased angular movement by reduced abutment.

a large number of tension members in the structure of the joint. A simple hinge joint consisting of a strip of ligament between two rigid members (Fig. 6.14a) can resist tension up to the limiting stress which the ligament will stand. But it cannot resist compression unless the ligament is short enough to preclude buckling (Fig. 6.14b). Conversely the more sophisticated ball joint can withstand compression (given adequate lubrication) but may disarticulate in tension. Prevention of disarticulation in tension is less of a problem for endoskeletal forms (which can hang great quantities of muscles and tendons around the joint) than for exoskeletal forms which must accommodate the entire control and limit system inside the joint.

If we take these two simplified examples, we see that in general we might expect that:

(a) the simple flexible hinge would exist where the forces are tensile and/or low.

(b) the ball and socket archetype is to be expected where forces are likely to be compressional and high.

On this basis we may classify joints as 'flexible' and 'sliding'.

6.10.3 Flexible Joints

In a flexible joint two rigid elements are connected by a more pliant region. In its simplest form this is found in many aquatic crustaceans where there is often no clearly defined joint but the limb is flattened and the cuticle is thinner and more pliant. The great majority of arthropod joints are of this type as, also, are the joints between shell plates of polyplacophoran molluscs and between the valves of inarticulate brachiopods.

As an aid, consider again Fig. 6.14a where two rigid members are joined by a flexible strip. Movement is restricted by the potential abutment (Fig. 6.14c) of the ends of the rigid elements and load transfer in compression is limited by the

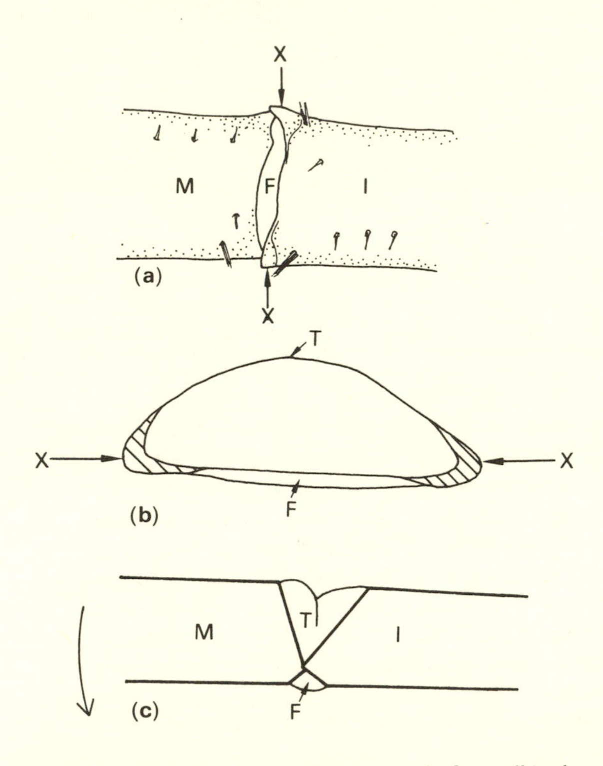

Fig. 6.15 Ischiopodite-meropodite joint in the first walking leg of a lobster. (**a**) Surface view from inside; (**b**) transverse section through points X–X in (**a**); (**c**) diagram of external view as seen from above. F, thick flexible cuticle; I, ischipodite; M, meropodite; T, thin flexible cuticle. Arrow indicates direction of muscular flexure.

buckling tendency of the strip. The movement can be increased by tapering the ends of the abutting members–the gentler the taper, the greater the potential excursion-(Fig. 6.14d). Various modifications exist–one of the more notable is the ischiopodite-meropodite joint in the cheliped of the lobster (Fig. 6.15). This has but one degree of freedom provided by a straight hinge of elastic cuticle and it has about 25° of rotation about one axis. The only muscle required is a flexor–the extension force is feeble and is provided by the elasticity of the cuticle (Currey, unpublished). In order to increase the excursion, a combination flexible/sliding joint is needed. Other simple 'flexible' joints exist in the sclerotized abdomens of many insects and this type of joint is, perhaps, most commonly found in arthropods.

But whatever, and wherever it is used, the essential conditions are clear. Such a joint can only be used under conditions of light loading, being limited (a) by the strength of the material in tension and (b) by liability for local buckling in compression (and torsion). It has a limited number of degrees of freedom which are set by both material and geometry.

6.10.4 Sliding Joints

We define these here as joints in which two surfaces move relatively to each other–the surfaces may be lubricated or not so that sliding joints can vary from the essential simplicity of some brachiopod hinges to the structural complexity of snakes' vertebrae and the mechanical subtlety of the bovid ankle.

Essentially the basic requirement is that of a bearing surface on each of the abutting members. The shape of the bearing surface determines, to a large extent, the number of degrees of freedom. Echinoid spines have essentially ball and socket joints–with movement in some directions limited by a cone of collagen (Fig. 6.16). At the other, more sophisticated, end of the scale, we have the lubricated synovial joint of the vertebrates in which the bones are capped by hyaline cartilage and are lubricated by synovial fluid (see Chapter 4). Joint excursions are limited by ligaments which may be on the outside of the joint or may develop within the synovial cavity. But, compared with other joints, the synovial joint is extremely free moving and appears in many forms such as the hinge (the temporomandibular joint of many mammals, the elbow), the saddle (carpometacarpal joints of the human thumb, the neck vertebrae of birds), the ellipsoid–which in its extreme form becomes the ball and socket–(human metacarpophalangeal joint, radiocarpal joint, and, ultimately scapulohumeral and hip joints).

But, notwithstanding the many variations, sliding and synovial joints have one requirement in common–that of a large bearing surface. The need for increasing surface area goes up as the lubrication requirements become more demanding. A second requirement is the need for some limitations on movement–at all costs, tension across the bearing surfaces must be minimized or the joint will disarticulate–compression loads of quite considerable magnitude can be tolerated.

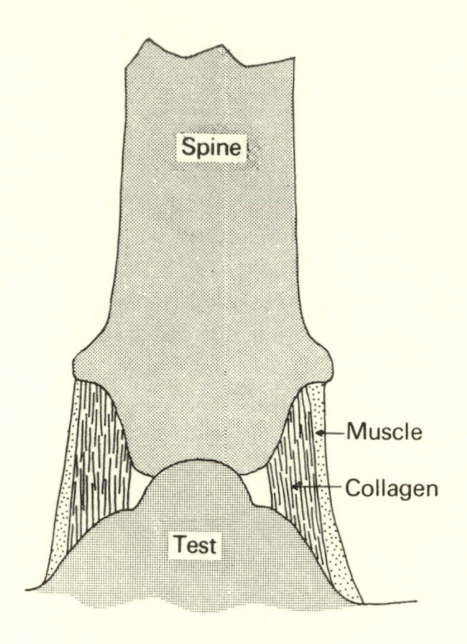

Fig. 6.16 The ball and socket joint at the base of an echinoid spine (drawn from information given in TAKAHASHI, 1967).

6.11 Adaptation of Shape

We have already noted a number of adaptations—the shape of shells, the provision of stiffening ribs in the thorax of flying insects and the adoption of particular cross-sectional shapes which maximize I or J or both. But probably the clearest examples of adaptation arise in the limbs of active animals and these provide the main topic here.

A simple adaptation is found at many epiphyseal plates in the bones of mammals. The function of these is to act as growth planes and, when growth ceases, it is converted from a layer of hyaline cartilage into a fused joint (or synostosis). It is, however, a relatively weak plane in an otherwise strong structure and fracture through the epiphyseal plate is, unfortunately, not an uncommon accident in children. Epiphyses are arranged so that tensile stress across the plane of the plates is rare and accidents are usually the result of shearing stresses. To counteract this, epiphyseal plates are often rather wavy—this prevents shearing stresses from acting across the whole plate at once. SMITH (1962) has shown how the epiphyseal plates of many bones are arranged so that they run at as large an angle as possible to the planes of maximum shearing stress. This is well shown in the plate of the human calcaneus where the epiphysis runs a zigzag course in a region where the geometry makes it impossible to avoid having the plate roughly in the direction of the shear stress (Fig. 6.17).

The major adaptations in limb shape are those which are associated with the need for movable joints and, with the earlier comments in mind, we can now begin to see what forms these may take. The type of adaptation is determined,

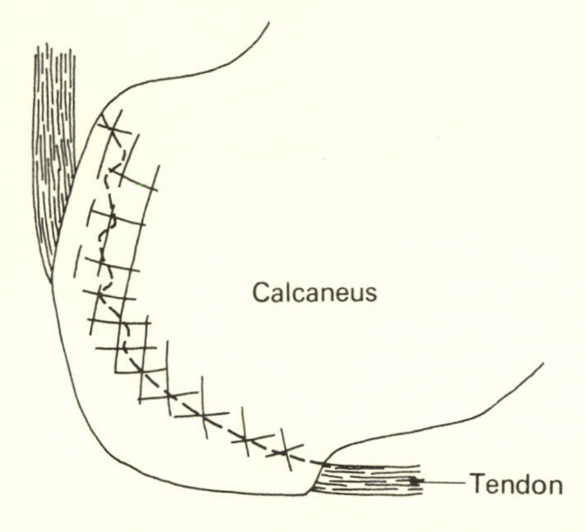

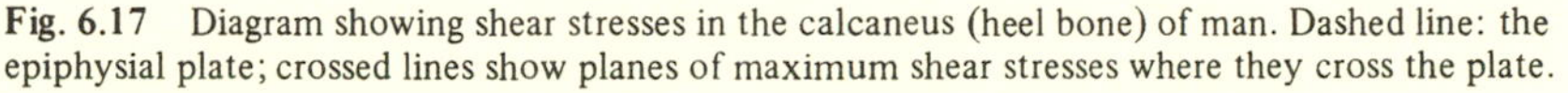
Fig. 6.17 Diagram showing shear stresses in the calcaneus (heel bone) of man. Dashed line: the epiphysial plate; crossed lines show planes of maximum shear stresses where they cross the plate.

in the first instance, by the number of degrees of freedom. The selection of the joint to be used–either flexible or sliding–is determined by the loads to be transmitted and by the design of the elements to be joined.

Consider first a one-degree joint. This can either displace in one plane *only* or it can rotate about one axis *only*. Where excursion is limited and the bones abut, two planes can move across each other as in the roughly cubical bones of the ankle where a limited displacement is possible. As the need for more excursion increases, especially at the ends of long members, the simplest rotating joint is a hinge involving displacement about a line (Fig. 6.18a). An initially cylindrical member must taper if the situation shown in Fig. 6.14c is to be avoided. Two solutions are shown in Fig. 6.18, one using a flexible joint, the other using a sliding joint (Fig. 6.18d). But, in tapering the member, the amount of material transmitting the stress in Fig. 6.18b is reduced so that the cross-sectional area must be increased (to minimize stress) either by increasing the wall thickness, if a tube, or by becoming spatulate or both (Fig. 6.18c). As we shall see in the next section material adaptations are also possible to accommodate to high bearing stresses.

The extreme case of unrestricted excursion requires a shaft rotating in a bearing. Since this seems nowhere to have been adopted by any living organism we conclude either that unlimited excursions are never required or that the complexity of organization which is required to stabilize and to control such a joint is so great that no economic solution has evolved. This seems reasonable when we remember that mechanical systems are limited to applying simple tensions and compressions and that complex mechanical linkages to control freely rotating systems are mechanically inefficient and occupy a fair amount of space.

A limited amount of rotation about a second axis can be provided if the members reduce to a point and the flexible strip of Fig. 6.18c is replaced by a flexible rod (Fig. 6.18e). Excursions can now be quite large but the joint is still constrained by an inability to withstand tensile forces in excess of the strength of the flexible rod and by the need to keep the rod short in order to preclude

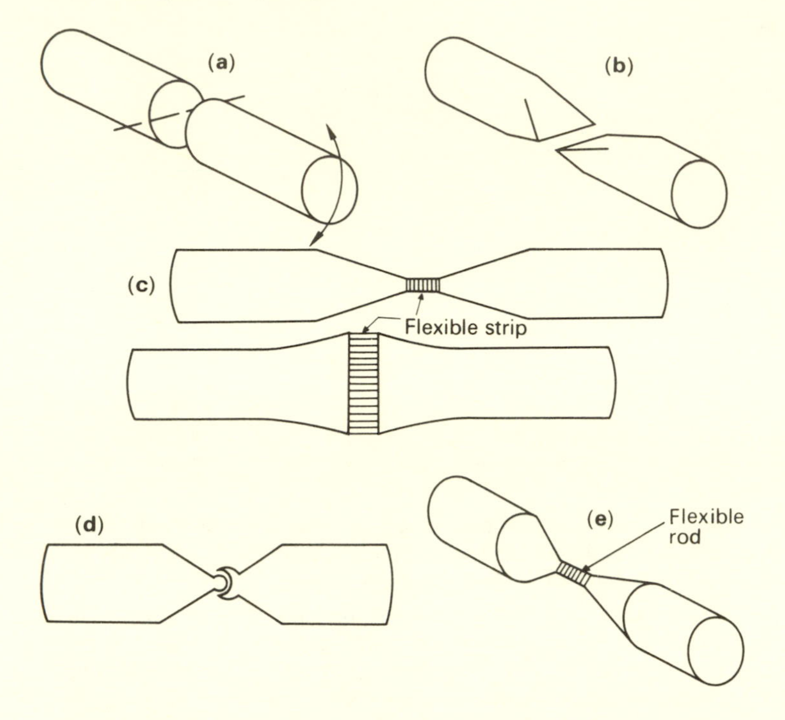

Fig. 6.18 Adaptations to shape by a simple hinge joint. (**a**) Two rigid elements and the axis (dashed line) around which they are to bend. (**b**) The elements' ends are shaped and joined (**c**, two views) by a flexible strip. (**d**) Sliding joint, side view. (**e**) Flexible rod allowing free rotation.

buckling. Reducing the section to a point further intensifies the problems of stresses at the bearing tips.

As a fairly general observation, then, we expect flexible joints to involve some reduction in external dimensions of the rigid element and to be used when forces are small or unidirectional. At the same time they provide an economic method of jointing for animals with exoskeletons whose elements, being by definition tubular, may be used to carry within them the necessary flexible material without exposing it unduly to the world outside.

The other solution represents the opposite extreme–it is to make the jointing surface as large as possible and to permit relative motion of two *surfaces*–the constraints now being set by the shape of the surfaces and by the number of stabilizing members necessary to prevent disarticulation of the joint. Thus the above situation is reversed here–in principle, the easiest joint to design has all six degrees of freedom available to it–the excursions being limited by the shape of the opposing members. The problem in these joints then is one of restraint requiring the provision of stops and controls and the provision of mechanical linkages to maintain continuity. Thus we expect such systems, generally speaking, to be associated with endoskeletons where the bulk of muscle, tendon, etc., can be easily accommodated on the outside of the joint and the most general shape adaptation to be in terms of increasing the surface area of the joint in order to improve both mechanical stability and also to provide design constraints which help to reduce the size and complexity of the external control system.

Nevertheless, sliding joints are quite often found in exoskeletons. A good example is the body-coxa joint of insects, where a single ball and socket joint allows three degrees of freedom. Many joints in Crustacea have two hinges, which may be either sliding or deforming in type. These hinges lie on opposite sides of the exoskeletal cylinder, and so restrict the joint to one degree of freedom. The sliding surfaces are small, and so tend to be stiffer than the surrounding material.

6.12 Adaptation of Material

We now consider some of the ways in which the materials may adapt in order to meet a given loading situation. We consider plants in the first instance, since here there are no problems which involve relative movement of one element past another.

We noted that one way of minimizing the stresses set up by bending, torsion or buckling is to change—often quite drastically—the second moment of area. But another strategy is, of course, to use a stronger material. Stronger material, however, may require a different level of atomic or molecular organization and we may suspect that, in biology as in engineering, a factor of two on the properties may involve a factor of ten on the cost. Thus really good material (either stiffer or stronger or whatever) must be used sparingly and the way to do this is to place it at those points in the element where it will do the most good. Teeth are a good and obvious example. The major forces involved are those of crushing—since the tooth must not be indented by the nutshell, we need great hardness. But intrinsic hardness carries along with it the penalty of brittleness and a tooth that shatters is of little use. So—the hard enamel is placed on the surface where it can resist indentation and do its proper job of cracking nuts or munching T-bones and it is underlain with a layer of softer, more pliant dentine whose only job is supportive and load spreading. A tooth made wholly of enamel would be a disaster in brittle fracture. And, by now, it will not surprise you to learn that engineers and materials scientists have caught on to this strategy of placing the properties where they will do the most good—your car windscreen, the valve seats in the engine, reinforced concrete and armour plate are all examples.

But the nicest examples of a simple strategy come in plant stems where the strong sclerenchyma and collenchyma are, as often as not, placed where the stresses will be highest—or at least most unfortunate. One excellent example is the stem of *Mentha*—this is roughly square in section and therefore has equal resistance to bending about the principal axes. But bending about a corner-to-corner axis would mean that a weak unsupported region at the 'corner' could fail by local buckling. This is precluded by the concentration of collenchyma in the corners. Many other examples can be found. K. Muzik has pointed out (personal communication) that the sclerenchyma in the petiole of the tree fern is arranged as shown in Fig. 6.19. Since this is subjected primarily to bending, we expect an I section or a T section. But given that the external shape is

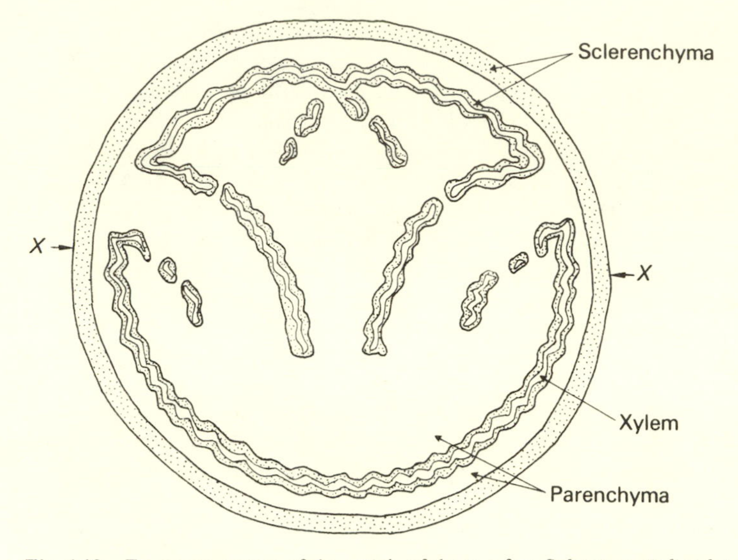

Fig. 6.19 Transverse section of the petiole of the tree fern *Sphaeropteris lunulata* showing the distribution of rigid sclerenchyma and spongy parenchyma. The distance *X–X* is 13 mm.

roughly circular, it seems entirely reasonable to suppose that the sclerenchyma should be disposed in a manner which is, at least, reminiscent of the I section.

Bone and cuticle offer two excellent examples of adaptations. Typical load bearing bone is solid except for blood channels and spaces for cells, but in cancellous bone the structure is much more open. It is found, typically, in two rather different situations—the first is at places where a load, distributed over a large area must be channelled into some distant place. The second is where a fairly flat sheet of bone is subjected to bending loads. We shall call the first 'load distributing' bone and the second 'sandwich bone'.

Load distributing bone is exemplified by the vertebrae and by the ends of many long bones. Examples of sandwich bone are the bones of the top of the human skull (Fig. 6.20), the iliac crests and the bony part of the chelonian shell. But, in both bone types the ultimate function is the same—that of carrying a small load over a large area while using little weight.

Sandwich bone consists, essentially, of two thin sheets of compact bone with

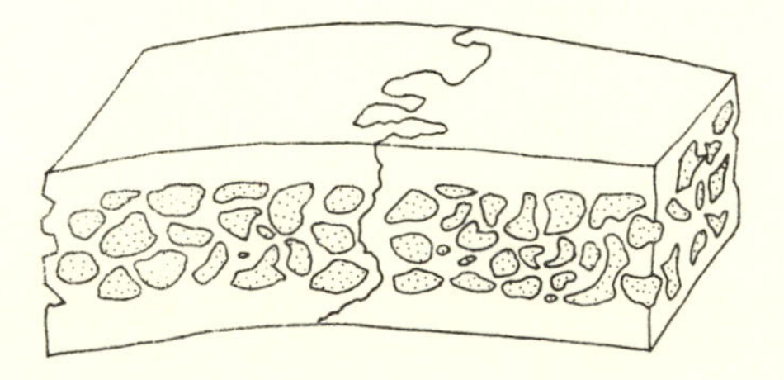

Fig. 6.20 Diagram of a piece of sandwich bone from the human cranium.

an intermediate layer of cancellous bone and it comes close to the ideal for resistance to bending in that, like the I beam, the load bearing material is displaced from the neutral axis and the 'web'—in this case the cancellous bone—acts to keep the outer surfaces separate.

The principle here seems to be that, when forces are small, little material is needed to resist them. There is also another possible advantage—that sandwich bone absorbs more energy per unit weight of material than compact bone. There is no direct experimental proof of this, though experiments by McELHANEY *et al.* (1964) provide a strong indication—sandwich bone from a human cranium when loaded radially (i.e., with the cancellous bone in series with the compact bone) absorbed 2.5 times as much energy as the same bone loaded tangentially (i.e., with the cancellous bone in parallel). And, certainly, the energy absorbing capacity of open, porous structures is a widely used technique in engineering practice—modern packaging in styrofoam provides an immediate (if sometimes irritating) analogy.

The use of cancellous bone in the joints of vertebrates is, however, easier to understand. If we accept the suggestions made in Section 6.10.4, namely that a sliding joint requires the development of a large bearing surface so that the ends of limbs tend to be expanded, it is at once clear that this expansion, if the same internal geometry were maintained, would lead to end fixings of ridiculously high weight. Furthermore since the area of the bearing surface has been increased, the stresses are reduced so that less material is needed. Thus the wall thickness may be reduced in the bearing area but, at the same time, the stresses which are distributed all over the bearing surface, must be channelled into the shaft wall. Cancellous bone here provides a solution. The trabeculae act as short struts, each of which is stiffened against buckling by interaction with short sideways struts and the distributed load on the bearing surface is concentrated into the sidewall of the shaft. And, since the side wall must be thick anyway in order to resist bending, there is no point in extending the trabeculae too far and, indeed, the bone becomes free of trabeculae about two shaft diameters from the end.

This is an elegant, least weight solution to the problem of concentrating a distributed load into the shaft. Arthropod joints have the opposite problem—the distributed load in the shaft must be concentrated at the joint where Z is lower than in the middle length of the limb. Thus, the stresses at the joint will be higher and the need for stiffer, stronger material will be most marked there. Even where the ends are slightly larger than the middle—as in the tibia of the locust—there will be local high load bearing forces which, because of the exoskeleton, cannot be transmitted by the strategy described above.

One immediately obvious solution is to increase the wall thickness of the cuticle near the joint. This probably occurs, but it can only be done to a limited extent because of the need for maintaining a central cavity of given size.

Arthropod cuticle may be sclerotized—this generally darkens it, though this is by no means invariable (ANDERSEN and BARRETT, 1971). Little is known about the properties of tanned cuticle, but Dr. A. C. Neville informs us that in a patch of cuticle on the cockroach *Blaberus*, which has a dark central region and clear periphery, there is a definite though small difference in *hardness*, with the darker region being harder, as the whole cuticle becomes tanned.

It is most noticeable that in a multicoloured cuticle the dark-tanned regions tend to be those subjected to large loads, and where stiffness will be particularly important. In the locust, for example, most of the cuticle is yellow or amber, but there are dark regions: these are on the spines, the grinding parts of the jaw, and in regions round joints. In some places as at the bottom of the femur, the darkened cuticle is clearly associated with thickened struts that support the femoro-tibial joint. Presumably this dark cuticle is stiffer than the clearer, tanned cuticle.

We suggest then that the tanned cuticle provides a stiffer, and probably stronger bearing surface at these points where loads are likely to be concentrated and the topography of the joint precludes increasing the cross-sectional area.

Why, if stiffness and strength are so desirable, is the whole cuticle not tanned? We do not know, but an interesting natural experiment concerning this is described by RICHARDS (1958). *Ephestia kühniella,* a moth, has a resilient pupal cuticle that springs back unharmed if compressed with a pair of forceps. A mutant form has a cuticle that will crack and stay indented if compressed. The only obvious difference between the normal and the mutant cuticle is that the mutant has virtually no endocuticle, though the exocuticle seems the same. Perhaps, then, the function of the more pliant endocuticle is to prevent the exocuticle from fracturing. It could do this by preventing the exocuticle from being bent into a curve with a very small radius of curvature. Such a curve would be associated with very high strains.

Many other adaptations exist—some important ones will be discussed further in Chapter 7. But we should note that adaptation of material is not always in the direction of increased strength. For instance, where epiphyseal plates of vertebrates are necessarily subjected to tension, the material becomes fibrous rather than hyaline cartilage (SMITH, 1962). In the same way the strong, though flexible, symphysis between the pubic bones of the mammalian pelvis, which are normally united by a block of fibrocartilage, relaxes in the later stages of pregnancy with much of the collagen being replaced by elastic tissue. This, of course, allows for easier widening of the birth canal.

Here again the arthropods demonstrate the wide variety of adaptation which are possible with cuticle. The blood sucking bug *Rhodnius*, for instance, changes its cuticle reversibly—plasticizing it when feeding to allow expansion and reversing the process when the cuticle has returned to its normal size. This is probably brought about by hydration of the cuticle (BENNETT-CLARK, 1962; MADRELL, 1966). Finally we suggest that one additional reason for the existence of flexible joints in arthropods is the comparative simplicity of leaving some cuticle untanned, thereby producing a ring of cuticle of lower stiffness. In the Crustacea this must be accompanied by a lack of calcification in the ring.

Chapter 7
Support in organisms

7.1 Introduction to Rigid and Flexible Systems

An understanding of the complexity and diversity of organisms is a central goal of biological research. Since the shape of most organisms is largely an expression of their mechanical support systems, comparative studies of mechanical function should help us understand some of the observed complexity and diversity of form and function in terms of selective processes in evolution. A goal of this chapter is to interpret the structure of those plants and animals for which we have some mechanical information in terms of principles of mechanical engineering. By applying mechanical design principles to the analysis of support systems, we hope to demonstrate advantages in and explanations of the complexity and diversity.

The mechanical support systems of multicellular organisms are complex in that each consists of more than one type of structural element, each type serving a different function. Furthermore many elements must perform functions other than support—transmission of nutrients in plant stems, conversion and storage of chemical materials in bone marrow, and so on. These often make it difficult to separate the purely supportive duties of a member from its other requirements. The design of both the individual member and of the system as a whole is, clearly, more complex than the design of an engineering structure where the duties of each element can be fairly precisely defined.

But even the simplest support system must resist tension, compression and bending—three very different kinds of forces. As has been described in detail in previous chapters, the specialization of macromolecular materials to resist these different kinds of forces in organisms is as marked as is any of the better known non-mechanical specializations of biomacromolecules (e.g., oxygen-binding of haemoglobin, specific catalysis of enzymes). Consideration of the resistance to compression and tension provides the immediate explanation of the necessity for speaking of the 'supportive system' rather than the 'skeleton'. As visualized in many textbooks of general biology, 'skeleton' refers to the bony structures in the vertebrate but, in fact, bones alone support no body without the aid of tension-resisting tendons and muscles.

The first organisms of Earth lived in the sea. One may conceive of the early organism as a tensile pattern of carbon-containing molecules surrounding an internal fluid. Whether this fluid was mostly sea water or a hydrophobic fluid (FOX and DOSE, 1972; KEOSIAN, 1964), it was the compression-resisting component and, right from the start, the surrounding skin of molecules and intermolecular bonds served the tensile role it has never relinquished. It is believed that from some such simple system as this has evolved the diversity of functional shapes in the bacteria, the algae, the fungi, the higher plants and over twenty phyla of animals. A sea watery fluid has proved to be handy and useful as a compression material for many species throughout evolution, but structural evolution of both plants and animals has involved the elaboration of stiff materials that resist bending as well as compression. Such a step has permitted arthropods, vertebrates and the higher plants to deploy numerous species on land, while the lack of this step in the evolution of the molluscs may help to account for their relative lack of success in exploiting terrestrial habitats where low humidity and high wind velocity promote water loss.

All terrestrial bodies must, ultimately, support their weight on the Earth's surface by compression. The simplest composite support system for a large static body is that seen in the flagstaff or the mast of a racing dinghy where a central compressive element is held resistant to buckling and bending by tensile wires. By an extension of the flagstaff, the problem of enclosing a volume is most effectively achieved in the simple bell tent where the tensile wires of the flagstaff are replaced by a tensile membrane. Compared with the simplicity and effectiveness of these two systems—one a support device, the other an enclosure device—everything else is a compromise. At the other end of the scale the pyramid, built of wholly compression members, is a hefty, costly and immobile way of holding up a flag (even though it encloses a space as it does so)—and it is clumsy not because all of its elements are in compression (which they are) but rather because the materials used mean that the compression stresses have to be borne by large blocks. The single-celled Radiolaria and multicellular glass sponges are evidence that better, lighter compression structures can be built even in a brittle—and therefore non-tensile material.

What we must now consider is that the supportive system of an organism is a set of discrete elements. Each element is a subset in that it either consists of components that are different (bone = collagen + apatite) or that it is made up of the same components arranged in different ways (silk = oriented + disoriented molecules). But the properties of each supportive element are functions of the properties of each constituent, of the relative proportions of each, and of the geometrical arrangements they adopt. Finally each of these individual elements must be combined into a structure that may have properties different from those of the individual elements.

Thus consider a tall tree: in a good wind the top may deflect by several feet and, when loaded with snow, it may bend to half its height. The accumulation of a whole lot of limiting strains over a great enough length produces a situation which cannot be visualized from a mere consideration of the behaviour of a small element.

Thus, having decided the optimal design of the individual elements, we are now faced with the much more difficult problem of deciding how the whole structure behaves when these optimally designed elements are fitted together in series or in parallel. At this level of sophistication almost anything we say becomes an over-simplification of the real situation and so we must, in our ignorance, content ourselves with two principal topics to point up the principles involved. The first is, essentially, the 'space framework' adopted by terrestrial animals to transmit forces from one point to another. The second is the 'membrane' by which an animal subjected to pretty uniformly distributed forces balances their effect. Thus the first case is more likely to be associated with a 'point' loading situation, the second with a 'distributed' force situation. And, by not improbable semantics, we may conceive that the first is associated primarily with terrestrial situations with gravity and muscles pulling on tendons acting as the major 'point loads' and that the second applies more to aquatic situations where the hydrostatic pressure forms the 'distributed' load.

7.1.1 The Optimization of Space Frames

In Chapter 6 we discussed the principles underlying the optimal design of struts, ties and beams—the elements that make up rigid structures. We now face the more difficult problem of optimizing a whole structure composed of an assembly of these simple elements. Once again the objective depends upon the requirements—do we need a structure involving the least weight, the least cost or what? Since we know very little about metabolic cost, we shall assume that, for rigid terrestrial structures, least weight is a good starting point since least weight implies the use of minimal material in order to perform a given function. While this does not necessarily imply least cost (since the best material may be very expensive), it does provide a convenient baseline against which other structures and other materials may be assessed.

For the engineer the design of a framework of minimum weight is an exercise in structural synthesis—the forces and their directions are known and it is an exciting exercise for him to predict the perfect framework. But how is it relevant to biology where the structure already exists? The exercise has, as it were, been done.

We believe that it is important that the biologist should start to look at structures in this way—and if the actual structure differs significantly from the theoretically best one, the reasons must lie either in some factors which he has ignored in his analysis or in the fact that his initial assessment of function was wrong. In either case he had better look for the missing factors.

The basic theorem used in the synthesis of ideal frameworks is due to CLERK MAXWELL (1890) and is often known as Maxwell's lemma. We shall not prove the lemma here—the reader is referred to PARKES (1965) for a clear exposition of the proof for a two-dimensional framework which can be extended to the three-dimensional case. In a series of elements joined to make a framework, we can describe any one RS (Fig. 7.1) by its coordinates $x_r y_r$ and $x_s y_s$. The forces acting at R and S can be resolved into forces $X_R Y_R$ and $X_S Y_S$. Now if each tension element supports a stress $+\sigma_t$ and each compression element supports $-\sigma_c$ (note that these are not necessarily numerically equal), then it can be shown that, if the total volume of all the tension members is V_t and of the compression members is V_c

$$\sigma_t V_t - \sigma_c V_c = \sum_R \{X_R x_r + Y_R y_r\} = const. \tag{7.1}$$

This statement makes three important points:

(a) If all the tension members are at stress $+\sigma_t$ and all the compression members are at $-\sigma_c$ the expression $(\sigma_t V_t - \sigma_c V_c)$ is a constant *regardless of the shape of the framework.*

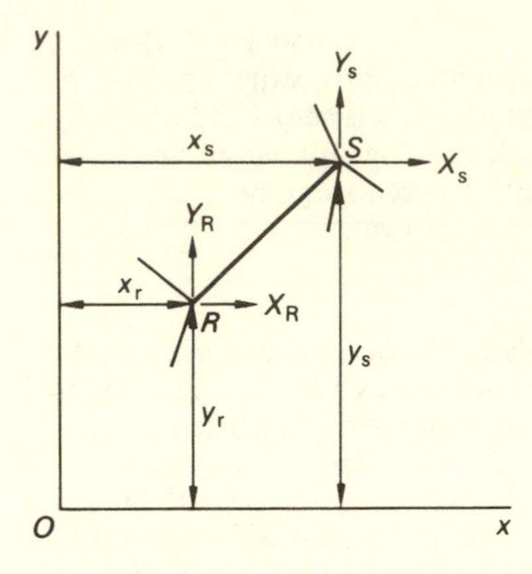

Fig. 7.1 Diagram of bar RS—a component of a rigid framework—showing coordinates and resolved forces. (From PARKES, 1965, *Braced Frameworks*, Pergamon Press, Oxford).

(b) For minimum weight $(V_t + V_c)$ must be as small as possible. *Equation 7.1* shows that

$$V_c = \frac{\sigma_t V_t}{\sigma_c} - const.$$

so that *if we minimize V_t we simultaneously minimize V_c.*
(c) The smallest value of either V_t or V_c is zero. Thus if we can devise a structure in which *all members are in tension or all are in compression*, then *that structure will be of minimum weight.*

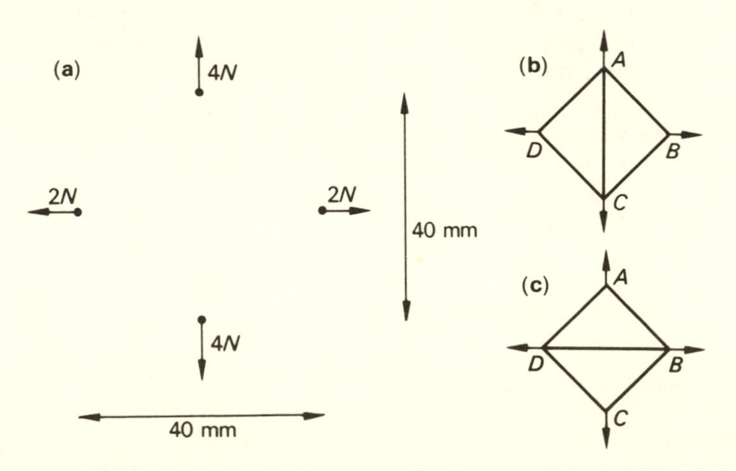

Fig. 7.2 (a) Distribution of forces to be equilibrated by a space frame. (**b** and **c**). Two possible solutions. (From PARKES, 1965, *Braced Frameworks*, Pergamon Press, Oxford).

7.1.1 The Optimization of Space Frames

A simple example may help–it is essentially the same one used by Parkes. The system of forces is shown in Fig. 7.2a and two solutions immediately suggest themselves (Fig. 7.2b, c). Note that each uses the same number of members and each has the same length of bracing member AC or BD but in (b) all members are in tension whereas in (c) BD is in compression. Let us assume that the material used is all the same and that in both tension and compression the maximum safe working stress is 5 N mm^{-2}. We resolve forces to get the force in each member and from $\sigma = F/A$ we obtain the areas and hence the volumes.

Case (b)					Case (c)				
Member	Force N	Area A mm^2	Length mm	Vol.	Member	Force N	Area A mm^2	Length mm	Vol.
AB	$+\sqrt{2}$	0.28	28.3	8	AB	$+\sqrt{8}$	0.566	28.3	16
BC	$+\sqrt{2}$	0.28	28.3	8	BC	$+\sqrt{8}$	0.566	28.3	16
CD	$+\sqrt{2}$	0.28	28.3	8	CD	$+\sqrt{8}$	0.566	28.3	16
DA	$+\sqrt{2}$	0.28	28.3	8	DA	$+\sqrt{8}$	0.566	28.3	16
AC	+2	0.4	40	16	DB	−2	0.4	40	16
				48					80

Now if we apply Maxwell's lemma to the force situation, the right-hand side of *Eq. 7.1* becomes

$$2 \times 20 + 2 \times 20 + 4 \times 20 + 4 \times 20 = 240 \text{ N mm}$$

whence, for a working stress of 5 N mm^{-2} the minimum volume is 240/5 = 48 mm^3 as obtained in case (b). Notice that in case (c) the difference between the volume of tensile material and the volume of compression material is 64 – 16 = 48 mm^3, which is as it should be. Of course we would have obtained different solutions had we used, say, a different material for the compression member or had we been prepared to allow higher compression stresses than tension in the same material.

In the above example it was possible to devise a structure with all its members in tension but for some systems of forces this is not possible. Minimum weight structures for these systems are solved by a theorem due to MICHELL (1904) and are often known as Michell structures. A full treatment of some simpler Michell structures is given by Parkes but, basically, the theorem considers a domain of space that contains all possible frameworks in which all tension members are under stress σ_t and all compression members have stress $-\sigma_c$ and then proceeds to eliminate those which do not keep the deflections within a specified limit. The theorem then shows that such a framework must have a volume which is less than any other framework lying within that domain of space. For systems that are all tensile or all compressive the solutions reduce, as they should, to Maxwell's lemma *Eq. 7.1.*

What is particularly interesting and what has not, so far as we know, been applied to biological situations is the sort of structures which result. Essentially

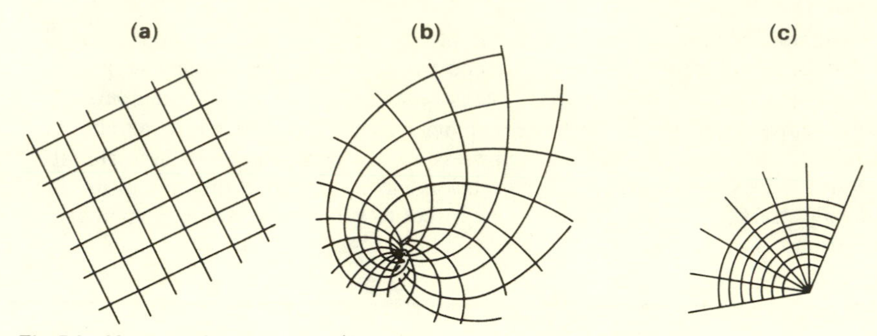

Fig. 7.3 Nets extending into space. (From PARKES, 1965, *Braced Frameworks,* Pergamon Press, Oxford).

these reduce to orthogonal nets. If the nets are such that they extend to the whole of space, they are especially important because they require no additional support and must therefore occupy the absolute minimum volume. The simplest of these is the rectangular net, but Parkes also points out that the equiangular spiral and the circular fan come very close to absolute minimum weight except for restrictions at the origin of the spiral and at the closure of the fan where strains may be incompatible (Fig. 7.3).

But it is when we get into space frames supporting concentrated loads that the situation becomes biologically intriguing. Parkes shows that the optimal solution for a three-point loading system where a force W must be reacted at A, B is best achieved, and at lower weight by a Michell framework. Figure 7.4 shows the weight (volume) saved. The similarity of this sort of pattern to, say, the distribution of trabeculae in the cancellous bone in the head of the femur is irresistible.

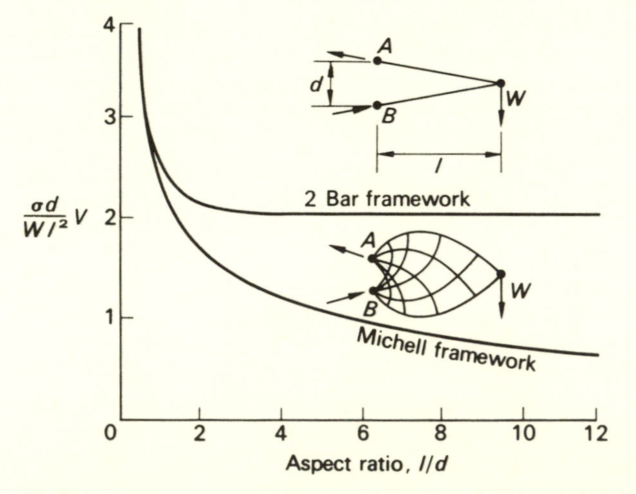

Fig. 7.4 Curves comparing a bar frame (upper curve) with a Michell framework. The ordinate is a measure of volume and the abcissa is the aspect ratio of the framework. (From PARKES, 1965, *Braced Frameworks,* Pergamon Press, Oxford; after CHAN, 1960).

But there is one more important conclusion from the Michell theorem. It is that the optimum framework of minimum weight is also *the stiffest of all possible frameworks* whose members sustain stresses σ_t and $-\sigma_c$ in that domain of space. Now this seems to us to be of immense importance–if the lightest framework is also the stiffest, i.e. has the least deflections per unit stress in the members of the framework, it would seem to suggest that any framework which is not so designed must be intentionally less stiff, i.e. it must be designed in order to accommodate deflections. This may indeed be the true situation in the head of the femur and probably in most biological frameworks–they may not in fact be designed to resist deformation but rather to accommodate deformation.

7.1.2 Fibre-wound Cylinders as Reinforced Membrane Systems
Another approach to support is the pneumatic or hydrostatic one wherein a central volume of fluid under pressure is surrounded by a container whose material is in tension. Most man-made pressurized cylinders and the walls of

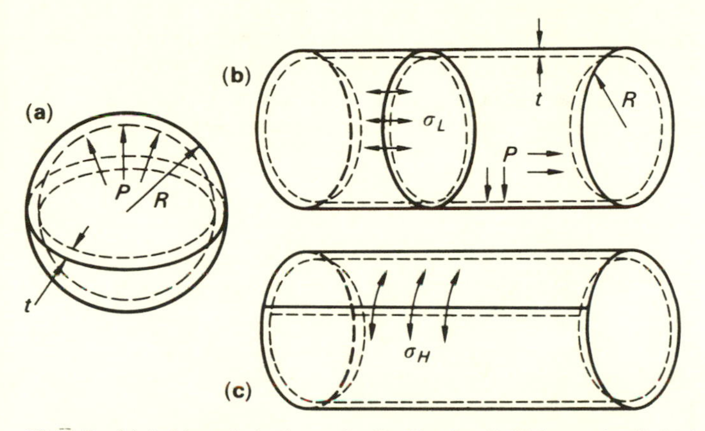

Fig. 7.5 (a) A thin-walled sphere of radius ***R*** and wall thickness ***t*** is split by internal pressure (***p***) along a circumference. A thin-walled cylinder is similarly split circumferentially in (b) by longitudinal stress, σ_L. Hoop stresses, σ_H, causing a longitudinal split are shown in (c).

plant cells are rigid enough to support the cylinder at zero pressure. Animal pneumo- and hydrostats are notably flabby and collapse without internal pressure. They are thus membrane space structures surrounding fluid as the compression component of the structure.

Consider a thin-walled sphere, of radius R and thickness t under internal pressure p. (The same argument applies to a vessel under external pressure, but the internal one is easier to visualize.) If we split the sphere into halves as in Fig. 7.5a, the total force separating the two halves is $F = \pi R^2 p$, this is resisted by the area of vessel wall $A = 2\pi R t$. Thus the stress in the wall is

$$\sigma = \frac{F}{A} = \frac{pR}{2t} \qquad (7.2)$$

and it is uniform in all directions.

Now consider a cylindrical vessel which is closed at its ends. Split it across (Fig. 7.5b) and we see that the situation is as above, i.e., the stress in the longitudinal direction is

$$\sigma_L = \frac{pR}{2t} \tag{7.3}$$

To find the stress in the circumferential direction, take an element of unit length and split as shown in Fig. 7.5c. The force normal to the diameter is now $F = 2pR$, the area $A = 2t$ so that the circumferential or hoop stress is

$$\sigma_H = \frac{pR}{t} = 2\sigma_L \tag{7.4}$$

This means that if slight increases in internal pressure are not to cause disproportionate increases in body diameter, the cylinder wall must be reinforced so as to control the change of shape with changes in pressure. Man-made thin-walled cylinders are reinforced by being wound with a relatively inextensible fibre of high tensile strength and stiffness (FAUPEL, 1964). Cell and body walls of pressurized organisms are similarly reinforced by helically oriented fibres. CLARK

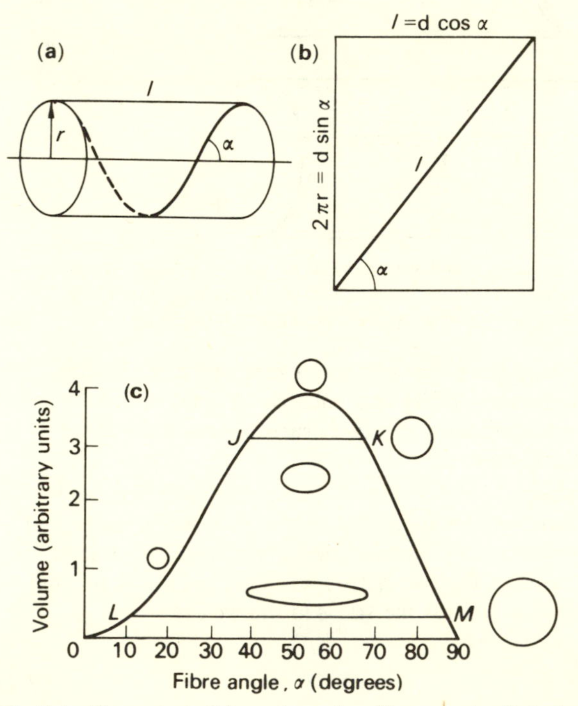

Fig. 7.6 The control of shape change in a fibre-wound cylinder by the fibre angle, α (from CLARK and COWEY, 1958; courtesy of the *J. exp. Biol.*). This figure is explained in the text.

and COWEY (1958) and HARRIS and CROFTON (1957) have indicated the degree that crossed-helical fibre systems in the body walls of some worms control the changes in shape undergone by various cylindrical and flattened worms.

In an open fibre-reinforced cylinder, the volume is maximum when the fibre angle (the angle between the fibre and the cylinder's long axis) is 54° 44′ (Fig. 7.6c). Figure 7.6a shows a length of such an open cylinder of length l and radius r subtended by one turn of a helically wound, stiff, reinforcing fibre making a fibre angle α with the cylindrical axis. Figure 7.6b shows this length of cylinder cut longitudinally and laid out flat and it shows that the fibre describes the diagonal bisecting the surface area of the cylinder. The volume of this cylinder is

$$V = \pi r^2 l$$

and when r and l are expressed in terms of the fibre angle,

$$V = \pi \left(\frac{d \sin \alpha}{2\pi} \right) \cdot \cos \alpha$$

and

$$V = d^3 \cdot \frac{\sin^2 \alpha}{4\pi} \cdot \cos \alpha$$

An open cylinder whose wall contains inextensible fibres in a crossed helical array can have shapes varying from very long and thin with a small fibre angle to very short and fat with a large fibre angle. At both extremes, volume will approach zero and the maximum volume will be included by a helical system with a fibre angle of 54° 44′, as shown in Fig. 7.6c. A closed cylinder of constant volume can change its shape along any horizontal line such as *JK* or *LM* according to the ability of the cylinder to be elliptical in cross-sectional shape at fibre angles near 55°. In other words, *JK* would represent a closed cylinder that deviated very little from perfectly circular, whereas a cylinder represented by *LM* would have a ratio of major axis to minor axis of its sectional area of 20. Further, if the fibres are relatively inextensible and cannot slide past one another in the wall material, the length of the cylinder will be given by $d \cos \alpha$.

In a closed, fibre-wound cylinder, the hoop and longitudinal stresses will be balanced when the fibre angle is 54° 44′ as will now be shown. Figure 7.7 shows such a closed cylinder. Internal pressure applies a force to each fibre that is given by

$$F_\alpha = \sigma_o A_f$$

where σ_o is the breaking stress of the fibre and A_f is the cross-sectional area of the fibre. The hoop force will be

$$F_H = F_\alpha \sin \alpha$$

and the hoop stress

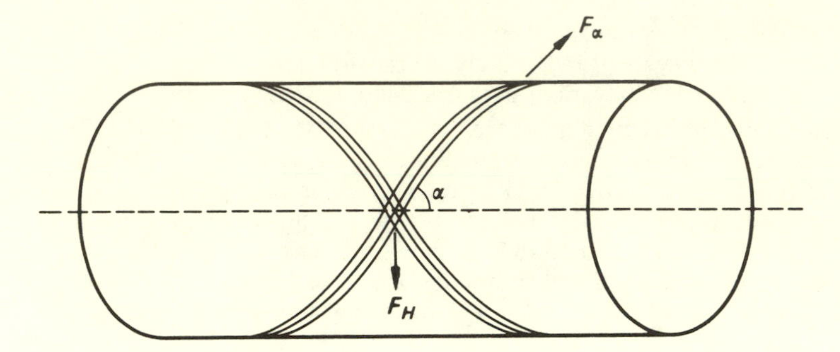

Fig. 7.7 Fibre-wound cylinder. α, fibre angle; F_α, force in a fibre; F_H, hoop force in the cylinder wall. This figure is explained in the text.

$$\sigma_H = \frac{F_H}{A} = \frac{\sigma_o A_f \sin \alpha}{(A_f/\sin \alpha)} = \sigma_o \sin^2 \alpha$$

where A is the sectional area of the cylinder wall. In a similar way, longitudinal stress can be shown to be

$$\sigma_L = \sigma_o \cos^2 \alpha.$$

Balancing these stresses

$$\frac{\sigma_H}{\sigma_L} = \frac{pr/t}{pr/2t} = 2 = \frac{\sigma_o \sin^2 \alpha}{\sigma_o \cos^2 \alpha} = \tan^2 \alpha$$

and

$$\alpha = 54° \, 44'.$$

These notes on rigid frameworks and reinforced membrane space-enclosing systems are not meant to represent the structural systems of all macroscopic organisms. For example, NACHTIGALL (1971) briefly treats other aspects of quadrupedal mammalian systems. The models discussed here are, in fact, the only general models we have found. Few whole organisms are rigid frameworks. Most animals have complex support systems with both flexible and rigid parts. We cannot state at this time any design principles that may apply to this combination. Here is an intriguing field for study. And what models are truly appropriate for massive live oaks, hollow cacti, supple palm trees and grasses, wilting and succulent herbs, filamentous fungi and algae, and giant kelps?

Since we have no further models of whole systems to offer, we will introduce the discussion of design in whole organisms by stating the following design principles for structural systems.

7.2 Design Principles for Biological Structural Systems

Principle 1 The lightest weight structures will be those consisting entirely of tension members or of compression members.

(a) Ability to resist tension is independent of both length and cross-sectional shape.
(b) Ability to resist compression decreases with increase in length and is dependent on shape.
(c) For a structure of a given size containing both tension and compression elements:

(i) Compression elements should be short and (due to the need to increase I) of large external dimensions. Tension members may be long and slender.
(ii) The most efficient least weight solution will maximize the amount of tension material and minimize the amount of compression material.
(iii) The most appropriate arrangement of the elements in rigid frameworks will be that in which they are at right angles to each other.
(iv) A non-orthogonal array of tensile reinforcing fibres lying in helices around the body is the most appropriate arrangement for highly flexible systems.

Principle 2 Whereas most biological structures have complex loading patterns involving tension, compression and bending, purely tensile elements resist tensile forces only. Bending and compression elements also resist tension and shear.

(a) Tension elements may be reduced to wires or cables. All other elements may develop resistance to bending by maximizing I. This will be achieved by the use of hollow sections and/or a distribution of strong stiff material in the periphery of the section.

Principle 3 It is advantageous to minimize the use of compression materials and to maximize the use of tensile materials.

(a) For a material of given density, a compression member must have a larger cross-section and must therefore be heavier than a tension member per unit of load carried.
(b) Disproportionate use of compression materials makes organisms heavier and puts greater demands on metabolism for synthesis, support and locomotion. To reduce these demands, the compressible materials should be:

(i) used sparingly,
(ii) cheap and easy to obtain, or
(iii) readily synthesized and
(iv) of low density.

(c) In a complex support system the number of compression elements should be as small as possible and they may be connected by the use of continuous tensile elements.

(i) In an optimal design, the compressible elements are small and isolated from one another, suspended by a continuous tensile component.

Principle 4 In a composite material the amount and distribution of the stiffer component may be used to control strain and to maximize the resistance to stress.

(a) The degree of anisotropy in mechanical properties in a composite material reflects the degree of anisotropy (preferred orientation) of the stiffer component.
(b) In a two-dimensional system the reduction in properties associated with random orientation of the stiffest component plus the associated weight penalties of using more material imply that random orientation will not be used as a solution in cases of multiaxial stressing. The solution will generally be found by a superimposition of differently oriented layers.
(c) The orientation of fibres will be affected (and may be determined) by the directions in which the forces act during the formation of the material.
(d) In a block or sheet of material containing parallel reinforcing fibres, the change in width per unit change in length is inversely related to the angle between the fibres and the longitudinal axis to the sample.

Principle 5 Thin-walled cylinders are most effectively reinforced against explosion and buckling by crossed-helically wound fibres.

(a) As a cylinder, wound helically with stiff fibres, changes shape (e.g. becomes short and fat), the fibre angle changes (e.g. increases).
(b) In pressurized fibre-wound cylinders, hoop and longitudinal stresses will be balanced when the fibre angle is $54° 44'$.
(c) Open fibre-wound cylinders will attain maximum volume when the fibre angle is $54° 44'$.
(d) Fibre-wound cylinders whose nonfibrous material is viscoelastic can sustain very large controlled deformation/flexures.
(e) Cylinders, especially flexible ones, are most effectively reinforced against implosion by internal circular wall thickenings.

Principle 6 Accommodation to and survival in a stressful environment may be achieved by resisting forces and/or deflections or by permitting them:

(a) Resistance to forces and deflections demands generally high strength and/or modulus; compliance with forces and deflections involves primarily low modulus.
(b) Structural forms consisting of rigid members distribute the load between all members and allow point loads to be sustained without excessive deflection. Structures built of flexible members can carry uniformly applied loads but may deform greatly under them. Point loads produce large deflections.
(c) Elements built of pliant materials may gain strength by reinforcement (change in E), whereas rigid elements may gain strength by changing cross-sectional shape or area (change in I).

7.3 Real Organisms: an Overview

The design principles listed above are now to be applied to diverse plant and animal structures and their mechanical function. The plethora of plant and animal species can be conveniently classified according to their mechanical life style: (a) some organisms are attached to solid substrata or are otherwise nonmotile while the others are motile and (b) some organisms live surrounded by water while others live in air.

7.3.1 Symmetry

Most macroscopic plants and many of those aquatic animals that are attached to a solid or semisolid substratum face their surrounding fluid environment, air or water, very much equivalently in every direction except toward the substratum. They have some degree of radial symmetry and hemispherical shape. Leaves, sensory receptors, and food catchers are extended equally in all directions. Animals that move actively through their environment and those attached plants and animals faced with persistent directional flow of the fluid environment past them, are not spherically shaped. Such animals usually have an elongated body with head and tail (upstream and downstream). When such 'polarized' organisms relate their movement via gravity or friction to a solid substratum, they have bilateral symmetry as well.

7.3.2 Reaction to Force

Gravity may govern distribution of nonmotile aquatic organisms in the water column, but organisms in water may be large and heavy with deposited mineral without great danger of failure by compression or buckling. These organisms may be large by being jellylike—a weak hydrostatic system reinforced by a dispersed three-dimensional network of collagen and other proteins and polysaccharides. They may also be large by the accretion of $CaCO_3$ in such great concentration that they are dense, absolutely heavy (the skeleton of a single reef coral colony may weigh several tons), rigid and brittle.

Animals that locomote through water or sediment have supportive systems in which compression-resisting materials and structures are sharply definable from tension-resisting ones. The body wall of the hydrostatic worm or mollusc is a tension-resisting system of fibre-reinforced connective tissue and active muscle. The solid skeletons of swimming and burrowing fishes and arthropods are systems of short compressive struts (Principle 1ci) articulated by a variety of tensile ligaments and membranes of varying degrees of rigidity and allowable strain. Large size correlated with more rapid locomotion has been attained by fishes, whales, turtles, medusae, cephalopods and colonial pelagic tunicates.

The evolution of structures of least weight has come about in terrestrial plants and animals. Their bodies are not supported by an ambient medium whose density is near that of cellular material, and gravity alone renders dangerous a calcified branch that is a 'safe' structure under calm water (Principle 3b). Weight reduction comes with volume reduction, and volume reduction of either tensile or compressive components to a minimum leads to optimization (Principle 1). The longer a compression element is, the greater must be its diameter to avoid failure by buckling (Principle 1ci). It is not surprising that those organisms spanning the greatest spaces or sustaining the highest relative wind velocities have greatly reduced proportions of compressive materials (Principle 3). At their best, these systems have design features that accentuate this optimization: the fibre reinforcement of the palm frond; the hollow bones and feathers of birds; the thin, light exoskeletons of insects (Principle 2a). Here again, running or flying speed may vary with size, but because the pull of gravity increases proportionally to mass, while muscular strength increases only proportionally to cross-sectional area, the size of running and flying animals is limited: the mass to be moved increases more rapidly than the strength to move it. Non-mineralized insect cuticle apparently supported its greatest loads in the dragon-fly-like beasts with half-metre wing-span of the Carboniferous Period. Specialization of other properties has led insects to remarkable manoeuvrability in rapid flight and has led some millipedes to a slow but powerful wedge-type of burrowing. In both these situations, the reduction of compressive elements is such that these elements do not touch one another but are suspended in a tensile web or membrane (Principle 3c). This describes a useful concept in engineering and architectural design promulgated by R. Buckminster Fuller who calls the phenomenon 'tensegrity' (MARKS, 1960).

The most extreme case of weight minimization among land animals is exactly described by the limiting case in Maxwell's lemma (Principle 1cii) that calls for the complete elimination of one mode, compression in this case. The orb-weaving spider relies on rigid elements in her habitat to support her purely tensile web. Silk threads have the strength to support the spider and her prey and the visco-elasticity to resist the shock of impact of flying or wind-blown bodies larger in size and weight than prey (Principle 6a). What is perhaps most remarkable in this extreme case is that rather than this animal being the fleetest, farthest ranging one, it is equivalent to the sessile aquatic plankton catchers that remain motionless and depend on the flow of water to bring food and gametes to them. Because of the lack of buoyant support by an aerial medium, a suspension-filtering net on land requires a totally competent neuromuscular host to perceive the compressive structure of the environment, to lay the net appropriately and to capture and kill the large-bodied animal prey (aerial phytoplankton are perhaps too scarce for large animals to live on). So a highly tuned 'higher' arthropod, able to secrete some of the best designed fibre systems in the natural world, is seen to be equivalent, ecologically, to a sea fan.

Multicellular macroscopic plants have remained largely nonmotile but have come to occupy nearly all the same habitats in the lighted biome that animals do. In rushing waters and pounding surf, algae rely on tensile properties of congruent cell walls in resisting the drag of the water (Fig. 7.8c). They also avoid high drag forces generally by being flexible and bending with the current (Principle 6). On land, as plants replaced hydrostatic water with all-purpose composite materials composed of cellulose and lignin, they have attained great height via slender trunks in tropical forest trees (Fig. 7.8a) and great horizontal breadth in the live oak (Fig. 7.8b) and thus, command of large volumes of the environment. Optimization in plant structures is best seen in the fibre reinforcement of individual cell walls (Principle 5) and whole organs and even organisms (most large monocotyledons), and in the placement of reinforcing materials in the plant (Principle

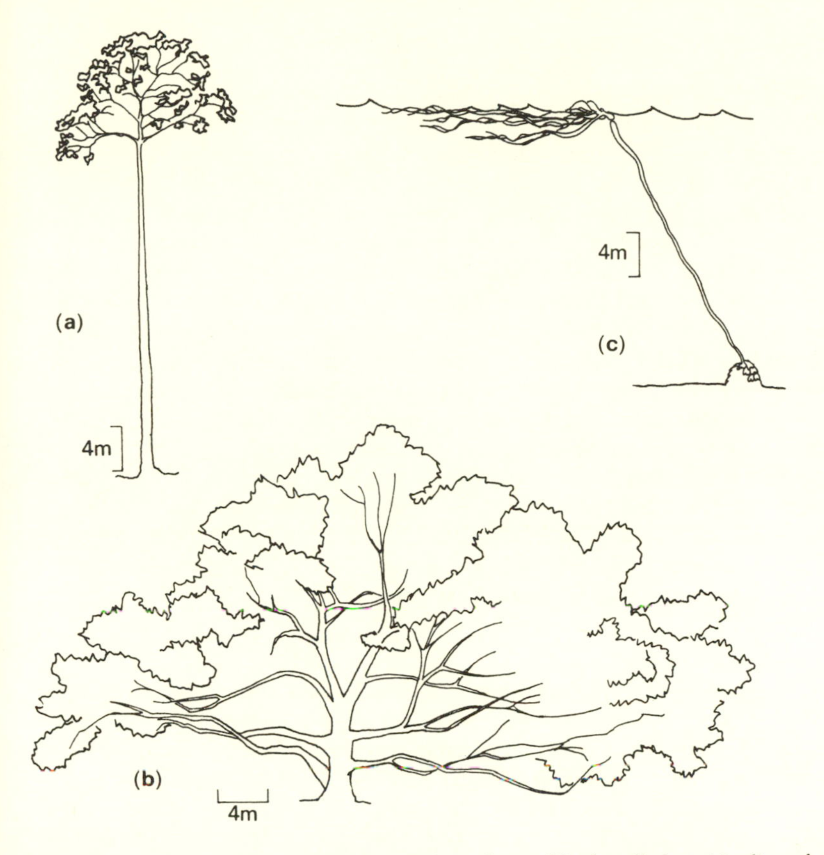

Fig. 7.8 (a) A tall, slender tree from a tropical rain forest, (b) a broadly-branching live oak, *Quercus virginiana*; (c) a giant kelp, *Nereocystis*, whose gas-filled float keeps the fronds afloat thus putting the stipe in continuous tension.

2a) and the orientation of macromolecules in the materials themselves (Principle 4a). Wind resistance is discussed in Chapter 8.

In this pocket survey, the emphasis has been on the large and the swift. This is justified by the intuition that organisms in performing extreme mechanical tricks can be expected to have the most extreme–and therefore most easily perceived and studied–mechanical features. The giant dragon-fly-like insects of 0.5 m wing span, mentioned above, have been extinct for a hundred million years, the great dinosaurs died off 70 million years ago and the largest birds and land mammals disappeared one million years ago. It appears that evolution has led most plant and animal taxa in other directions as well. The world of so-called meiofauna is among the most diverse and least understood. Tiny animals, less than 1 mm diameter, that live in soil or among sand grains in the sea floor, and small plants that colonize new areas of exposed earth have their own special mechanical problems and design features that are every bit as extreme as those of the great, the light and the fleet.

7.4 Fluid Support Systems in Plants and Animals

Throughout all organismic supportive systems there are two compressive modes: one based on fluid material and the other on solid material.

Organisms whose compression resistance is fluid are cylindrical or subcylindrical (Fig. 7.9). Some can vary their volume and some cannot. Some have great longitudinal extensibility and some have not. Some operate at high internal pressures and some do not. Obviously there is not a single set of optimum design features that can prescribe the structural requirements of all these systems. What all fluid systems have in common is a centrally located fluid volume under pressure. The pressure is generated by tensile muscular forces in animals and by compressive turgor forces in plants. In systems to be discussed here, the fluid is water, so the discussion will be one of the patterning of tensile materials.

The major structural proteins (collagen, keratin) and polysaccharides (cellulose, chitin) of living things are straight, long-chain polymers. These are commonly organized into bundles of parallel macromolecules that are recognizable as filaments, fibrils or fibres in light and electron optical instruments. For this reason we think of most structural biomaterials as fibre-reinforced composite materials, as discussed in Chapter 4. The bodies made up of these materials will thus be treated hereafter as fibre-wound cylinders.

7.4.1 High-pressure Worms

The most nearly cylindrical animal body with the highest internal pressure is that of the nematode *Ascaris*, parasitic in the guts of man and some of his dearest domesticates. Nematodes generally are round in section and have thick, multilayered cuticles containing fibrillar collagen. BIRD and DEUTSCH (1957) described nine layers in the cuticle of *Ascaris lumbricoides.* Three of these layers are composed of collagen fibres. In each layer the fibres are parallel to each other and wrap helically around the animal. Successive layers are alternately right and left-handed helices. This array of fibres will hereafter be referred to as a crossed-fibre array.

Internal pressures measured in *Ascaris* by HARRIS and CROFTON (1957) during rhythmic locomotory movements varied from 6.5 to 20 kN m^{-2}. A maximum value of 30 kN m^{-2} was attained and held for a few seconds when the worm's front end was tightly coiled and the tail strongly contracted. They recorded changes in length of the worms up to 10–15% of resting length. In discussing the possible means by which this cuticle could play a part in the control of body shape in *Ascaris*, Harris and Crofton considered collagen fibres to be 'practically inextensible'. They observed that the fibre angle in *Ascaris* cuticle at rest is $75°30'$ and, in appreciation of Principle 5b above, they realized that in a closed cylinder wound at such an angle, the contraction of longitudinal muscles would tend to make the cylinder short and fat, thus tending to decrease the volume (to the right of the volume maximum in the curve in Fig. 7.6c). Actual changes in volume can only be regional: a contraction in the tail end will cause an increase in the volume of the relaxed head end of the worm. Since even resting pressures are high, the worm is always circular in section. The collagen fibres in the cuticle put strong limitations on direction and magnitude of strain due to increase in

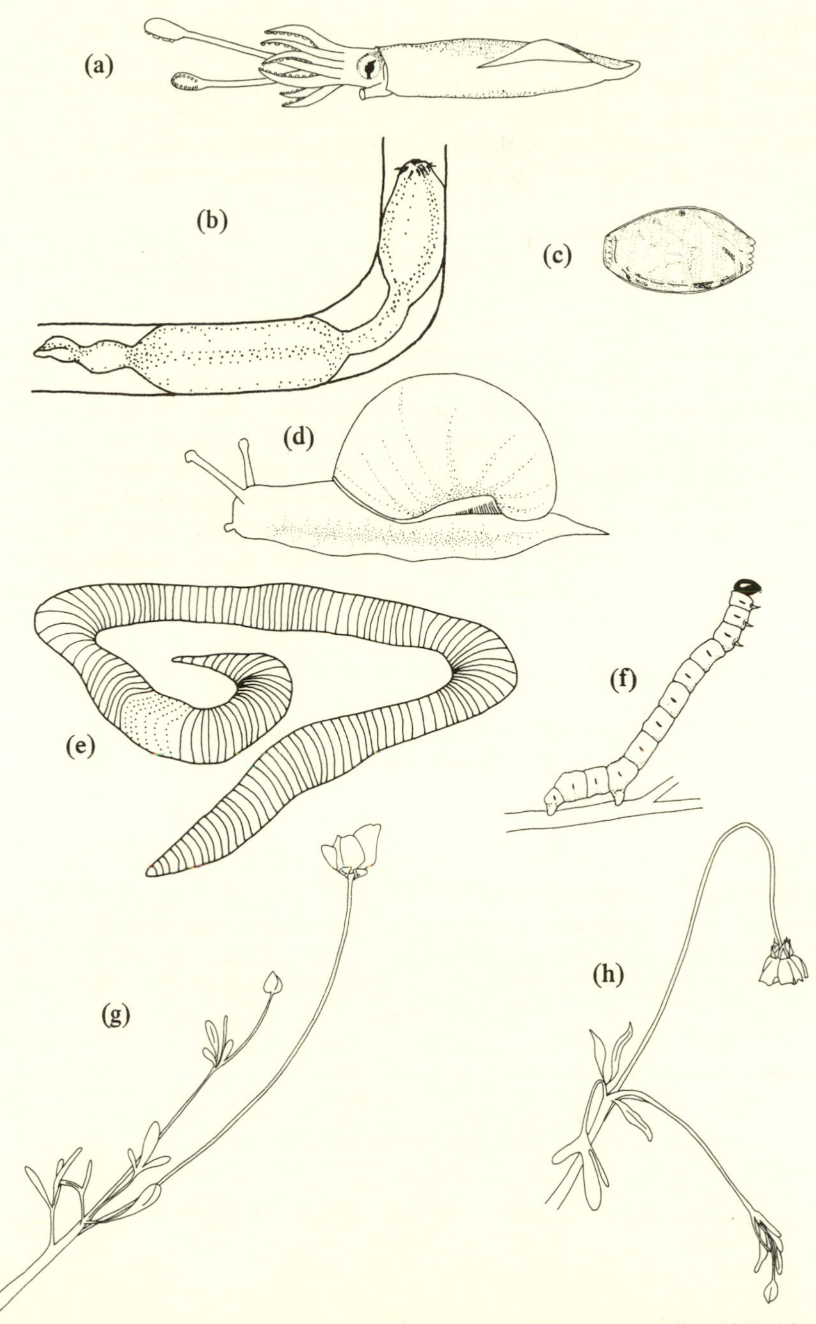

Fig. 7.9 Some organisms with fluid compression elements in their support systems. **(a)** Squid, **(b)** echiuroid worm, *Urechis caupo* in its burrow, **(c)** salp, **(d)** land snail, **(e)** earthworm, **(f)** caterpillar, **(g)** buttercup, **(h)** wilted buttercup. Not to scale.

internal pressure. Following Principle 5b as illustrated in Fig. 7.6c a fibre angle above 55° will cause-elastic recovery from increased pressures, and make the worm longer and thinner, thus antagonizing the action of longitudinal muscles. Tensile patterning in nematodes involves their only set of locomotory muscles, longitudinal ones, and a passive elasticity in the cuticle whose direction of action is controlled by the fibre angle of the stiff, helically wound collagen fibres.

7.4.2 Low-pressure Worms

Unlike the thrashing, high-pressure nematodes, turbellarian and nemertean worms move about by crawling on a mucus trail on a broad, flat, ventral surface or by muscular swimming that depends on the dorso-ventral flattening that is characteristic of these worms. The worms lack a cuticle but they do have a basement membrane to which their epithelial cells as well as their longitudinal and circular muscles attach. The basement membrane consists of a fabric of layers of 'reticulin' fibres that are in crossed fibre array around the worm's body. CLARK and COWEY (1958) in a classic paper that generated much of the argument for Principle 5b, studied the changes of length, fibre angle and cross-sectional shape of three turbellarian and six nemertean species, and they sought evidence bearing on the possible function of helical fibres in the control of body shape in these worms.

They showed that for a closed, fibre-wound cylinder of elliptical cross-section at fibre angles near 55°, the theoretical range of length that the cylinder may have is related to fibre angle and to the ellipticity $\left(n = \dfrac{\text{major axis}}{\text{minor axis}}\right)$ of the cross-section when the cylinder is at zero pressure (n_r). They then show that the maximum volume of a cylinder bound by one turn of a helix is

$$V = \frac{2n_r}{(n_r)^2 + 1}$$

By measuring the major and minor axes of worms' cross-sections, they could assign a value for the volume of worms of each species that could be represented by a horizontal line drawn on the curve in Fig. 7.6c. The abscissal values of the points where the horizontal line intersects the curve determine the maximum

Table 7.1 (from CLARK and COWEY, 1958; courtesy of the *J. exp. Biol.*)

Species	Theoretical extensibility	Measured extensibility
Rhynchodemus lineatus	3	3
Geonemertes dendyi	3-4	3-4
Amphiporus lactifloreus	6-7	5-6
Lineus gesserensis	6-7	5-6
L. longissimus	10	9
Cerebratulus lactens	10	2-3
Dendrocoelum lacteum	20	2-3
Malacobdella grossa	20	2-3
Polycelis niger	20	2

and minimum fibre angles and thus the maximum and minimum lengths of the worm.

They then measured the maximum length of an anaesthetized worm stretched to its limit and the minimum length of another, unanaesthetized, specimen after it had been dropped into formalin and allowed to contract. The actual extensibility of each species differed from the theoretical extensibility (Table 7.1) and the authors identified those design features observed in worms of each species that may account for the discrepancies. *Geonemertes* and *Rhynchodemus* are terrestrial genera whose theoretical extensibility is low because they depart very little from a circular shape, hence their position near the maximum volume on the curve in Fig. 7.10. Since their observed extensibility is the same as their theoretical extensibility, one is justified in saying that the limits to the range of

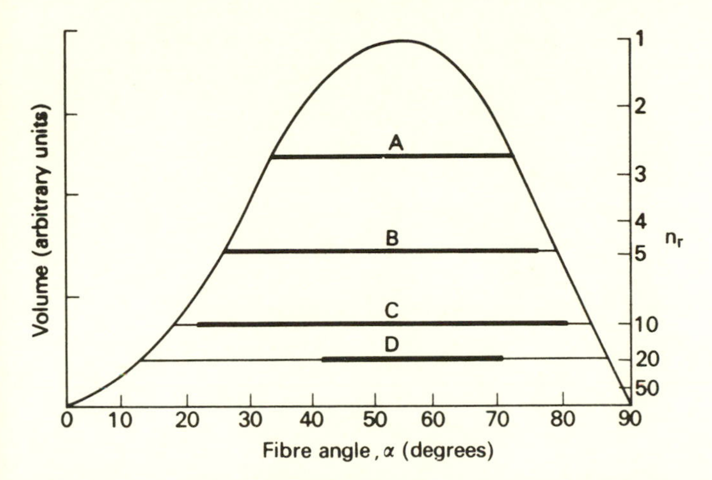

Fig. 7.10 Ability of worms to acquire elliptical shape (ordinate) versus their ability to attain maximum or minimum length (abcissa) (from CLARK and COWEY, 1958). A, *Geonemertes, Rhynchodemus*; B, *Amphiporus, Lineus gesserensis*; C, *Lineus longissimus*; D, *Malacobdella, Cerebratulus, Polycelus* and *Dendrocoelum.* Explained in the text. Courtesy of the *J. exp. Biol.*

length and sectional shape in these worms are set by the pattern of tensile fibres in their basement membranes.

Amphiporus and *Lineus gesserensis* are marine nemerteans of elliptical cross-section at rest, whose extended lengths are 5 to 6 times their contracted lengths but, because they never attain a circular sectional shape in the contracted state, they are not able to attain the maximum longitudinal contraction or, therefore, maximum theoretical extensibility. *Lineus longissimus* is similar but even more elliptical at rest and attains circular cross-section at neither the fully extended nor the fully contracted state. Clark and Cowey suggest that these worms fail to attain maximum extension because the decrease in area of the basement membrane that accompanies extreme changes in length compresses the bases of epidermal cells to such an extent that the cell bodies hydrostatically oppose the attainment of extreme shortening or extension. And finally, the marine nemerteans, *Malacobdella* and *Cerebratulus* and the fresh water turbellarians

Polycelis and *Dendrocoelum* are the flattest of all, thereby having the greatest theoretical extensibility. However, dorso-ventral musculature keeps them flat, and they have the smallest extensibilities of all the worms studied. As Clark and Cowey state, creatures that depend on cilia for locomotion benefit from a high surface to volume ratio such as is attained in these permanently flattened worms, perhaps at the expense of the benefit of high extensibility.

7.5 Open, Extensible Cylinders: Sea Anemones

An open cylinder whose wall is flexible enough to collapse of its own weight sounds more like an empty stocking than it does an organism. Nevertheless the gut cavities of sea anemones and some other cnidarians and synaptid holothurians allow, by controlled opening of the mouth (or the anus of the holothurian) and ingress or egress of sea water, an enormous range of body sizes and shapes in these organisms that are not available to creatures with hydrostats of constant volume.

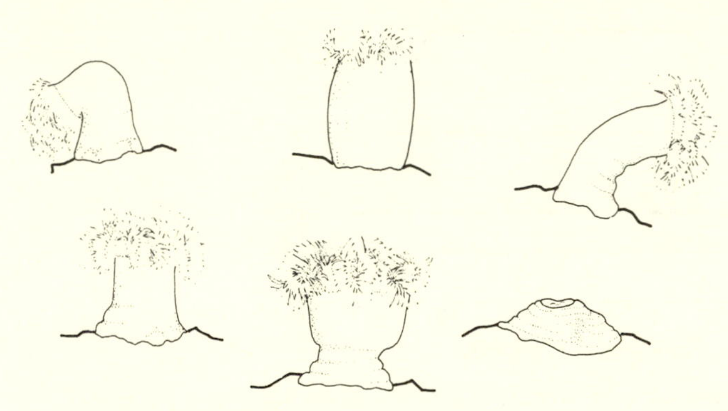

Fig. 7.11 Changes in volume and shape in the normal posturing behaviour of the anemone *Metridium senile* (drawn from photographs in BATHAM and PANTIN, 1950).

The anemone *Metridium senile* can open its mouth, contract its muscles and shrink down to a wrinkled blob, or it can take in sea water and expand into a vertical column four times taller than wide (BATHAM and PANTIN, 1950). At sizes between these extremes, the columnar polyp can sway and sweep the sea floor with its tentacular crown; it can contract only its crown, or contract or expand any portion of the column (Fig. 7.11). It ingests food and egests undigested materials by passing peristaltic waves along the column. The hermit crab-riding anemones (e.g. *Calliactis parasitica* (ROSS, 1960)) somersault from rock to shell. *Stomphia coccinea* swims by bending its column smartly from side to side while *Boloceroides* swims by flapping its many long tentacles simultaneously in a medusa-like fashion (ROBSON, 1966). *Peachia* and some other anemones burrow into sediments. All of these activities are accomplished by an open, cylindrical body with circumferential, radial and longitudinal muscles on the walls and on longitudinal mesenteries protruding radially inward from the walls. Postural movements, aside from quick withdrawal contractions and swimming are very slow—a single sweep may take hours. These movements are supported by very low hydrostatic pressures, rarely exceeding 110 N m^{-2} (BATHAM and PANTIN, 1950).

The supportive material in the wall and mesenteries of *Metridium* is mesoglea—a visco-elastic composite of fibrillar collagen in an aqueous matrix of undetermined structure. In what follows, 'fibres' are distinguished with the light microscope and are thought to comprise a large number of parallel 'fibrils', resolved by the electron microscope. Cells make up less than 1% of its volume. GOSLINE (1971*a*) has described the mesoglea of large specimens of *Metridium* (20 cm diameter) on the basis of polarized light and electron microscopic study as being a two-layered system of collagen fibrils. There is a crossed-helical array of fibres in the outer layer and a denser array of circumferentially and radially oriented fibres in the inner layer (Fig. 7.12). Electron micrographs show large electron-lucent spaces between fibrils, while polarized light micrographs show that there are alternating zones of lower and higher concentrations of fibres that appear as concentric rings (growth increments?) in the outer layer.

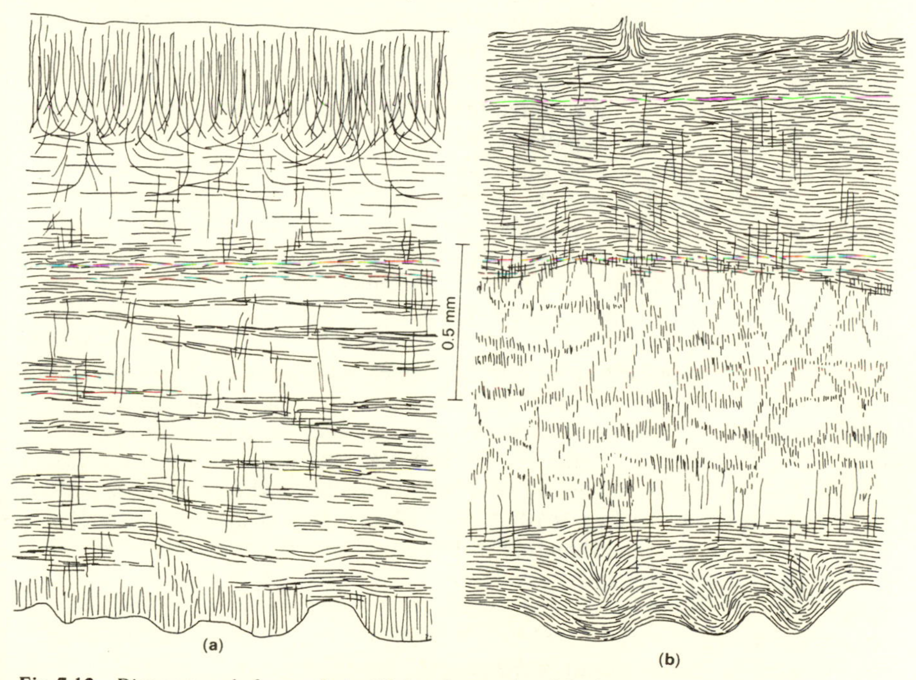

Fig. 7.12 Diagrams made from polarized light micrographs of the fibre array in mesoglea of *Metridium*. (**a**) longitudinal section; (**b**) transverse section. The inner layers are at the top of the figures. Contrast is reversed; birefringent fibres are depicted as lines (after GOSLINE, 1971*a*).

In his study of mechanical properties of mesoglea of *M. senile*, ALEXANDER (1962) measured creep and creep recovery. He found that the body wall could recover from long-range postural deformations by the elastic properties of the mesoglea to a 'unique set of dimensions' without the aid of muscular force. The very long time (10^5 s) required for the complete recovery from such deformations—the viscous interaction—is the property of mesoglea that determines its function in slowly posturing, low pressure anemones.

The densely packed inner layer of circumferentially oriented fibres is responsible for mechanical anisotropy of the mesoglea. Results of stress-relaxation experiments (GOSLINE, 1971*b*) show that at short times of 10 s or less there is little difference between the elastic moduli in the two directions (Fig. 4.10) but that at longer times of 0.5 to 2 hours, the modulus in the circumferential direction is more than an order of magnitude higher than that in the longitudinal direction. Also, the equilibrium modulus at times greater than 24 h is significantly higher in the circumferential direction. Although these curves can be expected to be shifted lower and displaced to the right (longer times) at the 13°–15°C temperatures the animals normally experience, there is no reason to expect the differences in magnitude of modulus to change. Gosline interprets this difference in terms of the shear stresses that should occur when circumferentially oriented fibres are caused to slide past one another by stress in that direction. He concludes from the fact that collagen affects the equilibrium modulus of mesoglea that the collagen fibrils are not free to slide past one another unimpeded, but that they are linked directly to long, randomly coiled matrix molecules. The functional expression of this anisotropy is understood in terms of the effect it has on the force required to expand the column in the two directions. Hoop stresses being twice longitudinal stresses, as in Principle 5b, one would expect a cylinder with a wall of isotropic properties to expand radially at twice the rate of longitudinal expansion. By having a higher circumferential modulus due to the preferred orientation of collagen fibrils in this direction, the column wall of large *Metridium* is allowed to expand in length faster than in radius for any pressure increment without the use of muscular force.

CHAPMAN (1953) recorded a higher breaking stress for mesoglea of *Metridium* in circumferential (4.4 N m^{-2}) than in longitudinal (2.9 N m^{-2}) direction, and this is consistent with Gosline's observation of higher modulus in the hoop direction.

While there is ample evidence that the mechanical properties of mesoglea collagen are similar to those of rat-tail tendon collagen, the properties of the materials as a whole are remarkably different. Rat tendon has an elastic modulus of 10^9 N m^{-2}, whereas the plateau modulus of mesoglea, wherein it takes part in the slow posturing behaviour, is 10^5 N m^{-2}. The message is clear: one cannot predict the properties (and therefore the function) of a material simply by knowing that it is predominantly collagen. We currently appreciate too little of the range of possible cross-linking mechanisms and the nature and role of solutes and solvents in the interfibrillar matrix of collagenous connective tissues. Here lie exciting discoveries to be made in the next few years.

7.5.1 Hydra and Other Polyps

There are polyps that are simpler in form than sea anemones in that they lack internal structures such as mesenteries and have fewer tentacles. Hydra's linear dimensions are a hundredth of those of large *Metridium* and its mesoglea is similarly populated with collagen fibrils that are oriented parallel in areas (DAVIS and HAYNES, 1968) that may be the orthogonal array of reinforcing fibres reported by HAUSMAN and BURNETT (1969).

7.5.1 *Hydra and Other Polyps*

Hausman and Burnett isolated mesoglea from hydra by rapidly freezing and thawing the animals and then washing away the lysed cells with distilled water. They note that the isolated mesoglea can be stretched up to 3 times resting length. Although they did not estimate mechanical properties quantitatively, they noted that a stretched mesoglea snaps back to resting length instantly as it is released from the deforming force. This instantaneous return from long range deformation is a property unknown to date in collagen-dominated materials, but it indicates that hydra, as well as sea anemones, has an equilibrium shape and size that can be maintained without muscular or ciliary work. Another advantage of having a particular resting shape is that it does not require proprioceptors and other neuromuscular apparatus in order to identify and reattain a basic shape following deformation.

But what of the tentacles of hydra? In brown hydra (e.g. *Pelmatohydra oligactis*) the distal half of each tentacle can extend to 20 times its contracted length. Preliminary observations (Wainwright, unpublished) with the electron microscope of sections of extended tentacles show that (1) the lumen is circular in cross-section, (2) longitudinal muscles in the ectoderm are closely packed, (3) no circular muscles are visible, (4) there are fibrils in the mesoglea that are thicker than those shown by DAVIS and HAYNES (1968) of the collagen fibrils in the mesoglea of the body column, and (5) these fibrils lie in helical array at very low angles, probably less than 10° to the tentacular axis. At the moment there is no information about the orientation of these fibrils in the contracted tentacle nor about the volume or internal pressure in the tentacular lumen in either extended, relaxed or contracted state.

It is generally supposed that, hydra being hydrostatic, tentacles may be extended either by the contraction of circular muscles in the tentacles so that they become long and thin, or that contraction of any muscles in the column can force water from the coelenteron into the tentacle, blowing it out. The former explanation is unlikely to be true if there are no circular muscles in the tentacles. The latter, seen in terms of hoop and longitudinal stresses (Principle 5b), predicts that increased internal volume and/or pressure should blow the tentacle out in diameter twice as fast as it extends longitudinally. Because this does not happen, one may suppose the crossed-helical array of fibres in the mesoglea play a role in maintaining the cylindrical shape of the extending tentacle. The curve in Fig. 7.6c predicts that such a constraint is possible in a cylinder that is becoming longer and thinner only so long as the fibre angle exceeds 55°. At lower fibre angles, such as the observed one, less than 10°, the volume must be decreasing And, for all we know, it does so in hydra tentacles.

One assumption inherent in the model portrayed in Fig. 7.6c is that the fibres are inextensible. Although we do not know the range of fibre angles or the extensibility of the fibrils in mesoglea of hydra tentacles, the 20-fold extension of the tentacle itself indicates that mesogleal fibres must either be tremendously extensible themselves or else they must slide past one another. This reduces the restraints that helically wound fibres have on changes in diameter and still allows that the fibrils in hydra tentacles have a shape-constraining function to oppose the imbalance of hoop over longitudinal stresses.

The source of the force causing tentacles to elongate is still not known and the concept of 'resting shape' may be applicable. When the mesoglea of a tentacle is isolated by rapid freezing and thawing, the resting shape it takes on is an extended shape. If these preparation methods have not altered the mechanical properties of the mesoglea, it is a plausible hypothesis that the elastic strain energy of the mesoglea, whose resting shape is a long, thin tube, provides the restoring force for the tentacle following the relaxation of the longitudinal muscles.

The tentacles of some hydrozoan and scyphozoan polyps have a single row of large, vacuolated endodermal cells completely filling the central core of the tentacle. D. M. CHAPMAN (1970) has ascribed the extending force in the filled tentacles of the larva of *Aurelia aurita* to viscoelastic properties residing in the tentacle. Tentacles of these polyps have longitudinal but no circular muscles and an array of minute mesogleal fibres (visible in electron micrographs) that has no obvious degree of preferred orientation at any degree of tentacular extension. Following contraction of the longitudinal muscles, and presumably when the muscles are totally relaxed, the tentacles extend by 2 to 4 times the contracted length in 10 s and then more slowly, taking 150 s to extend another contracted-length increment. If the tentacle is held contracted for more than a few seconds, it extends only at the slower rate.

As Chapman points out, there seems to be a viscoelastic component with an initially high elastic modulus acting at short times after deformation, whose modulus declines at longer times. This is similar to the phenomenon observed in mesoglea of *Metridium* by ALEXANDER (1962) and GOSLINE (1971*b*). Gosline's model is described and discussed in Chapter 4, and we feel that the model can explain the extension of scyphistoma tentacles observed by Chapman. It presumes the presence of an amorphous polymer, so far unobserved in electron micrographs, that has the requisite properties. It allows agreement with Chapman's speculation that the strained polymer that accounts for this behaviour ". . . might be a carbohydrate bridging and coating the collagen fibrils".

7.5.2 Medusae

Medusae are cnidaria that swim by the pulsatile contractions of circumferential and sometimes radial muscles on the lower surface of the bell at the rate of about one beat per second. The muscular contraction bends the bell (Fig. 7.13), stretching the outer surface and compressing the bulk of the material in the bell.

G. CHAPMAN (1953, 1959) describes the optically visible fibre systems in the mesoglea of some medusae. There is a concentration of fine fibres oriented randomly in a plane parallel to the upper surface in the mesoglea of *Chrysaora mediterranea* and *Pelagia noctiluca* (Fig. 7.14). These fine fibres appear to be branches of much thicker fibres or fibre aggregates that are oriented at or near right angles to the network near the upper surface and that connect it with a less well developed network near the lower surface. GLADFELTER (1972) describes in *Polyorchis montereyensis* an array of fibres, visible by light microscopy, that extends radially between basement membranes underlying the exumbrellar and subumbrellar epithelia. Each fibre divides into a fan or cone of tapering, further dividing, branches toward the subumbrellar surface. A radial tensile force on one microscopic fibre will thus be dissipated over a wide area of the subumbrellar surface.

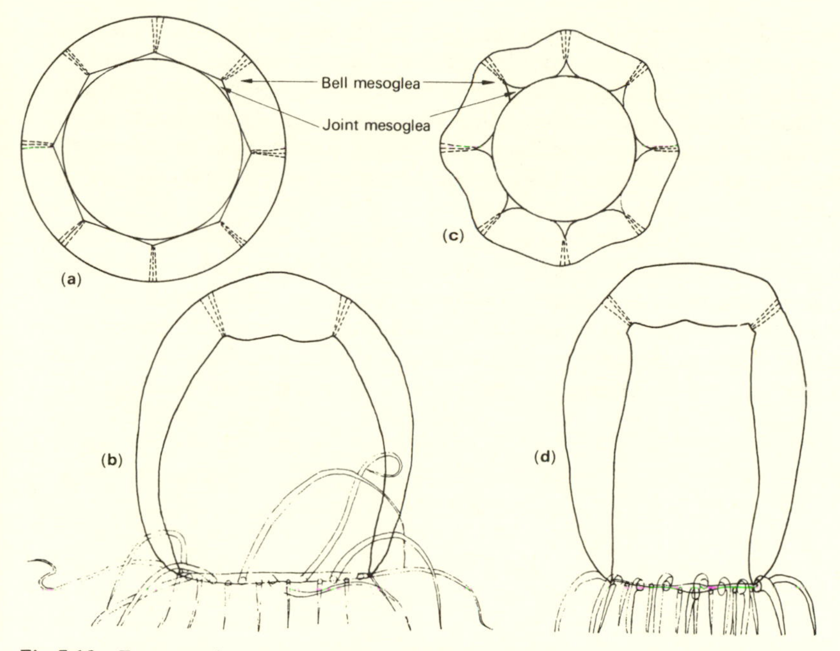

Fig. 7.13 Transverse (**a** and **c**) and longitudinal (**b** and **d**) sections of the bell of the hydromedusan *Polyorchis montereyensis* at rest (**a** and **b**) and at the end of a power stroke (**c** and **d**). Longitudinal wedges of highly deformable, nonfibrous joint mesoglea provide fulcra around which the more rigid, fibrous bell mesoglea is bent. Areas of bell mesoglea with the greatest density of fibres (dashed lines) undergo least thickening in a power stroke (redrawn from GLADFELTER, 1972).

Although we lack measurements of the tensile properties of individual fibres, they show moderate elastic extensibility. Gladfelter shows that the zones of bell mesoglea with the greatest density of these fibres (Fig. 7.13) undergo the least thickening during a power stroke. Gladfelter concludes that compressive cir-

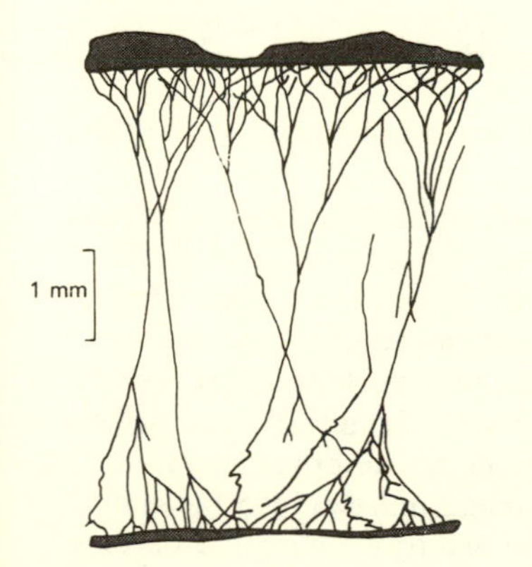

Fig. 7.14 Array of microscopic fibres in a hand-cut section of the bell of the medusa *Pelagia noctiluca*. Outer (exumbrellar) surface is at the top of the diagram (from CHAPMAN, 1953).

cumferential forces are more important than radial tensile forces in the return to resting shape. From the added information about the highly oriented microscopic elastic fibres, we suggest that the function of these fibres is to transmit radial forces. Also, their elastic modulus, whatever it is, and extensibility in tension during muscular contraction probably play a greater role in the return stroke, in terms of work done, than do the circumferential compression forces on the mesogleal matrix.

At the ultrastructural level, a highly dispersed three-dimensional network of fibrils, presumably collagen, has been found in *Aequorea* mesoglea (D. CHAPMAN, 1970) and in mesoglea of the siphonophore *Hippopodius hippopus* (MACKIE and MACKIE, 1967). G. CHAPMAN (1959) studied the shrinkage of mesoglea from *Pelagia* when heated in sea water and concluded from the 80% shrinkage in the bell diameter, from the change to a more gelatinous consistency and from electron

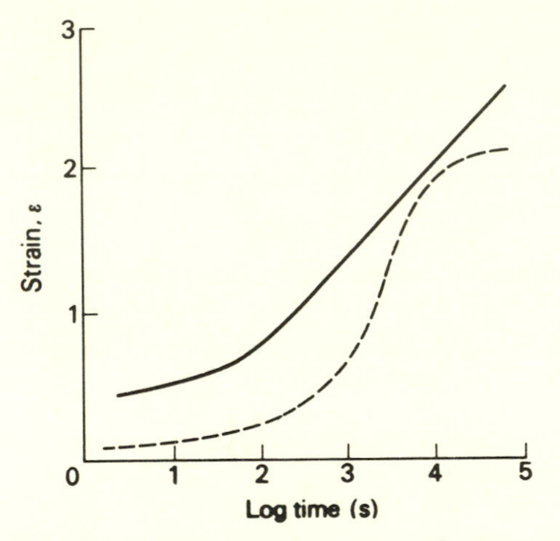

Fig. 7.15 Creep curve for times up to 10^5 s for mesoglea of the medusa *Cyanea* (solid line) and the anemone *Metridium* (dashed line). Note that the curve for the anemone levels off at long times indicating an equilibrium modulus and a maximum extension that is to be expected from a polymeric gel in which the molecules are in a cross-linked network (ALEXANDER, 1964; courtesy of the *J. exp. Biol.*

micrographs of normal material, that collagen is the most abundant solid in this predominantly fluid mesoglea.

Creep curves of longitudinal strips of mesoglea from the medusa *Cyanea* and the anemone *Metridium* are shown in Fig. 7.15 over times from less than 10 to 10^5 s (ALEXANDER, 1964). He notes that at longer times and at strains greater than 2 the strain against the log of time never attains an equilibrium modulus or a final length that a cross-linked material would be expected to have, and he concluded that the properties he measured were those of simple polymeric gels.

Jellyfish mesoglea has a resting shape that is probably never realized in the life of the organism. CHAPMAN (1959) cut the circular muscles by 8 radial cuts on the lower surface of *Pelagia* and noted that the bell took on a more flattened form than that of the relaxed, intact animal. This observation indicates that medusa mesoglea in living animals is a stressed material and that the residual stress in this material functions in the maintenance of the resting shape of the animal throughout life. The mechanisms that maintain resting shape are specialized macromolecular design features that have evolved via natural selection from materials that were presumably even simpler, more homogeneous and less mechanically specialized.

Some medusae do not retain the smooth bell-like outline throughout swimming. From ciné films of swimming *Polyorchis*, GLADFELTER (1972) reported a 16 to 34% increase in thickness of the bell mesoglea concurrent with a 56% decrease in diameter of the inner surface of the bell during each power stroke. He found no measurable change in length of the bell. Therefore, there must be a structural mechanism resisting change in length but allowing change in thickness of the bell.

The contractions of circumferential muscles in the power stroke of *Polyorchis* results in a regularly folded shape of the mesoglea (Fig. 7.13c). This folded shape comes about because of 8 adradial columns, having triangular cross-sections, of joint mesoglea (GLADFELTER, 1972). Joint mesoglea has a very dispersed array of fibrils seen with electron optics and has no microscopically visible fibres. The material is a gelatinous solid with a high degree of elastic extensibility and an apparently very low modulus. Since the circular muscles are attached to the bell mesoglea only at the eight interradii, the eight interradii are pulled closer together, causing the bell mesoglea to bend around the columns of joint mesoglea rather than to undergo circumferential compression only. The advantage of this folding of the bell mesoglea is that less force is required to bend the bell mesoglea and to decrease the subumbrellar radius than would be required to compress the mesoglea. Since the force of a power stroke is proportional to the volume of water expelled, and since the volume of this cylinder of water varies with the square of the radius or circumference, any such economy of muscular work per unit decrease of radius is obviously advantageous.

7.5.3 *Tube Feet*

The hydraulic, contractile tube feet and radial canals in the water-vascular systems of echinoderms perform a wide variety of locomotory, burrowing and feeding activities. Tube feet of some starfish (Asteroidea) and cake, heart, and sea urchins (Echinoidea), and sea cucumbers (Holothuroidea) have terminal suckers. These tube feet function primarily in tension, allowing stepping in locomotion and the forceful grip and pull that some species are known to exert in opening bivalved molluscs. Tube feet of brittle stars (Ophiuroidea) are tapering and bear no suckers (Fig. 7.16). These function in burrowing by bending and by pushing sand grains in a backward direction, promoting forward locomotion. The long, slender tube feet of feather stars (Crinoidea) and some ophiuroids are extended perpendicularly to ambient flow from an extended arm. From this position they catch suspended food particles by flicking inwards toward the ambulacral groove (NICHOLS, 1960).

WOODLEY (1967) has described the hydrostatic tube foot system in some brittle stars. From the ring canal surrounding the gut in the disc, a radial canal extends into each arm where it is embedded in, and protected by, each of the large 'vertebral ossicles' that comprise the compression component of the arm skeleton. From the radial canal extend pairs of lateral canals and each of these connects to

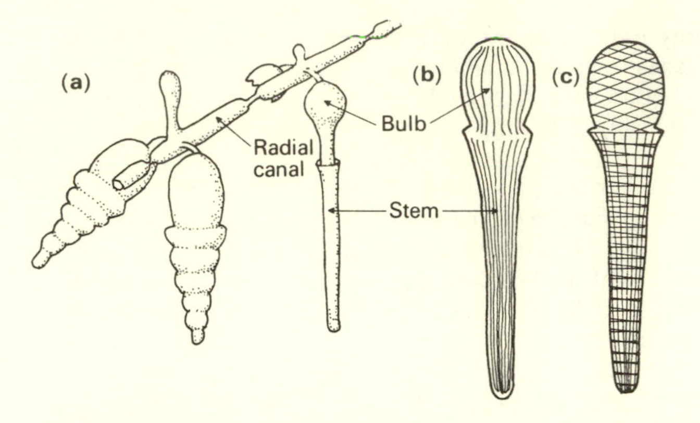

Fig. 7.16 (a) Ophiuroid tube feet and the connecting radial canal. The two at the left are contracted; the one on the right is fully extended. Orientation of collagen fibres in the (**b**) outer and (**c**) inner connective tissue layers of the tube foot of *Amphiura filiformis.* Fibre angle in the bulb is ca 73°; in the stem it is ca 85°. (**a**, adapted from BUCHANAN and WOODLEY, 1963; courtesy of *Nature*; **b** and **c**, drawn from information in WOODLEY, 1967).

the bulb of a tube foot (Fig. 7.16a). There is a muscularly operated valve that, when closed, hinders the flow of fluid from the bulb to the radial canal. The tube foot has an extended conical shape with the bulb at its base, and it is a closed tube of connective tissue lined with longitudinal muscles. The muscle layer of the bulb is thicker than and discrete from that of the stem.

The tube foot protracts when bulb muscles contract (with the valve closed), forcing fluid into the stem. The tube foot is retracted when stem muscles contract, inflating the bulb. Extreme retraction occurs when all bulb and stem muscles contract with the valve open: the fluid is displaced into the radial canal. Protracted tube feet bend and wave when stem muscles on the concave sides contract. Throughout all movements, there is no discernible change of diameter of the tube foot.

The integrity of the tube foot lies in the connective tissue sheath. Woodley isolated the sheaths from the feet of *Amphiura filiformis* by partial maceration in KOH and glycerine solution. Between crossed polaroids, the fibrous sheath appears as a double layered structure in the stem whose inner layer is continuous with the single layer over the bulb. The outer stem layer contains longitudinal fibres (Fig. 7.16b), whereas the inner stem layer and bulb sheath contains crossed helical fibres (Fig. 7.16c). In the bulb, at rest, the helical fibre angle is about 73°, and in the stem it is 85°. The fibres can be removed by persistent trypsin treatment from a non-birefringent matrix material that is more collapsible than the entire sheath.

In an anaesthetized animal, or in a tube foot isolated from one, the relaxed, resting shape is the protracted shape. Seen in the context of fibre-reinforced cylinders (Fig. 7.6c), these large fibre angles become even larger as longitudinal muscle contraction causes shortening of the tube foot. During retraction with the valve closed, the modulus of the tube foot wall that resists explosion (hoop stresses) will rise with the sine of the fibre angle. Simultaneously the sheath-enclosed volume will decrease. In the retracted state, the fibre angles will be virtually 90°. Steeply angled helical fibres in the resting state confer control of changes of the tube foot's length that circumferentially oriented fibres could not. Presumably the longitudinal fibres limit the protracted length of the tube foot, but the crossed helical fibre system will help retain the smooth, non-buckled shape of the tube foot during shortening. The lower fibre angle in the stem leads to the prediction that the stem may undergo a greater amount of shortening than the bulb, but at these generally high fibre angles, the difference is slight (see Fig. 4.7c).

The radial canal is not only contractile, but its connective tissue sheath confers the property of long range elasticity on the canal. Radial canal material isolated by partial maceration returns instantly to resting shape after release from being stretched to twice resting length. The material is isotropic at rest and, like rubberlike proteins, it is insoluble in weak acids and alkalis. It resembles these proteins also in shrinking reversibly in low pH media and swelling in high pH.

We do not have comparable information on the connective tissue support of water-vascular systems of other echinoderms, but it is clear that there are important differences. All tube feet have internal longitudinal musculature and all maintain constant diameter throughout protraction and retraction. Tube feet and radial canals of the crinoid *Antedon bifida* have a single layer of circular connective tissue fibres (NICHOLS, 1960). Each tube foot of asteroids and echinoids has an ampulla at the proximal end of the tube foot and most species have a disc-shaped sucker at the distal end. The ampulla acts in an analogous manner to the bulb in ophiuroids, but it has a larger volume than the bulb relative to the stem volume.

In the asteroid *Asterias rubens*, SMITH (1946) reports from the study of paraffin sections that the connective tissue sheath is bilayered. The inner layer contains circumferentially oriented, inextensible fibres that have the staining characteristics of collagen. No elastin was found. These fibres are in rings that are separated in the protracted stem. During retraction, the stem shortens until the rings are close-packed. Thereafter the stem wall buckles into a wrinkled shape. The outer layer contains longitudinally oriented fibres that are presumed to limit the protracted length of the tube foot and to give tensile strength and tensile stiffness to the protracted foot as it acts as a pulling member in stepping locomotion and especially in pulling open bivalved molluscs.

The detailed comparative studies by NICHOLS (1959*a*, *b*; 1961) have shown a variety of tube foot types among echinoids that have highly specialized mechanical roles in burrowing and in stepping locomotion. Stepping echinoids *Cidaris cidaris* and *Echinus esculentus* have 3-layered connective tissue sheaths in their suckered tube feet in which the outer layer contains diffusely arrayed fibres and curved calcareous spicules. The middle layer contains a denser array of longitudinally oriented fibrils whose description suggests they may describe helices of very small fibre angle. The inner layers have the densest arrays of fibres, all circularly oriented. Both these species also have non-suckered buccal tube feet that lack the outer spiculiferous layer found in the suckered tube feet.

The heart urchin burrows more than its height below the surface of the sediment, and it maintains respiratory contact with the water above by the action of funnel-building tube feet that extend over this long distance. The sheath of these remarkable tube feet has two layers: an outer one of diffuse fibre array and an inner one with a dense, circumferential fibre array. It seems likely that the circumferential fibre arrays reported in asteroid and echinoid tube feet may be helical arrays with a high fibre angle. If so, they could exercise a greater degree of control over change of tube foot shape than could a purely orthogonal system.

Tube feet are semi-autonomous compartments of a large, complex hydraulic system. Such a system has the advantage of the diversity of functions and multiplicity of effectors to aid in locomotion, feeding and other activities. The inherent danger of puncture causing loss of pressure and thus loss of function in a pressurized fluid system is to a great extent averted by having many small, isolatable fluid filled compartments instead of one large one. The water vascular, haemal and perihaemal systems of the Echinodermata are the most complex and diversely functioning set of fluid systems known in invertebrates. A simpler way to arrange compartments of a larger fluid filled system is to have all compartments alike and in a serial array. This condition is called metamerism.

7.5.4 Metamerism

In the locomotion of long, thin animals the division of the musculature and the body cavity into a series of identical segments amounts to a strategic advantage over the nonsegmented state. Identical nervous messages can be transmitted along the segments causing muscles in each segment to perform a power stroke. A wave of power strokes thus controlled and generated is a simple and effective way to produce swimming, burrowing, crawling, walking or running. In an evolutionary and a developmental sense it saves time and information to have one basic plan that is repeated, rather than having a different plan for each set of legs. For those legless worms, such as the earthworm, that use their whole bodies in locomotion, it is similarly advantageous to have separate identical blocks of circular and longitudinal muscles, controlled by a segmented central nervous system.

The coelom of some polychaetes is effectively compartmentalized by complete muscular septa, but most species have incomplete septa that play some part in the operation of parapodia, the segmented appendages (CLARK, 1962). These parapodia are also diverse but in general are broad, flat paddle-shaped organs supported from within by stiff bristles (chaetae) that serve as endoskeletal rods. The coelom of each segment extends into each parapodium and there is no separation of parapodia from trunk coelom. Parapodia are waved back and forth and up and down by the action of muscles pulling the chaetae and attached to the body wall. Since each parapodium has a broad base that extends around a third of the circumference of the worm, there is extensive interruption in the generally cylindrical body plan of these worms. Figure 7.17e shows the cross-section of the polychaete *Nereis*. Note the lack of circular muscles at the level of the section diagrammed. The longitudinal muscles are restricted to cords dorsal and ventral to the parapodial bases. Polychaetes are hydrostatic animals–some much more than others–but they have developed further than the earthworm in the direction of the use of nonhydrostatically operated appendages.

To get a bit ahead of the story, the vertebrates have solid compression elements rather than fluid ones, and they are also metameric with regard to their locomo-

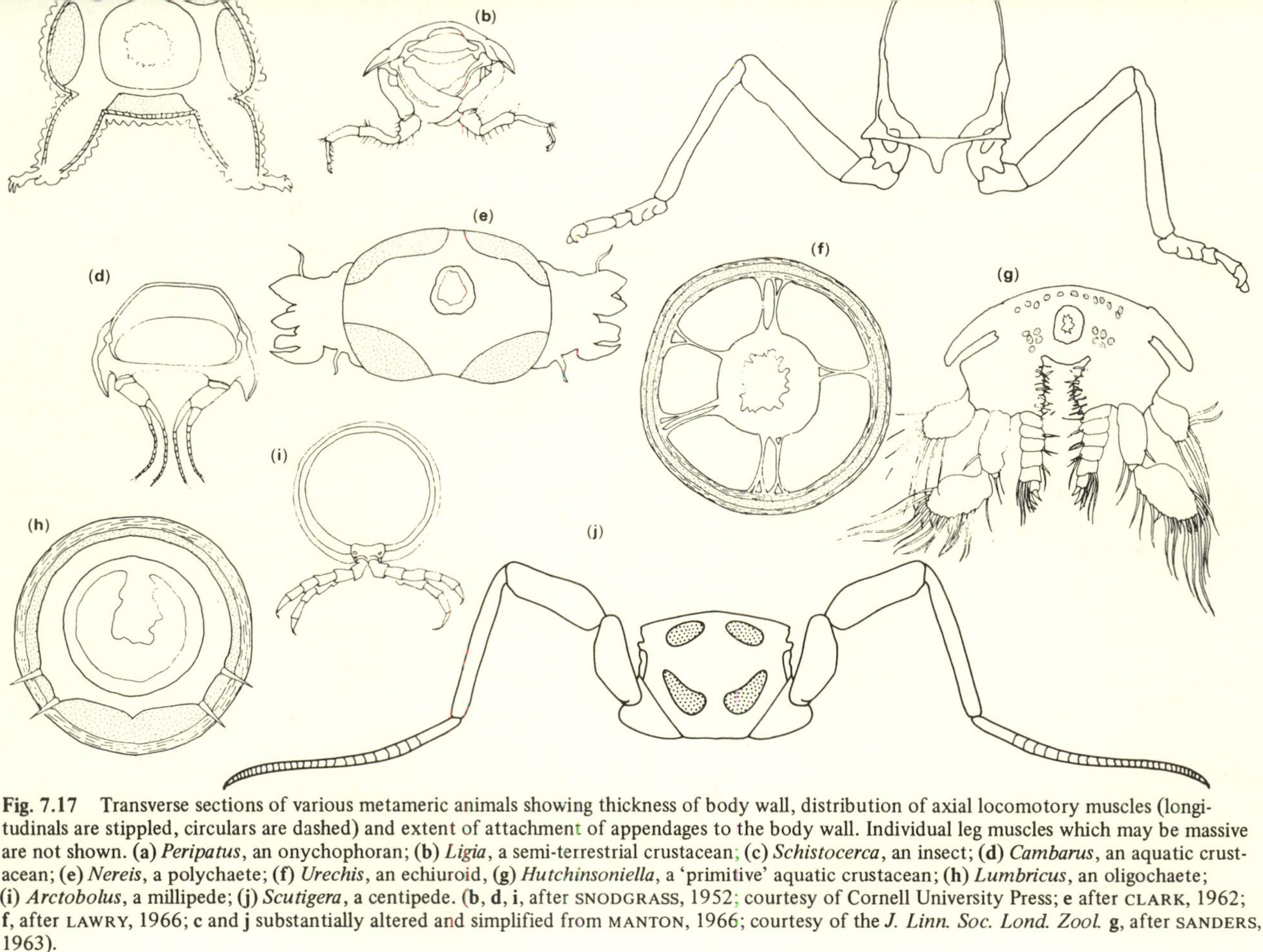

Fig. 7.17 Transverse sections of various metameric animals showing thickness of body wall, distribution of axial locomotory muscles (longitudinals are stippled, circulars are dashed) and extent of attachment of appendages to the body wall. Individual leg muscles which may be massive are not shown. **(a)** *Peripatus*, an onychophoran; **(b)** *Ligia*, a semi-terrestrial crustacean; **(c)** *Schistocerca*, an insect; **(d)** *Cambarus*, an aquatic crustacean; **(e)** *Nereis*, a polychaete; **(f)** *Urechis*, an echiuroid, **(g)** *Hutchinsoniella*, a 'primitive' aquatic crustacean; **(h)** *Lumbricus*, an oligochaete; **(i)** *Arctobolus*, a millipede; **(j)** *Scutigera*, a centipede. (**b**, **d**, **i**, after SNODGRASS, 1952; courtesy of Cornell University Press; **e** after CLARK, 1962; **f**, after LAWRY, 1966; **c** and **j** substantially altered and simplified from MANTON, 1966; courtesy of the *J. Linn. Soc. Lond. Zool.* **g**, after SANDERS, 1963).

tory apparatus. They swim by propagating a lateral (or dorso-ventral, in case of whales, dolphins and some seals) undulatory wave along the body. The simplest design for such a system is a series of identical muscles controlled by a single nervous conduit with segmental sensory and motor branches. Each segment does exactly what the one before it did, and each one acts at the correct time, due to the progressively later arrival of the control message as it passes along the linear central nervous system.

The evolutionary and mechanical implications of metamerism have been critically discussed by CLARK (1964) in one of the most succinct and thorough accounts ever made of the diversity of animal form and function. The reader is urged to consult this important work.

7.6 On Being Surrounded by Air

Organisms on land are subject to greater gravitational forces per unit mass than are those in water. The ability to draw sustenance from a greater volume of the environment by speedy locomotion or by static reaching is size-dependent: the larger the organism, the faster the locomotion and the greater the reach. Among organisms surrounded by air, we will see the greater advantage of structures of minimum weight.

Hydrostatic support mechanisms are convenient for organisms that live in or by reliable water sources. As plants and animals evolved into terrestrial habitats, any means of reducing the amount of water required for support has meant an increased efficiency of support and has permitted even drier environments to be inhabited. By having an exoskeleton that could be made water-proof and rigid, aquatic arthropods were well suited to exploit the possibilities of support and locomotion by non-hydrostatic means. Presumably the aquatic ancestors of modern onychophorans and arthropods were segmented, and segmentation appears to have been no drawback in becoming terrestrial.

7.6.1 Wilting Plants

Let us make a technical digression from self-bending hydrostatic animals and examine the hydrostatically supported terrestrial plants. These are plants that wilt when deprived of water and then recover their upright posture after water supply is restored. They are different from hydrostatic worms, snails and caterpillars in that the fluid-filled cavity of wilting plants is subdivided into thousands of single cells. All the cells of all types in, for example, the wilting petiole of *Nasturtium* (Fig. 7.18), are full of a very watery fluid. Water may diffuse from one cell to the next through cell membranes stretched over pits in the adjacent walls of two cells. Water moves up the stem to the leaves by mass flow in the xylem vessel members. When water supply to the roots is restricted, the water in all the non-conducting cells will, by diffusion, be entrained into the water moving up the xylem to be evaporated away through the leaves.

A hydrostatic mechanism, we have said, is basically a two-component structure: a tension-resisting container surrounding a compression-resisting fluid under pressure. Pressure in the fluid stretches the container to occupy a maximum volume. According to this model, there should be centrally placed fluid under

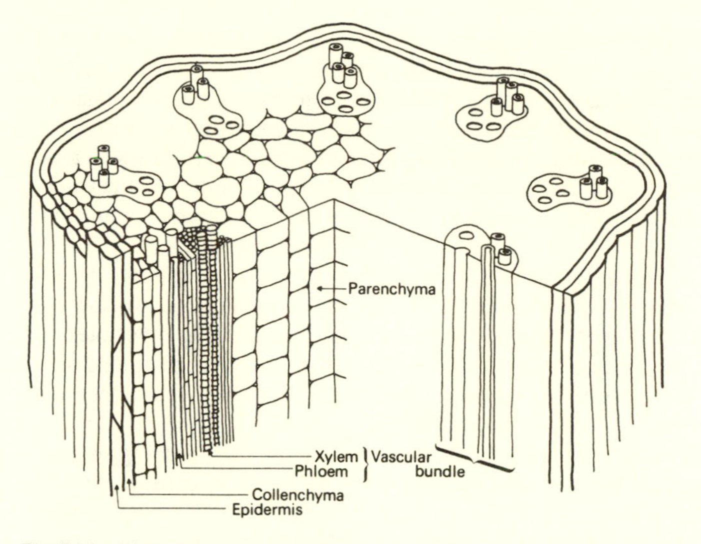

Fig. 7.18 Three-dimensional diagram of a cut-away view of a *Nasturtium* petiole. Petiole diameter, ca 3 mm.

pressure and peripherally placed material in tension in the wilting stem. In fact, this is so. A fine glass needle poked into any of the parenchymal cells of a turgid, upright *Nasturtium* petiole will allow the escape of a drop of cell sap when the needle is withdrawn (WAINWRIGHT, 1970), If the needle penetrates a xylem vessel, the drop appears as the needle is withdrawn and then, in a few seconds, disappears. Presumably the drop of fluid escaped from punctured parenchymal cells and it disappears as it is entrained in the transpiration stream up the xylem.

Parenchyma cells are typically thin-walled and approach being isodiametric. The analogy of the shapes of animal fat cells and plant parenchyma cells to the shape of soap bubbles and tightly packed plastic spheres has been demonstrated (MATZKE and NESTLER, 1946). Certainly these plant cells have a shape and wall thickness that best allow them to resist their own turgor pressure.

In seeking the tensile elements around the periphery of the wilting stem (Principle 2a), one's attention is drawn to the fibro-vascular bundles: supportive fibres occur in both xylem and phloem. Fibrovascular bundles of wilting stems are located one third or less of the radius from the outer cuticle. However, fibres are not common in primary xylem and phloem, and wilting stems have only primary tissues. The chief tension-resisting tissue in wilting stems is collenchyma that lies just under the epidermis and may be 2 to several layers thick according to the diameter of the stem.

In the *Nasturtium* petiole, for example, collenchyma forms an unbroken cylinder two cells thick. The tangential walls are fused and are thicker than the radial walls (Fig. 7.18). This structure gives the collenchyma the sandwich structure described in Chapter 5, that has such superior bend-resistant properties combined with a design of least material. It is interesting that collenchyma is the only supportive tissue of higher plant stems whose cells are known to continue growth and to maintain their living protoplasts throughout the life of the plant (ESAU,

1965). A wilting stem grows much faster in length than in diameter, and the cellulose microfibrils in collenchyma walls have an orientation that is virtually parallel to the stem's long axis (CHAFE, 1970). Since fibres and other elongate plant cells have predominantly transverse orientation of cellulose, we may expect a striking deviation from this general pattern to be an adaptation to a special function. We suggest that it is the special function of collenchyma walls to resist tensile forces in bending stems to which the axially oriented cellulose is an adaptation.

The principles of design tell us that in pressurized cylinders, hoop stresses are twice longitudinal stresses, and we must therefore discover the hoop-stress-resisting component of the stem. Hoop stresses in turgid *Nasturtium* petioles cause the round holes made by fine glass needles in the epidermis to spread as longitudinal slits. We suggest that tangential epidermal and collenchyma cell walls may play a tensile role in resisting hoop stresses.

To appreciate fully the concept of collenchyma as a tension-resisting tissue and the unsuitability of fibrovascular bundles to resist tensile stretching, one has only to consider the celery stalk. Pull it or bend it till it breaks. You will observe that the bulk of the material is parenchymatous and breaks crisply, leaving two sorts of strands unbroken. The thin, flat strands that occupy the ridges on the outside of the intact stem are samples of pure collenchyma. The round strands from within the stem are vascular bundles. Even though the collenchyma strands are thinner, they are stronger than the bundles (ESAU, 1936). Furthermore, the bundles stretch a considerable extent before they break, whereas collenchyma does not.

Seeing vascular tissue as not being very good at the rigid resistance of tension brings us to another mechanical design feature of these cells that is perhaps not widely understood. This is that xylem vessel members have circular or helical thickenings of their cell walls, shown diagrammatically in Fig. 7.18, that are often said to reinforce the strength of these cell walls. Unfortunately, the implication is made that the reinforcement allows the cell to help in the upright support of the stem.

Ringed vessels in plant stems carry water that is under strong negative pressure (KRAMER, 1949). Vessels in animals that are similarly reinforced by internal rings of rigid materials are the tracheae of insects and the bronchi of mammalian lungs. These vessels are gas-filled tubes that require strengthening against collapse from hydrostatic pressures deep within the body and local buckling stresses of flexing joints. Xylem vessels are also in danger of implosion. The engineering solution to reinforcing a cylinder against collapse is internally placed rigid reinforcement. So we suggest that the extensibility of vascular bundles of celery and the ubiquitous internal ring-reinforcement of xylem vessel members should remove them from the realm of general stem support and should see them ideally adapted to resist the implosive negative pressures attendant to their major function as water channels up the stem.

7.6.2 Woody Plants

Presumably a hydrostatic green plant (an alga) has given rise to all terrestrial green plants. Wood evolved and liberated the early air-breathing plants from dependence on standing pools of water or water-saturated soils for the upkeep of rigid upright support. Just as chitin-based cuticle and apatite-and-collagen-based bone pre-adapted the arthropods and vertebrates for support in low density air, so cellulose-based wood pre-adapted the green plants for life on land. We have already discussed wood as a material (Section 5.18) and we have presented the tree trunk (Section 6.2) and the petiole of the palm frond (Section 6.5) as examples of column and cantilever design.

Mechanically speaking, cellulose is the best tensile material for its weight available to organisms; wood is as effective per unit weight in resisting bending as any biomaterial; and plant parts are as simple and elegantly designed as any organisms on Earth. What more can we say?

7.6.3 Reaction Wood

In Section 6.2, we discussed the problem of improving bend-resistance by maximizing EI. We then acted as right-thinking design engineers and, reckoning that once the material is chosen, the properties are set, we discussed clever ways of designing I. Animals and plants are full of clever I's. But animals and plants are also able to redesign material properties. The gorgonian, *Plexaurella*, deposits more or less $CaCO_3$ in the sides of its collagenous axis according to the direction of predominating wave force. Trees make reaction wood. This general class of phenomena is the local alteration of material within an organism to ensure compatibility of stress and strain.

Considering wood as described in Chapter 5 as typical, atypical wood (secondary xylem) called reaction wood is formed in tree trunks and limbs according to the direction in which they are displaced and to their phylogenetic position. The result is that the tree recovers its original posture through subsequent growth. Softwood species (gymnosperms) form so-called compression wood on the side where growth in length will recover original posture. Hardwood species (dicotyledonous angiosperms) form so-called tension wood on the side of the stem where shortening will restore the original posture. Figure 7.19 shows that both types of reaction wood occur as eccentric growth.

In a recent review, SCURFIELD (1973) discusses his suggestion that lignification is the mechanism by which reaction wood actively restores posture. He cites the evidence that reaction wood plays an active role: (1) Boards cut from compression wood (tangential and parallel to the underside of a softwood stem) expand longitudinally. Boards cut from tension wood (tangential and parallel to the upperside of a hardwood stem) contract longitudinally (HEJNOWICZ, 1967). (2) Eccentric growth evoked by application of the plant growth hormone indolacetic acid has the same structure as reaction wood (WARDROP and DAVIES, 1964). (3) If an upright stem is held in a vertical loop (Fig. 7.19b) while reaction wood grows and is then cut horizontally as shown, the upper chord expands and the lower chord contracts. Transverse sections of the chords indicate the presence of compression wood in the softwood and tension wood in the hardwood as shown in Fig. 7.19c and f (JACCARD, 1938). (4) When aerial roots of the fig tree *Ficus* reach the ground, they penetrate the soil, produce a layer of tension wood and contract. If the root is allowed to grow into soil in a container, the container will be lifted off the ground (ZIMMERMAN *et al.*, 1968).

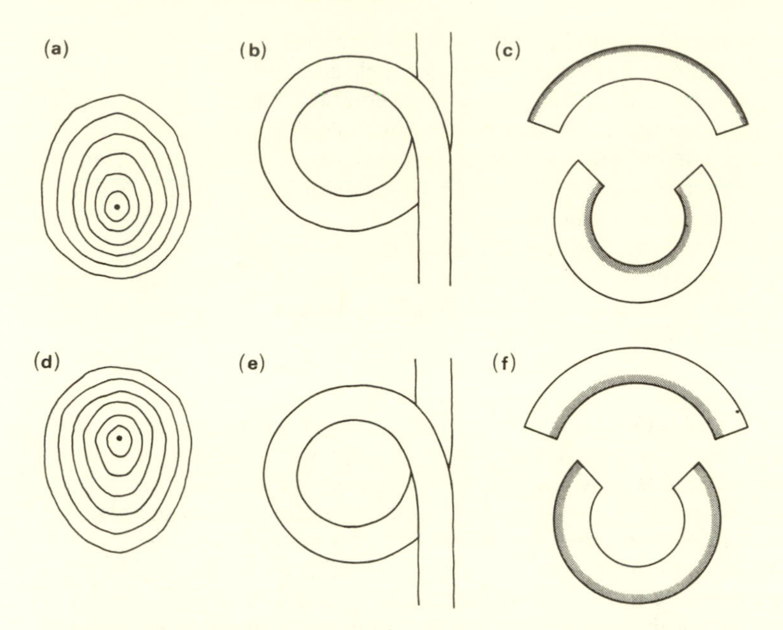

Fig. 7.19 Eccentricity of reaction wood in (**a**, **b**, **c**) hardwood (angiosperm) and (**d**, **e**, **f**) softwood (gymnosperm). (**a** and **d**) are transverse sections of horizontal branches showing eccentric growth rings. (**b** and **e**) are saplings bent into loops and restrained. After a period of growth, the saplings were cut and the loops sprung to the shapes in (**c**) and (**f**). Grey areas indicate the position of eccentric wood produced during the experiment (**b**, **c**, **e**, **f** after JACCARD, 1938).

Structurally, the two types of reaction wood cell walls are different from normal wood and from each other (Fig. 7.20). Compression wood cells (Fig. 7.20a) are thick-walled and notably round in cross-section. Roundness leads to large intercellular spaces. The secondary wall consists of S1 and S2, both thicker

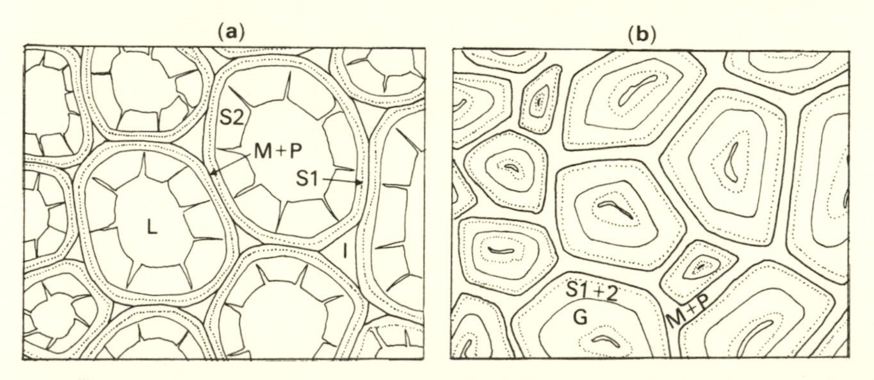

Fig. 7.20 Histological preparations (transverse sections) of reaction wood cell walls. (**a**) Compression wood: rounded cells, intercellular spaces, checked secondary layers. (**b**) Tension wood: G layer occluding cell lumen. I, intercellular space; G, G layer; L, cell lumen; M & P, middle lamella plus primary layer; S1, S2, secondary layers; (**a**, after CÔTÉ, 1968; **b**, after WARDROP & DADSWELL, 1955).

than normal. Cellulose microfibril angle is normal in S1 but is 30° to 45° greater than normal in S2. Deep slits extend from near S1 through S2 to the lumen: they are parallel to the microfibrils. Lignin content of S2 is higher in compression wood than in normal wood; lignin content of S1 is the same in both. There is less cellulose in compression wood than normal wood. S3 is missing.

Tension wood cells will develop an inner 'G' layer of a different sort, adding it to whatever layer was last formed when force was applied. The G layer delaminates from S2 readily and comes to occlude the lumen as shown in Fig. 7.20b. Layer G is distinctly lamellated and predominantly cellulose, though it does contain peroxidase and phenolic compounds. Its cellulose microfibrils are axially oriented. Initially G is highly convoluted, and birefringent in cross-section but convolutions and birefringence disappear with time. It is presumed these changes occur as cellulose becomes axially aligned. Cracks also appear in G parallel to cellulose microfibrils. Layers external to G become highly lignified while G remains low in lignin and high in cellulose. Fibres are more abundant in tension wood than in normal wood and vessels tend to be collapsed due to tight-packing. There are no intercellular spaces, and tension wood is harder, denser and darker in colour than normal wood. Extensive mechanical analyses like those of MARK (1967) have not been performed on reaction wood cells, so we are left, with SCURFIELD (1973), to conjecture the mechanical implications of reaction wood cell wall structure.

Compression wood cells lose contact with each other where intercellular spaces are formed. Less cellulose in the wall may increase porosity allowing greater concentrations of lignin to accumulate. PRESTON and MIDDLEBROOK (1949) and ALEXANDROV and DJAPARIDZE (1954) have shown that lignin deposition can cause cell walls to swell. If this occurs in the formation of compression wood and if there is a mechanism by which swelling could be limited to the longitudinal direction, the cells could extend and cause the observed curvature in stem growth. Perhaps deposition of bulk material between steeply helical cellulose microfibrils could provide a lengthening force.

Design principles of fibre-wound cylinders tell us that the decrease in volume of a cylinder wound with fibres at angles less than 55° will cause the cylinder to shrink in diameter but to increase in length. If water from the wall or the lumen is pumped out of wood cells, they would predictably appear as compression wood cells do: they would become rounder and thinner but longer.

Tension wood cells fit their role by being mostly longitudinally oriented cellulose. They would need only to put down a lot of cellulose (as in the convoluted G layer) and then crosslink, dehydrate and crystallize it, causing the wall to shorten. Shortening of tension wood cells would provide the restoring force for recovery of posture.

There are many 'if's' in this scheme at the moment, but active alteration of material properties by an organism is important to the organism. Man is inclined to see and use wood as primarily a rigid material produced by a tree that is considered as passive as a great organism can be. It is as important to us as it is to the tree that we understand the mechanisms by which it actively produces and modifies material that efficiently resists tension, compression, bending and impact, conducts water and acts to recover externally caused posture changes.

7.6.4 *Fibre-reinforced Palm Trees*

The entire coconut palm (*Cocos nucifera*) is a fibre-reinforced column, the trunk, from which each leaf develops to hang as a curved, tapered, fibre-reinforced cantilever supporting the photosynthetic tissues in the sun. Petioles, leaflets and the husks of the nuts are fibre-reinforced materials. The fibres are the fibrovascular bundles, homologous and analogous with the same-named structures in celery and all other monocotyledonous angiosperms. As described in Chapter 6, optimization of I and J is nowhere in biology more cleverly achieved than in the palm petiole. Fibrovascular bundles that provide the tensile integrity of the entire plant are present throughout the section of the petiole, but the number of bundles per unit area is greatest (they are close-packed) around the periphery, thus further optimizing sectional design. Palm roots and trunk wood bear careful further study to complete this picture. Furthermore, resistance to wind forces is accomplished in the leaves by a delicate balance of strain control in bending and torsion by changes in sectional shape and by streamlining of petiole and leaflets and leaflet attachments.

This subject was first appreciated by Tomlinson who analysed the fibrous anatomy of stems and leaves in a very wide range of palm species (TOMLINSON, 1962; 1964). We can add very little to his analysis at this time other than the treatment of the petiole as a bend- and torsion-resisting element in Chapter 6. It is very hard to convey, in a neat package of two or three words, that the palm frond is made of rigid materials but because it is designed to spread strains over

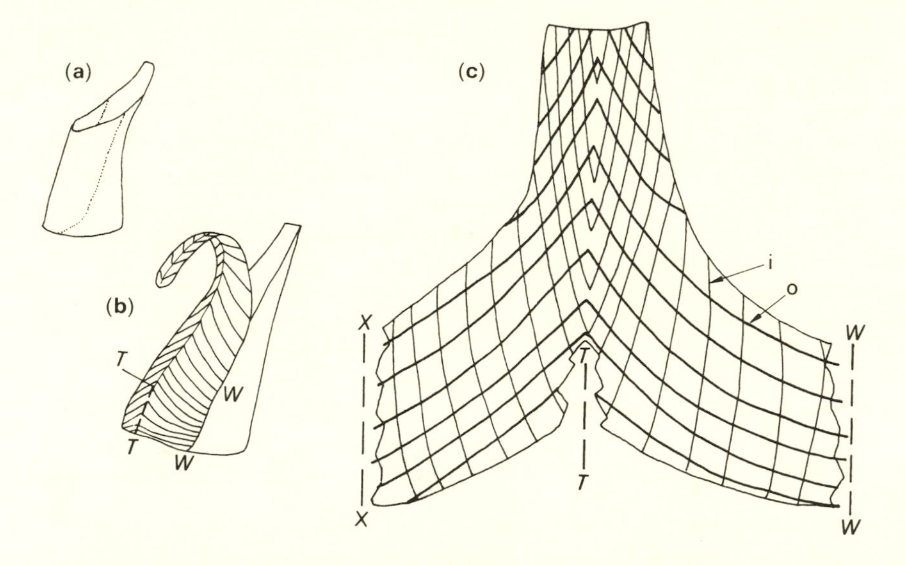

Fig. 7.21 Leaf base of the coconut palm. (**a**) Shape of the leaf base at an early stage. (**b**) A later stage following growth in girth of the trunk and death of the leaf base tissue on the side away from the petiole: only the fabric remains. (**c**) The fabric from a leaf that has just fallen due to failure of the fabric along the torn lines *X-X*, *W-W*, *T-T*. The two layers of fibres are shown: heavy lines represent fibres in the outer layer (o) and finer lines represent fibres in the inner layers (i). (**a** and **b**, after TOMLINSON, 1962; courtesy of *J. Arnold Arboretum*; **c**, Wainwright and Biggs, unpublished results).

long distances via its rigid petioles–and even the trunk in hurricanes–it will undergo very large total deformations to allow avoidance of large stresses due to wind.

Dissection of the growing tip of a coconut palm reveals that the base of each petiole develops as a cylinder completely surrounding the stem (Fig. 7.21a). The leaflets develop on one side and fibrovascular bundles develop in two layers as shown in the top of the diagram of fabric in Fig. 7.21c. As the leaf matures, the tissue on the side away from the petiole dies leaving the fabric of fibres. The fibres of the two layers form a crossed helical pattern with a line up the middle where the helical direction reverses (Fig. 7.21b and c) without interrupting integrity of the fibres. With increased girth, the fibre angle in the outer layer becomes greater, as predicted by Principle 5a until, at maturity, the fibres in the outer layer are virtually circumferential. The fibres of the inner layer have a very low fibre angle throughout and, at the 'seam' in the mature leaf base, they have been spread apart by growth in girth of the trunk.

The fabric splits away from the leaf base gradually. When a leaf is one of the two or three oldest leaves left on the tree, the circumferential fibres of the fabric play a major role in holding the leaf on the tree. At this time these fibres are taut and when cut one at a time with a knife, they snap with a sound like a breaking banjo string. When enough of them are cut, the leaf will fall: there is an abscission layer, a zone of weakness, where the leaf base meets the tree that ensures a clean break. A take-home lesson here for keepers of tropical estates: do not remove the unsightly fabric from the leaf bases of your coconut palms, else the leaves will fall on your head. Citizens of the temperate zone may need to be reminded that each leaf weighs 6 kg.

7.7 The Hydrostatic Onychophora

Before web building, ballooning and flight became ways of life, terrestrial animals inhabited land surfaces, leaf litter and whatever sediments they could burrow in. MANTON (1961*b*) has documented the design features of onychophorans and terrestrial arthropods that suit them for the special mechanical tasks they each perform. The accounts of her researches in this field (MANTON, 1950; 1952*a*; 1952*b*; 1953; 1954; 1956; 1958*a*; 1958*b*; 1961*a*; 1961*b*; 1964; 1965; 1966; 1972) constitute a monument in the study of mechanical design of the most mechanically diverse groups of organisms that have ever lived. Taking onychophorans and arthropods as a whole, they are walking and burrowing wormlike animals. Some swim either by walking movements or by a new specialization and, of course, some can fly, and many no longer look like worms, though some of these have very wormlike larvae. The most wormlike of these animals is the onychophoran *Peripatus* that lives in leaf litter on tropical forest floors (Figs. 7.17a and 7.22d). The surface layers of such a 'sediment' are structured and small things can move through the continuous spaces between structural elements. This *Peripatus* does by being capable of extreme changes in cross-sectional shape. MANTON (1958*b*) watched while *Peripatus* took 20 minutes in stepping slowly through a round hole in a card that was one-ninth the diameter of the resting animal. The only hard parts of *Peripatus* are a pair of chitinous jaws and a pair

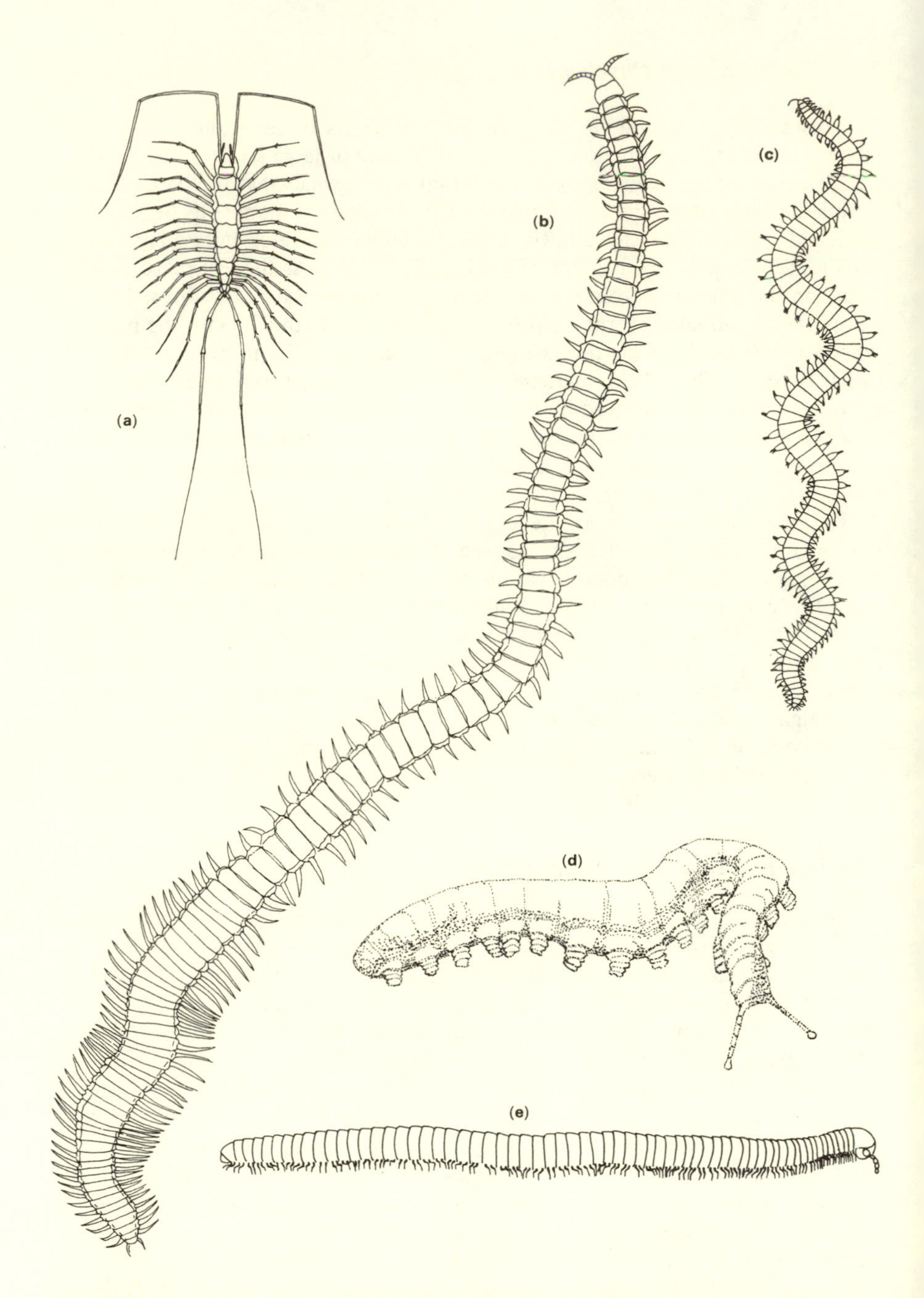

Fig. 7.22 Some segmented invertebrates. (**a**) a running centipede, *Scutigera coleoptrata*; (**b**) a burrowing centipede, *Orya barbarica*; (**c**) a swimming polychaete, *Nereis*; (**d**) litter-burrowing *Peripatopsis moseleyi*; (**e**) a crawling millipede. NB: *Orya* has just awakened and walking movements that began at the head have only reached about half way down the body. Segments that are walking become elongated and narrow (**a**, from SNODGRASS, 1952; courtesy of Cornell University Press; **c**, after WELLS, 1968; **d**, adapted from ROBSON, 1964).

of small chitinous claws on each leg. For the rest, this animal has a single internal fluid-filled cavity, the haemocoel, supported on metamerically arranged, paired, short legs controlled by a metameric central nervous system. Body muscles are smooth muscles and the cuticle is furrowed and folded but unstretchable (ROBSON, 1964). That is, it has great flexibility and probably a high tensile elastic modulus.

This capacity for great local changes in body diameter and length is the basis for the three gaits of *Peripatus*. One gait is properly called a 'low gear' since it is the slowest and most powerful. For this gait, the animal's body is short and as it walks by passing a wave of stepping movements forward along the body, most of the legs are on the ground at any one time, giving maximum power to forward movement. In changing 'gears' the animal's body lengthens and each leg spends a proportionally longer time on its recovery stroke and less on the power stroke. In top gear, length of body is maximum and at any one instant most legs are off the ground and power of movement is sacrificed for speed. The legs are short and unjointed and the speed in top gear is not very impressive when compared with fleet, long-legged arthropods, but the versatility of having three gaits represents a high degree of sophistication for a hydrostatic terrestrial animal.

7.8 Jointed Frameworks of Solid Materials

Many animals, including aquatic ones, have evolved away from dependence on pressurized fluid for compression resistance. The addition of rigid solid materials to the worm-like body wall is the basis of body form and mechanical function in such different but highly successful groups as arthropods, echinoderms and vertebrates. We have already seen in Section 7.6 that the functional replacement of turgor water by more rigid cell wall material allowed especially the terrestrial plants to diversify and to colonize the land.

In our treatment here, we have not considered design in the vertebrate body. We feel that this has been done effectively by GRAY (1968) and NACHTIGALL (1971) and that we could not significantly add to these accounts. We omit such phyla as Bryozoa and Brachiopoda because we have no mechanical information. We offer the reader a few concepts about arthropods, echinoderms and molluscs.

7.8.1 Running and Burrowing Myriapods

Forceful movements are accomplished by muscles of great cross-sectional area. High speed locomotion can be achieved by taking longer, faster steps. A long limb carrying thick strong muscles will be heavy and difficult to move rapidly: long slender muscles can take faster but less forceful steps. A long muscle can move its insertion over a greater distance or through a wider angle than a short one can, and this is another design feature of long-legged running creatures: each step may be quite long relative to the body length.

The most wormlike adult arthropods are the Myriapoda, the centipedes (Chilopoda) and the millipedes (Diplopoda) (Figs. 7.17i and j and 7.22a, b and e). These animals, like polychaetes and *Peripatus*, are metameric in plan and have one or two pairs of walking legs on each body segment. Some of the centipedes have retained some hydrostatic function of the haemocoel, but the main differences between true arthropods and their soft-bodied relatives is the inclusion of rigid, compression-resisting materials into the cuticle. The rigid areas of the cuticle are called sclerites and are surrounded by highly flexible cuticle called arthrodial membrane. Insect cuticle is discussed in detail in Chapter 5.

The adaptations that enhance the speed and agility of running centipedes include very long, very slender legs joining the body at the ventro-lateral corner (see Fig. 7.17j). The arching leg and underslung body allow the centre of gravity to remain low. With 14 pairs of long legs moving swiftly over wide arcs, the potential danger of tripping on one's own feet is high, and the precise timing of the movement of each leg becomes important. The movements of each leg of the fleet centipede *Scutigera* are controlled by no less than 33 muscles originating from sites on the exoskeleton of the body—analogous leg movements of a slow powerfully burrowing millipede (discussed below) are controlled by two muscles per leg. Speed and agility in locomotion demand more neuromuscular units and greater complexity of coordinating processes than does slow but powerful locomotion.

The rigid sclerites of arthropods are in linear series along the body: tergites down the dorsal surface, sternites ventrally and pleurites laterally. These scutes are attachment sites for both body and leg musculature. Sclerites of fleet centipedes are relatively small and appear to float as islands in the flexible arthrodial cuticle. This gives the animal the ability to bend its body into lateral S-shaped curves or to coil tightly dorso-ventrally into a ball. At the same time, the more rigid the body is, the less it undulates from side to side during locomotion, and the more muscular energy can go into the power stroke. The floating sclerites of

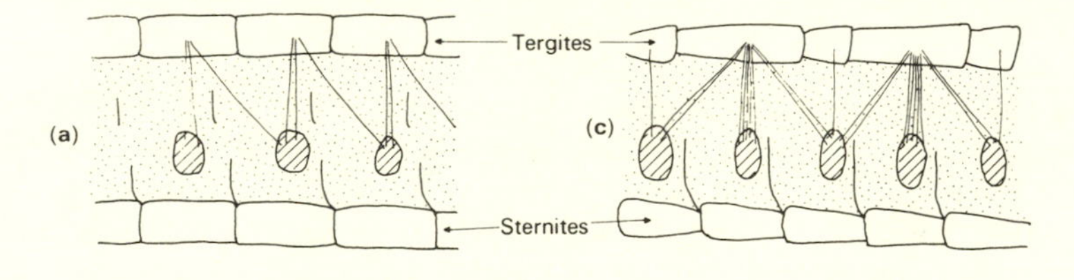

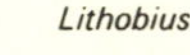

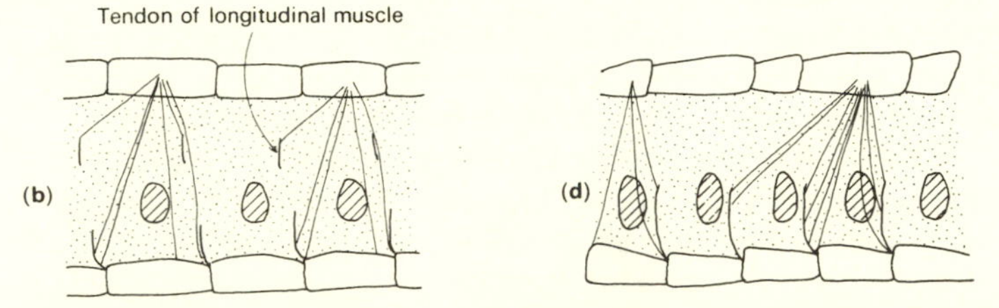

Fig. 7.23 Diagrams, from side view, head to the left, showing the number of muscles from tergites to leg base (**a** and **c**) and from tergites to tendons of longitudinal muscles (**b** and **d**) in the burrowing centipede *Haplophilus subterraneus* (**a** and **b**) and the running centipede *Lithobius forficatus* (**c** and **d**). Hatched ovals are sites of leg attachments; stippled area is flexible arthrodial membrane; enclosed blank areas are rigid sclerites (redrawn and altered from Manton, 1965; courtesy of the *J. Linn. Soc. Lond. Zool.*).

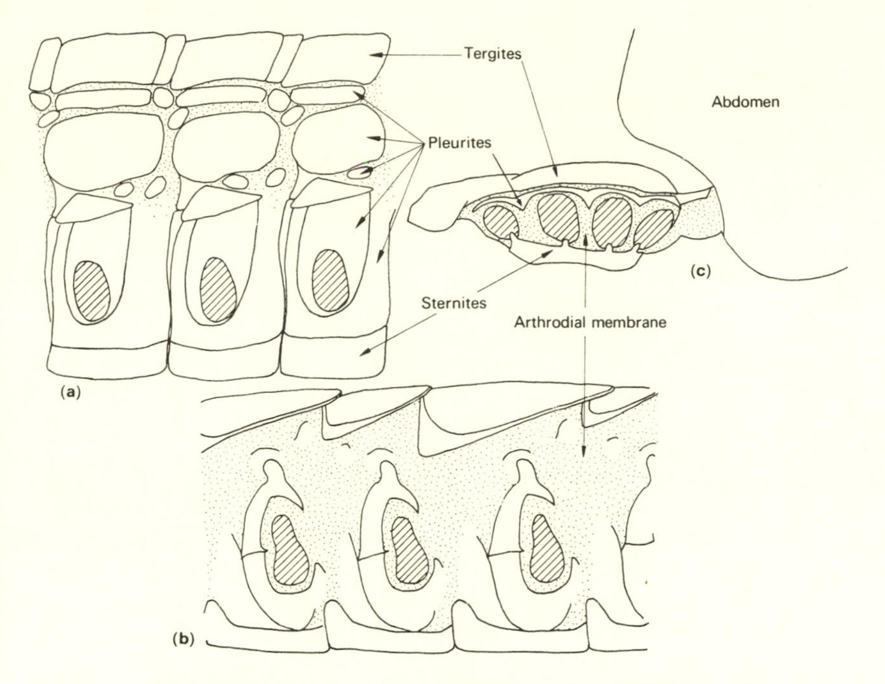

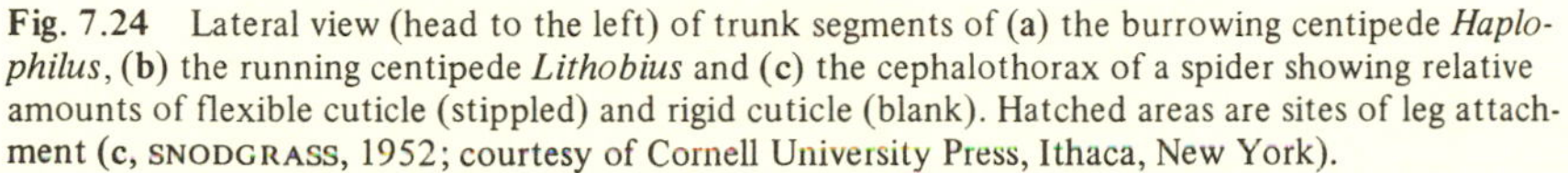

Fig. 7.24 Lateral view (head to the left) of trunk segments of (**a**) the burrowing centipede *Haplophilus*, (**b**) the running centipede *Lithobius* and (**c**) the cephalothorax of a spider showing relative amounts of flexible cuticle (stippled) and rigid cuticle (blank). Hatched areas are sites of leg attachment (c, SNODGRASS, 1952; courtesy of Cornell University Press, Ithaca, New York).

running centipedes are anchored by muscles to each other in an extraordinary tensile suspension system: in *Lithobius* each tergite bears muscles for 3 successive pairs of legs and is united to 5 successive sternites (Figs. 7.23 and 7.24). In a slow, powerful burrowing centipede such as *Haplophilus* (discussed below), sclerites occupy proportionally more of the body cuticle and have fewer separate tensile attachments: each tergite has muscular connections with only one pair of legs and is linked by muscles to two successive sternites.

The most striking cuticular difference between fleet centipedes and slow, powerful centipedes and millipedes is the extent of flexible vs. rigid cuticle on the sides and ventral surface where the legs join the body (see Fig. 7.24). This is related to the ability to change shape locally, to the angular swing of the legs and to lateral or ventral bending of the body. Centipedes generally have alternate large and small tergites (Fig. 7.25) whose function is best seen in the burrowers, but the possession of the smaller intercalary sternites as muscle insertions enhances the lateral bending movements of all centipedes.

Running centipedes have arrived at an interesting compromise: long thin legs require many muscles of body origin per leg, and they have proportionally more of this appendage-controlling musculature than of longitudinal body musculature (Fig. 7.23). However, lateral undulations do not contribute to forward thrust, and rigidity against lateral undulations is enhanced by strong longitudinal body

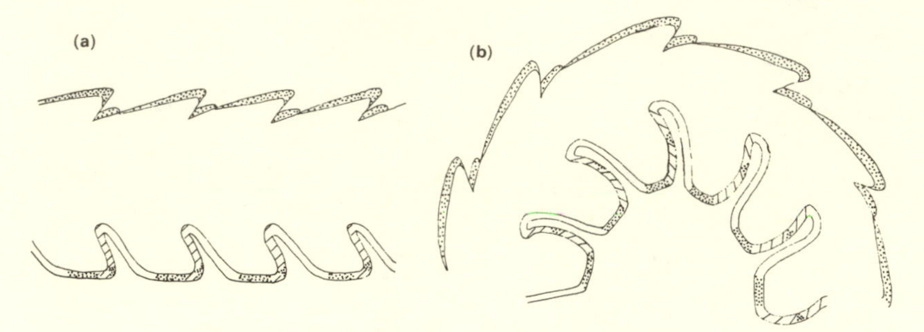

Fig. 7.25 Diagrammatic sagittal section through a burrowing centipede showing the distribution of cuticle of graded rigidity relative to flexure points. Most rigid cuticle is stippled, least rigid cuticle is blank and cuticle of intermediate rigidity is hatched. **(a)** before and **(b)** after ventral bending.

musculature. The compromise seems to be that fleet centipedes have fewer segments, and the large 7th and 8th tergites are fused. What they gain in speed they lose in flexibility. Flexibility is of more interest to burrowers and slower creatures that squeeze and locomote in tight places. Fleet centipedes race about in open places, actually pounce on their prey and are the only myriapods to have well developed compound eyes that see far enough ahead to allow speedy pursuit of prey.

Centipedes that burrow do so by wedging their dorso-ventrally flattened bodies into a crevice and by forcing the crevice to become wider. The shape of *Orya* (Fig. 7.22b), for example, tapers from the widest, longest, and deepest segments in the middle of the body to narrowest, shortest and shallowest segments at the head and tail ends. Hydrostatically driven extension movements and stout anterior legs push the head into the soil or leaf litter or tree-bark crevice. Once the head is in, the legs maintain a hold while strong longitudinal body musculature pulls the next segments into the space. Since these next segments are deeper than the first, the crevice is made wider by this powerful action and the head can be hydrostatically extended further into the widened crevice.

Such burrowing could not be done with long, spindly legs, and the speed of leg motion in burrowing is less important than its force. Burrowing centipedes have legs so short that they present none of the potential tripping problems of the running centipedes. Precision of step-placing is less crucial in burrowing than in running and there are many fewer extrinsic leg muscles in burrowers (Fig. 7.23).

Burrowers do not run swiftly and one reason may be that lateral and dorso-ventral flexibility that is useful to a burrower is not compatible with running. *Orya* can be seen in Fig. 7.22b to be very flexible indeed, and such flexibility is based both on critically placed flexible cuticle but also on the fact that the rigid sclerites have soft edges that furl and, in some cases, a flexible line across the sclerite along which a fold may be made (Fig. 7.25). Also, when the animal is maximally contracted, the large tergites come to overlap the shorter intercalary tergites. These mechanisms allow shortening to half fully extended length, and while this degree of length change is not difficult for a helically wound, soft body, it requires complex design features in a cylindrical shell that carries its own rigid compression-resisting materials.

Millipedes move slowly compared to centipedes and they burrow by simply forcing their head end forward into the soft soil and leaf litter that is their habitat (MANTON, 1961*b*). The millipede head is bent rather sharply downward (see Fig. 7.22e) and its dorsal surface is protected by the large rounded shield-like tergite, called the collum, of the first body segments. The collum faces forward and upward and together with the top of the head is the surface that meets the soil most forcefully. Since this bull-dozing method of burrowing requires great forward thrust, the body axis is put under longitudinal compression and thus could be expected to have design features that resist buckling. Millipedes do not have a powerful hydrostatic mechanism. Instead, millipedes have an exoskeleton whose sclerites are made rigid by densely deposited calcium salts. Typically the

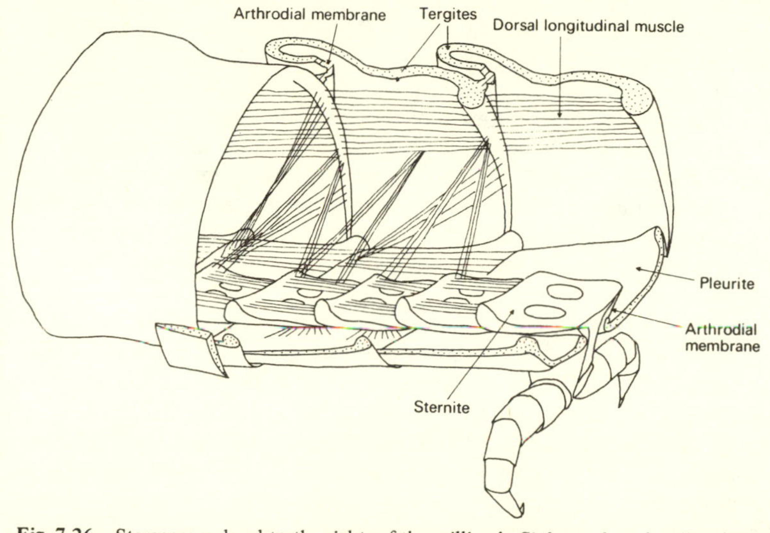

Fig. 7.26 Stereogram, head to the right, of the millipede *Siphonophora hartii* with viscera removed showing the 'tensegrity' suspension of the leg-bearing sternites by muscles and flexible cuticle. Note that each segment bears two pairs of legs—only one pair is shown, but the oval attachment sites for others can be seen on the inside of each sternite. Note also the overlapping tergites that 'telescope' as the animal bends (redrawn and substantially altered from MANTON, 1961*a*; courtesy of the *J. Linn. Soc. Lond. Zool.*).

tergite is at least a half-ellipsoidal arc, the pleurites are then ventral and sternites bear the legs (Fig. 7.26). There are two pairs of legs per segment and there are many segments. The legs join the body close to the midventral line and project first downward and then straight laterally as shown in Fig. 7.17i. They are seldom longer than the radius of the cylindrical body and therefore are invisible from above.

In locomotion, both legs of a pair move together and the wave of steps move forward as the animal progresses (see Fig. 7.22e). Each leg swings fore and aft over a small arc and therefore takes small steps. Speed of stepping is not rapid and the time occupied by the power stroke is about twice that for the return stroke. As a result, most legs are in power stroke at any instant, and the total forward thrust of all thrusting legs is potentially very great. In order that the thrust of legs at the rear be applied to forcing the head through soil, a mechanism for making the body a compression-resisting axis is required.

In fact, the hard exoskeleton of millipede segments is so shaped that the narrow front end of each tergite fits into the flared back end of the tergite in front of it (Fig. 7.26). When longitudinal musculature in the body contracts, tergites are pressed on each other and, because tergites are joined to each other by flexible cuticle, they can rotate on each other. This freedom of intertergal rotation is not the same in all directions: they can curl forward ventrally tightly into a ball, but lateral and dorsal flexure are restricted.

During walking, when longitudinal compression is not so important, the tergites separate and flexible cuticle can be seen between tergites. Since step length is restricted by how far a leg may swing forward before it bumps into the leg in front, any longitudinal separation of legs allows for a longer step and, as a consequence, the potential for greater speed.

Another feature of diplopod design that promotes forceful, head-on burrowing is the ability of the leg-bearing sternites to move completely independently of the tergites. In forceful burrowing by *Siphonophora hartii* the legs swing forward and contact the ground. Then, instead of the entire segment being pulled forward by the relatively weak leg musculature, only the tension-supported sternite is pulled forward (Fig. 7.26). Then, with leg muscles contracted, the powerful (thick) longitudinal tergo-sternal muscles contract and forcefully pull the bulk of the segment forward.

This two-phase action has interesting mechanical advantages. There are relatively few muscles attached to the sternites that effect the power stroke of the millipede leg, and muscles extending into the leg must be slender. However, reflexed edges or, indeed, the broad flat planes of sclerites can be attachment sites for muscles of large cross-sectional area and these can be used in forcefully moving the sclerites. The ability of the leg-bearing sternite to move independently of the tergite allows the massive body musculature to provide the force for forceful forward movement and the more weakly muscled legs have only to make the forward step, fix contact with the ground and then resist the tension put on them by the action of body muscles. This is mechanically equivalent to the sliding seat that allows an oarsman to use his leg muscles as well as trunk and arm muscles in his power-stroke.

This sophisticated mechanism for forceful burrowing is truly a tensegrity system, defined by Principle 3ci, in which the compression-resisting sclerites are connected together by a force-dispersing web of tension-active muscles and tension-passive flexible cuticle (Fig. 7.26).

7.8.2 Insects

Insects, especially flying ones, have fewer, generally longer legs. Their neuromuscular complexity is not one of great serial repetition of units, but of refinement of control over few legs. There is a general stiffening of the leg-bearing thorax that forever eliminates the problem of undulation and, in most advanced forms, also eliminates the ability for either dorso-ventral or lateral flexion.

To consider the thorax of flying insects as a rigid box is to over-simplify badly a deliciously complex mechanical system. It is not uniformly rigid as is the exoskeleton of a millipede's leg-bearing segment, but it is a complex system of rigid sclerites and flexible cuticle that permit the various wide-angled motions of legs and wings. For a detailed description, see PRINGLE (1957). Wings can be flapped up and down in a figure-8 pattern and can be twisted at the same time. To accommodate these movements, the sclerites in the rigid box are able to move relative to one another, suspended in the tensile web of flexible cuticle.

Also involved in insect flight are tendons consisting mainly of resilin. These tendons undergo long-range deformations during each stroke of the wings and provide elastic force that helps to accelerate the wing. The extra acceleration made possible by these remarkable rubber-like tendons at both the beginning and the end of each upstroke and downstroke allows the striated muscle that powers the wings to supply flight power in each stroke at wing-beat frequencies up to several hundred per second. (See ALEXANDER, 1968, for a review of insect flight.)

The tensile muscles and tendons of insects have been much studied. The nature of the rigid compressive cuticle is less well known, but clearly, if the high-speed, high frequency, powerful and precisely controlled movements of wings are to effect aerodynamically stable flight, the compressive parts of the system must be as precisely constructed as the tensile parts. Deforming polymeric materials at high frequency raises questions of the possibility of fatigue due to viscous interaction of the macromolecular components. The door is open here for some interesting research on mechanical properties of biopolymeric compressive materials whose chemistry and microstructure is well known.

7.9 Complex Support Systems: Molluscs and Echinoderms

Molluscs and echinoderms are animals whose many species represent alternatives to the linear body plan of polyps, worms and arthropods. Every species of both phyla possesses at least one hydrostatic system and many species have more than one hydrostatic system plus a solid-based system of calcium carbonate elements.

Molluscan locomotory mechanisms have been reviewed by MORTON (1967), GRAY (1968), TRUEMAN (1968*a*, *b*, *c*) and JONES and TRUEMAN (1970). TRUEMAN (1969) has summarized aspects of the fluid skeletons for the three major classes as typified by their different locomotory fortes.

Bivalves make use of internal hemocoelic spaces in the foot in burrowing. They also use water in an external space, the mantle cavity in burrowing and in swimming. Figure 7.27 shows the burrowing cycle of bivalves. In part (a) the adductor muscle is relaxed, allowing the shells to be forced apart by the elastic hinge ligament. This action holds the shell against the sediment while transverse muscles in the foot contract, forcing blood from the visceral mass into the foot and extending it. Then the powerful adductor muscles contract (Fig. 7.27b), more blood is forced into the extended foot causing it to expand laterally, enhancing frictional contact with the sediment (Fig. 7.27b). This is the highest pressure phase of the cycle: in the razor clam *Ensis arcuatus*, TRUEMAN (1967) has recorded pressure pulses of 1.18 kN m^{-2} lasting for 0.5 s in this phase of the cyclic burrowing activity. The

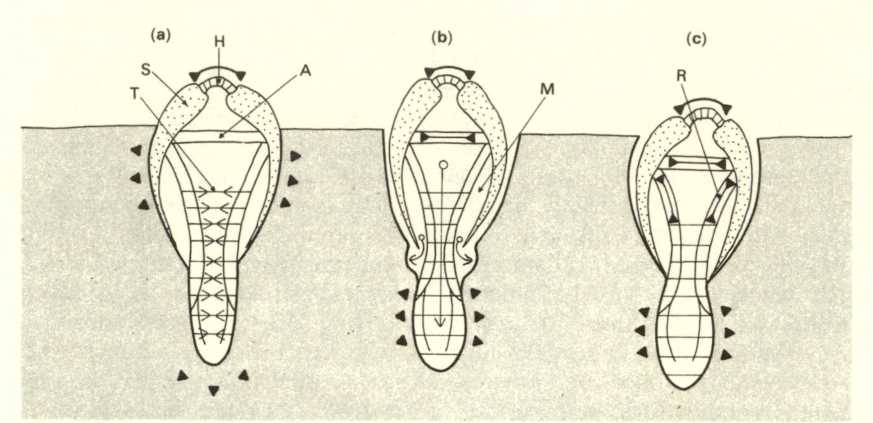

Fig. 7.27 Diagram of the burrowing cycle in bivalved molluscs. (**a**) probing phase, (**b**) high pressure phase during which sediment is loosened by water jet (2 small hollow arrows) from mantle cavity and more blood is forced into the foot (long thin arrow). In the retraction phase (**c**) the retractor muscles contract, pulling the shell down into the sand. Solid arrowheads indicate force directions. A, adductor muscle; H, hinge ligament; M, mantle cavity; R, retractor muscle; S, shell valve; T. transverse muscle. The cycle is described in the text (TRUEMAN, 1968*a*; reproduced from *Symp. Zool. Soc. Lond.*).

pressure of the adducting shells forces water out between the mantle-shell and the foot (Fig. 7.27b). This water jet loosens the sediment and thus aids the thrusting phase of the burrowing cycle. Then the longitudinal pedal retractor muscles contract, pulling the animal and its shell down a step further into the sediment and completing the cycle (Fig. 7.27c).

Crawling by the limpet (JONES and TRUEMAN, 1970) and burrowing by a number of bivalve species (TRUEMAN, 1968*a*) are effected by the use of internal and external fluid compression systems. The bivalves have the additional use of rigid shells that close, forcing blood into the foot where pressure peaks allow this soft tissue to act as a firm frictional anchor for a short time.

A few gastropod species such as *Polynices* burrow (TRUEMAN, 1968*b*) into sediments by the action of an internal hydraulic system that works much like the foot of burrowing bivalves and the anterior end of the polychaete *Arenicola* (TRUEMAN, 1966).

Swimming bivalves use only the external mantle cavity and the adduction of shell valves to generate locomotory power. The fast acting adductor muscle of swimming scallops is antagonized by the rubber-elastic inner layer of the ligament (ALEXANDER, 1966). This ligament is composed largely of abductin, a rubber-like protein so far known only from the ligaments of bivalves (Chapter 6).

7.10 Squid Locomotion

The fastest swimming invertebrates are the squids whose locomotion is famous as the prime biological example of jet propulsion. Circular muscles in the mantle contract against the water in the mantle cavity—technically an external space—producing the highest pressure peaks ever recorded in invertebrates (39.2 kN m^{-2} for 150 ms: TRUEMAN and PACKARD, 1968). This pressure forces water out of the siphon and provides the power stroke for rapid locomotion. Several power strokes can be delivered at a rate of 2 per second and they constitute an effective escape reaction. From studying ciné-films of swimming squid, WARD (1972) determined that during an average jet power stroke, the outer diameter of the mantle decreases by 20%, the mantle tissue thickens by about 20%, and there is no discernible change in length.

Less obvious and still incompletely understood is the mechanism by which the squid mantle, having contracted in a power stroke, effects its return to resting shape prior to the next power stroke. The return stroke mechanism has been studied by WARD (1972) who tested the hypothesis of YOUNG (1938) and WILSON (1960) that radial muscle fibres scattered among the circular muscles of the mantle provide the restoring force. Certainly radial muscles are correctly placed, such that their contraction would antagonize the deformation caused by circular muscles. Circular muscle contraction causes mantle tissue thickening, and radial muscle contraction should cause the tissue to become thin again. Each of these authors has observed that the two muscle systems are innervated by branches of the same nerve fibres, and that action potentials in both sets of muscles are simultaneous. As yet there is no direct recording showing the separate time course of the development and relaxation of tension in the two systems, so radial muscle function in return stroke is neither established nor denied.

The mantle is organized into a thick muscular layer bounded by inner and outer tunics of collagen fibres (Fig. 7.28) (WARD and WAINWRIGHT, 1972). The muscle layer is organized into thick blocks, 10 to 30 muscle cells wide, of circular muscles along the length of the mantle with alternating blocks, 2 to 4 cells wide, of radial muscles. Each radial muscle cell attaches to both inner and outer tunics.

Each tunic consists of layers of collagen fibres running in helical courses round the mantle. Each layer contains a single layer of large (7 μm diameter) collagen fibres that are parallel and close-packed and the layers themselves are close-packed. The fibre angle in the 7-to-10 layered outer tunic is $28^\circ \pm 1^\circ$ (s.e.) and in the 3-to-4 layered inner tunic is $24^\circ \pm 1^\circ$. This acute angle causes the collagen fibres to resist increase in length more forcefully than they resist increase in girth. The compactness of these tunics and their collagenous nature contribute to the crisp cartilaginous 'feel' of living or freshly killed squid mantle tissue.

The other connective tissue fibre system consists of fibres of average diameter 2.6 μm and unknown composition that lie scattered in radial planes throughout the muscle layer. These fibres are birefringent, straight and unbranching and each one inserts at an oblique angle onto both inner and outer tunics. The angle these fibres make with the tunics (i.e., the angle they make with the longitudinal axis of the animal: the fibre angle) varies widely around 28°. Elastic fibres having small angles to the long axis will resist increase in length more forcefully than they resist increase in thickness of the tissue: at a 30° fibre angle, resistance to longitudinal stretch will be 3.5 times the resistance to increase in thickness. Results of preliminary experiments by Gosline, on whole squid mantle tissue with the skin removed, show that the modulus rises sharply with longitudinal extension, whereas circumferential extension of *ca.* 25% is necessary before modulus begins to rise.

Once again within the Mollusca we see the interaction of an external (mantle cavity) and an internal (mantle tissue) fluid system in the support of locomotion.

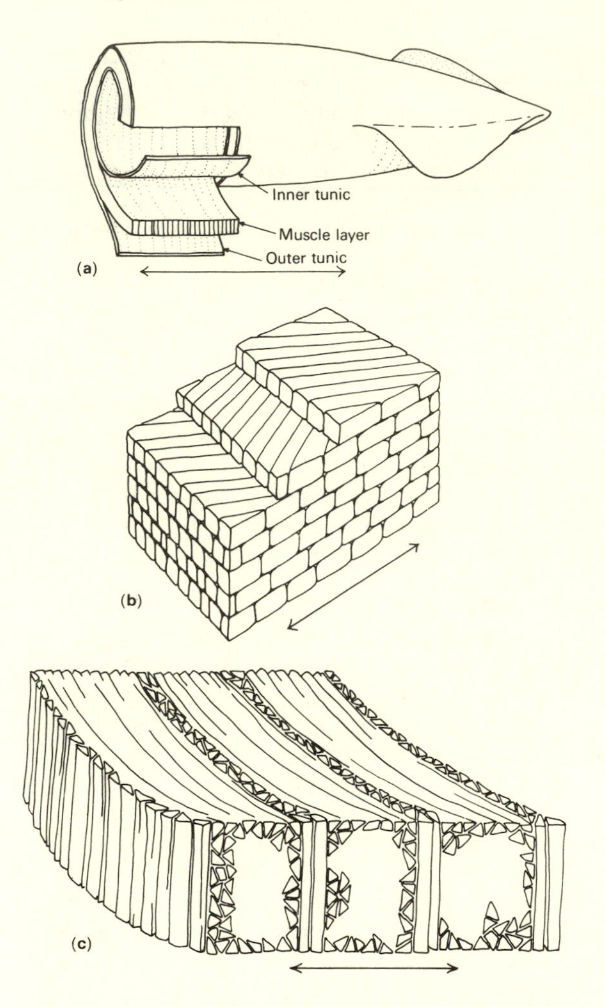

Fig. 7.28 Design features of the squid mantle. (**a**) Mantle (head end at left) tissue delaminated to show relationships of the tunics and the muscle. The inner and outer skin have been omitted. (**b**) Diagram of a block cut from the outer tunic showing the crossed-fibrillar arrangement of the large collagen fibres (7 μm diameter). (**c**) Diagram showing the arrangement of muscle in blocks of circular muscle cells separated by layers of radial muscle cells. In *Lolliguncula brevis*, the muscle layer is 20 times the thickness of the outer tunic. Arrows indicate the longitudinal axis of the animal (WARD and WAINWRIGHT, 1972; courtesy of *J. Zool., Lond.*).

WARD and WAINWRIGHT (1972) have suggested the following mode of operation of the various components. Power stroke is accomplished by contraction of circular muscles that decrease the diameter of the mantle and increase the thickness of the mantle tissue. The tendency for mantle tissue to lengthen during power stroke appears to be opposed by three mechanisms: the rigid chitinous pen, the low fibre angle, crossed-helical collagen fibre system in the tunics, and the low fibre angle intramuscular fibres.

Further, just as the graded series of giant neurons ensure simultaneous contraction of circular musculature throughout the mantle in a power stroke (YOUNG, 1938), the pattern of helically-wound, highly elastic collagen fibres in the tunics ensures that any local perturbation in this contraction is resisted by dispersing its force over the largest possible area of the mantle.

The squid mantle is a contractile open hydraulic locomotory system whose substance is a closed hydrostatic tissue. The deformation of the tissue is controlled both by the active muscles and by two tensile fibre systems. The integration of its function is controlled physically by the crossed helical array of collagen fibre in the tunics and the rigid chitinous pen. It is controlled physiologically by the stellate ganglion through graded sizes of giant axons. These mechanisms ensure a smooth, rapid contraction of all circular muscles at once in power stroke. The return stroke may be largely driven by radial muscles, but the elastic recoil in both low angle fibre systems will aid in driving and in controlling the uniformity of return stroke.

7.11 Backbones

The notochord of amphioxus (*Branchiostoma*) is unique among axial supportive structures in that it contains striated muscle whose contraction increases the notochord's bending modulus (FLOOD *et al.*, 1969). The notochord consists of a row of disc-shaped cells close-packed like a stack of coins (Fig. 7.29), wrapped in a fibrous sheath. The locomotory muscle segments are attached to the sheath. From outside in, the sheath consists of a layer of longitudinally oriented collagen fibrils, a layer of circumferentially oriented collagen fibrils and an 'elastic interna' of unknown composition. The muscle in the notochordal cells is in the form of sarcomeres of thick and thin filaments extending from right- to left-hand sides of each cell. The striation repeat distance is about 33 μm, and the thick filaments resemble paramyosin filaments of the anterior byssus retractor muscle of the mussel *Mytilus*

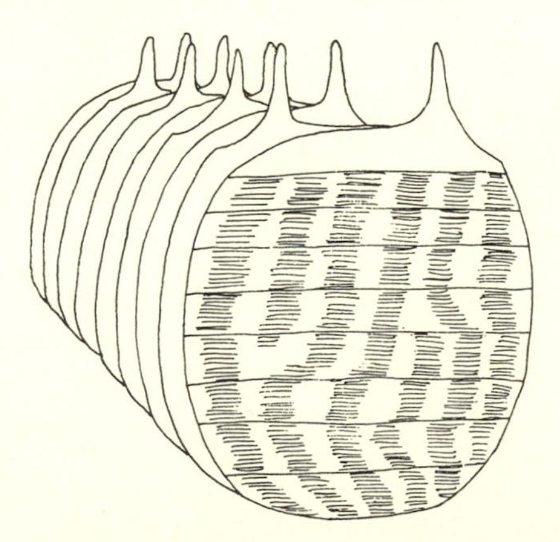

Fig. 7.29 A section of the notochord of amphioxus. Each segment is a single vacuolated cell. Striated muscle fibres within each segment extend from side to side of the notochord. A dorsal extension of each segment forms a typical myoneural junction with the spinal cord that lies directly above the notochord.

edulis in their 14.5 nm axially repeated structure, their solubility and the structure of the precipitate. When the notochord is bent and then stimulated electrically, active movements of the muscle-containing cells are visible and the force tending to straighten the notochord increases. FLOOD *et al.* (1969) determined a pulse duration (4 ms) and frequency of stimulus (4 Hz) that evoked the greatest tension and noted that the maximum tension occurred at moderate flexion. Since no dimensions were reported, one cannot calculate stress or modulus from the published figures. The behaviour of excised bits of notochord in response to electrical stimulus was characterized by summation and a physiological relaxation time of 75 s. The suggestion was made that this long relaxation time makes it unlikely that the changes in stiffness of the notochord produced by its own muscle are synchronous with the tail beat that is driven by the body musculature at frequencies of several beats per second during swimming.

No lateral contraction or other change in cross-sectional dimensions were noticed in the notochord while its muscles were active. If body musculature and sheath are pulled away, the notochord flattens laterally, and Flood suggested that body musculature prevents lateral flattening by creating a hydrostatic force.

The vertebral column of quadrupeds is generally viewed as an arched member whose structure and attached muscle and ligaments transmit the force of gravity of the supported body weight as compressive stresses along the spine to the shoulder and pelvic girdles (GRAY, 1968; NACHTIGALL, 1971; SLIJPER, 1946). From there, in an animal standing motionless, the force is resisted by compressive stresses in the legs.

Clearly this is an oversimplification and accounts functionally for only a part of the total spine. The spine gives rigidity to the body, and the tendency in larger terrestrial animals is to increase the amount and proportion of bony material in the spine. However, the spine, consisting of vertebrae and intervertebral discs cannot be thought of as a rigid structure in the sense used in this book. Every spine (except those of turtles and some fishes and birds) has some flexibility due to metamerically arranged elements of pliant material: cartilaginous intervertebral discs give the spine as a whole a rather low stiffness over the range of deformation it experiences in the intact animal. The spinal columns of many small animals have very weak elastic components to the first few degrees of bending strain. This flexibility is controlled in the intact animal by the muscles in antagonistic sets on opposite sides of the spine that bend it back and forth. In this way mechanical properties of the spine will vary according to muscle tensions on opposite sides.

So the function of 'backbones', whether or not they contain mineralized vertebrae, is complex. Every spine has a measure of compression-resisting rigidity and bending flexibility. The pattern of compression-resisting bony elements is stereotyped for modern vertebrates, but it has apparently not always been so. PARRINGTON (1967) discussed this in terms of the spinal columns of early tetrapods in which each vertebra is represented by a cluster of bones. Since it is impossible to experiment with long extinct animals, Parrington turned to model building and constructed a possible representation of the spine (Fig. 7.30) that he compared to a model of a spine of a modern vertebrate in which each vertebra is a single bone. Figure 7.30a shows that the model in which each vertebra is represented by several bones with adjacent faces at angles other than 90° to the

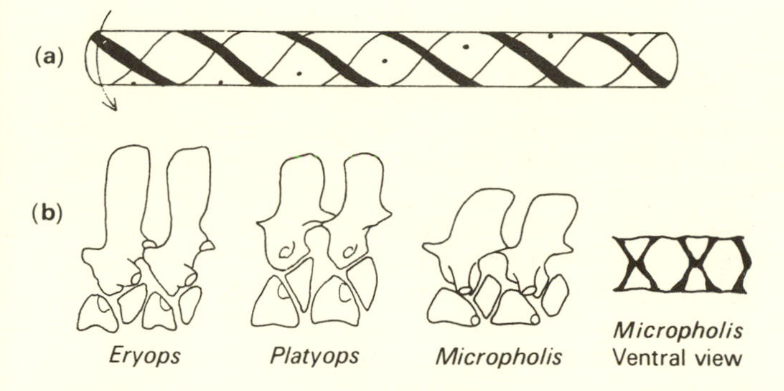

Fig. 7.30 (a) A model for a column of rhachitomous vertebrae. The model is made by fastening rigid blocks (vertebral components) to a solid rubber rod (notochord). Torque (arrow) causes differential separation. Dots are at the centres of blocks that lay on a line parallel to the rod's axis before torsion. (b) Diagrams of arrangements of vertebral components from some early tetrapods (PARRINGTON, 1967).

long axis is more readily twisted through greater torsional angles than are the rectilinear vertebral columns of modern vertebrates, thus allowing roll between the head and the tail. Even if this exercise may one day be shown to be irrelevant to the behaviour of labyrinthodonts, it does illustrate another possible mechanism for flexibility of vertebral columns.

In the swiftest running quadrupeds such as cheetahs and greyhounds, the spinal column flexes with each stride in full gallop. This effectively increases the length of each stride over that possible without body flexure (HILDEBRAND, 1959, 1962). The force exerted by the trunk musculature in straightening the body is added to that of the legs in producing forward motion.

Is it conceivable that the hydrostatic abdominal cavity whose internal pressure is under control of the muscular diaphragm and the external abdominal muscles may play an active role in cycles of flexure and straightening in galloping and leaping quadrupeds? As the hind legs push off in a power stroke, the body must be held rigidly straight if the forefeet are to be advanced in the largest possible stride. A high internal abdominal pressure asserted by the diaphragm and peripheral muscles would provide a compression-resisting component to contribute to that of the spine held rigid by its musculature. Further, the dorso-ventral flexing of the body could be accomplished in part by ventral abdominal muscles stretching the backbone over the viscera that would be acting as a hydrostatic compression element. If this were so, then the entire complex of spine-plus-dorsal musculature would be under tension.

The possibility that running and jumping mammals may either use their backbones in tension or use tension mechanisms as shock absorbers for the axially compressed spine is being investigated by Vogel (personal communication).

7.12 Stressed Tissues

Living tissues that generate mechanical force play a significant role in the design of organisms. Muscle tissue in animals is the most obvious one and accounts of its structure and function fill many books. Less widely known is the reaction wood in trees discussed in Section 7.6.3 and the ability of the roots of many plants to contract (GALIL, 1958). Another tissue that is constantly stressed is human nasal septal cartilage that divides the nasal cavity. This cartilage is stressed, and a lesion in one side of it will allow tensile stresses in the intact side to curl the septum toward the intact side.

FRY and ROBERTSON (1967) have studied the source of this stress. Since samples of nasal cartilage did not lose their contractility upon being repeatedly frozen in liquid nitrogen and thawed to room temperature, cellular metabolic activity was ruled out. Incubation of samples in papain and hyaluronidase together completely abolished the contractility. Neither enzyme alone had a measurable effect on elastic properties, but both treatments caused the tissue to lose most of its sulphated glycosaminoglycan (chondroitin sulphate). Since neither enzyme alone caused any loss of collagen, it is probable that protein affected by papain is part of a protein-polysaccharide matrix surrounding the collagen. The effects of collagenase treatment were an initial loss of 'elasticity' whereby the samples became more 'plastic'. The residual stresses and the elasticity, however, were not completely abolished until the enzyme had rendered the sample an amorphous mass with only 34% of the original dry weight remaining. The reader can now appreciate the need for quantitative data on clearly defined mechanical properties.

Since both matrix and collagen framework are apparently required for tissue contractility, and since elasticity is characteristic of the framework, it would appear that the matrix is acting as a compression-resisting component. The high affinities for water of dispersed protein–polysaccharide complexes in connective tissues provide an organism with a source of compression resistance, i.e., water that is accumulated and held by the investment of a very small amount of polymer.

Surgeons have learned that living cartilage is not to be regarded as a mechanically inactive material. For example, GIBSON and DAVIS (1958) describe the stressed outer fibrous layer of intercostal cartilage that can be used, in carefully carved grafts made from this tissue, as a mechanical agent that will actively reshape a face.

The presence of stressed fibres in a composite material will give a higher tensile modulus in the direction of stress, and presumably this is effected at no sacrifice of density. It seems to be an economic way for organisms to maximize EI and E/ρ wherever it is appropriate.

This opens the consideration to the many different elastic mechanisms that oppose muscle contraction in animals: rubber-like proteins in insects, bivalves and mammals; cartilaginous tubes that extend the brachioles (gills) of fan worms (sabellid polychaetes); salp and ascidian tests of unknown materials; medusa mesoglea; etc. All of these are adaptations providing a strain-recovering function that does not require ATP for its operation.

7.13 Safety Factors

When an engineer speaks of a 'safety factor' he implies a factor of ignorance which allows for situations he cannot take into account. His design is formulated on the basis of a number of simplifying assumptions that he must make in order that it be amenable to mathematical analysis. Some of these assumptions involve the material properties, others involve a simplification of the mathematics, and yet others involve assumptions about the state of stress in the member.

Now, in principle at least, no assumptions need be made and, for complex cases, very detailed and mathematically accurate analyses can be done if the cost is not too great–the increased use of computer techniques is providing the designer with a tool of great power which enables him to perform ever more sophisticated calculations. But even the computer cannot supply all the answers; particularly it is not usually possible to get useful information on the statistical chances of either overload or accident.

In order to cope with all these imponderables the engineer normally designs his structure on the basis of a 'safety factor'–a numerical factor is built into the calculations so that, under reasonable loading conditions, the maximum stress or strain which the material will withstand is never attained. Thus, say, if it is the yield stress of the material that represents the critical loading condition, one might decide that the design will be based on one half the measured yield stress, giving a safety factor of two. Each of the analyses of bending, compression, buckling, etc., described earlier would be so treated. In this way the engineer provides a 'reserve' (of strength, deflection, energy absorption, etc.) against the possibility that an unknown, and hence uncalculable, effect will arise.

Clearly, living organisms have safety factors built into them. In their case, however, the safety factor is not arbitrary but is of such a value that it is favoured by natural selection over any other value. This is a complex point, and we shall develop it slightly.

Apart from the concept of a safety factor, engineers make use of the idea of the 'life expectancy'. This is the mean (or mode or some other statistical parameter according to requirements) life of a population of supposedly similar structures exposed to real conditions of service. For the most part methods of assessing life expectancy depend on the assumption that damage caused during service is irreparable. This is true of most engineering materials and structures, but is not always the case in biology. At the same time biological structures have one disadvantage not shared by engineering artifacts: they must grow and therefore pass through weaker stages before they reach the fully operational reproductive stage. In fact organisms often adopt different ways of life at different stages in their life history, and what we shall say will notionally refer to the adult stage, but can clearly be extended, with appropriate modification, to other stages.

In trying to describe the idea of a global safety factor, which takes account of all the characteristics of the individual, we must try to consider the statistics, in particular the probability that an overload (causing failure) will occur during the stage we are considering. Clearly, such an overload cannot occur often, probably never if the organism is to survive. At the same time it is obvious that the structure may be able to withstand a large number of underloads.

Taking a hypothetical population of animals, we can suppose that the probability, for an individual, of one of its limbs being subjected to loads of various severities will be something like Fig. 7.31. (Note that this is on probability paper, so the ends of the distribution are drawn in.) This shows that the chance of a limb being subjected

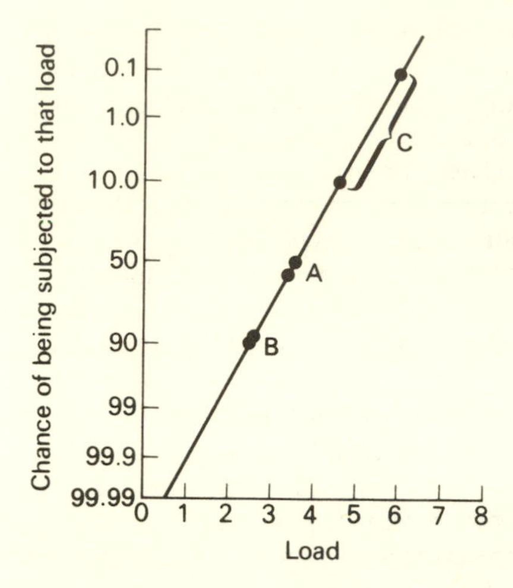

Fig. 7.31 Graph of the hypothetical relationships between load (abcissa) and chance of being subjected to a load at least as great (ordinate). Pairs of points at A, B and C refer to pairs of organisms of which the weaker (lower left in each case) has a selective disadvantage of 0.1. Note how the difference in loads that can be borne increases as the probability of being loaded that much decreases.

to a low load, of one unit, is high, 99.9% for the line we have drawn. The chance of being subjected to a high load, six units, is conversely low, 0.1%. Now if an animal's limb breaks when subjected to a particular load, and this causes the death of the animal, we wish to know what load, in life, the limb would be designed to withstand. It is obvious from the diagram that the animal must be able to withstand a load of 2 units, because it has a 97% chance of being loaded to this extent. On the other hand it is improbable that selection would increase its strength until it could withstand loads of 7 units, because it has less than a 0.01% chance of being loaded that much. Thus insuring against the rare event would make the animal weigh more, use more metabolic energy in moving around and making its skeleton, and take longer to produce offspring.

Suppose we have an individual of a particular genotype, such that the interaction between the genotype and the environment makes it have limbs that fail at some load. We can state the probability that it will be loaded to this load in some period of time (say before reproducing). Call this probability P. Suppose we have two kinds of animals in the species, whom we call weakies and toughies, with different values of P. We assume for convenience that all weakies have the same value of P, and all toughies the same value. Suppose $P_{\text{weakie}} = 0.55$, and $P_{\text{toughie}} = 0.50$. They are represented by the points A on Fig. 7.31, and their limbs will fail at loads of 3.45 and 3.55 units. Clearly, for every hundred toughies that survive only 90 weakies will. Weakies are said to have a selective disadvantage compared to toughies of 0.1. It is the proportion of survivors that counts. Values of $P_{\text{weakie}} = 0.91$ and $P_{\text{toughie}} = 0.90$ will produce the same selective difference, but now the slaughter will be much greater (Fig. 7.31B). Conversely, if $P_{\text{weakie}} = 0.01$, and $P_{\text{toughie}} = 0.001$, then ten times as many weakies will be killed as toughies, but because so few animals are killed anyhow, the proportions of the survivors (0.99:0.999) are so similar that the selective disadvantage of being a weakie is only 0.009.

Would genes making toughies out of weakies spread through a population of weakies? It depends, of course, on the *disadvantages* of being a toughie. We have itemized some of the disadvantages of being strong already. If the disadvantages of being strong outweigh the advantages of not being loaded excessively once in several lifetimes, then clearly the gene will not spread, and indeed a population of toughies would be expected to turn into one of weakies in the course of time.

In general the smaller P, the probability of meeting too large a load, the less likely that changes will take place in a population to decrease P still further. This is because the cost of doing so, in terms of increased weight, etc., will at least remain the same, or even increase, while the advantage of being stronger diminishes. The table shows this latter point.

P_{weakie}	$P_{toughie}$	Selective disadvantage (because of accidents) of being a weakie
0.5	0.05	0.4737
0.4	0.04	0.3750
0.3	0.03	0.2784
0.2	0.02	0.1837
0.1	0.01	0.0909
0.05	0.005	0.0452
0.01	0.001	0.0090
0.001	0.0001	0.0009

Note that in the conditions chosen here, always ten times as many weakies are killed as toughies.

The figure shows the converse of this: if the same selective advantage or disadvantage is chosen (0.1) then the amount stronger than the weakies that the toughies have to become for the same selective advantage becomes much greater as they become stronger (Fig. 7.31A, B, C).

Unfortunately, we do not yet know anything about the actual shapes of the curve we have drawn, or of the disadvantage of being stronger. However, this qualitative analysis does emphasize one point: by the time that accidental death is not very common, the selective advantage in being stronger is not large, and therefore the compensating disadvantages will become as important. We would not expect, therefore, the majority of animals to have very large safety factors. We should not be surprised, for instance, that bones get broken quite often in the wild; even the small cost of making the bones stronger outweighs the advantages of not breaking the bones so often. SCHULTZ (1939) showed how fine the balance can be. He shot 118 wild gibbons, mainly *Hylobates lar.* Among these 118 animals he found the following bones with healed fractures: humerus 11, radius 3, ulna 5, femur 21, tibia 5 and fibula 3. There were also 26 other bones with fractures. Even if gibbons with healed fractures are somewhat easier to shoot than others, these figures show that many gibbons break a bone, and presumably many die from so doing. Of course, gibbons must be particularly prone to violent accidents. (When a gibbon misjudges his landing, he is in real trouble, but a bird merely makes a heavy landing.)

The following example (CONNELL, 1961) shows, how in a particular species, the selective pressure for a strong skeleton will be greater in some circumstances than in others. The whelk *Nucella lapillus* lives, amongst other places, in Ireland. Whelks living in an exposed, wave battered shore (Carrigthorna) had shells markedly thinner-walled than did those from a quiet shore 1 km away (The Rapids). In experimental cages the crab *Carcinus*, a common local predator of *Nucella*, preferentially chose the Carrigthorna animals, and in fact had difficulty in crushing the shells of Rapids animals. When exposed on a slate in a current, on the contrary, the Carrigthorna animals could remain in place much more easily than the Rapids animals. More natural experiments confirmed these findings. At Carrigthorna, Carrigthorna animals, having been moved a few tens of metres and allowed to re-attach, mostly remained in place after several stormy high tides, whereas Rapids animals, allowed to re-attach at Carrigthorna, tended to wash off. Conversely, when animals from both populations were transported to a quiet bay, many more of the Carrigthorna animals were eaten by crabs. Crabs cannot reach the *Nucella* at Carrigthorna usually, because the rocks are too exposed and wave washed. The Carrigthorna animals are able to hold on better than the Rapids animals both because they have a lighter shell and because they have a rather larger foot. We have then, two strategies that allow these two populations to live in particular areas that would be unsuitable for the other population.

Part III
Ecomechanics

Chapter 8
Ecological mechanics

8.1 Introduction

On our tour through the 'innards' of organisms, as we have been, we may lose sight of a bigger, more important view: that plants and animals evolved in so-called natural habitats, not in laboratories. Structural materials, elements and systems in organisms operate over a narrow range of conditions in the life span of an individual. With mutation and recombination as a source of design change and with natural selection as the means of testing the changes and rejecting unfit designs, the individual can be said to be designed to live in a particular set of conditions.

The understanding of the interactions of organisms and environments is the goal of ecology. Because organism–environment interactions are complex and interdependent, important and answerable questions are elusive in ecology. An important question is one whose answer explains a phenomenon or a function. Often, in biology, function at one level of organization (e.g. cellular) is satisfactorily explained by knowledge of changing structure at a lower level (e.g. molecular). It is important to make explanations over as many levels of organization as possible.

In biology today, major areas of research include all levels of integration, but the ecological area is in for unusually strict scrutiny. A common demand is for predictions of the effects of a phenomenon (e.g., an oil spill, the operation of a nuclear power plant, a flood) on the activities of organisms residing in a target area. A common but expensive and unsatisfactory approach is for a team of ecologists to enter the area and attempt to measure every physical and biological factor, to identify every species and count its individuals, and so on. Then, *after* the event, they remeasure and recount and present a profit and loss statement whose entire difference from the first survey is too often judged as undesirable.

The feeling behind this book, and especially this chapter, is that by understanding the physical bases of a particular phenomenon, one can, by applying general principles, predict how any given organism is likely to respond to a particular perturbation in its environment. In the case developed here, the phenomenon is the reliance of mechanical function upon structure, the environmental perturbations are changes in mechanical forces, and the response is either continued function or failure. With terms defined operationally at every step and with the solid background of materials science, this seems to us one good way to approach ecological predictions.

No ecologist would claim that gravity, wind and water flow are unimportant, but as yet no general account of their implications for organisms exists. We give a brief account here to demonstrate the range of important questions that can be illuminated by measuring these forces and their effects on organisms as they occur in nature.

Another supra-organismic topic that is amenable to some structure–function correlations is the reconstruction, by hypothesis, of paleoenvironments. If we can learn what mechanical properties arise from identifiable structures in extant organisms, then we have some reason to suppose fossil structures had similar properties. Having correlated extant properties of organisms with flow conditions in the environment, we may hypothesize that the fossil organism lived in such conditions.

We are simply saying that the biomechanical viewpoint and technology are available to contribute accurate and precise information towards the answers to questions concerning organisms and their environments, past, present and future.

8.2 The Stressful Environment

Environmental forces include gravity, fluid flow, pressure and surface tension. The importance of each of these to an organism depends on its body size. Thus, a small organism will not be in danger of structural failure due to gravity but its life may be governed by surface tension. Atmospheric or external water pressure is a compressive force and structural adaptations to it are not well known. Gravity causes failure by bending in stems and legs. Surface tension is a tensile force that will keep an organism afloat if its mass is small enough or, in the case of swans and other large water birds, if the organism has a suitably great 'non-wettable' surface. Surface tension can also collapse a wet gill in air by tension if the laminae of the gill are not strongly supported. Air and water flow present potentially dangerous shear forces to organisms with a high ratio of surface to volume. For organisms of great mass, flow amounts to a great pressure from one direction: trees are bent and non-rooted animals are simply pushed around.

Gravity and surface tension are constant. Pressure pulses on marine organisms vary with waves and tides. The duration of air and water flow pulses varies from a few seconds (waves), to 2×10^4 s (tides) to 10^7 s (seasonal currents). Structural materials may be adapted to function at any of these periods of stress, and it is clear that viscoelastic properties have always been very important to organisms. Further, animals have behavioural responses that help them cope with forces lasting up to about 10^4 s long, and plants and animals have physiological and developmental responses that can adjust resistance or compliance to stress of from 10^3 s (16+ min) to 10^8 s (3+ years) duration. The important parameters of fluid flow are its velocity, direction, duration and frequency. After reviewing what we have already said about adaptations of self-weight to gravity, we will treat only fluid flow.

8.2.1 Adaptations to Gravity (Mass)

Gravity is a threat chiefly to those larger organisms whose bodies are surrounded by air. Reduction of mass of the supportive system is the critical feature allowing mammals, birds and large insects to fly. Some plants, notably bamboo and other grasses, and large fleet animals such as cats, dogs and gazelles have gone in for extremes in the specialization and use of tensile materials, the reduction in the amount and density of compressive materials and optimization of the design of elements (*EI*) to allow further reduction of weight. Since speed on land is achieved by longer legs or wings (Chapter 7) and since resistance to buckling requires disproportionate increase in thickness with increase in length (Chapter 6), and since weight increases with thickness, there is a compromise on land between size and speed: the largest are not the fastest as they are in the water (GRAY, 1968).

8.2.2 Adaptations to the Velocity of Flow (Strength and Rigidity)

The property of flowing water that can cause failure of biomaterials is *drag*. In very slow flow at Reynolds numbers less than 10 (see ALEXANDER, 1968), the viscosity of the fluid interacting with the surface of the object is the most important component of drag. In rapid flow and Reynolds numbers above 400, the viscous interaction is less important than the pressure gradient around the object. This 'pressure drag' has been determined empirically to be roughly proportional to $(\text{velocity})^2$ whereas 'skin drag' is proportional to velocity. Organisms may oppose drag by being rigid or they may avoid excessive drag either by being flexible and bending with the flow, thus reducing the area perpendicular to flow and pressure drag, or by operating to keep pressure drag within functional limits by a suitable and usually viscoelastic tolerance of strain.

8.2.3 Rigid Stony Corals

The rigid opposition to wind and waves is best seen in trees on windy mountain tops and in the stony corals (Class Anthozoa, Order Scleractinia). Most trees bend in high winds but coral skeletons are so highly mineralized (90% to 99.6% by weight of aragonite: SILLIMAN, 1846; WAINWRIGHT, 1963) that they have no possibility of being flexible. The great branching elkhorn coral, *Acropora palmata*, lives on surge-swept reef tops around Caribbean islands. Stony coral skeletons are exceedingly brittle and elkhorn coral branches are broken to rubble by hurricanes, but it grows up from the bases very rapidly and is an important component of the community on seaward margins of the reef-top. The elkhorn coral thrives only in the areas where water motion is great, its strategy for survival is that of orienting its rigid branches to the direction of the current (SHINN, 1966). This will be discussed in Section 8.2.7.

Most stony corals are massive, subspherical lumps of porous aragonite. In spite of their mass (thousands of kg), these will be torn from the sea floor before they will be broken. Some are encrusting and avoid drag by living entirely within the boundary layer of the reef. Delicately branched corals, including the brittle precious octocoral *Corallium rubrum*, grow in deeper water where surge forces are not important.

8.2.4 Compliant and Tensile Grasses, Seaweeds and Spider Webs

The opposite strategy to being rigid is to comply—to bend and stream in the direction of flow. Leaves and flower stems of grasses bend in the wind. This either reduces pressure drag by reducing the surface area that is normal to the wind, or the plant may be pushed far enough into the boundary layer that wind speed is insufficient to cause dangerous stresses. Although bending stresses are important here, there will be an overall tensile stress in these stems and blades. Principles 1a, 1cii, 2a and 4a in Section 7.2 tell us that this is the easiest sort of stress for which to design resistance in materials and elements: align structural macromolecules in the direction of stress, and ignore the cross-sectional shape. We therefore believe that the round, hollow cross-section of grass stems is an adaptation to resist failure by local buckling in bending.

Most of the large benthic algae, or seaweeds, comply with water motion. As do the grasses, they stream out with the flow, thus putting their holdfasts and thalli into tension. Seaweeds see a very different mechanical world from that of terrestrial plants. Sea water is more dense and has a higher viscosity than air. Because they have a density near that of sea water, the slightest wave, tide, or current will spread the fronds and expose them to the sun. Kelps, the largest of the seaweeds, occupy all the known zones of water movement near rocky temperate shores (NEUSHUL, 1972). In flow past a solid surface, velocity is zero at the surface and increases sharply through the next few millimetres. Spores attach to solid surfaces where velocity is lowest. Sporelings germinate and grow within a zone of low velocities (10^{-2} to 10^{-1} m s^{-1}). Juvenile plants grow through the surge zone (1 m s^{-1} maximum; reciprocal flow) and continue to grow until their tops contain gas-filled floats and occupy surface layers in the current zone (1 to 10 m s^{-1} flow where frequency of reciprocal flow cycles is 0.5 to 1.0/day or *ca.* 10^5 s). Released spores, of course, must retrace these steps, and the mechanical world of spores and sporelings is one of the few mechanical features of marine environments that has been looked into (see NEUSHEL, 1972, for review).

Results of DELF's (1932) tensile tests on seaweeds indicate clearly that there is a range of tensile, time-dependent properties in the stipes of various genera. Because tensile loads were added stepwise and intermittently, breaking strengths and elastic moduli from Delf's work are not strictly comparable to values from either static loading tests or creep tests. Nevertheless Delf found that tensile breaking strength of stipes was independent of plant size or whether the plants grew in sheltered or exposed coastal areas. The exception to this was that tensile strength of young *Laminaria* (stipe cross-sectional area = 30 mm^2) was about one fifth of that of older plants (area = 75 mm^2).

Figure 8.1a shows that *Fucus serratus* was found to fracture at strains approaching 10%, while *Ascophyllum* showed a lengthy plastic region and did not fail below 25% strain and *Laminaria* failed at intermediate strains. Delf slowly loaded an *Ascophyllum* thallus beyond its elastic limit and recorded the gradual attainment of a final strain between 0.25 and 0.30 (Fig. 8.1a, curve B, and Fig. 8.1b).

In tensile tests, DELF (1932) noted that stipes of *Laminaria* and *Ascophyllum* broke cleanly in a transverse plane, while two species of *Fucus* broke along irregular, conical surfaces. The microscopic structure of *Ascophyllum* and *Fucus* are similar in having a central core of filaments, but because they fracture differently, Delf suggested that mucilaginous substances in the cell walls may play an important role in determining the elastic properties of the stipe. The polarized light photomicrographs of sections of *Laminaria* by G. ANDERSEN (1956) show that the thick, transparent cell walls in the centre of the stipe have positive axial birefringence and that cell walls in more peripheral tissue show radially positive birefringence. Material of low birefringence, believed by its solubility

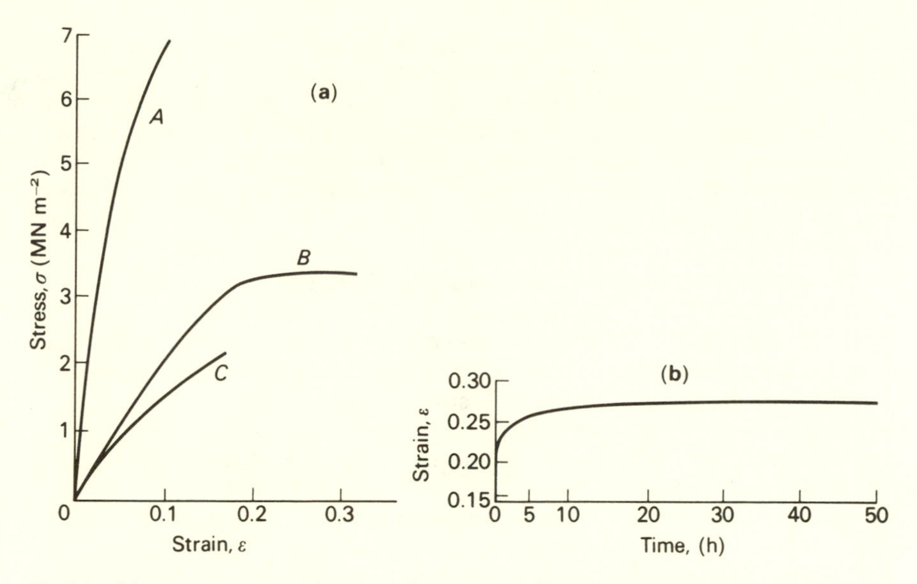

Fig. 8.1 (a) stress–strain curves for stipes of seaweeds. *A*, *Fucus serratus*; *B*, *Ascophyllum nodosum*; *C*, *Laminaria digitata*. All tests shown ended in fracture. Loads were added stepwise. (**b**) Graph showing the slow extensions of *Ascophyllum* loaded over a few minutes to 1.8 MN m^{-2} (a and **b** after Delf, 1932). Courtesy of the *J. exp. Biol.*

characteristics to be polymerized alginic acid, may be selectively removed, and the highly birefringent cellulosic cell walls that remain are easily separated. We believe Delf's hypotheses regarding the functional integrity of seaweeds are well worth pursuing.

In spite of Delf's findings that strength of stipe in *Fucus* is independent of the degree of wave action in the plant's habitat, her other information indicates an important range of viscoelastic properties in various genera. Further information on variation of mechanical properties of kelp stipes in habitats with different wave activity is discussed below.

The fronds of a giant kelp plant 20 m long are floated by gas-bladders in the alongshore currents from a depth of 5 to 15 m (Fig. 7.8c). The buoyant force oscillates due to surface waves and creates purely tensile stresses in the kelp stipe. Tensile structures are interesting in that the principles of their design are few and simple. A simple cable of any cross-sectional shape will do. If reinforcing filaments are involved, they should be oriented longitudinally, and they need be joined to one another to assure that tensile forces pull along the molecular chains.

The morphology of kelp stipes has not been extensively studied, but they are circular or elliptical in section and composed of elongated cells that are aligned longitudinally—a thick outer cortex grading into a central medulla. In the medulla of the stipe of *Laminaria* (OLTMANNS, 1922) the cells are the most elongate and form a three-dimensional network with transversely oriented cells. These cells are joined end-to-end by widened perforated ends, and it is suspected that this medullary tissue functions to translocate materials in solution. Cell walls are thicker toward the centre of the stipe where they are separated by an immensely thickened middle lamella-plus-primary wall layer. Cellulose appears to be confined to the secondary wall layers next to the cell and the bulky $M + P$ layer's main constituent is the polyuronic acid called alginic acid (G. ANDERSEN, 1956). Because the bulk of the extracellular material in the stipe is an alginic acid gel, we may predict that the mechanical properties of the stipe will have strong time dependence. It will be especially interesting to learn the correlation between the values of these properties and the flow velocities over wave and tidal cycles. We would expect the kelp's properties to be closely tuned to the environmental force-time pattern.

The stipe of *Eisenia arborea* has been analysed as a nonlinearly bending thin rod (MITCHELL, 1957) by CHARTERS *et al.* (1969) who also estimated the Reynolds number (10^4–10^5) and the skin and friction drag coefficients for this kelp. (ALEXANDER, 1968, defines these terms in a lucid introduction to the relevant fluid dynamics.) Charters *et al.* studied stipes of plants from a bed exposed to water velocities estimated to reach greater than 4 m s^{-1} and from a bed protected from swift currents. The stipes studied were 40 cm long and had elliptical cross-sections.

Average values for I and E for swift water plants were found to be 1.78 cm^4 and 2.34 MN m^{-2} respectively, and for calm water plants 1.29 cm^4 and 1.42 MN m^{-2}. Over the 30° to 70° bends tested, the curvature of the stipe was that predicted by Mitchell's slender bent rod formula. Skin drag was seen to be much smaller than pressure drag, and all drag decreased with increasing velocity. As flow velocity increased the flexible stipe and the blades bent further in the flow direction. The bending stipe lowers the crown of the plant further into the zone over the substratum where baffling by other organisms and rocks causes decreased flow velocity. By bending with the flow, the fronds reduce their projected area exposed at high angles to the flow. Since drag is reduced by reduction both in velocity and in projected area, the ability of the entire plant to bend and its strain-related modulus are adaptations to the flow regime the plant inhabits.

It is indeed encouraging to see a study of this nature where functional mechanical differences are correlated with mechanical features of the habitat and dimensions of the stress-resisting element. We may wonder what further morphological adaptations are shown by kelps living on the Anglesey coast where JONES and DEMETROPOULOS (1968) have recorded water velocities up to 14 m s^{-1}.

Without more detailed information from cyclic stress–strain tests at various frequencies and stress–relaxation tests we cannot assess the roles of the cortical and medullary tissues of these plants in their resistance of tensile stress. Especially intriguing is the question of the role of the alginic acid-rich outer cell walls in the medulla. It may take a second or more for a large storm wave to develop its maximum force by pulling on the floats and fronds of large kelps such as *Nereocystis* and *Macrocystis*, and a means of absorbing this strain energy would be advantageous. If alginic acid is present as a polymer in a viscoelastic gel with a high modulus at times up to a few seconds, it could well be the shock absorbing mechanism.

The structural carbohydrates are more diverse in the cell walls of the Brown Algae than in any other group of plants. Since these algae include plant forms from tiny filaments of cells (*Sphacelaria*) and crusts of leathery tissue (*Colpomenia*) to the floating bushes of *Sargassum* and the giant kelps discussed here, it is reasonable to suppose that some of the carbohydrate diversity displayed by the Brown Algae (see ANDERSEN *et al.*, 1969) constitutes evolutionary experiments with different materials in the solution of mechanical problems. If this is not so, how else are we to view the restriction of truly giant and wave-resistant plant forms to this one Division of the Plant Kingdom?

It is predictable that elastic moduli of kelp stipes are very much time-dependent and that cell walls such as those described here will have different master curves from *Nitella* cells or *Juniper* tracheids (Chapter 5). The fact that all plant bodies are built up as aggregates of their filament-reinforced cylindrical cells makes it imperative that we learn, both by theoretical development and by looking at more plant species, the range of mechanical solutions attainable by this mechanism.

Orb webs of spiders exemplify design principles of least weight structures. Mark Denny has unselfishly shared the following information from his unpublished work. The strategy of using a minimum weight of material to build a food catching structure of large surface area is achieved in the orb web by the tactics of (1) using materials in tension only, (2) ensuring that the adhesive and energy-absorbing properties of the materials include the ranges of strain and time duration of stresses that are set up by flying and struggling prey, (3) distributing silk in threads of varying size in such a way that stresses are equilibrated throughout the web and (4) building the catching area as a plane that thus equals the projected area of the catching structure in the directions from which prey are expected to arrive.

The materials used are two kinds of silk (see Sections 3.2–3.2.3): viscid silk that is coated with an adhesive substance and framework silk that supports the viscid silk. Denny found that framework threads of *Araneus seracatus* break at stresses up to 1.6 GN m^{-2} and strains up to 0.3, whereas viscid threads are weaker (0.5 GN m^{-2}) but sustain strains up to 4. These dramatically different properties occur in two materials that are chemically quite similar. Table 3.3 shows that the most profound difference between these silks is that much of the alanine of framework silk is replaced by proline in viscid silk (Section 3.2.3).

The web is a viscoelastic structure that may be anchored to rigid or to highly flexible objects. The web must absorb the impact of an airborne insect and accommodate to its struggles without breaking. Most insects the size of a honey bee can struggle free of the sticky threads if given the minutes necessary to do this. The spider, however, usually arrives to further ensnare the prey well before a minute has elapsed, so the adhesive is adequate. As a large insect such as a locust struggles in a web, the greatest length of a forceful thrust it can give to the viscid thread is the length of the thrusting leg, and since there is nothing to thrust against except the weight of the insect's body and the resistance of the threads, such thrusts do not impart enough force over a great enough distance to break the highly extensible threads. The higher strength and modulus of the framework threads resist impact of flying insects while the extensibility of the sticky viscid threads allows them to absorb the energy of struggling.

Figure 8.2a shows the geometric features of the web of *Araneus*. The framework of this web is further abstracted in Fig. 8.2b which shows the magnitude of forces set up in the framework if one of the guys is subjected to an axial tension of 2 units in the plane of the web. Since the number of silk fibres (each

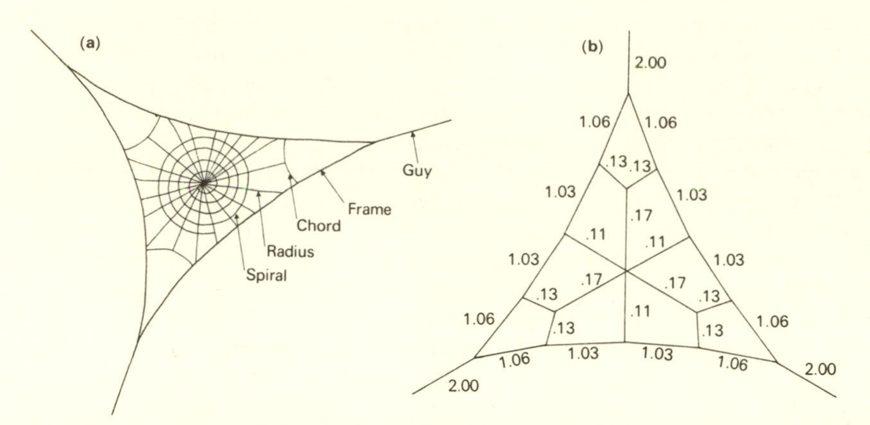

Fig. 8.2 Diagrams of the web of *Araneus*. Radii, frame and guy threads contain about 4, 10 and 20 strands respectively. (**a**) Principal features of the web. (**b**) Simplification of the web in (**a**) showing only the main support threads and the distribution in these threads of forces when a force of 2 is pulling on one guy.

fibre is 3 μm in diameter) that constitute each radius, frame and guy is 2, 10 and 20 respectively, this approximately equilibrates stresses throughout the framework. Maxwell's lemma (Section 7.1.1) predicts that a least volume system occurs when

(1) all members are either in tension or in compression and
(2) all members are equally stressed near their breaking stresses.

Since an orb web satisfies these criteria, it may be regarded as a minimum volume (therefore also minimum weight) structure for resisting forces in the plane of the web.

Wind and wind-blown debris and flying insects exert forces perpendicular to the plane of the web. Anyone who has tried to pull straight a long rope tied to a tree knows it cannot be done. This is because the tensile stress σ_t in a stretched line will be proportional to the force bowing the line divided by sin α where α is the angle of sag. If $\alpha = 0°$, sin $\alpha = 0$ and σ_t is infinite. Individual threads in orb webs therefore must each sag with gravity, and the whole web will bow laterally in light breezes or when struck by flying insects.

In order that a web structure of rigid biomaterials might resist the forces applied to it, the structure would have to be bulky and heavy. But a highly flexible material such as silk allows reduction in the tensile stress by $1/\sin \alpha$ by bowing just as its tensile function contributes to its use in minimum volume. The extensibility of the web is just as important to web survival with wind and gravity as it is against the struggling prey. This is another example of a biological material that, in its functioning system in a mechanically stressful environment, avoids breaking stresses by tolerating large strains.

8.2.5 Drag Control in Air: Trees

The most predominant and interesting strategy for avoiding critical stresses in the face of high winds, waves and currents is that of controlled flexibility. This is an ill-defined category including everything between the rigid coral and the floppy seaweed. The structural materials of organisms in this category are rigid enough to hold the trophic organs (leaves of plants and tentacles of suspension-feeding animals) into the flow that brings their nutrients and allows gas exchange and waste dispersal, but flexible enough to give with unusually high winds or currents and thus to reduce pressure drag.

Trees bend in the wind. Their trunks and branches bend and their leaves fold. This flexibility allows reduction of drag by reducing the projected area acted on by wind force. FRASER (1962) put spruce, hemlock, pine and Douglas fir trees 8.2 m tall into a wind tunnel and recorded the drag forces evoked by wind speeds up to 26 m s^{-1} (50.5 knots). Drag around a rigid body varies with the square of the wind velocity, but for these flexible, stream-lining trees, drag was roughly proportional to wind speed and to the weight of the tree. These relationships hold even when the geometry of the crown of the tree varies: a tree with a small dense crown has a similar drag to a tree with a much larger, more open crown, if their weights are the same. Conversely, two trees with similar crowns have different drag if their weights are different. One tree was tested a second time with all branches of one side removed. Its drag was reduced by the amount predicted by the loss of the weight of the branches.

8.2.6 Drag Control in Water: Passive Suspension Feeders

Passive suspension feeders are those animals that live attached to solid substrata and that rely on ambient water currents for food, gas exchange, sanitation and gamete dispersal. They feed by catching particles out of the water stream as it flows by. We suppose, and lack evidence either for or against, that there is an optimum range of ambient velocity in which each species feeds most effectively. We further suppose that waves and tides routinely have velocities exceeding this optimum range. It is probable that the mechanical properties of the structural materials of these animals are so well tuned to the flow regime occurring in the habitat that the velocity past the food catcher is kept within functional limits.

In Chapter 6 we saw how the cross-sectional shape of the coconut palm petiole is adapted both as a gravity-resisting cantilever maximizing I_x, I_y and J at the base (Table 6.3) and I_y and J away from the base (A in Fig. 6.10a). Torque comes to the palm tree with the wind. The great reduction of cross-sectional area along the petiole from A through B and C in Fig. 6.10a will account for the reduction of I_x and J. In a light breeze, the frond tips (at point C) bend laterally in the *XX* direction. In stronger wind, fronds twist (Fig. 6.10b) and bend in the *YY* direction. In calm air or high wind, the petiole is held firmly onto the tree trunk by tensile fibres in the petiole base (o in Fig. 7.21c) that reinforce the tension-resisting top and corners of the petiole section near the base (A in Fig. 6.10a). Thus the frond is designed to perform different mechanical functions along its length as required by environmental forces:

(1) to allow bending and twisting with the wind in its distal 2/3,

(2) to resist torsion due to wind forces and bending due to gravity in its basal 1/3 and
(3) to hold firmly to the trunk by fibres in tension.

There are other trees that resist high winds, but none that is regularly exposed to gale force winds or that grows as fast or as tall as the coconut palm. It appears that, for the following mechanisms the coconut palm has the most intricate design for wind compliance of any large organism yet studied: (a) by maximizing I and J in the basal part of the petiole, (b) by reducing the sectional area of the distal 2/3 of the petiole, (c) by splitting the broad leaf into narrow, pointed leaflets that can stream with the wind and (d) by making use of the tensile properties of fibres in reinforcing the entire structure of the wood, the petiole and the leaflets.

For example, consider the plume-like gorgonian *Pseudopterogorgia americana* that lives in surge-swept zones of Caribbean reefs (see KINZIE, 1973). Ciné films taken on a calm day show that the entire area undergoes constant to-and-fro

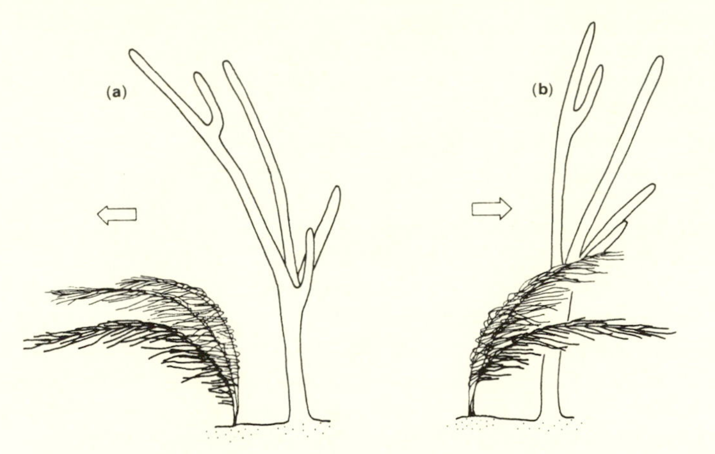

Fig. 8.3 To and fro motion of gorgonians in surge. *Plexaurella nutans* is tall and rigid; *Pseudopterogorgia americana* is short and flexible. Note that shoreward flexure (**a**) is greater than seaward flexure (**b**). Images are traced from ciné film.

motion and that the many gorgonian species living there sway with the flow. Most do so rather rigidly, only bending slightly in the main axis. *P. americana* also bends slightly in the main axis, but the terminal branches are completely compliant and flop like seaweeds (Wainwright, unpublished). So at each change of flow direction, the colony is bent and the 10 to 20 cm long terminal branches stream with the current (Fig. 8.3), and no matter what direction the ambient current comes from, each polyp faces it with its upstream face. In a colony 30 cm tall, the total to-and-fro travel of a single branch tip or feeding polyp may be 30 cm. During the travel from maximum 'to' to maximum 'fro' positions, the flow velocity experienced by a polyp will be much less than the ambient current. This slowing down of ambient current should provide the feeding polyp with more time to catch potential food particles.

8.2.6 Drag Control in Water: Passive Suspension Feeders

Many supple gorgonians such as the sea fan *Gorgonia* live among the rigid elkhorn corals in the surge zone. Sea fans may look delicate, but the strategy of controlled flexibility combined with the strengthening tactic of anastomosing the fine branches allows sea fans to survive the hurricanes that break up the rigid and brittle branches of elkhorn coral.

Typically a gorgonian colony's height is 20 to 50 times its diameter, and most colonies are so flexible that a branch can easily be tied in a tight knot while still wet, and it will return to its unstrained shape if it is released soon. It is not easy to break them or to cut their axial skeletons in order to collect them from the reef. It is easier to chip the brittle limestone reef away around the holdfast with a geologist's pick than it is to cut the axis with a knife or a pick.

Holaxonians dominate the reef top and slopes in many Caribbean reefs, but they are a relatively insignificant component of the reef communities in the tropical Pacific. However, in the Pacific, living in more protected but still shallow places, there is a diversity of scleraxonian gorgonian sea feathers, whips and fans. The axes of these species are composites of microscopic calcareous spicules and organic material whose chemical and mechanical nature is wholly unknown. *Melithaea* is so rigid and brittle that the 4 cm diameter axis of a fan 50 cm across can be broken off by hand by a diver. In contrast, *Subergorgia* axis is tough and 'woody' but still partially mineralized and requires cutting with a blade. Primary and secondary branches of *Melithaea* are jointed, with alternating rigid and flexible sections, and the joints are shaped so that they prefer to bend in the plane of the fan: the major axis of the elliptical cross-section is in the plane of the fan (Muzik and Wainwright, personal observations). Much remains to be learned about mechanical design of these animals.

Two sorts of alcyonacean corals that inhabit shallow Pacific reefs have still other tactics by which they resist ambient flow up to a certain velocity, and thereafter avoid high velocity flow by bending. (1) Colonies of some nephtheids are hydrostatically supported, and muscles in the colony wall can contract causing the entire colony to shorten to a fifth or less of its extended length. (2) The alcyoniid genera *Lobophytum*, *Sarcophyton*, and *Sinularia* are dominant on some reefs. The basic support of these colonies is a voluminous transparent mesoglea that contains isolated calcareous spicules ranging in size from 10 μm to 10 mm in length. The mesoglea is perforated by a network of endodermal tubes. Some species have flexible, waving branches 20 cm long and are able to contract the colony by as much as one half extended height. Most species have less contractile colonies and, instead, a larger volume fraction of spicules in their mesoglea. This makes a very tough material that feels like vulcanized rubber. It responds elastically to rapidly applied and released forces, but recovers slowly from very forcefully made strains or strains resulting from stresses of long duration. It is not known what role hydrostatics play in this.

If the strategy of soft corals is to keep drag from reaching high values by elastic bending, there are clearly at least three tactical approaches to this solution as exemplified by the skeletal materials: (1) the rigid central axis of gorgonians, (2) the hydrostatic support of nephtheid colonies and (3) the spicules-in-a-gel support of various alcyoniid genera. From what we have seen in Chapter 5 about the range of possible structural arrangements and resultant properties in different spicular materials, we can predict that each of these three tactical approaches may include structural material having a wide range of macromolecular and microscopic structure.

8.2.7 Adaptations to the Direction of Flow (Anisotropy)

That the postures and activities of plants and animals are oriented to the direction of wind and water flow is widely appreciated. Particular attention has been paid to planar or fan-shaped seaweeds, hydroids, soft and hard corals, bryozoans and crinoids (THÉODOR and DENIZOT, 1965; MEYER, 1971; RIEDL, 1966, 1971*a*; RIEDL and FORSTNER, 1968; SVOBODA, 1970; GRIGG, 1972; KINZIE, 1973; WAINWRIGHT and DILLON, 1969; VELIMIROV, 1973). The reader should refer to the reviews by RIEDL and SCHWENKE (1971) for a more broadly based view of the implications of water movements.

In a study of fan-shaped objects, WAINWRIGHT and DILLON (1969) argued that planar objects with centrally placed holdfasts will be in equilibrium with directional flow in only two positions: parallel and perpendicular to the flow. Where populations of oriented planar organisms occur, the azimuth of flow varies a few degrees according to tidal phases and shifts of the wind. Small deviations in flow direction will render parallel orientation difficult to maintain, and this reasoning is advanced to explain the observed orientation of planar organisms perpendicular to prevailing current.

Wainwright and Dillon also report that while populations of large fans of *Gorgonia* show preferred orientation parallel to reef edges and perpendicular to wind-driven surge, the smaller fans showed no preferred orientation. They suggested that flow is turbulent and therefore random in direction, a few cm off the reef surface where small fans grow amongst a clutter of attached seaweeds, sponges and corals. Only when the fans are tall enough and extend into the directional flow are they exposed to the orienting force. On the basis of growth increments visible in sections of the fans' axes, they concluded that small fans initially oriented randomly to the directional current become reoriented passively more or less perpendicular to the current. According to this hypothesis, the new orientation achieved passively is accompanied by creep in the axial material and by new, unstrained growth increments that bury and preserve the old pattern.

The information to be gleaned about flow conditions in the environment from measuring the preferred orientation of planar organisms is useful in several ways. It is possible to predict flow conditions from photographs taken by unmanned cameras at the bottom of the sea. These photographs can be taken anywhere, even in the oceanic depths. Where the orientation of planar organisms is preserved in fossil beds, at least the direction, if not the velocity, of currents can be guessed.

Certain species adapt to either a more-or-less constantly directional flow or to conditions where there is little or no preferred flow direction. RIEDL (1966) has observed that the featherlike *Aglaophenia* species (*A. septifera, A. tubulifera, A. pluma octodonta*) have a planar shape in directional flow (Fig. 8.4). When grown in currents from several directions, the planes of the colonies of closely related *A. pluma dichotoma, A. elongata* and *Lytocarpia myriophyllum* are twisted (Fig. 8.4). In another hydroid genus, *Eudendrium rameum* grows in unidirectional currents and its branches are arranged in a plane, whereas *Eudendrium ramosum*

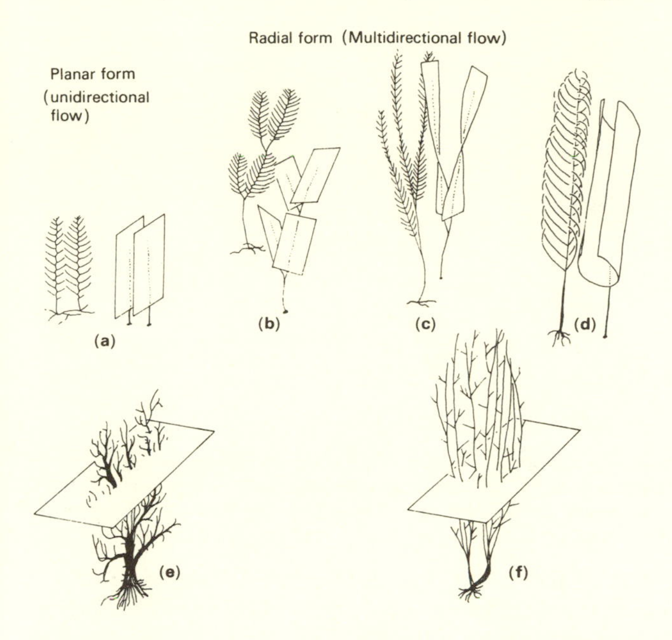

Fig. 8.4 Colony shape of hydroids correlated with the degree of preferred directionality of ambient flow. (a) *Aglaophenia septifera, A. tubulifera, A. pluma octodonta.* (b) *A. pluma dichotoma.* (c) *A. elongata.* (d) *Lytocarpia myriophyllum.* (e) *Eudendrium rameum.* (f) *Eudendrium ramosum* (after RIEDL, 1966).

grows in more turbulent water and has a bushy form. The tendency for individual colonies of bushy organisms to take on planar form and to orient the plane perpendicular to directional flow is known also in gorgonians such as the Caribbean *Plexaurella*, the tropical Pacific *Plexauroides* and *Ellisella* (Muzik and Wainwright, personal observations), and the temperate *Leptogorgia* (LEVERSEE, 1972).

Most supportive materials of oriented plants and animals are anisotropic. Macromolecular and mechanical anisotropy should be characteristic of any material that is adapted to resist directional stresses. Reaction wood in trees (see Section 7.6.3) is one such material whose structure is modified and oriented to the direction of environmental forces. The basal axis of the gorgonian genus *Plexaurella* that grows in directional flow contains polycrystalline calcareous bodies that are distributed in a plane perpendicular to the flow direction, thus giving preferential flexibility in the flow direction and increased rigidity in other directions (Fig. 8.5). This mechanical anisotropy assures a particular spatial separation of the branches of the suspension-feeding network that is held into the passing current. Again, controlled flexibility allows for the functioning of a rigidly supported structure in slow flow and the avoidance of high drag values by bending when the flow is intermittently too fast.

The sea pen, *Scytaliopsis djiboutiensis* has a rigid axis and the plane of its featherlike array of branches is passively oriented by the flow to a position perpendicular to the direction of flow (MAGNUS, 1966). Change in orientation is apparently accompanied by shear in the mesoglea and allows a total range of

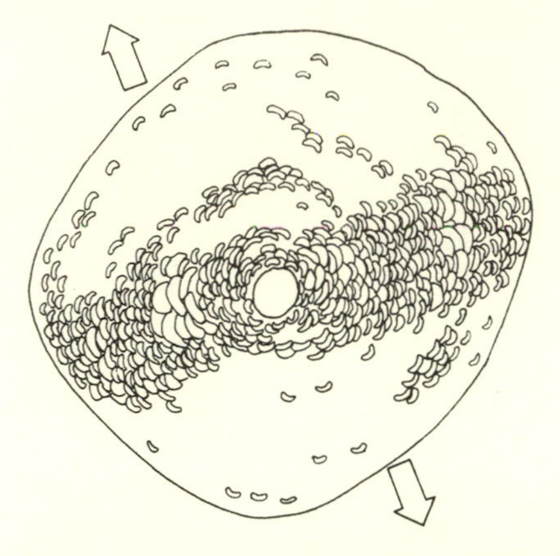

Fig. 8.5 Diagram of transverse section of the axial skeleton of *Plexaurella nutans*, a gorgonian from the Caribbean. Most of the axis is organic but it is reinforced with polycrystalline aggregates of $CaCO_3$ shown as bean-shaped bodies in section. As the colony grows in height into a predictable direction of flow, the mineral is preferentially deposited in a plane perpendicular to the flow direction (arrows).

reorientation through 180°. Given current changes of greater angles than ±90° from resting, the rootlike base of the colony slowly turns in the sediment. This is a very slow reaction requiring many minutes or hours per degree of turn. It is not known if this whole-colony rotation is active or passive.

Comatulid crinoids extend a fan-shaped array of featherlike arms perpendicular to the current. Most species orient themselves with the convex aboral side of the arms upstream. The arms are uncurled and held as semirigid beams across the current for hours at a time. MEYER (1971) has pointed out that the large ligament running down the aboral side of each arm presumably holds the arm erect and is, in fact, not a muscle. It belongs to a category aptly called 'problematical ligaments' that occur in various echinoderms. These ligaments appear by their histochemical characteristics and the longitudinally periodic banding seen in electron micrographs to be collagenous. Yet there is evidence (TAKAHASHI, 1967) that they contract and can remain contracted without consuming energy against applied force for long periods of time. Hence, it is called a *catch mechanism*. If the dorsal ligament of the crinoid arm is indeed a catch mechanism, it is in just the right place for a contractile, low-energy stiffening mechanism to keep the suspension-catching arms spread into the current. The mechanism does not contract rapidly or often.

SVOBODA (1970) collected rocks bearing oriented populations of the fan-shaped hydroid *Aglaophenia pluma*. He returned the rocks to their original places after rotating them 90°. He observed that fully grown fans were cast off and new fans grew from their broken bases to have a new orientation 90° to the orientation of the original fans. Fans that had only begun to develop continued to grow, but with a helical twist as shown by *Aglaophenia elongata* in Fig. 8.4c.

In experiments in which this species was grown in a tank and exposed to a bidirectional flow (maximum velocity 0.3 to 0.5 m s^{-1} and one swash cycle every 7 s), similar results were obtained: fully grown fans degenerated and fell off and were replaced by new fans, all of which were perpendicular to the flow.

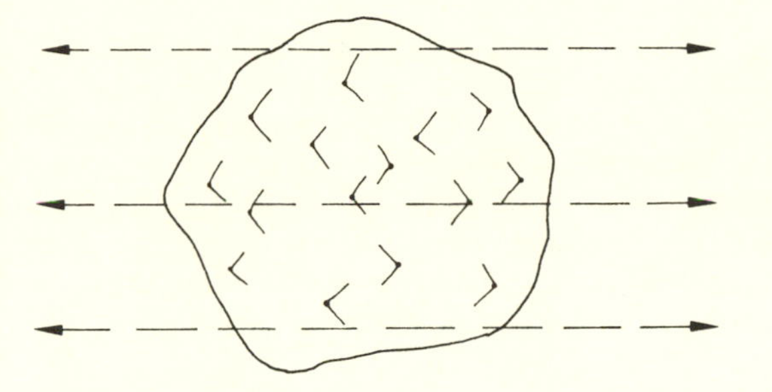

Fig. 8.6 Orientation of the fan-shaped hydroid *Aglaophenia pluma* that grew on a rock in bidirectional flow (arrows) (modified from SVOBODA, 1970).

While many species of attached colonial organisms show a planar form that is an adaptation to bidirectional currents, a few are adapted to unidirectional flow. Among these are the holaxonian genus *Melithaea* and the hydroid genus *Aglaophenia*. The two sides of each fan are different: seen from above, the fan is shallowly concave, and the polyps extend into the concave side. This is called the ventral side of the colony and in fans grown in unidirectional flow, the ventral side faces downstream. In the populations of *Aglaophenia* grown on rocks in bidirectional flow by SVOBODA (1970), half the fans grew up oriented facing one way and half the other (Fig. 8.6).

The concern about orientation of planar suspension feeders is based on the oft stated assumption that a suspension feeding apparatus will catch a larger proportion of the particles from the ambient stream if it is perpendicular to the stream than if it has some other orientation. At this time there is no information about how or in what conditions of flow velocity or direction the polyps or tentacles of any organism actually catch food particles. RIEDL and FORSTNER (1968), in analysing the flow pattern around branches and polyps of a hypothetical sea fan (Fig. 8.7), note that as water flows between branches, its velocity increases because

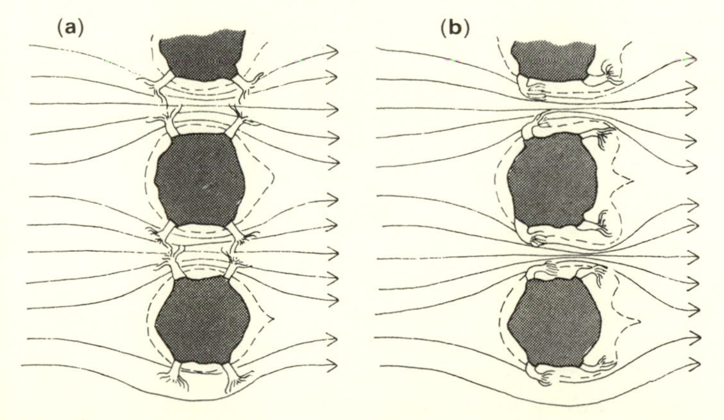

Fig. 8.7 Hypothetical flow velocity profiles and behaviour of polyps of a fan-shaped gorgonian. Solid lines with arrows represent stream lines; dashed lines enclose a region of minimum flow velocity. (a) Slow flow; (b) fast flow (RIEDL and FORSTNER, 1968).

the diameter of the flow passage is reduced. Immediately past the branch, there is an expanded area where there is eddying and a reduced flow velocity. It is known that the polyps of many planar colonial cnidarians and bryozoans are arranged biserially on the edges of branches such that they extend into the space between branches. This posture appears to place the particle-catching tentacles where the particles are moving fastest. As Riedl and Forstner point out, currents too fast for particle catching may well bend the polyps so that tentacles are in the low velocity eddy region on the downstream side of the branch. The point being made here is that the flexibility of the expanded polyp—a property depending on mesogleal structure and hydrostatic pressure in the polyp—is the property that allows the polyps to feed in a narrow range of flow velocity while the colony as a whole is exposed to a wide range of ambient flow velocities. We presume that the polyps could not feed effectively if they were rigid and experienced the high ambient flow velocities.

The only data on the effect planar colony orientation has on feeding is from a study by LEVERSEE (1972) on the temperate, bushy gorgonian *Leptogorgia virgulata*. In the field, he observed that where water currents are highly turbulent and variable in direction throughout tidal cycles, 64% of the colonies show no planar form (thickness: width = 1:1). In a tidal channel where currents are clearly bidirectional, 64% of the colonies are strongly planar in form (thickness: width = 1:4 or 1:5). Leversee placed planar colonies in a recirculating unidirectional flow tunnel with a velocity of 40 mm s^{-1} and added nauplii of *Artemia salina* to a concentration of 20 per litre. Colonies perpendicular to the flow caught 50% of the nauplii in 1 hour while colonies parallel to the flow took 2 hours to accomplish this. Perpendicular colonies regularly caught more ($>$ 90%) of the nauplii than did parallel colonies (80–90%). This allows us to say that it may indeed be advantageous to suspension feeding to be planar and therefore perpendicular to flow, but we do not know what aspect of colony orientation allows this greater catching rate.

If resistance to bending is as important to passive suspension feeders as we suggest it is, the time course of the directionality of water flow is important to those attached species that have directionally oriented form and behaviour. The fact that planar organisms such as the hydroid *Aglaophenia pluma* and the gorgonian *Melithaea* have polyps extending from one surface only leads to the supposition that they may have a greater prey catching rate when the flow is in one direction (with the polyps on the downstream side of the fan) than when it is in the reverse direction.

The growth form of elkhorn coral colonies on the Florida reef tract was observed by SHINN (1966) to vary in different but predictable ways with the distance from the sea surface and with the intensity of surge action. First, Fig. 8.8b shows that, viewed from the side, colonies growing near the sea surface tend to be flat—their branches extend horizontally from the colony base. The deeper the site of growth, the more acute is the angle included between seaward and shoreward branches (Fig. 8.8c, d, e). Second, as seen from above, colonies growing in more sheltered areas tend to branch equally in all directions, whereas colonies in exposed areas have branches extending seaward and landward but not perpendicular to the surge.

8.2.7 Adaptations to Direction Flow (Anisotropy)

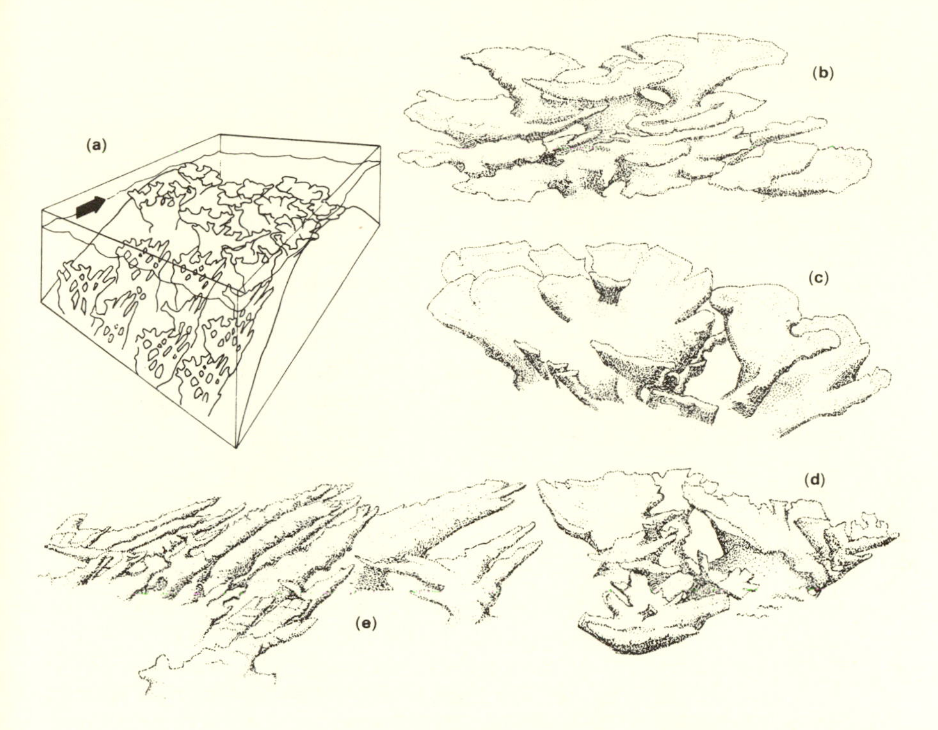

Fig. 8.8 The growth form of the elkhorn coral, *Acropora palmata*, varies with depth and intensity of surge action. (**a**) Stereogram of colonies and water conditions on the reef: depth at seaward edge, 5 m. Arrow: direction of onshore surge. (**b**) Colony form in shallow, protected area is flat and branches in all directions. (**c**) Colony form in deeper protected area shows a smaller branching angle. (**d**) At the same depth but more exposed to surge, the colony shows a preferred orientation of branches up- and downstream. (**e**) In deeper, surge-swept water, all branches are oriented up- and downstream and the branching angle is constant. In (**b**) through (**e**) the surge direction is from left (offshore) to right (onshore). Drawn from photographs.

Of these two aspects of form correlated with a forceful habitat, the second is most obviously advantageous. Where surge occurs, the branches are oriented in such a way that they expose a minimum area to the wave direction, and that they take wave impact as compressive forces along the long axis of the branch. In view of the known brittleness of stony coral skeleton, it makes sense not to put energy into building brittle branches that will be bent in impact.

The tendency to take wave forces as compressive forces may also explain the first aspect of form described. At the sea surface, the maximum velocity of a wave is directed nearly horizontally, whereas at the 2 to 6 m depths where elkhorn coral's vertical branching angle is acute, the maximum wave velocity is directed at a downward angle. We suggest that both these aspects of growth form in corals are adaptations to the direction of maximum wave action. That they depend on a rigid and brittle material being put into compression is further evidence of a rather high degree of specific mechanical design. Purely compressive and purely tensile structures are uncommon in biological systems. That they are found in the equally uncommon situations where the direction of force is highly predictable cannot have been arrived at by chance.

8.2.8 Adaptations to Duration and Frequency of Flow (Stress Rate and Fatigue)
As we have noted above, one advantage of being flexible is that, while a structural element is bending, the stress rate is less than if the element were rigid. This is a mechanism for sustaining impact by absorbing the energy as strain energy. The capability of sustaining a deformation for hours or days and then sustaining a similar deformation in the opposite direction when the tide or current changes is also advantageous. It is difficult to separate adaptations to or effects of frequency of applied force from those of the duration. Perhaps in the example given below, the interstitial animals feed most effectively at zero flow velocity and their neuromuscular systems are tuned to start and stop feeding on the irregular but limited range of flow pulse frequency they experience.

8.2.9 Meiofauna and the 'Stormy Interstices'
The most serious shortcoming in our knowledge of the mechanical behaviour of marine epibenthic animals and plants in flow is that we have little reliable data on the velocity of flow at the level of the organism. We have a little information on the direction of flow and none on the duration and frequency of flow pulses shorter than tidal cycles. In sharp contrast to this, RIEDL and MACHAN (1972), RIEDL *et al.* (1972) and RIEDL (1971*b*) have provided us with a detailed account of flow velocities and directions in the spaces among sand grains in the intertidal beaches and shallow subtidal sea of Onslow Bay, North Carolina.

These workers have shown that from the beach, where waves are pushing water into the sediment, to the edge of the continental shelf, oxygenated water is in constant motion in the interstices of the surface sediments. The 63 predominant species of animals living in this habitat belong to 7 major taxa: nematodes, turbellarians and harpacticoid copepods are the most abundant. These animals have the following average diameter/length ratios (given in μm) 25/1500; 100/1500; 50/400 respectively.

On the beach where the waves enter the sand, mean downward flow velocity in the interstices is 780 $\mu m\ s^{-1}$ and the mean seaward velocity is 300 $\mu m\ s^{-1}$. Thus most interstitial animals are exposed to flow rates that are at least as great as their own body length per second with pulses up to 4 $mm\ s^{-1}$ or twice the body length of large animals per second. It is not surprising that these animals are variously arrayed with claws, jaws, suckers and adhesive pads, papillae and bristles by which they attach to sand grains. Adaptations of these animals in the many other aspects of their life style are as yet unknown, but in their behaviour they must be opportunistic and able to move about in the intervals of a few seconds each when flow is least.

The entire question of design principles and structural limitations to mechanical function of very small systems and their materials is virtually untouched. The stormy interstices provide an accessible habitat of a known range of flow velocities with a distinctive fauna of creatures adapted to it: this could provide a new window on the world of the consequences of body size and the interactions between organisms and environmental force.

8.3 Active Suspension Feeders

Active suspension feeders are animals that create their own flow from which they catch suspended food particles. Bivalved molluscs, brachiopods and other lophophorates, tubicolous polychaetes, endoprocts, bryozoans, barnacles, ascidians, cephalochordates and sponges all do this while attached or burrowed into *terra firma*. Larvae of many phyla as well as adult medusae, siphonophores, Crustacea, motile tunicates, planktotrophic fishes and baleen whales cause water to flow through their fishing apparatus by swimming through the water. Since most of the forces they meet are generated with their bodies, they receive little attention here.

8.4 The Informative Environment

A specialty of biological systems is the transduction of information from one mode to another. Sensory cells transduce selected chemical, mechanical, thermal, and electromagnetic signals from the environment to electrical impulses. Muscles reconvert the electrical message to a mechanical response. Hormones are the chemical means by which the information that 'days are getting longer in the Spring' is transduced as the order that gonads must ripen in birds and the order that 'long day' plants should produce flowers. These final sections seek to point out the range of nonmechanical environmental information that is transduced into altered mechanical properties in the organism's support system. No attempt is made here to evaluate these changes in properties in terms of the organisms' well being. This important aspect of each change in each organism must be carefully evaluated by the responsible investigator.

8.4.1 Chemical Information

There are several ways in which chemicals in the environment can become involved in mechanically important phenomena when they are taken up by organisms. (a) 'Impurities' can compete with normal constituents and be secreted into skeletal materials. This can possibly affect mechanical properties of the materials. Sr and Mg compete with Ca with different effectiveness in various species of bone- and shell-secreting animals. The effects of these impurities on mechanical properties are potentially important. (b) Metabolic interference of any sort may overstimulate or inhibit any part of normal synthesis of structural materials. The changes in mechanical properties of these materials that would result from skewed synthesis would be easily measured. (c) Some chemicals that are resistant to biodegradation and that do not directly affect oxygen or nutrient concentrations or light penetration may have profound effects on the viscosity of the water thus upsetting organisms that pump it, swim in it, and filter or trap food from it. Detergents lower the surface tension of water and cause insects and birds that normally float due to their buoyancy (induced by labile wettability-control compounds) to fall through the water surface.

8.4.2 Thermal Information

Changes in environmental temperature have profound effects on the synthetic processes of plants and cold-blooded animals. The nature and rate of locomotion, feeding and other mechanical behaviours are affected by ambient temperature. This is partly due to the fact that nerve and muscle function involve chemical reactions whose rate is temperature dependent, but there is another mechanism, purely physical in nature, by which organismic biology may be affected by temperature. This is described in Chapters 2 and 4 as the temperature (and time) dependent elastic behaviour of polymeric materials. This phenomenon has been studied over an environmental range of temperatures in very few structural materials and, for the most part, organisms and their materials are found to be 'tuned' in such a way that mechanical properties do not change much over the temperature range they experience. All studies to date have dealt only with the dynamic testing of excised bits of materials: no one has yet analysed the behaviour of an organism to learn what role may be played in its behaviour by the temperature-dependent properties of its constituent materials.

Lenhoff (personal communication) has noted that hydra in a defined culture medium at 20°C are notably stickier and harder to dislodge from glass surfaces than they are at 15°C. Is a temperature dependent stimulus to secretion of mucus responsible for this? Or is it a change in the adhesive properties of the mucus? The implication of the latter mechanism is of interest in the seasonal (temperature) control of settling (adhesion) of larvae of sessile organisms.

The possibility exists that abnormal or unexpected temperature changes in an environment may cause structural materials in organisms to behave in a part of their retardation time spectrum that is totally inappropriate.

The frequency of rhythmic muscular activities in many poikilothermic animals is known to vary with ambient temperature. SASSAMAN and MANGUM (1970) recorded the frequency, duration and amplitude of spontaneous longitudinal contractions of the column in anemones, *Diadumene leucolena*, at 10°C, 17.5°C and 27.5°C. Prior to experiments, animals were kept at 10°C. The amplitude increased ten-fold from 10°C to 27.5°C but, in four of the five animals, the frequency of contraction of the animals at 27.5°C decreased to 19% of that at 10°C. By comparison with other cnidarian muscular rhythms the increase in amplitude was expected, but the decrease in frequency with temperature was not. As pointed out by ALEXANDER (1964), muscles of cnidarians pull less against other muscles than they do against rather massive viscoelastic mesoglea. Therefore, physical conditions of the environment that affect the properties of the support system thereby affect the behaviour of the whole organism.

GOSLINE (1971*b*) showed that the modulus of mesoglea isolated from *Metridium senile* dropped an order of magnitude over the temperature range 3°C to 24°C and then rose again by a factor of 5 from 27°C to 32°C. A decrease in modulus will mean that the resistance to a given muscular force will decrease, and perhaps the increased amplitude of contraction in intact *Diadumene* depends in part on decreased mesogleal modulus. It is well to realize that neither of these anemones may ever experience the higher experimental temperatures used in these studies, and that if they could not survive at these temperatures, the altered mechanical properties of their support system may be partially to blame. Whatever the cause(s) may be, the mechanical behaviour of these animals has changed considerably over the stated temperature range. How such behavioural changes over environmental temperature ranges normally affect feeding rates or other aspects of survival of these or other poikilothermic animals is not known at this time.

8.4.3 Rheological Information

Flowing fluid is a source of mechanical force and has been discussed earlier in this chapter. We presented the idea that each organism lives in a habitat that can be characterized by a range of the velocity, direction, duration and frequency of air or water flow. In this section we consider changes in the environment that may result in a new range of flow parameters. Changes in the ambient current regime may render a habitat totally unsuited to an organism. This may result in the organism leaving the habitat if it can, or it may result in decline in normal function and death if it cannot.

Here we wish to point out the value of measurable mechanical phenomena to the business of monitoring changes in the environment. All changes are not detrimental. In most cases where change has caused one species to depart or die out, one or more different species may expand or immigrate to take its place. The kinds of changes man proclaims as detrimental are those that reduce the number of species and those that reduce species diversity.

Change in flow velocity may have mechanical or non-mechanical effects. Decrease in flow velocity is likely to have effects unrelated to mechanics that are detrimental to the existing community. Stagnation means starvation to passive filter feeders and may mean anoxia for all organisms that cannot leave the area. Increase in flow velocity is more likely to become limiting in the mechanical sense, but may be as detrimental to some predators as it is to some prey species. The altering of water current direction and magnitude by solid refuse or natural sediments in streams, rivers, estuaries, tidal flats, reefs, etc., is only one aspect of geographical changes wrought by man and nature. Just how important the rheological change is in the resulting changes in distribution of plant and animal species in the habitat is information we badly need.

We are concerned about the environment and its fragility. It is desirable that we learn enough about the tolerances of species to environmental change so that we may avoid making harmful changes. Studying lethal limits of temperature, light, flow factors and concentrations of various chemicals is one approach, and tells us how much species can tolerate. It is equally important to understand the mechanism by which a harmful change renders a habitat lethal or otherwise intolerable for a species. If this mechanism is understood then the specific feature of the pollutant that is harmful may be amenable to treatment to nullify its harmful effects.

It is here proposed that the study of the structure and mechanical properties of supportive materials and systems in macroscopic plants and animals is an operationally feasible and theoretically sound means of acquiring knowledge of criteria by which effects of environmental changes on organisms can be observed, measured and, ultimately, understood.

8.5 The Next Few Years

So impressive is our ignorance of the sensitivity and response of plants and animals to mechanical information in their environment, that it is easy to predict a decade of exploration in this field. Environmental forces and organismic responses must be recorded, and then mechanisms of the responses are to be explained. Perhaps we can predict the source of thoroughly new concepts to most molecular, cellular and organism biologists: we have taken directional and reciprocal flow for granted for so many years that we are a long way from thinking analytically about materials that withstand these forces for years on end. Medical engineers have solved a few problems about how to make replacement parts of the human body that will survive the forces they must for the time they must. But materials that are adapted to ocean wave forces lasting a few seconds and those adapted to tidal frequencies of several hours and those that are designed to cope with seasonal changes in direction, velocity, and duration of flow have properties we have not yet measured. We will find the time-dependent mechanical properties of structural biomaterials to be the most challenging phenomena to explain. These explanations will surely lead us to new concepts of macromolecular structure, especially the means of cross-linking the long, straight-chain structural proteins and polysaccharides. How do sliding linear molecules retain their integrity and what is the nature of their immediate surroundings? What roles are played by water and nonaqueous solvents and their ionic or nonionic solutes?

We look forward to being able to study the structure of a shell, a bone, or what have you, and to predict something about the mechanical conditions of the habitat in which its owner lived. Allied to this sense of hindsight is a predictive sense we should have about what will happen to the organisms along a bit of coastline if conditions are altered. If we can identify pathological structures in terms of environmental conditions, we should even be able to suggest steps to take to improve or avoid those conditions. We do not believe that biomechanics holds the secrets to all important problems, but we do believe that it embodies a useful point of view.

References—Author index

Titles of journal articles given in parentheses are abbreviated descriptions of the articles. Italic numbers in parentheses refer to citations in this text.

ABENDSCHEIN, W. and HYATT, G. W. (1970). Ultrasonics and selected physical properties of bone. *Clin. Orthop.*, **69**, 294–301. *(174, 183)*

ABRAHAMS, M. and DUGGAN, T. C. (1964). Mechanical characteristics of costal cartilage. In *Biomechanics and Related Bioengineering Topics*, R. M. Kenedi, ed. Pergamon Press, Oxford, pp. 285–300. *(143)*

AKLONIS, J. J., MACKNIGHT, W. J. L. and STEIN, M. (1972). *Introduction to Polymer Viscoelasticity*. John Wiley, New York. *(28, 29, 31, 54)*

ALEXANDER, P. and EARLAND, C. (1950). Structure of wool fibres. *Nature, Lond.*, **166**, 396–7. *(189)*

ALEXANDER, R. McN. (1962). Viscoelastic properties of the body wall of sea anemones. *J. exp. Biol.*, **39**, 373–86. *(127–9, 307, 310)*

ALEXANDER, R. McN. (1964). Viscoelastic properties of the mesogloea of jellyfish. *J. exp. Biol.*, **41**, 363–9. *(312, 366)*

ALEXANDER, R. McN. (1966). (pectinid inner ligament) *J. exp. Biol.*, **44**, 119–30. *(52, 114, 116, 334)*

ALEXANDER, R. McN. (1967). *Functional Design in Fishes*. Hutchinson University Library, London. *(3)*

ALEXANDER, R. McN. (1968). *Animal Mechanics*. Sidgwick and Jackson, London. *(3, 15, 118, 333, 349, 352)*

ALEXANDER, R. McN. (1971). *Size and Shape*. Edward Arnold, London. *(3, 248)*

ALEXANDROV, W. G. and DJAPARIDZE, L. I. (1954). (effects of lignin on cell wall structure) *Planta*, **4**, 306. *(323)*

ANDERSEN, G. (1956). (alginic acid and birefringence of kelp tissue) *Second Int. Seaweed Symp.* Pergamon Press, Oxford, pp. 119–24. *(350, 352)*

ANDERSEN, S. O. (1967). (phenolic cross-links in bivalve ligament protein) *Nature, Lond.*, **216**, 1029–30. *(116)*

ANDERSEN, S. O. (1970). Amino acid composition of spider silk. *Comp. Biochem. Physiol.*, **35**, 705–11. *(74, 79, 81)*

ANDERSEN, S. O. (1971). Resilin. In *Comprehensive Biochemistry*. M. Florkin and E. H. Stotz, eds., **26-C**, 633–57, Elsevier, Amsterdam. *(111–13, 165)*

ANDERSEN, S. O. (1973). (sclerotization in larval and adult cuticle of *Schistocerca gregaria*) *J. Insect Physiol.*, **19**, 1603–14. *(163)*

ANDERSEN, S. O. and BARRETT, F. M. (1971). (isolation of ketocatechols from insect cuticles) *J. Insect Physiol.*, **17**, 69–83. *(158, 160, 285)*

ANDERSON, F. R. (1964). (structure of polyethylene) *J. appl. Physics.*, **35**, 64–70. *(68)*

ANDERSON, N. S., CAMPBELL, J. W., HARDING, M. M., REES, D. A. and SAMUEL, J. W. B. (1969). (structure of some polysaccharides) *J. molec. Biol.*, **45**, 85–99. *(122)*

ANDREWS, E. H. (1964). (strain crystallization of natural rubber) *Proc. Roy. Soc.*, **A277**, 562–70. *(70)*

ANDREWS, E. H. (1968). *Fracture in Polymers.* Oliver and Boyd, London. *(40, 42, 177)*

ARNOLD, J. M. and ARNOLD, K. O. (1969). Hole-boring by octopus. *Am. Zool.*, **9**, 991–6. *(273)*

ASCENZI, A., BONUCCI, E. and BOCCIARELLI, D. S. (1965). (EM of osteon calcification) *J. Ultrastr. Res.*, **12**, 287–303. *(169, 172)*

ASTBURY, W. T. (1933). *Fundamentals of Fibre Structure.* Oxford University Press, London. *(188)*

ASTBURY, W. T. and WOODS, H. J. (1933). (structure and properties of hair keratin) *Phil. Trans. R. Soc. Lond.*, **A232**, 333–94. *(191)*

ATKINS, E. D. T. (1967). (structure of some insect fibrous proteins) *J. molec. Biol.*, **24**, 139–41. *(81)*

AUSKERN, A. and HORN, W. (1971). Some properties of polymer-impregnated cements and concretes. *J. Am. Ceram. Soc.*, **54**, 282–5. *(228)*

BAILEY, A. J. (1967). Labile intermolecular cross-links in tendon collagen. *Biochem. J.*, **105**, 34P–35P. *(88)*

BAILEY, A. J. (1968a). The nature of collagen. In *Comprehensive Biochemistry*, M. Florkin and E. H. Stotz, eds., **26-B**, 297–424. Elsevier, Amsterdam. *(82, 83, 85, 88)*

BAILEY, A. J. (1968*b*). (intermolecular cross-links in collagen) *Biochim. Biophys. Acta*, **160**, 447–53. *(87, 88)*

BAILEY, A. J., PEACH, C. M. and FOWLER, L. J. (1970*a*). Biosynthesis of intermolecular cross-links in collagen. In E. A. Balazs, ed. (1970), *The Chemistry and Molecular Biology of the Intercellular Matrix.* Academic Press, New York, Vol. 1, pp. 385–404. *(86, 88)*

BAILEY, A. J., PEACH, C. M. and FOWLER, L. J. (1970*b*). Chemistry of the collagen cross-links. *Biochem. J.*, **117**, 819–31. *(88)*

BAILEY, K., and WEIS-FOGH, T. (1961). (amino acid composition of resilin) *Biochim. Biophys. Acta*, **48**, 452–9. *(112, 113)*

BAILEY, S. W. (1954). Hardness of arthropod mouthparts. *Nature, Lond.*, **173**, 503. *(221)*

BAILEY, T. L. W., TRIPP, V. W. and MOORE, A. T. (1963). Cotton and other vegetable fibres. In *Fibre Structure*, J. W. S. Hearle and R. H. Peters, eds. Butterworths, London, pp. 422–54. *(100)*

BALAZS, E. A., ed. (1970). *Chemistry and Molecular Biology of the Intercellular Matrix*, 3 vols. Academic Press, New York. *(119, 138)*

BALAZS, E. A. and JEANLOZ, R. W. (1965). *The Amino Sugars.* Academic Press, New York. *(119)*

BARNES, D. J. (1970). (growth of coral skeleton) *Science*, **170**, 1305–8. *(210)*

BARRETT, A. J. (1968). Cartilage. In *Comprehensive Biochemistry*, M. Florkin and E. H. Stotz, eds., **26-B**, 425–74. Elsevier, Amsterdam. *(138)*

BARTH, F. G. (1969). (spider cuticle) *Z. Zellforsch.*, **97**, 137–59. *(162)*

BASSETT, C. A. L. (1965). Electrical effects in bone. *Sci. Am.*, **213** (Oct.), 18-25. *(185)*

BASSETT, C. A. L. (1971). Biophysical principles affecting bone structure. In *Biochemistry and Physiology of Bone*, G. H. Bourne, ed. Academic Press, New York, Vol. 3, pp. 1–76. *(185, 187)*

BATHAM, E. J. and PANTIN, C. F. A. (1950). Muscular and hydrostatic action in the sea anemone *Metridium senile. J. exp. Biol.*, **27**, 264–89. *(127, 306, 307)*

BECKER, G. W. (1961). (stress relaxation in polyethylene) *Kolloid Z.*, **175**, 99–110. *(71, 92)*

BENDIT, E. G. (1957). The α–β transformation in keratin. *Nature, Lond.*, **179**, 535–6. *(191)*

BENEDICT, J. V., WALKER, L. B. and HARRIS, E. H. (1968). (stress-strain characteristics of unembalmed human tendon) *J. Biomech.*, **1**, 53–63. *(88)*

BENNET-CLARK, H. C. (1962). (active control of mechanical properties of insect cuticle) *J. Insect Physiol.*, **8**, 627–33. *(286)*

BERGEL, D. H. (1961*a*). Static elastic properties of the arterial wall. *J. Physiol., Lond.*, **156**, 445–57. *(135–137)*

BERGEL, D. H. (1961*b*). Dynamic elastic properties of arterial wall. *J. Physiol., Lond.*, **156**, 458–69. *(138)*

BERGEL, D. H. (1966). Stress-strain properties of blood vessels. *Lab. Pract.*, **158**, 77–81. *(135)*

BERGEL, D. H. and SCHULTZ, D. L. (1971). Arterial elasticity and fluid dynamics. *Prog. Biophys. molec. Biol.*, **22**, 1–36. *(118, 138)*

BEVELANDER, G. and NAKAHARA, H. (1969). (formation of the nacreous layer in the shell of bivalved molluscs) *Calc. Tiss. Res.*, **3**, 84-92. *(229)*

BHIMASENACHAR, J. (1945). Elastic constants of calcite and sodium nitrate. *Proc. Indian Acad. Sci.*, **22A**, 199–208. *(226)*

BIGGS, W. D. (1966). Theoretical background. In *Composite Materials*, L. Holliday, ed. Elsevier, Amsterdam, pp. 28–64. *(144)*

BIRD, A. F. and DEUTSCH, K. (1957). The structure of the cuticle of *Ascaris lumbricoides. Parasit.*, **47**, 319–28. *(302)*

BITTIGER, H., HUSEMANN, E. and KUPPEL, A., (1969). (EM of fibril formation). *J. Polymer Sci.* **28-C**, 45–56. *(97)*

BLACKWELL, J. (1969). (structure of chitin) *Biopolymers*, **7**, 281–98. *(105)*

BLANTON, P. L. and BIGGS, H. L. (1970). Ultimate strength of fetal and adult human tendon. *J. Biomech.*, **3**, 181–9. *(89)*

BLOWER, G. (1951). A comparative study of chilopod and diplopod cuticle. *Quart. J. Microsc. Sci.*, **92**, 141–61. *(164)*

BOCCIARELLI, D. S. (1970). Morphology of crystallites in bone. *Calc. Tiss. Res.*, **5**, 261–69. *(169)*

BOEDETKER, H. and DOTY, P. (1956). Native and denatured states of soluble collagen. *J. Am. Chem. Soc.*, **78**, 4267–80. *(84, 85)*

BØGGILD, O. B. (1930). (the shell structure of the mollusks) *K. danske Vidensk. Selsk. Skr.*, **2**, 232–325. *(213)*

BOULIGAND, Y. (1965). (architecture torsadée répandue dans cuticules d'arthropodes) *Compt. rend. hebd. Séanc. Acad. Sci., Paris*, **261**, 3665–8. *(161, 162)*

BOULIGAND, Y. (1971). Les orientations fibrillaires dans le squelette des arthropodes. *J. Microscopie*, **11**, 441–72. *(161)*

BOULIGAND, Y. (1972). Twisted fibrous arrangements in biological materials and cholesteric mesophases. *Tiss. Cell*, **4**, 189–217. *(161)*

BOUTEILLE, M. and PEASE, D. C. (1971). (structure of native collagen fibrils) *J. Ultrastr. Res.*, **35**, 314–38. *(87)*

BOYDE, A. and HOBDELL, M. H. (1969). Scanning electron microscopy of lamellar bone. *Z. Zellforsch.*, **93**, 213-31. *(171, 172)*

BROOKES, M. (1971). *The Blood Supply of Bone.* Butterworths, London. *(172)*

BRUNET, P. C. J. (1967). Sclerotins. *Endeavour,* **26**, 68–74. *(158)*

BRYAN, W. H. and HILL, D. (1941). (hexacoral growth as spherulitic crystallization) *Proc. roy. Soc. Queensland,* **52**, 78–91. *(210)*

BUCHANAN, J. B. and WOODLEY, J. D. (1963). Extrusion and retraction of the tube-feet of Ophiuroids. *Nature (Lond.),* **197**, 616-7. (*314*)

BUECHE, F. (1962). *Physical Properties of Polymers.* Interscience, New York. (*54, 59*)

BUNN, C. W. (1955). The melting point of chain polymers. *J. Polymer Sci.,* **16**, 323–43. (*66*)

BURSTEIN, A. H., CURREY, J. D., FRANKEL, V. H. and REILLY, D. T. (1972). (ultimate properties of bone tissue) *J. Biomech.,* **5**, 35–44. (*8, 174, 178*)

CAIN, A. J. (1964). The perfection of animals. *Viewpoints in Biology,* **3**, 36–63. (*235*)

CARLSTRÖM, D. (1957). Crystal structure of α-chitin. *J. biophys. biochem. Cytol.,* **3**, 669–83. (*106*)

CARNIGLIA, S. C. (1965). Petch relation in single-phase oxide ceramics. *J. Am. Ceram. Soc.,* **48**, 580–3. (*225*)

CARNIGLIA, S. C. (1966). Grain-boundary and surface influence on the mechanical behaviour of refractory oxides. *Mat. Sci. Res.,* **3**, 425–71. (*225*)

CARRIKER, M. R. (1961). (gastropod boring mechanisms) *Am. Zool.,* **1**, 263-6. (*273*)

CARRIKER, M. R. (1969). (shell-boring by a gastropod) *Am. Zool.,* **9**, 917-33. (*273*)

CARRIKER, M. R., SMITH, E. H. and WILCE, R. T., eds. (1969). (lime-boring by lower plants and invertebrates) *Am. Zool.,* **9**, 629–1020. (*273*)

CHAFE, S. C. (1970). Fine structure of the collenchyma cell wall. *Planta,* **90**, 12–21. (*320*)

CHAN, A. S. L. (1960). (braced frameworks) *Coll. Aeronaut., England. Rep. No. 142.* (*292*)

CHANZY, H. D., DAY, A. and MARCHESSAULT, R. H. (1967). (structure of polyethylene) *Polymer,* **8**, 567–88. (*98*)

CHANZY, H. D. and MARCHESSAULT, R. H. (1969). (structure of polyethylene) *Macromolecules,* **2**, 108–10. (*98*)

CHAPMAN, D. M. (1970). Re-extension mechanism of a scyphistoma's tentacle. *Canad. J. Biol.,* **48**, 931–43. (*310, 312*)

CHAPMAN, G. (1953). (histology of mesogloea) *Quart. J. microsc. Sci.,* **94**, 155–76. (*310, 311*)

CHAPMAN, G. (1959). Mesogloea of *Pelagia noctiluca. Quart. J. microsc. Sci.,* **100**, 599–610. (*310, 312*)

CHAPPELL, T. W. and HAMANN, D. D. (1968). Poisson's ratio and Young's modulus for apple flesh under compressive loading. *Trans. Am. Soc. Agric. Engin.,* **11**, 608–610. (*12*)

CHARTERS, A. C., NEUSHUL, M. and BARILOTTI, C. (1969). Functional morphology of *Eisenia arborea. Sixth Int. Seaweed Symp.* Madrid, pp. 89–105. (*249, 352*)

CHATTERJI, S. and JEFFERY, J. W. (1968). Changes in structure of human bone with age. *Nature, Lond.,* **219**, 482–4. (*171*)

CHAVE, K. E. (1964). Skeletal durability and preservation. In *Approaches to Paleoecology,* J. Imbrie and N. Newell, eds., John Wiley, New York, pp. 377–87. (*273*)

CHIA, F. S. (1973). (heavy sand grains in echinoid juveniles) *Science,* **181**, 73–4. (*262*)

CIFERRI, A. (1961). Present status of the rubber elasticity theory. *J. Polymer Sci.,* **54**, 149–73. (*50, 57*)

CLARK, G. L. and SMITH, A. F. (1936). X-ray diffraction studies of chitin, chitosan and derivatives. *J. phys. Chem.,* **40**, 863–79. (*107*)

CLARK, R. B. (1962). On the structure and function of polychaete septa. *Proc. zool. Soc. Lond.,* **138**, 543–78. (*316, 317*)

CLARK, R. B. (1964). *Dynamics in Metazoan Evolution.* Oxford Univ. Press, London. (*3, 317*)

CLARK, R. B. and COWEY, J. B. (1958). (factors controlling the change of shape of some worms) *J. exp. Biol.,* **35**, 731–48. (*294, 295, 304, 305*)

CLOUDSLEY-THOMPSON, J. L. (1950). (cuticle and water relations of a millipede) *Quart. J. microsc. Sci.,* **91**, 453–64. (*216*)

COBLE, R. L. and KINGERY, W. D. (1956). (voids in ceramics) *J. Am. Ceram. Soc.,* **39**, 379. (*157*)

CONNELL, J. H. (1961). (predation on whelks by crabs) *Ecol. Monog.,* **31**, 61–104. (*344*)

COOK, J. and GORDON, J. E. (1964). A mechanism for the control of cracks in brittle systems. *Proc. roy. Soc.* **A282**, 508–20. (*154–6, 227*)

COOPER, G. A. (1966). Orientation effects in fibre-reinforced materials. *J. Mech. Phys. Solids,* **14**, 103. (*156*)

COOPER, R. R., MILGRAM, J. W. and ROBINSON, R. A. (1966). (morphology of the osteon) *J. Bone Joint Surg.,* **48-A**, 1239–71. (*172*)

CÔTÉ, W. A. Jr. (1968). The structure of wood and the wood cell. In Kollman and Côté (1968), **1**, 1–54. (*322*)

CÔTÉ, W. A. Jr. and DAY, A. C. (1965). (structure of reaction wood) In *Cellular Ultrastructure of Woody Plants,* W. A. Côté, ed., Syracuse Univ. Press, Syracuse, New York. (*100*)

COTTRELL, A. H. (1964). *Mechanical Properties of Matter.* John Wiley, New York. (*14, 17*)

COX, H. L. (1952). Elasticity and strength of fibrous materials. *Brit. J. appl. Phys.,* **3**, 72–9. (*149, 151*)

CREWTHER, W. D., FRASER, R. D. B., LENNOX, F. G. and LINDLEY, H. (1965). Chemistry of keratins. *Adv. Protein Chem.,* **20**, 191–346. (*188, 190, 191*)

CRICK, F. H. C. (1953). (packing of α-helices) *Acta Cryst.,* **6**, 689–97. (*188*)

CURREY, J. D. (1960). (blood supply of different types of bone) *Quart. J. microsc. Sci.,* **101**, 351–70. (*172*)

CURREY, J. D. (1962). Histology of the bone of a prosauropod dinosaur. *Palaeontol.,* **5**, 238–46. (*172*)

CURREY, J. D. (1964*a*). Metabolic starvation as a factor in bone reconstruction. *Acta Anat.,* **59**, 77–83. (*172*)

CURREY, J. D. (1964*b*). Stress concentrations in bone. *Quart. J. microscop. Sci.,* **103**, 111–33. (*182*)

CURREY, J. D. (1964*c*). (bone as a composite). *Biorheol.,* **2**, 1–10. (*175*)

CURREY, J. D. (1967). (exoskeletons versus endoskeletons) *J. Morph.,* **123**, 1–16. (*258*)

CURREY, J. D. (1968). Adaptations of bones to stress. *J. Theoret. Biol.*, **20**, 91–106. (*187*)

CURREY, J. D. (1969*a*). Mechanical consequences of varying the mineral content of bone. *J. Biomech.*, 2, 1–11. (*183, 272*)

CURREY, J. D. (1969*b*). The relation between stiffness and the mineral content of bone. *J. Biomech.*, 2, 477–80. (*183*)

CURREY, J. D. (1970). The mechanical properties of bone. *Clin. Orthopaed.*, **73**, 210–31. (*174, 257*)

CURREY, J. D. (1970*a*). *Animal Skeletons.* Edward Arnold, London. (*229*)

CURREY, J. D. and BREAR, K. (1974). Tensile yield in bone. *Calc. Tiss. Res.*, **15**, 173–9. (*177*)

CURREY, J. D. and NICHOLS, D. (1967). Absence of organic phase in echinoderm calcite. *Nature, Lond.*, **214**, 81–83. (*234*)

CURREY, J. D. and TAYLOR, J. D. (1974). (mechanical behaviour of mollusc shells). *J. Zool. Lond.* **173**, 395–406 . (*230, 231*)

CURREY, J. D. (1975). A Comparison of the strength of Echinoderm spines and Mollusc shells. *J. Mar. Biol. Ass. U.K.* **55**, 419–424. (*234*)

DAVIS, L. E. and HAYNES, J. F. (1968). (ultrastructure of hydra mesogloea) *Z. Zellforsch.*, **92**, 149–58. (*308, 309*)

DAVIS, S. S. (1970). Saliva is viscoelastic. *Experientia,* **26**, 1298–1300. (*122*)

DAVIS, S. S. and DIPPY, J. E. (1969). Rheological properties of sputum. *Biorheol.*, **6**, 11–22. (*122*)

DELF, E. M. (1932). Experiments with the stipes of *Fucus* and *Laminaria. J. exp. Biol.*, **9**, 300–13. (*350, 351*)

DENHAM, W. S. and LONSDALE, T. (1933). (tensile properties of silk) *Trans. Faraday Soc.*, **29**, 305–16. (*78, 79*)

DENNELL, R. (1947). (phenolic tanning in decapod crustacean cuticle) *Proc. roy. Soc.*, **B134**, 485–503. (*160, 216*)

DENNELL, R. (1960). Integument and exoskeleton. In *Physiology of Crustacea,* T. H. Waterman, ed. Academic Press, New York, Vol. 1, pp. 449–72. (*216*)

DENNELL, R. (1973). Structure of the cuticle of the shore crab *Carcinus maenas* (L). *Zool. J. Linn. Soc.*, **52**, 159–63. (*161, 162*)

DIAMANT, J., KELLER, A., BAER, E., LITT, M. and ARRIDGE, R. G. C. (1972). (collagen structure and function in aging) *Proc. roy. Soc.*, **B180**, 293–315. (*87*)

DIGBY, P. S. B. (1968). (lime in the cuticle of the shore crab *Carcinus maenas*) *J. Zool.*, **154**, 273–86. (*216*)

DISALVO, J. and SCHUBERT, M. (1966). Interaction during fibril formation of soluble collagen and cartilage proteinpolysaccharide. *Biopolymers,* **4**, 247–58. (*140*)

DISMORE, P. F. and STATTON, W. O. (1966). Chain folding in oriented nylon 6-6 fibers. *J. Polymer Sci.*, **13C**, 133–48. (*69, 70*)

DOBB, M. G., FRASER, R. D. B. and MACRAE, T. P. (1967). (structure of silk) *J. Cell Biol.*, **32**, 289–95. (*75*)

DOBRIN, P. B. and ROVICK, A. A. (1969) (arterial wall mechanics) *Am. J. Physiol.*, **217**, 1644–51. (*135*)

DOUGILL, J. W. (1962). (upper and lower bound constants of two-phase materials) *Proc. Am. Concrete Inst.*, **59**, 1363. (*146*)

DOW, N. F. (1963). (stresses near a discontinuity in a filament) GEC Rep. TIS R63S D61, Reinforced Composite Metal. (*149*)

DUDICH, E. (1931). (Kalkeinlagerungen des Crustaceenpanzers) *Zoologica* (Berlin), **30**, 1-154. *(216)*

EASTOE, J. E. (1971). Dental enamel. In *Comprehensive Biochemistry,* M. Florkin and E. H. Stotz, eds., **26C**, 785–834. Elsevier, Amsterdam. *(223)*

EINSTEIN, A. (1906). Ein neue Bestimmung der Moleküldimensionen. *Annalen der Physik.,* 4te Folge., **19**, 289–306. *(145)*

ELDEN, H. R. (1968). Physical properties of collagen fibers. In *Int. Rev. Connective Tiss. Res.* **4**, 283–348, D. Hall, ed. Academic Press, New York. *(82)*

ELLIOTT, D. H. (1965). Structure and function of mammalian tendon. *Biol. Rev.,* **40**, 392–421. *(88, 89)*

ELLIOTT, G. F., HUXLEY, A. F. and WEIS-FOGH, T. (1965). On the structure of resilin. *J. molec. Biol.,* **13**, 791-5. *(112)*

ENGEL, J., KURTZ, J., KATCHALSKI, E. and GERGER, A. (1966). (polymers of tripeptides) *J. molec. Biol.,* **17**, 255–72. *(84)*

ENLOW, D. H. and BROWN, S. O. (1957). (histology of fossil and Recent bone). *Texas J. Sci.,* **9**, 186–214. *(172)*

ENLOW, D. H. and BROWN, S. O. (1958). (histology of fossil and Recent bone). *Texas J. Sci.,* **10**, 187-230. *(172)*

ESAU, K. (1936). (properties of collenchyma in celery petioles) *Hilgardia,* **10**, 431–76. *(320)*

ESAU, K. (1965). *Plant Anatomy.* John Wiley, New York. *(320)*

EVANS, F. G. and KING, A. I. (1961). (properties of human spongy bone) In *Biomechanical Studies of the Musculoskeletal System,* F. G. Evans, ed. C. C. Thomas, Springfield, Ill., pp. 49–67. *(167, 174)*

FAUPEL, J. H. (1964). *Engineering Design.* John Wiley, New York. *(16, 245, 249, 256, 294)*

FELTHAM, P. F. (1955). On the representation of rheological results. *Brit. J. appl. Phys.,* **6**, 26–31. *(39)*

FERRY, J. D. (1970). *Viscoelastic Behaviour of Polymers,* 2nd edn. John Wiley, New York. *(28, 31, 59, 72, 73)*

FESSLER, B. F. (1960). A structural function for mucopolysaccharides in connective tissue. *Biochem. J.,* **76**, 124–32. *(123)*

FILSHIE, B. K. and ROGERS, G. E. (1961). Fine structure of α-keratin. *J. molec. Biol.,* **3**, 784–6. *(188)*

FILSHIE, B. K. and ROGERS, G. E. (1962). (structure of feather keratin) *J. Cell Biol.,* **13**, 1–12. *(191)*

FINNEY, E. A. and HALL, C. W. (1967). Elastic properties of potatoes. *Trans. Am. Soc. Agric. Engin.,* **10**, 4–8. *(12)*

FISCHER‘ E. W. (1957). (growth form of high polymers) *Z. Naturf.* **12-A**, 753–4. *(68)*

FJERDINGSTAD, E. J. (1970). Ultrastructure of the spicules of *Spongilla lacustris.* In *Biology of the Porifera.* Edited by W. G. Fry. *Symp. zool. Soc. Lond.* No. 25, 125–33. Academic Press, London. *(209)*

FLOOD, P. R., GUTHRIE, D. M. and BANKS, J. R. (1969). Paramyosin muscle in the notochord of *Amphioxus. Nature, Lond.,* **222**, 87–8. *(337, 338)*

FLORY, P. J. (1953). *Principles of Polymer Chemistry,* Chap. X. Cornell University Press, Ithaca, New York. *(50)*

FLORY, P. J. (1969). *Statistical Mechanics of Chain Molecules.* Interscience, New York. *(50)*

FLORY, P. J. and GARRETT, R. R. (1958). Phase transitions in collagen and gelatin systems. *J. Am. Chem. Soc.*, **80**, 4845. (*83*)

FLORY, P. J., HOEVE, C. A. J. and CIFERRI, A. (1959). Bond angle restrictions on polymer elasticity. *J. Polymer Sci.*, **34**, 337–47. (*57*)

FORSYTH, P. J. E. (1965). Fibre strengthened materials. In *Composite Materials.* Iliffe Books, London. (*144*)

FOX, S. W., and DOSE, K. (1972). *Molecular Evolution and the Origin of Life.* W. H. Freeman, San Francisco. (*287*)

FRANZBLAU, C. (1971). Elastin. In *Comprehensive Biochemistry.* M. Florkin and E. H. Stotz, eds., **26-C**, 659–712. Elsevier, Amsterdam. (*166, 118*)

FRASER, A. I. (1962). Wind tunnel studies of the forces acting on the crowns of small trees. *Rep. For. Res., HMSO, London,* 178–83. (*355*)

FRASER, R. D. B., MACRAE, T. P. and ROGERS, G. E. (1962). Molecular organization in α-keratin. *Nature, Lond.*, **193**, 1052–5. (*188, 189*)

FRASER, R. D. B., MACRAE, T. P., STEWART, F. H. C. and SUZUKI, E. (1965). Poly-1-alanylglycine. *J. molec. Biol.*, **11**, 706–12. (*75*)

FRASER, R. D. B., MACRAE, T. P. and STEWART, F. H. C. (1966). (a model for crystalline silk) *J. molec. Biol.*, **19**, 580–2. (*75*)

FRASER, R. D. B., MACRAE, T. P., PARRY, D. A. D. and SUZUKI, E. (1969). Structure of β-keratin. *Polymer,* **10**, 810–26. (*189*)

FRASER, R. D. B., MACRAE, T. P., PARRY, D. A. D. and SUZUKI, E. (1971). Structure of feather keratin. *Polymer,* **12**, 35–56. (*191*)

FREEMAN, M. A. R., DAY, W. H. and SWANSON, S. A. V. (1971). Fatigue fractures in subchondral bone. *Med. Biol. Engin.*, **9**, 619–29. (*184*)

FRETTER, V. and GRAHAM, A. (1962). *British Prosobranch Molluscs.* Ray Society, London. (*273*)

FREY-WYSSLING, A. (1952). Deformation of plant cell walls. In *Deformation and Flow in Biological Systems,* A. Frey-Wyssling, ed. North-Holland, Amsterdam, pp. 194–254. (*100, 101*)

FREY-WYSSLING, A. (1955). Structure of cellulose. *Biochim. Biophys. Acta,* **18**, 166–8. (*95, 96*)

FREY-WYSSLING, A. (1959). *Die Pflanzliche Zellwand.* Springer Verlag, Berlin. (*96*)

FREY-WYSSLING, A. and MÜHLETHALER, K. (1963). Die Elementarfibrillen der Cellulose. *Makromol. Chem.* **62**, 25–30. (*96, 97*)

FREY-WYSSLING, A. and MÜHLETHALER, K. (1965). *Ultrastructural Plant Cytology.* Elsevier, Amsterdam. (*194*)

FRIEDMAN, E. (1967). (whisker-reinforced composites) Rep. No. AFML-TR-66–362, Air Force Materials Lab, Dayton, Ohio. (*168*)

FRÖHLICH, H. and SACK, R. (1946). Theory of rheological properties of dispersions. *Proc. roy. Soc.*, **A185**, 415–20. (*146*)

FROST, H. M. (1964). *The Laws of Bone Structure.* C. Thomas, Springfield, Ill. (*186*)

FRY, H. and ROBERTSON, W. V. B. (1967). Interlocked stresses in cartilage. *Nature, Lond.*, **215**, 53–4. (*340*)

FRY, P., HARKNESS, M. and HARKNESS, R. D. (1963). (effect of age on rat skin collagen) *Am. J. Physiol.*, **206**, 1425–9. (*133*)

GALIL, J. (1958). (physiology of contractile roots) *Bull. Res. Council Israel, D, Botany*, **60**, 221–36. (*340*)

GARONNE, R. (1969). Collagène, spongine et squelette minéral chez l'éponge *Haliclona rosea. J. Microscopie*, **8**, 581–98. (*209*)

GEBAUER, J., HASSELMAN, D. P. H. and LONG, R. E. (1972). (polymer impregnation of ceramic bodies) *Bull. Amer. Ceramic. Soc.*, **51**, 471-3. (*228*)

GEBHARDT, W. (1906). (structure and function of Haversian systems) *Arch. Entwickl.-Mech. Org.*, **20**, 187–334. (*171*)

GEDDES, A. J., PARKER, K. D. and BEIGHTON, E. (1968). Cross-β conformation in proteins. *J. molec. Biol.*, **32**, 343–58. (*80, 81*)

GERARD, G. (1956). *Minimum Weight Analysis of Compression Structures.* New York University Press, New York. (*267*)

GIBBS, D. A., MERRILL, E. W., SMITH, K. A. and BALAZS, E. A. (1968). Rheology of hyaluronic acid. *Biopolymers*, **6**, 777–91. (*121*)

GIBSON, T. and DAVIS, W. B. (1958). (stressed cartilage grafts) *Brit. J. Plastic Surg.*, **10**, 257–74. (*143, 340*)

GIBSON, T., STARK, H. and EVANS, J. H. (1969). (extensibility of human skin) *J. Biomech.*, **2**, 201–4. (*134*)

GILLIS, P. P. (1969). (hydrogen bonds and stiffness in cellulose) *J. Polymer Sci.*, **7A**, 783–94. (*102*)

GILMAN, J. J. (1960). Direct measurements of the surface energy of crystals. *J. appl. Phys.*, **31**, 2208–18. (*226*)

GILMORE, R. S., POLLACK, R. P. and KATZ, J. L. (1970). Elastic properties of bovine dentine and enamel. *Arch. Oral Biol.*, **15**, 787–96. (*167*)

GJELSVIK, A. (1973*a*). Bone remodelling and piezoelectricity, I. *J. Biomech.*, **6**, 69–77. (*187*)

GJELSVIK, A. (1973*b*). Bone remodelling and piezoelectricity, II. *J. Biomech.*, **6**, 187–93. (*187*)

GLADFELTER, W. B. (1972). (bell design in the medusa *Polyorchis*) *Helgoländer wiss. Meeresunters.*, **23**, 38–79. (*310, 311, 313*)

GLASSTONE, S. and LEWIS, D. (1962). *Elements of Physical Chemistry*, 2nd edn. MacMillan, London. (*45*)

GOLDBERG, W. (1973). Chemistry, structure and growth of gorgonian and antipatharian coral skeleton. Dissertation, University of Miami. (*191–3*)

GOODIER, J. N. (1933). (voids in composites) *J. appl. Mech.*, **55**, A-39. (*157*)

GORDON, J. E. (1952). (optimal design in fibrous composites) *J. aeronaut. Soc.*, **56**, 710. (*151*)

GORDON, J. E. (1968). *The New Science of Strong Materials.* Penguin Books, Harmondsworth, U.K. (*3*)

GORDON, J. E. (1970). The design of materials. *Proc. roy. Soc.*, **A139**, 137–43. (*229*)

GORING, D. A. I. and TIMMELL, T. E. (1962). Molecular weight of native cellulose. *Tappi*, **45**, 454–9. (*100*)

GOSLINE, J. M. (1971*a*). (structure and composition of mesogloea in *Metridium*) *J. exp. Biol.*, **55**, 763–74. (*128, 129, 307*)

GOSLINE, J. M. (1971*b*). (viscoelastic properties of mesogloea in *Metridium*) *J. exp. Biol.*, **55**, 775–95. (*128–30, 308, 310, 366*)

GOULD, B. S., ed. (1968). *Treatise on Collagen,* 3 vols. Academic Press, New York. *(2)*

GOW, B. S. and TAYLOR, M. G. (1968). (viscoelastic properties of dog arteries) *Circ. Res.,* 23, 111–22. *(138)*

GRAY, J. (1968). *Animal Locomotion.* Weidenfeld and Nicholson, London. *(3, 327, 333, 338, 349)*

GREEN, P. B. and CHAPMAN, G. B. (1955). (cell wall in *Nitella*) *Am. J. Bot.,* **42**, 685–92. *(196)*

GRIFFITH, A. A. (1921). The phenomena of rupture and flow in solids. *Phil. Trans. roy. Soc.,* **A221**, 163–198. *(18)*

GRIGG, R. W. (1972). Orientation and growth form of sea fans. *Limnol. Oceanog.,* **17**, 185–92. *(358)*

GRIMSTONE, A. V., HORNE, R. W., PANTIN, C. F. A. and ROBSON, E. A. (1959). (fine structure of *Metridium* mesenteries) *Quart. J. microsc. Sci.,* **99**, 523–40. *(127)*

GROGG, B. and HELMO, D. (1958). (stress relaxation of wheat dough) *Cereal Chem.,* 36, 260–73. *(39)*

GROSS, J., HIGHBERGER, J. H. and SCHMITT, F. O. (1954). Collagen structures considered as states of aggregation of a kinetic unit. *Proc. Nat. Acad. Sci.* **40**, 679–88. *(85)*

GUTH, E. and MARK, H. (1934). (statistics of polymer chains) *Monatsch. Chem.,* **65**, 93. *(53)*

HACKMAN, R. H. (1953). (water insoluble proteins of beetle cuticle) *Biochem. J.,* **54**, 367–70. *(158)*

HACKMAN, R. H. (1960). (complexes containing covalently linked chitin and protein) *Austral. J. biol. Sci.,* **13**, 568–77. *(158, 161)*

HACKMAN, R. H. and GOLDBERG, M. (1971). (sclerotization of insect cuticle) *J. Insect Physiol.,* **17**, 335–47. *(158–60)*

HALL, R. H. (1951). Changes in length of stressed collagen fibers with time. *J. Soc. Leather Trade Chem.,* **35**, 11–17. *(92)*

HAMMERMAN, D. (1970). (structure and function of synovial joints) In E. A. Balazs (1970), Vol. 3, pp. 1259–77. *(122)*

HARE, P. E. and ABELSON, P. H. (1965). Amino acid composition of some calcified proteins. *Ann. Rep. Geophysics Lab., Carnegie Inst.,* 223–32. *(213, 231)*

HARKNESS, M. L. R. and HARKNESS, R. D. (1959). Changes in the physical properties of the uterine cervix of the rat during pregnancy. *J. Physiol.,* **148**, 524–47. *(131)*

HARKNESS, M. L. R. and HARKNESS, R. D. (1961). Properties of rat cervix *post partum. J. Physiol.,* **156**, 112–20. *(131)*

HARKNESS, R. D. (1968). Mechanical properties of collagenous tissues. In *Treatise on Collagen,* B. S. Gould, ed. Academic Press, New York, Vol. 2A, pp. 247–310. *(89, 132, 134, 135)*

HARKNESS, R. D. (1970). (connective tissues of skin) In E. A. Balazs (1970), Vol. 3, pp. 1309–40. *(133, 134)*

HARKNESS, R. D. and NIGHTINGALE, M. A. (1962). (extensibility of rat cervix during pregnancy) *J. Physiol.,* **160**, 214–20. *(131)*

HARRINGTON, W. F. (1964). (H bonds in collagen structure) *J. molec. Biol.,* **9**, 613–17. *(84)*

HARRIS, J. E. and CROFTON, H. D. (1957). (internal pressure and cuticular structure in *Ascaris*) *J. exp. Biol.,* **34**, 116–30. *(93, 295, 302)*

HARTMAN, W. D. and GOREAU, T. F. (1970). (Jamaican coralline sponges). In *Biology for the Porifera.* Edited by W. G. Fry. *Symp. Zool. Soc. Lond.* No. 25, 205–43. Academic Press, London. (*207*)

HASCALL, V. C. (1972). (untitled discussion) In *Comparative Molecular Biology of Extracellular Matrices,* H. C. Slavkin, ed. Academic Press, New York, pp. 170–88. (*140*)

HASHIN, Z. (1955). (elastic constants of components of composites) *Bull. Res. Council Israel,* **5-C**, 46. (*146*)

HAUGHTON, P. M., SELLEN, D. B. and PRESTON, R. D. (1968). (dynamic properties of *Nitella* cell walls) *J. exp. Bot.* **19**, 1–12. (*36, 38, 104, 199*)

HAUSMAN, R. E. and BURNETT, A. L. (1969). (physical and histochemical properties of hydra mesoglea) *J. exp. Zool.,* **171**, 7–14. (*308*)

HAUT, R. C. and LITTLE, R. W. (1969). (rheology of canine ligament) *J. Biomech.,* 2, 289–98. (*90, 91*)

HEJNOWICZ, Z. (1967). (mechanism of orientation in woody stems) *Am. J. Bot.,* **54**, 684–9. (*321*)

HENDERSHOT, O. P. (1924). Thermal expansion of wood. *Science,* **60**, 456–7. (*206*)

HENISCH, H. K. (1970). *Crystal Growth in Gels.* Pennsylvania State University Press, College Park. (*217, 230*)

HEPBURN, H. R. and BALL, A. (1973). (structure and properties of beetle shells) *J. Materials Sci.,* 8, 618–23. (*167*)

HERP, A. and DIGMAN, W. (1968). (carbohydrates of rat skin) *Biochim. Biophys. Acta,* **165**, 76–83. (*133*)

HERRING, G. M. (1968). Chemical structure of tendon, cartilage, dentine and bone matrix. *Clin. Orthopaed.,* **60**, 261–99. (*169*)

HERZOG, R. O. (1926). Fortschritte in der Erkenntnis der Faserstaffe. *Z. angew. Chem.,* **39**, 297–302. (*108, 109*)

HILDEBRAND, M. (1959). Motions of the running cheetah and horse. *J. Mammal.* **40**, 481–95. (*339*)

HILDEBRAND, M. (1962). Walking, running and jumping. *Am. Zoologist,* **2**, 151–5. (*339*)

HIRAMOTO, Y. (1962). Mechanical properties of the protoplasm of the sea urchin egg. *Exp. Cell Res.,* 56, 101–218. (*33*)

HOEVE, C. A. J. and FLORY, P. J. (1958). Elastic properties of elastin. *J. Am. chem. Soc.* **80**, 6523–6. (*116*)

HÖHLING, H. J., SCHOLZ, F., BOYDE, A., HEINE, H. and REIMER, L. (1971). (nucleation and growth of crystals in dentine matrix) *Z. Zellforsch.,* **117**, 381–93. (*171*)

HÖHLING, H. J. and SCHÖPFER, H. (1968). (apatite nucleation in hard tissues) *Naturwiss.* 55, 545. (*171*)

HOLLISTER, G. S. and THOMAS, C. (1966). *Fibre Reinforced Materials.* Elsevier, Amsterdam. (*144*)

HOMANN, H. (1949). (growth and mechanics preceding spider moulting) *Z. Vergleich. Physiol.,* 31, 413–40. (*108*)

HUNT, S. (1970). *Polysaccharide-protein Complexes in Invertebrates.* Academic Press, London. (*119*)

HYMAN, L. H. (1955). *The Invertebrates: Echinodermata.* McGraw-Hill, New York. (*229*)

IFJU, G. (1964). Tensile strength as a function of cellulose in wood. *For. Prod. J.* **14**, 366–72. (*167, 205*)

IIZUKA, E. (1965). (modulus and crystallinity in silk) *Biorheol.*, 3, 1–8. (*78, 79*)

IIZUKA, E. (1966). Mechanism of fiber formation in silkworm, *Bombyx mori. Biorheol.*, 3, 141–52. (*76, 77*)

ILER, R. K. (1963). Strength and structure of flint. *Nature, Lond.*, **199**, 1278–9. (*225*)

INGLIS, C. E. (1913). Stresses in a plate due to the presence of cracks and sharp corners. *Trans. Inst. Naval Arch.*, **55** (1), 219–30. (*16*)

IRWIN, G. R. (1958). Fracture. *Encyclopedia of Physics*, S. Flugge, ed., Springer, Berlin, Vol. 6 pp. 551–90. (*42*)

JACCARD, P. (1938). (reaction wood in bent saplings) *Ber. Schweiz. bot. Ges.*, **48**, 491–537. (*321, 322*)

JAMES, W. L. (1962). (dynamic tensile properties of wood) *For. Prod. J.*, **12**, 253–8. (*167*)

JEFFREY, G. B. (1923). The motion of ellipsoidal particles immersed in a viscous fluid. *Prod. roy. Soc. Lond.* **A-102**, 161–179. (*146*)

JENSEN, M. and WEIS-FOGH, T. (1962). (mechanical properties of locust cuticle) *Phil. Trans. roy. Soc. Lond.*, **B245**, 137–69. (*109, 113, 114, 165, 167, 168, 229*)

JEUNIAUX, C. (1971). Chitinous structures. In *Comprehensive Biochemistry.* M. Florkin and E. H. Stotz, eds. **26-C**, 595–632. Elsevier, Amsterdam. (*104, 106*)

JONES, H. D. and TRUEMAN, E. R. (1970). Locomotion of the limpet *Patella vulgata. J. exp. Biol.*, **52**, 201–16. (*333, 334*)

JONES, W. C. (1970). Composition, development, form and orientation of calcareous sponge spicules. In *Biology of the Porifera.* Edited by W. G. Fry. *Symp. Zool. Soc. Lond.* No. 25, 91–123. Academic Press, London. (*209*)

JONES, W. E. and DEMETROPOULOS, A. (1968). (wave action on a rocky shore) *J. exp. mar. Biol. Ecol.*, 2, 46–63. (*352*)

JOPE, M. (1971). Constituents of brachiopod shells. In *Comprehensive Biochemistry,* M. Florkin and E. H. Stotz, eds. **26-C**, 769–86. Elsevier, Amsterdam. (*215*)

JUSTUS, R. and LUFT, J. H. (1970). A mechanochemical hypothesis for bone remodelling induced by mechanical stress. *Calc. Tiss. Res.*, **5**, 222–35. (*185, 187*)

KATCHALSKY, A. (1964). Polyelectrolytes and their biological implications. *Biophys. J.*, **4**, 9–41. (*140*)

KATZ, J. L. (1971). Hard tissue as a composite material. *J. Biomech.*, **4**, 455–73. (*176*)

KAUFMAN, K. W. (1971). (avicularia of the ectoproct *Bugula*) *Postilla,* No. 151, 1–26. (*238*)

KELLER, A. (1957). (evidence for folded chains in polymer crystals) *Philos. Mag.*, Ser. 8, 2, 1171–5. (*68*)

KELLER, A. and MACHIN, M. J. (1968). (strain crystallization in polymers) In *Polymer Systems, Deformation and Flow,* R. E. Wetton and R. W. Whorlow, eds. MacMillan, London, pp. 97–102. (*70*)

KELLY, A. (1964). Strengthening of metals by dispersed particles. *Proc. roy. Soc. Lond.*, **A282**, 63–79. (*150*)

KELLY, P. G., OLIVER, P. T. P. and PAUTARD, F. G. E. (1965). The shell of *Lingula unguis. Proc. 2nd Europ. Symp. Calcified Tissues.* University of Liège, pp. 337–45. (*215*)

KELLY, R. E. and RICE, R. V. (1967). (abductin) *Science,* **155**, 208–210. (*113, 116*)

KENEDI, R. M., GIBSON, T., DALY, C. H. and ABRAHAMS, M. (1966). (mechanics of human skin and cartilage) *Proc. Fed. Am. Soc. exp. Biol.,* 25, 1084–7. (*133, 142*)

KENNAUGH, J. (1959). (cuticle structure in two scorpions) *Quart. J. microsc. Soc.,* **100**, 41–50. (*164*)

KENNAUGH, J. (1968). (cuticle structure in three Ricinulei) *J. Zool. Lond.,* **156**, 393–404. (*163*)

KENNEDY, W. J., TAYLOR, J. D. and HALL, A. (1969). Environmental and biological controls on bivalve shell mineralogy. *Biol. Rev.,* **44**, 499–530. (*211, 213, 233*)

KEOSIAN, J. (1964). *The Origin of Life.* Reinhold, New York. (*287*)

KING, A. I., and EVANS, F. G. (1967). (fatigue strength of human compact bone) *Digest 7th Int. Conf. Med. Biol. Engin.,* Stockholm, p. 514. (*184*)

KINZIE, R. A. (1973). Zonation of West Indian gorgonians. *Bull. mar. Sci.,* **23**, 93–155. (*356, 358*)

KLAUDITZ, W. (1957). (role of cellulose and hemicellulose in strength of wood) *Holzforsch.,* **11**, 110–16. (*101*)

KLEIN, L. and CURREY, J. D. (1970). Echinoid skeleton: absence of a collagenous matrix. *Science,* **169**, 1209–10. (*217*)

KOLLMAN, F. (1951). *Technologie des Holzes und der Holzwerkstoffe.* Springer Verlag, Berlin, Vol. I, 2nd edn. (*206*)

KOLLMAN, F. (1962). (rheology of wood) *Materialprüfung,* **4**, 313–19. (*205*)

KOLLMAN, F. and CÔTÉ, W. A., Jr. (1968). *Principles of Wood Science and Technology.* Vol. 1, *Solid Wood,* Springer Verlag, New York. (*167, 196, 201, 203, 204, 206*)

KORATKY, O., WAWRA, H., PILY, I., SEKORA, A. and VAN DEINSE, A. (1964). (X-ray studies of silk in solution) *Monatsch. Chem.,* **95**, 359–72. (*75*)

KRAMER, P. J. (1949). *Plant and Soil Water Relationships.* McGraw-Hill, New York. (*320*)

KRENCHEL, H. (1964). *Fiber Reinforcement.* Akademisk Vorlag, Stockholm. (*151, 152*)

KRISHNAN, G. and RAJULU, G. S. (1964). (cuticle structure of a symphylid) *Z. Naturforsch.,* **196**, 640–5. (*164*)

KUHN, W. (1934). Fadenförmiger Molekule in Lösungen. *Kolloid Z.,* **68**, 2–15. (*53*)

KUHN, W. (1939). (molecular conformation and crystalline orientation in rubber elasticity) *Kolloid Z.,* **87**, 3–12. (*53*)

LAFON, M. (1943). (tégument des arthropodes) *Ann. Sci. nat., Ser. 11, Zool. Biol. anim.,* **5**, 113–46. (*164, 216*)

LANG, S. B. (1969). Elastic coefficients of animal bone. *Science,* **165**, 287–8. (*174*)

LAURENT, T. (1966). Physicochemical characteristics of the acid glycosaminoglycans. *Proc. Fed. Am. Soc. exp. Biol.,* **25**, 1037. (*121*)

LAURENT, T. (1970). Structure of hyaluronic acid. In E. A. Balazs (1970), Vol. 3, pp. 703–32. (*122*)

LAWRY, J. V. (1966). (burrow irrigation by *Urechis caupo*) *J. exp. Biol.,* **45**, 343–56. (*317*)

LEAROYD, B. M. and TAYLOR, M. G. (1966). (viscoelastic properties of human arterial wall) *Circ. Res.*, **18**, 278–92. (*138*)

LEVERSEE, G. J. (1972). (effect of water currents on morphology of *Leptogorgia*). *American Zoologist*, 12, 719. (*359, 362*)

LINDENMEYER, P. H. (1965). Crystallization and molecular folding. *Science*, **147**, 1256–62. (*68*)

LINDENMEYER, P. H. (1966). Dislocations in polyethylene crystals. *J. Polymer Sci.*, **15-C**, 109–27. (*69*)

LINN, F. C. (1968). Lubrication of animal joints. *J. Biomech.*, **1**, 193–205. (*122*)

LIŠKOVÁ, M. and HEŘT, J. (1971). (reaction of rabbit bone to intermittent loading) *Folia Morph.*, **19**, 301–17. (*185*)

LOCKE, M. (1967). (pattern development in insect cuticle) *Adv. Morphogen.* **6**, 33–88. (*161*)

LOTMAR, W. and PICKEN, L. E. R. (1950). (crystal forms of chitin) *Experientia*, **6**, 58–9. (*107*)

LUCAS, F. (1964). Spiders and their silk. *Discovery*, **25**, 20–6. (*78–80*)

LUCAS, F. and RUDALL, K. M. (1968). (silks) In *Comprehensive Biochemistry*, M. Florkin and E. H. Stotz, eds., **26-B**, 475–558. Elsevier, Amsterdam. (*74–5, 77, 80*)

LUCAS, F., SHAW, J. T. B. and SMITH, S. G. (1960). (amino acid analyses of silks from 74 arthropod species) *J. molec. Biol.*, **2**, 339–49. (*76*)

LUSCOMB, M. and PHELPS, C. (1967). Bovine nasal septum. *Biochem. J.*, **102**, 110–19. (*140*)

MACAN, T. T. (1963). *Freshwater Ecology.* Longmans, London. (*236*)

MACK, R. W. (1964). Bone–a natural two-phase material. *Tech. Mem. Biochem. Lab.*, University of California, Berkeley. (*175*)

MACKENZIE, J. K. (1950). Elastic constants of a solid containing spherical holes. *Proc. Phys. Soc. Lond.*, **63-B**, 2–11. (*146, 157*)

MACKIE, G. O. and MACKIE, G. V. (1967). (reversible opacity in siphonophore mesoglea) *Vie Milieu*, **A18**, 47–71. (*312*)

McBRIDE, O. W. and HARRINGTON, W. F. (1967*a*). (disulfied cross-links in *Ascaris* cuticle collagen) *Biochem.*, **6**, 1484–98. (*93*)

McBRIDE, O. W. and HARRINGTON, W. F. (1967*b*). (helix-coil transition in collagen) *Biochem.*, **6**, 1499–514. (*93*)

McCONNELL, D. (1963). (*Lingula* shell constituents) *Bull geol. Soc. Am.*, **74**, 363–6. (*215*)

McCRUM, N. G., READ, B. E. and WILLIAMS, G. (1967). *Anelastic and Dielectric Effects in Polymeric Solids.* John Wiley, New York. (*61*)

McCUTCHEN, C. W. (1966). (boundary lubrication by synovial fluid) *Proc. Fed. Am. Soc. exp. Biol.*, **25**, 1061–8. (*122, 142*)

McELHANEY, J. H. (1966). Dynamic response of bone and muscle tissue. *J. appl. Physiol.*, **21**, 1231–6. (*12, 175, 285*)

McELHANEY, J. H., FOGLE, J., BYARS, E. and WEAVER, G. (1964). Effect of embalming on the mechanical properties of beef bone. *J. appl. Physiol.*, **19**, 1234–6. (*285*)

MADDRELL, S. M. (1966). (nervous control of mechanical properties of *Rhodnius* body wall) *J. exp. Biol.*, **44**, 59–68. (*286*)

MAGNUS, D. (1966). (orientation to flow by a sea pen) *Veröff. Inst. Meeresforsch. Bremerhaven*, 2, 369–80. (*359*)

MANDELKERN, L. (1966). (molecular weight and properties of long chain molecules) *J. Polymer Sci.*, **C15**, 129–62. (*69*)

MANLEY, R. ST. J. (1963). (structure of cellulose acetate) *J. Polymer Sci.*, **1A**, 1875–92. (*97, 98*)

MANLEY, R. ST. J. (1971). Molecular morphology of cellulose. *J. Polymer Sci.*, **9A**-2, 1025–59. (*97, 98*)

MANTON, S. M. (1950). (locomotion of *Peripatus*) *J. Linn. Soc. Lond., Zool.*, **41**, 529–70. (*325*)

MANTON, S. M. (1952*a*). (locomotory mechanisms of arthropods) *J. Linn. Soc. Lond., Zool.*, **42**, 93–117. (*325*)

MANTON, S. M. (1952*b*). (locomotion of centipedes and pauropods) *J. Linn. Soc. Lond., Zool.*, **42**, 118–66. (*325*)

MANTON, S. M. (1953). Locomotory habits and the evolution of the larger arthropodan groups. *Soc. exp. Biol. Symp.*, **7**, 339–76. (*325*)

MANTON, S. M. (1954). (design in millipedes) *J. Linn. Soc. Lond., Zool.*, **42**, 299–368. (*237, 325*)

MANTON, S. M. (1956). (design in pselaphognath millipedes) *J. Linn. Soc. Lond., Zool.*, **43**, 153–87. (*325*)

MANTON, S. M. (1958*a*). (design in centipedes and millipedes) *J. Linn. Soc. Lond., Zool.*, **43**, 487–556. (*325*)

MANTON, S. M. (1958*b*). (body design in arthropods) *J. Linn. Soc. Lond., Zool.*, **44**, 58–72. (*325*)

MANTON, S. M. (1961*a*). (design in millipedes, especially Colobognatha) *J. Linn. Soc. Lond., Zool.*, **44**, 383–462. (*325, 331*)

MANTON, S. M. (1961*b*). Experimental zoology and problems of arthropod evolution. In *The Cell and The Organism*, J. A. Ramsay and V. B. Wigglesworth, eds. Cambridge University Press, London. (*325, 331*)

MANTON, S. M. (1964). Mandibular mechanisms and the evolution of arthropods. *Phil. Trans. roy. Soc. Lond.*, **B44**, 1–83. (*325*)

MANTON, S. M. (1965). (body design in centipedes) *J. Linn. Soc. Lond., Zool.*, **46**, 251–484. (*325, 328*)

MANTON, S. M. (1966). (design in Symphyla and Pauropoda) *J. Linn. Soc. Lond., Zool.*, **46**, 103–42. (*317, 325*)

MANTON, S. M. (1972). (body design in hexapod classes) *J. Linn. Soc. Lond., Zool.*. **51**, 203–400. (*325*)

MARK, H. (1932). *Physik und Chemie der Cellulose.* Springer Verlag, Berlin, (*72*)

MARK, H. (1933). Fine structure and mechanical properties of fibres. *Trans. Faraday Soc.*, 29, 6–13. (*78*)

MARK, R. E. (1967). *Cell Wall Mechanics of Tracheids.* Yale University Press, New Haven. (*97, 101, 102, 197–201, 229, 323*)

MARK, R. E., KALONI, P. N., RANG, R. and GILLIS, P. (1969). Cellulose: refutation of a folded chain structure. *Science*, **164**, 72–3. (*97*)

MÄRKEL, K. (1969). Morphologie der Seeigelzähne. *Z. Morph. Tiere*, **66**, 1–58. (*222, 223*)

MÄRKEL, K. (1970*a*) Morphologie der Seeigelzähne III. *Z. Morph. Tiere*, **66**, 189–211. (*222*)

MÄRKEL, K. (1970*b*). Morphologie der Seeigelzähne IV. *Z. Morph. Tiere*, **68**, 370–89. (*222*)

MÄRKEL, K. (1970*c*). Tooth skeleton of *Echinometra mathaei. Annot. Zool. Jap.*, **43**, 188–99. (*222, 223*)

MÄRKEL, K. and GORNY, P. (1973) Zur funktionellen Anatomie der Seeigelzähne. *Z. Morph. Tiere*, **75**, 223–42. (*223, 234*)

MÄRKEL, K. and TITSCHAK, H. (1969). Morphologie der Seeigelzähne I. *Z. Morph. Tiere,* **64**, 179–200. (*222*)

MÄRKEL, K., KUBANEK, F. and WILLGALLIS, A. (1971). Polykristalliner calcit bei Seeigeln. *Z. Zellforsch.,* **119**, 355–77. (*217*)

MARKS, M. H., BEAR, R. S. and BLAKE, C. H. (1949). (X-ray diffraction evidence of collagen in various phyla) *J. exp. Zool.,* **111**, 55–77. (*191*)

MARKS, R. W. (1960). *The Dymaxion World of Buckminster Fuller.* Reinhold, New York. (*300*)

MARSH, R. E., COREY, R. B. and PAULING, L. (1955*a*). (structure of silk) *Biochim. Biophys. Acta,* **16**, 1–34. (*74–6*)

MARSH, R. E., COREY, R. B. and PAULING, L. (1955*b*). Structure of Tussah silk fibroin. *Acta Cryst.,* **8**, 710–15. (*76*)

MARX-FIGINI, M. and SCHULZ, G. V. (1966). (biosynthesis of cellulose in higher plants) *Biochim. Biophys. Acta,* **112**, 84–101. (*97–8*)

MASON, P. (1965). Viscoelasticity and structure of keratin and collagen. *Kolloid Z.,* **202**, 139–47. (*90, 91*)

MATHEWS, M. B. (1965). (interaction of collagen and acid mucopolysaccharides) *Biochem. J.,* **96**, 710–16. (*139, 140*)

MATHEWS, M. B. and LOZAITYTE, I. (1958). (structure of chondroitin sulfate-protein complexes of cartilage) *Arch. Biochem. Biophys.,* **74**, 158–74. (*139*)

MATZKE, E. B. and NESTLER, J. (1946). Volume-shape relationships in foams. *Am. J. Bot.,* **33**, 58–80. (*319*)

MAXWELL, J. C. (1873). *Treatise on Electricity and Magnetism,* Vol. 1, p. 365. (*144, 145*)

MAXWELL, J. C. (1890). Scientific Papers, Cambridge Univ. Press, London, Vol. 2. (*289*)

MERCER, E. H. (1961). *Keratin and Keratinization.* Pergamon Press, Oxford. (*188*)

MEREDITH, R. (1946). Elastic properties of textile fibers. *J. Text. Inst.,* **37**, 469–80. (*101, 103*)

MEYER, D. L. (1971). (collagenous ligaments in crinoids) *Mar. Biol.,* **9**, 235–41. (*358, 360*)

MEYER, K. H. (1942). *Natural and Synthetic High Polymers.* Interscience, New York. (*73, 78, 101–3, 108*)

MEYER, K. H., and FERRI, C. (1936). (elastic properties of collagen and elastic fibers) *Pflüger's Arch. ges. Physiol.,* **238**, 78–90. (*116*)

MEYER, K. H. and LOTMAR, W. (1936). L'élasticité de la cellulose. *Helv. Chem. Acta,* **19**, 68–86. (*101*)

MEYER, K. H. and MISCH, L. (1937). (modèle spatial de la cellulose) *Helv. Chem. Acta,* **20**, 232–44. (*95*)

MICHELL, A. G. M. (1904). The limits of economy of material in frame-structures. *Philos. Mag.,* 8, 589–97. (*291*)

MILES, R. W. E., ed. (1967). *Structural and Chemical Organization of Teeth,* 2 vols. Academic Press, New York. (*223*)

MILLER, A. and PARRY, D. A. D. (1973). Structure and packing of microfibrils in collagen. *J. molec. Biol.,* **75**, 441–7. (*86*)

MILLER, A. and WRAY, S. S. (1971). Molecular packing in collagen. *Nature, Lond.,* **230**, 437–9. (*86*)

MITCHELL, T. P. (1957). (nonlinear bending in thin rods) *J. appl. Mech., Trans. Am. Soc. Mech. Engin.*, Pap. 58-A-50. (*249, 352*)

MORIIZUMI, S., FUSHITANI, M. and KABURAGI, J. (1973*a*). (temperature dependence of stress relaxation of saturated wood) *J. Japan Wood Res. Soc.*, **19**, 109–15. (*39, 207*)

MORIIZUMI, S., FUSHITANI, M. and KABURAGI, J. (1973*b*). (fine structure of tree trunk and stress relaxation properties) *J. Japan Wood Res. Soc.*, **19**, 81–8. (*39, 207*)

MORROW, C. T. (1960). *Nonlinear and transient dynamic behavior in bovine muscle.* Dissertation, Pennsylvania State University, University Park. (*39*)

MORTON, J. E. (1967). *Molluscs.* Hutchinson University Library, London. (*333*)

MOSS, M. L. (1961). Osteogenesis of acellular teleost bone. *Am. J. Anat.*, **108**, 99–110. (*174*)

MUGGLI, R., ELLIAS, H. and MÜHLETHALER, K. (1969). Feinbau der elementar Fibrillen der Cellulose. *Makromol. Chem.*, **121**, 290–4. (*97, 99*)

MÜHLETHALER, K. (1969). Fine structure of natural polysaccharide systems. *J. Polymer Sci.*, **28-C**, 305–16. (*97, 99*)

MUZIK, K. M. (1973). (morphology of octocorals) M.A. Thesis, Duke University, Durham, N. Carolina. (*220*)

NACHTIGALL, W. (1971). *Biotechnik.* Quelle and Meyer, Heidelberg. (*5, 255, 296, 327, 338*)

NADOL, J. B., GIBBINS, J. R. and PORTER, K. R. (1969). (collagen texture in basal lamella of fish skin) *Devel. Biol.*, **20**, 304–31. (*124*)

NAKAHARA, H. and BEVELANDER, G. (1971). (prismatic layer of *Pinctada radiata*) *Calc. Tiss. Res.*, **7**, 31–45. (*230*)

NEMETHY, G. and SCHERAGA, H. A. (1962*a*). (model for properties of liquid water in protein bonding) *J. chem. Phys.*, **36**, 3382–400. (*117*)

NEMETHY, G. and SCHERAGA, H. A. (1962*b*). (model for properties of hydrocarbons in aqueous solutions) *J. chem. Phys.*, **36**, 3401–17. (*117*)

NEMETHY, G. and SCHERAGA, H. A. (1962*c*). (properties of hydrophobic bonds in proteins) *J. Phys. Chem.*, **66**, 1773–89. (*117*)

NEUSHUL, M. (1972). (functional morphology of seaweeds) In *Contributions to the Systematics of Benthic Marine Algae of the North Pacific,* I. A. Abbott and M. Kurai, eds. Japan Soc. Physol. Kobe, pp. 47–74. (*350*)

NEVILLE, A. C. (1967). Chitin orientation in cuticle and its control. *Adv. Insect Physiol.*, **4**, 213–86. (*107, 161*)

NEVILLE, A. C. (1970). (cuticle structure in relation to the whole insect) In *Insect Ultrastructure,* A. C. Neville, ed. Roy. Ent. Soc., London , pp. 17-39. (*161*)

NEVILLE, A. C. and LUKE, B. M. (1969). (model for chitin-protein complexes in insect cuticle) *Tiss. Cell,* **1**, 689–707. (*161*)

NEWMAN, W. A., ZULLO, V. A. and WAINWRIGHT, S. A. (1967). (growth in Balanomorpha) *Crustaceana,* **12**, 167–78. (*216*)

NICHOLS, D. (1959*a*). (tube feet of *Echinocardium caudatum*) *Quart. J. microsc. Sci.*, **100**, 73–87. (*315*)

NICHOLS, D. (1959*b*). (tube feet of *Echinocyamus pusillus*) *Quart. J. microsc. Sci.*, **100**, 539–55. (*315*)

NICHOLS, D. (1960). (tube feet of a crinoid) *Quart. J. microsc. Sci.*, **101**, 105–17. (*313, 315*)

NICHOLS, D. (1961). (tube feet of two regular echinoids) *Quart. J. microsc. Sci.*, **102**, 157–180. (*315*)

NICHOLS, D. and CURREY, J. D. (1968). (echinoderm calcite) In *Cell Structure and its Interpretation.* S. M. McGee-Russell and K. F. A. Ross, eds. Edward Arnold, London, pp. 251–61. (*217, 218, 234*)

NIELSEN, L. E. (1962). *Mechanical Properties of Polymers.* Van Nostrand Reinhold, New York. (*72*)

NIELSEN, L. E. and CHEN, P. E. (1968). Young's modulus of composites filled with randomly oriented fibers. *J. Materials,* **3**, 352–8. (*164*)

NORTON-GRIFFITHS, M. (1967). Some ecological aspects of the feeding behaviour of the oystercatcher *Haematopus ostralegus* on the edible mussel *Mytilus edulis. Ibis,* **109**, 412–24. (*273*)

NORWIG, A. and HAYDUK, U. (1969). (invertebrate collagens) *J. molec. Biol.*, **44**, 161–72. (*128*)

ÖBRINK, B. and WASTESON, A. (1971). (interaction of glycosaminoglycans with collagen) *Biochem. J.*, **121**, 227–33. (*140*)

OGSTON, A. G. (1966). Protein-polysaccharide interaction. *Proc. Fed. Am. Soc. exp. Biol.*, **25**, 1039. (*121*)

OGSTON, A. G. (1970). Biological functions of the glycosaminoglycans) In E. A. Balazs (1970) Vol. 3, pp. 1231–40. (*140*)

OGSTON, A. G. and STANIER, J. E. (1950). The state of hyaluronic acid in synovial fluid. *Biochem. J.*, **46**, 364–76. (*121*)

OGSTON, A. G. and STANIER, J. E. (1953). (properties and function of hyaluronic acid in synovial fluid) *J. Physiol.*, **119**, 244–52. (*121, 122*)

OLSSON, R. (1964). Skin of *Amphioxus. Z. Zellforsch.*, **54**, 90–104. (*124*)

OLTMANNS, F. (1922). *Morphologie und Biologie der Algen.* Gustav Fischer, Jena, Vol. 2. (*351*)

OROWAN, E. (1950). *M.I.T. Symposium on Fatigue and Fracture of Metals,* John Wiley, New York. (*41*)

ORTNER, D. J. and VON ENDE, D. W. (1971). (sclerotic lamellae in human osteons) *Israel J. med. Sci.*, **7**, 480–1. (*172*)

ORTON, J. H. (1926). (growth of *Cardium edule*) *J. mar. biol. Assoc., U.K.*, **14**, 239–80. (*262, 273*)

ØRVIG, T. (1967). (phylogeny of tooth tissues) In *Structural and Chemical Organization of Teeth,* A. E. W. Miles, ed. Academic Press, New York, Vol. 1, pp. 45–110. (*173*)

OXNARD, C. E. (1971). Tensile forces in skeletal structures. *J. Morph.*, **134**, 425–36. (*187*)

PARKER, K. D. and RUDALL, K. M. (1957). Structure of silk of *Chrysopa* egg stalks, *Nature,* **179**, 905–7. (*80*)

PARKES, E. W. (1965). *Braced Frameworks.* Pergamon Press, Oxford. (*289, 290, 292*)

PARRINGTON, F. R. (1967). Vertebrae of early tetrapods. *Colloq. int. Centre nat. Rech. sci.*, No. 163, 269–79. (*338, 339*)

PARTINGTON, F. R. and WOOD, G. C. (1963). (role of non-collagenous components in mechanical behavior of tendon fibers) *Biochim. Biophys. Acta*, **69**, 485–95. (*91*)

PARTRIDGE, S. M. (1967). Diffusion of solutes in elastin. *Biochim. Biophys. Acta*, **140**, 132–141. (*117, 118*)

PARTRIDGE, S. M. (1970). Isolation and characterization of elastin. In E. A. Balazs (1970). Vol. 1, pp. 593–616. (*117*)

PASSAGLIA, E. and KOPPEHELE, N. P. (1958). (stress relaxation in cellulose) *J. Polymer Sci.*, 33, *281–9*. (*104*)

PAU, R. N., BRUNET, P. C. J. and WILLIAMS, M. J. (1971). (colleterial gland proteins in the cockroach) *Proc. roy. Soc. Lond.*, **B177**, 565-79. (*158*)

PAULING, L. and COREY, R. B. (1953*a*). Stable configurations of polypeptide chains. *Proc. roy. Soc. Lond.*, **B141**, 21–3. (*80*)

PAULING, L. and COREY, R. B. (1953*b*). (structure of α-helical proteins) *Nature, Lond.*, **171**, 59–61. (*188*)

PAUWELS, F. (1968). (cortex function in rachitic femora) *Z. Anat. Entwickl.*, **127**, 121–37. (*185*)

PEAKALL, D. B. (1964). (spider silk) *J. exp. Zool.*, **156**, 345–50. (*74*)

PEASE, D. C. and MOLNARI, S. (1960). (fine structure of muscular arteries) *J. Ultrastr. Res.*, 3, 447–68. (*135*)

PEASE, D. C. and PAULE, W. J. (1960). (fine structure of rat aorta) *J. Ultrastr. Res.*, 3, 469–83. (*134, 135*)

PERSON, P. and PHILPOTT, D. (1969). (invertebrate cartilages) *Biol. Rev.*, **44**, 1–16. (*139*)

PETCH, N. J. (1953). The cleavage strength of polycrystals. *J. Iron Steel Inst.*, **174**, 25–28. (*225*)

PETERLIN, A. (1969). Bond rupture in highly orientated crystalline polymers. *J. Polymer Sci.*, **7A-2**, 1151–63. (*70*)

PETERS, L. and WOODS, H. J. (1955). Protein fibres. In *The Mechanical Properties of Textile Fibres*, R. Meredith, ed. North Holland Publishing Company, Amsterdam, pp. 151–244. (*190*)

PETERSON, R. E. (1953). *Stress Concentration Design Factors.* John Wiley, New York. (*16*)

PICKEN, L. E. R. (1960). *The Organization of Cells and Other Organisms.* Oxford University Press, London. (*3*)

PICKEN, L. E. R., PRYOR, M. G. M. and SWANN, M. M. (1947). Orientation of fibrils in natural membranes. *Nature, Lond.*, **159**, 434. (*124*)

PIEKARSKI, K. (1968). Studies on the mechanical properties of bone. Dissertation, University of Cambridge. (*12*)

PIEKARSKI, K. (1970). Fracture of bone. *J. appl. Physics*, **41**, 215–23. (*177*)

PIEKARSKI, K. (1973). Analysis of bone as a composite material. *Int. J. Engin. Sci.*, **11**, 557–65. (*146*)

PIEZ, K. A. and GROSS, J. (1959). (comparative composition and structure of collagen) *Biochim. Biophys. Acta*, **34**, 24–39. (*85*)

POSNER, A. S.(1969). Crystal chemistry of bone mineral. *Physiol. Rev.*, **49**, 760–92. (*169*)

PRESTON, R. D. (1952). *Molecular Architecture of Plant Cell Walls.* Chapman and Hall, London. (*100*)

PRESTON, R. D. (1963). Observed fine structure in plant fibres. In *Fibre Structure,* J. W. S. Hearle and R. H. Peters, eds. Butterworths, London, pp. 235–68. (*103*)

PRESTON, R. D. and MIDDLEBROOK, M. (1949). (lignin and cell wall structure) *J. Text. Inst.*, **40**, T715. (*323*)

PRINCE, R. P. and BRADWAY, D. W. (1969). Shear stress and modulus of selected forages. *Trans. Am. Soc. Agric. Engin.*, **12**, 426–8. (*12*)

PROBINE, M. C. and BARBER, N. F. (1966). (plastic properties of *Nitella* cell wall) *Austral. J. biol. Sci.*, **19**, 439–57. (*196*)

PROBINE, M. C. and PRESTON, R. D. (1962). (structure and properties of *Nitella* cell wall) *J. exp. Bot.*, **13**, 111–27. (*196, 198, 199*)

PUHL, J. J., PIOTROWSKI, G. and ENNEKING, W. F. (1972). Biomechanical properties of paired canine fibulas. *J. Biomech.*, **5**, 391–7. (*174*)

RAMACHANDRAN, G. N. (1963). Molecular structure of collagen. *Int. Rev. Conn. Tiss. Res.*, **1**, 127–82. (*82, 83*)

RAMACHANDRAN, G. N. and SASISEKHARAN, V. (1961). Structure of collagen. *Nature, Lond.*, **190**, 1004–5. (*83*)

RAMSDEN, W. (1938). Coagulation by shearing and by freezing. *Nature, Lond.*, **142**, 1120–1. (*77*)

RAO, N. V. and HARRINGTON, W. F. (1965). (pyrrolidine residues in collagen stability) *J. molec. Biol.*, **21**, 577–81. (*84*)

RAUP, D. M. (1966).(analysis of shell coiling) *J. Paleont.*, **40**, 1178–90. (*261*)

RAUP, D. M. and STANLEY, S. M. (1971). *Principles of Paleontology*. W. H. Freeman, San Francisco. (*263*)

REYNOLDS, J. J. (1966). (ascorbic acid and growth of chick bone) *J. Exp. Cell Res.*, **42**, 178–88. (*140*)

RICE, R. W. (1972). (strength and grain size in ceramics) *Proc. Brit. Ceram. Soc.*, No. 20, 205–57. (*230*)

RICH, A. and CRICK, F. H. C. (1961). Molecular structure of collagen. *J. molec. Biol.*, **3**, 483–506. (*83*)

RICHARDS, A. G. (1958). (pupal cuticle of *Ephestia kühniella) Z. Naturforsch.*, **13**, 813–16. (*286*)

RICHARDS, A. G. (1967). Sclerotization and the localization of brown and black colours in insects. *Zool. Jb.* (*Anat.*), **84**, 25–62. (*163*)

RIEDL, R. (1966). *Biologie der Meereshöhlen.* Paul Parey, Hamburg. (*358, 359*)

RIEDL, R. (1971*a*). Water movement: "Introduction" (1085–8) and "Animals" (1123–56). In *Marine Ecology,* O. Kinne, ed. Wiley-Interscience, New York, Vol. 1, Part 2. (*358*)

RIEDL, R. (1971*b*). (how much water passes through intertidal interstices?) *Int. Rev. ges. Hydrobiol.*, **56**, 923–46. (*364*)

RIEDL, R. and FORSTNER, H. (1968). Wasserbewegung im Mikrobereich des Benthos. *Sarsia,* **34**, 163–88. (*358, 361*)

RIEDL, R. and MACHAN, R. (1972). (hydrodynamic patterns in intertidal sands) *Mar. Biol.*, **13**, 179–209. (*364*)

RIEDL, R., HUANG, N. and MACHAN, R. (1972). (the subtidal pump) *Mar. Biol.*, **13**, 210–21. (*364*)

RIGBY, B. S., HIRAI, N., SPIKES, J. D. and EYRING, H. (1959). Mechanical properties of rat tail tendon. *J. gen. Physiol.*, **43**, 265–83. (*89, 92*)

DE RIQLÈS, A. (1969). (les os longs des thériodontes) *Ann. Paléont.*, **55**, 1–52. (*172*)

RIVLIN, R. S. and THOMAS, A. G. (1953). Rupture of rubber. *J. Polymer Sci.*, **10**, 291-318. *(41)*

ROACH, M. R. and BURTON, A. C. (1957). (distensibility of arterial wall) *Canad. J. Biochem.*, **35**, 681-90. *(137)*

ROBERTS, A. D. (1971). Role of electrical repulsive forces in synovial fluids. *Nature, Lond.*, **231**, 434-6. *(122)*

ROBSON, E. A. (1964). Cuticle of *Peripatopsis. Quart. J. microsc. Sci.*, **105**, 281-99. *(326, 327)*

ROBSON, E. A. (1966). Swimming in Actiniaria. In *The Cnidaria and Their Evolution.* Edited by W. J. Rees. *Symp. Zool. Soc. Lond.* No. 16, 333-60, Academic Press, London. *(306)*

ROCHE, J., FONTAINE, M. and LELOUP, J. (1963). Halides. In *Comparative Biochemistry,* M. Florkin and H. Mason, eds. Vol. 5, pp. 493-547. Academic Press, London and New York. *(191)*

ROELOFSEN, P. A. (1959). *The Plant Cell Wall.* Gebr. Borntraeger, Berlin. *(194)*

ROELOFSEN, P. A. (1965). Fine structure of plant cell walls. *Adv. Bot. Res.*, **2**, 69-149. *(194)*

ROSEN, B. W. (1964). "Fibre Composite Materials", *Amer. Soc. Metals,* Bush, ed. Ohio. Chap. 3, p. 37. *(153)*

ROSEN, B. W. (1965). Mechanics of composite strengthening: fiber composites. *Am. Soc. Metals,* Ohio. *(153)*

ROSS, D. M. (1960). (behaviour of anemones on hermit crab shells) *Proc. zool. Soc. Lond.*, **134**, 43-57. *(306)*

ROSS, R. and BORNSTEIN, P. (1969). (macromolecular components of the elastic fibre) *J. Cell Biol.*, **40**, 366-81. *(113)*

RUDALL, K. M. (1947). (distribution of protein chain types in vertebrate epidermis) *Biochim. Biophys. Acta,* **1**, 549-62. *(191)*

RUDALL, K. M. (1955). Distribution of collagen and chitin. *Soc. exp. Biol. Symp.* **9**, 49-71. *(104)*

RUDALL, K. M. (1963). Chitin/protein complexes of insect cuticles. *Adv. Insect Physiol.*, **1**, 257-313. *(105-7)*

RUDALL, K. M. (1968). Intracellular fibrous proteins and the keratins. In *Comprehensive Biochemistry,* M. Florkin and E. H. Stotz eds.. **26-B**, 559-94. Elsevier, Amsterdam. *(188)*

RUDALL, K. M. (1969). (Chitin) *J. Polymer Sci.*, **28C**, 83-102. *(105, 107, 161)*

RUDWICK, M. J. S. (1964). (inference of function from structure in fossils) *Brit. J. Phil. Sci.*, **15**, 27-40. *(1)*

RUDWICK, M. J. S. (1970). *Living and Fossil Brachiopods.* Hutchinson Univ. Library, London. *(214)*

RUNHAM, N. W. and THORNTON, P. R. (1967). (mechanical wear of the gastropod radula) *J. Zool. Lond.*, **153**, 445-52. *(221)*

RUNHAM, N. W., THORNTON, P. R., SHAW, D. A. and WAYTE, R. C. (1969). Mineralization and hardness of the radular teeth of the limpet *Patella vulgata. Z. Zellforsch.*, **99**, 608-26. *(221, 222)*

RYSKEWITCH, T. (1953). (porosity in ceramics) *J. Am. Ceram. Soc.*, **36**, 65. *(157, 228)*

SAKURADA, I., NUKUSHINA, Y. and ITO, I. (1962). (modulus of oriented polymers) *J. Polymer Sci.*, **57**, 651-60. *(102)*

SANDERS, H. L. (1963). (functional morphology of Cephalocarida) *Mem. Connecticut Acad. Arts Sci.*, **15**, 1-80. *(317)*

SARKO, A. and MARCHESSAULT, R. H. (1969). Supermolecular structure of polysaccharides. *J. Polymer Sci.*, **28-C**, 317-31. *(98)*

SASSAMAN, C. and MANGUM, C. (1970). (temperature adaptation in sea anemones) *Mar. Biol.*, 7, 123–30. (*366*)

SCHMITT, F. O. (1956). Macromolecular interaction patterns in biological systems. *Proc. Am. Phil. Soc.*, **100**, No. 5, 476–86. (*84, 85*)

SCHOPF, T. J. M. (1969). Paleoecology of ectoprocts. *J. Paleontol.*, **43**, 234–44. (*238*)

SCHROEDER, J. H., DWORNIK, E. J. and PAPIKE, J. J. Primary protodolomite in echinoid skeletons. *Bull. geol. Soc. Am.*, **80**, 1613–16. (*222*)

SCHROEDER, W. A. and KAY, L. M. (1955). (amino acid composition of silks) *J. Am. chem. Soc.*, **77**, 3908–13. (*74*)

SCHUBERT, M. (1966). Structure of connective tissues, a chemical point of view. *Proc. Fedn. Am. Socs. exp. Biol.*, **25**, 1047–52. (*139, 140*)

SCHULTZ, A. H. (1939). (fractures in wild apes) *Bull. Hist. Med.*, **7**, 571–82. (*343*)

SCHWENKE, H. (1971). Water movements: Plants. In *Marine Ecology*, O. Kinne, ed. Wiley-Interscience, New York, Vol. 1, Part 2, pp. 1089–121. (*358*)

SCURFIELD, G. (1973). Reaction wood, its structure and function. *Science*, **179**, 657–59. (*321, 323*)

SEDLIN, E. D. (1965). Mechanical properties of bone. *Acta Orthoped. scand.*, Suppl. 83, 1. (*33*)

SEIFERT, G. (1967). (cuticle of a millipede) *Z. Morph. Ökol. Tiere*, **59**, 42–53. (*164*)

SERAFINI-FRACASSINI, A. and SMITH, J. W. (1966). (macromolecular interactions in bovine nasal septum) *Proc. roy. Soc. Lond.*, **B165**, 440–9. (*140*)

SHANLEY, F. R. (1957). *Weight-Strength Analysis of Aircraft Structures.* McGraw-Hill, New York. (*249, 264, 266*)

SHIMIZU, M., FUKUDA, T. and KIVIMURA, J. (1957). *The silk fibroins.* Sericulture Exptl. Station, Ministry of Agric. and Forestry, Tokyo. (*78*)

SHINN, E. A. (1966). Coral growth rate, an environmental indicator. *J. Paleontol.*, **40**, 233–40. (*349, 362*)

SIEGEL, S. M. (1968). (plant cell wall) In *Comprehensive Biochemistry*, M. Florkin and E. H. Stotz, eds. **26-A**, 1–51. Elsevier, Amsterdam. (*94*)

SILLIMAN, B. (1846). Chemical composition of calcareous corals. *Am. J. Sci. Arts*, **51**, 189–99. (*349*)

SIMKISS, K. (1964). Phosphates as crystal poisons of calcification. *Biol. Rev.*, **39**, 487–505. (*236*)

SIMPSON, G. G. (1953). *The Major Features of Evolution.* Columbia University Press, New York. (*235*)

SINGER, F. L. (1962). *Strength of Materials*, 2nd edn. Harper and Row, New York. (*249, 256*)

SLIJPER, E. J. (1946). (axial skeleto-muscular systems of mammals) *Verh. Kon. Ned. Ak. Wet.*, Ser. 2, **42**, 1–128. (*338*)

SMITH, D. W., BROWN, D. M. and CARNES, W. H. (1972). Preparation and properties of salt-soluble collagen. *J. biol. Chem.*, **247**, 2427–32. (*118*)

SMITH, J. E. (1946). (tube feet of starfish) *Phil. Trans. roy. Soc.*, **B232**, 279–310. (*315*)

SMITH, J. W. (1962). (structure and stress in fibrous epiphyseal plates) *J. Anat., London*, **96**, 209–25. (*280, 286*)

SMITH, J. W. (1968). Molecular pattern in native collagen. *Nature, Lond.*, **219**, 157–58. (*85–87*)

SNODGRASS, R. E. (1952). *A Textbook of Arthropod Anatomy.* Cornell University Press, Ithaca. (*317, 326, 329*)

SOBEL, A. E., LAURENCE, P. A. and BURGER, M. (1960). Nuclei formation and crystal growth in mineralizing tissues. *Trans. N. Y. Acad. Sci.*, **22**, 233–43. *(229)*

SOKOLOFF, L. (1963). (elasticity of articular cartilage) *Science*, **141**, 1055–9. *(140)*

SOKOLOFF, L. (1966). Elasticity of aging cartilage. *Fedn. Proc. Fedn. Am. Socs. exp. Biol.*, **25**, 1089–95. *(141)*

SOMMERHOFF, G. (1950). *Analytical Biology.* Oxford University Press, Oxford. *(1)*

SPIRO, B. F. (1971). (ultrastructure of the skeleton of *Tubipora musica*) *Bull. geol. Soc. Denmark*, **20**, 279–84. *(210)*

STANLEY, S. M. (1970). *Relation of Shell Form to Life Habits of the Bivalvia. Geol. Soc. Am. Mem.*, 125, 1–296. *(262, 263)*

STEVENS, W. C. and TURNER, N. (1948). *Solid and Laminated Wood Bending.* HMSO, London. *(206)*

STEVENSON, J. R. (1969). Sclerotin in the crayfish cuticle. *Comp. Biochem. Physiol.*, **30**, 503–8. *(160)*

SVOBODA, A. (1970). Oscillating flow and benthos. *Helgoländer wiss. Meeresunters*, **20**, 676–84. *(358, 360, 361)*

SWANSON, S. A. V. and FREEMAN, M. A. R. (1970). (mechanism of synovial joints) In *Modern Trends in Biomechanics.* D. C. Simpson, ed. Butterworths, London, Vol. 1, pp. 239–65. *(142)*

SZENT-GYÖRGI, A. G. and COHEN, C. (1957). (role of proline in protein chain structure) *Science*, **126**, 697–8. *(81)*

TAKAHASHI, K. (1967). (echinoderm catch mechanism) *J. Fac. Sci. Tokyo Univ.*, **11**, 109–35. *(280, 360)*

TANFORD, C. (1961). *Physical Chemistry of Macromolecules.* John Wiley, New York. *(58, 120)*

TAYLOR, J. D. (1973). The structural evolution of the bivalve shell. *Palaeontol.*, **16**, 519–34. *(232)*

TAYLOR, J. D., KENNEDY, W. J. and HALL, A. (1969). Shell structure and mineralogy of the Bivalvia. *Bull. Brit. Mus. (Nat. Hist.) Zool.*, Suppl. 3, 1–25. *(211)*

TAYLOR, J. D. and LAYMAN, M. (1972). Mechanical properties of bivalve shell structures. *Palaeontol.*, **15**, 73–87. *(232, 273)*

TERMINE, J. D. and POSNER, A. S. (1966). (age-dependent crystallinity in rat bone) *Science*, **153**, 1523–5. *(169)*

THÉODOR, J. and DENIZOT, M. (1965). (orientation to flow by benthos) *Vie Milieu*, **16**, 237–41. *(358)*

THINIUS, K. and HÖSSELBARTH, B. (1970). Fesigkeitseigenschaften gefüllter Thermoplaste. *Plaste Kautschuk*, 8, 567–576. *(219)*

THOMPSON, D. A. W. (1917). *On Growth and Form.* Cambridge University Press, London. *(3)*

THOR, C. J. B. and HENDERSON, W. F. (1940). (preparation of alkali chitin) *Am. Dyestuff Reptr.*, **29**, 461–4. *(108, 109)*

TIMOSHENKO, S. (1956). *Strength of Materials*, Part 2. Van Nostrand, Princeton. New Jersey. *(16)*

TOBOLSKY, A. V. (1960). *Properties and Structures of Polymers.* John Wiley, New York. *(50)*

TOMLINSON, P. B. (1962). (mechanical morphology of palm leaf base) *J. Arnold Arboretum*, **43**, 23–50. *(324)*

TOMLINSON, P. B. (1964). Vascular skeleton of the coconut leaf base. *Phytomorph.*, **14**, 218–30. (*324*)

TOOLE, B. P. (1969). (solubility of collagen with sulphated complexes) *Nature Lond.*, **222**, 872–3. (*133*)

TOOLE, B. P. and LOWTHER, D. A. (1966). (hexosamine compounds in bovine skin) *Biochim. Biophys. Acta*, **121**, 315–25. (*133*)

TOOLE, B. P. and LOWTHER, D. A. (1968). (dermatan sulfate-protein interaction with collagen) *Arch. Biochem. Biophys.*, **128**, 567–78. (*133*)

TOWE, K. (1972). (organic matrix concept and shell structure) *Biominer.*, **4**, 1–14. (*229*)

TOWE, K. and CIFELLI, R. (1967). (structure in calcareous foramenifera) *J. Palaeontol.*, **41**, 742–62 (*229*)

TOWE, K. and HAMILTON, G. H. (1968). Ultramicrotome-induced deformation artefacts in densely calcified material. *J. Ultrastr. Res.*, 22, 274–81. (*209*)

TOWE, K. and LOWENSTAM, H. A. (1967). (iron mineral in radular teeth of a chiton) *J. Ultrastr. Res.*, **17**, 1–13. (*221*)

TRAUB, W. and YONATH, A. (1966). (structure of tripeptides) *J. molec. Biol.*, **16**, 404–14. (*84*)

TRAVIS, D. F. (1970). (organization of five calcified tissues) In *Biological Calcification,* H. Schraer, ed. Appleton-Century-Crofts, New York. pp. 203–311. (*209, 217, 229*)

TRELOAR, L. R. G. (1958). *The Physics of Rubber Elasticity*, 2nd edn. Clarendon Press, Oxford. (*52–4, 57, 112*)

TRELOAR, L. R. G. (1960). (moduli of cellulose) *Polymer*, **1**, 209–303. (*102, 199*)

TRUEMAN, E. R. (1953). (mechanical properties of *Pecten* ligament) *J. exp. Biol.*, **30**, 453–67. (*115*)

TRUEMAN, E. R. (1966). (burrowing in *Arenicola*) *Biol. Bull.*, **131**, 369–77. (*334*)

TRUEMAN, E. R. (1967). Dynamics of burrowing in *Ensis. Proc. roy. Soc. Lond.*, **B166**, 459–76. (*333*)

TRUEMAN, E. R. (1968*a*). Burrowing activities of bivalves. *Zool. Soc. Lond. Symp.*, 22, 167–87. (*333, 334*)

TRUEMAN, E. R. (1968*b*). (burrowing in gastropods) *J. exp. Biol.*, **48**, 663–78. (*333, 334*)

TRUEMAN, E. R. (1968*c*). (burrowing in *Dentalium*) *J. Zool. Lond.*, **154**, 19–27. (*333*)

TRUEMAN, E. R. (1969). Fluid dynamics of molluscan locomotion. *Malacologia*, 9, 243–8. (*333*)

TRUEMAN, E. R. and PACKARD, A. (1968). Motor performances of some cephalopods. *J. exp. Biol.*, **49**, 495–507. (*334*)

TSAI, S. W. (1968). Strength theories of filamentous structures. In *Fundamental Aspects of Fiber Reinforced Plastic Composites,* R. T. Schwartz and H. S. Schwartz, eds. Wiley-Interscience, New York, pp. 3–11. (*152*)

TSCHANTZ, P. and RUTISHAUSER, E. (1967). (la surcharge mécanique de l'os vivant) *Ann. Anat. path.*, **12**, 223–48. (*178*)

TYLER, C. (1969). (avian egg shells) *Int. Rev. gen. exp. Zool.*, **4**, 81–130. (*218*)

VACELET, J. (1970). Les éponges pharétonides actuelles. In *Biology of the Porifera.* Edited by W. G. Fry. *Symp. Zool. Soc. Lond.* No. 25, 189–206. Academic Press, London. (*207*)

VELIMIROV, B. (1973). (sea fan orientation to flow) *Helgoländer wiss. Meeresunters.*, **24**, 163–73. (*358*)

VERONDA, D. R. and WESTMAN, R. A. (1970). (finite deformations of skin) *J. Biomech.*, 3, 111–24. (*132, 133*)

VINCENT, J. F. V. and WOOD, S. D. E. (1972). (mechanism of abdominal extension in *Locusta*) *Nature, Lond.*, **235**, 167–8. (*123*)

VISWANATHAN, A. and SHENOUDA, S. G. (1971). Helical structure of cellulose I. *J. appl. Polymer Sci.*, **15**, 519–63. (*96*)

VOLPIN, D. and CIFERRI, A. (1970). Elastic behavior of elastin. In E. A. Balazs (1970), Vol. 1, pp. 691–8. (*117*)

VOSE, G. P. and KUBALA, A. L. (1959). (bone strength and ash content) *Human Biol.*, **31**, 261–70. (*183*)

WAINWRIGHT, S. A. (1963). (structure of a coral skeleton) *Quart. J. microsc. Sci.*, **104**, 169–83. (*210, 349*)

WAINWRIGHT, S. A. (1964). (mineral phase of coral skeleton) *Exp. Cell Res.*, **34**, 213–30. (*210*)

WAINWRIGHT, S. A. (1970). Design in hydraulic organisms. *Naturwiss.*, **57**, 321–6. (*319*)

WAINWRIGHT, S. A. and DILLON, J. R. (1969). Orientation of sea fans. *Biol. Bull.*, **136**, 130–9. (*358*)

WARD, D. V. (1972). Locomotory function of squid mantle. *J. Zool., Lond.*, **167**, 487–99. (*335*)

WARD, D. V. and WAINWRIGHT, S. A. (1972). Locomotory aspects of squid mantle structure. *J. Zool., Lond.*, **167**, 437–49. (*335, 336*)

WARDROP, A. B. (1965). Formation and function of reaction wood. In *Cellular Ultrastructure of Woody Plants,* W. A. Côté, ed. Syracuse University Press, Syracuse. N.Y. pp. 371–90. (*100*)

WARDROP, A. B. and DADSWELL, H. E. (1955). (structure of reaction wood) *Austral. J. Bot.*, 3, 177. (*322*)

WARDROP, A. B. and DAVIES, G. W. (1964). (indole acetic acid and wood growth) *Austral. J. Bot.*, **12**, 24. (*321*)

WARWICKER, S. O. (1960). (fibroin crystal structure) *J. molec. Biol.*, 2, 350–62. (*76, 77*)

WEATHERWAX, R. C. and STAMM, A. J. (1946). (thermal expansion of wood) *U.S. For. Products Lab. Rep.*, No. 1487. Madison, Wisc. (*206*)

WEAVER, J. K. and CHALMERS, J. (1966). (cancellous bone) *J. Bone Jt. Surg.*, **48-A**, 289–98. (*174*)

WEBER, J., GREER, R., VOIGHT. B., WHITE, E. and ROY, R. (1969). (properties of echinoderm calcite) *J. Ultrastr. Res.*, 26, 355–66. (*167, 234*)

WEIS-FOGH, T. (1960). A rubber-like protein in insect cuticles. *J. exp. Biol.*, **37**, 889–906. (*111, 112*)

WEIS-FOGH, T. (1961*a*). (thermodynamic properties of resilin) *J. molec. Biol.*, **3**, 520–31. (*52, 54, 111, 112*)

WEIS-FOGH, T. (1961*b*). (molecular interpretation of resilin elasticity) *J. molec. Biol.*, 3, 648–67. (*52, 54, 112*)

WEIS-FOGH, T. (1970). Structure and formation of insect cuticle. In *Insect Ultrastructure,* A. C. Neville, ed. Blackwell Scientific Publications, Oxford, pp. 165–85. (*107, 161*)

WEIS-FOGH, T. (1972). (energetics in hummingbird and fruit fly flight) *J. exp. Biol.*, **56**, 79–104. (*111, 118*)

WEIS-FOGH, T. and ANDERSEN, S. O. (1970). (molecular model for elastin) *Nature, Lond.*, **227**, 718–21. (*117*)

WELLS, A. A. (1953). (brittle fracture in steel) *Welding Res.*, **7**, 34. (*42*)

WELLS, M. J. (1968). *Lower Animals.* World University Library, London. (*326*)
WILBUR, K. M. and SIMKISS, K. (1968). Calcified shells. In *Comprehensive Biochemistry.* M. Florkin and E. H. Stotz, eds. **26-A**, 229–95. Elsevier, Amsterdam. (*213, 218*)
WILDE, J. DE (1943). (properties of spider silk) *Arch. néerl. Sci.,* **27**, 118–32. (*78, 79, 81*)
WILLIAMS, A. (1968). (brachiopod shell structure) *Spec. Pap. Palaeontol.,* 2, 1–55 (*214*)
WILLIAMS, A. (1970). (brachiopod shell secretion) *Lethaia,* **1**, 268–87. (*214*)
WILLIAMS, A. and WRIGHT, A. D. (1970). (shell structure of calcareous inarticulate Brachiopoda) *Spec. Pap. Palaeontol.,* **7**, 1–51. (*215*)
WILLIAMS, G. C. (1966). *Adaptation and Natural Selection.* Princeton University Press, Princeton, New Jersey. (*1*)
WILSON, D. M. (1960). Nervous control of movement in cephalopods. *J. exp. Biol.,* **37**, 57–72. (*335*)
WOLINSKI, H. and GLAGOV, S. (1964). (structure and properties of aortive media) *Circ. Res.,* **14**, 400–13. (*135, 137*)
WOODLEY, J. D. (1967). (ophiuroid water-vascular system) In *Echinoderm Biology,* N. Millott, ed. Academic Press, London, pp. 75–104. (*313, 314*)
YAMADA, H. and EVANS, F. G. (1970). *Strength of Biological Materials.* Williams and Wilkins, Baltimore. (*174, 179*)
YOUNG, J. Z. (1938). Functioning of giant nerve fibers in the squid. *J. exp. Biol.,* **15**, 170–85. (*335, 337*)
YOUNG, S. D. (1971). (amino acids and glucosamine in coral skeleton) *Comp. Biochem. Physiol.,* **40B**, 113–20. (*210*)
ZIMMERMAN, M. H., WARDROP, A. B. and TOMLINSON, P. B. (1968). Tension wood in aerial roots of *Ficus benjamina. Wood Sci. Tech.,* 2, 95–104. (*321*)

Subject Index

Pages numbered in italics include a figure; pages numbered in boldface contain definitions or equations; pages whose numbers are marked with an asterisk contain values of the designated property.